STUDENT STUDY GUIDE

An Introduction to Concept Mapping for Campbell's BIOLOGY

Third Edition

MARTHA R. TAYLOR

Cornell University

The Benjamin/Cummings Publishing Company, Inc.

Redwood City, California ■ Menlo Park, California
Reading, Massachusetts ■ New York ■ Don Mills, Ontario
Wokingham, U.K. ■ Amsterdam ■ Bonn ■ Sydney
Singapore ■ Tokyo ■ Madrid ■ San Juan

Sponsoring Editor: Edith Beard Brady

Developmental Editor: Valerie Kuletz

Production Editor: Larry Olsen

Copyeditor: Lois Levitan

Composition: The Book Company

Cover Designer: Yvo Riezebos

ISBN 0-8053-1181-X

1 2 3 4 5 6 7 8 9 10--BA--98 97 96 95 94 93

The Benjamin/Cummings Publishing Company, Inc.
390 Bridge Parkway
Redwood City, California 94065

CONTENTS

Preface   *v*

Chapter 1
Introduction: Themes in the Study of Life   1

Unit I   The Chemistry of Life   5

Chapter 2
Atoms, Molecules, and Chemical Bonds   7

Chapter 3
Water and the Fitness of the Environment   13

Chapter 4
Carbon and Molecular Diversity   18

Chapter 5
Structure and Function of Macromolecules   22

Chapter 6
Introduction to Metabolism   29

Answer Section   36

Unit II   The Cell   45

Chapter 7
A Tour of the Cell   47

Chapter 8
Membrane Structure and Function   56

Chapter 9
Cellular Respiration: Harvesting Chemical Energy   63

Chapter 10
Photosynthesis   71

Chapter 11
Reproduction of Cells   78

Answer Section   83

Unit III   The Gene   91

Chapter 12
Meiosis and Sexual Life Cycles   93

Chapter 13
Mendel and the Gene Idea   98

Chapter 14
The Chromosomal Basis of Inheritance   107

Chapter 15
The Molecular Basis of Inheritance   114

Chapter 16
From Gene to Protein   120

Chapter 17
Microbial Models: The Genetics of Viruses and Bacteria   128

Chapter 18
Genome Organization and Expression in Eukaryotes   136

Chapter 19
DNA Technology   142

Answer Section   150

Unit IV   Mechanisms of Evolution   165

Chapter 20
Descent with Modification: A Darwinian View of Life   167

Chapter 21
How Populations Evolve   172

Chapter 22
The Origin of Species   179

Chapter 23
Tracing Phylogeny: Macroevolution, the Fossil Record, and Systematics   184

Answer Section   191

Unit V   Evolutionary History of Biological Diversity   197

Chapter 24
Early Earth and the Origin of Life   199

Chapter 25
Prokaryotes and the Origins of Metabolic Diversity   203

Chapter 26
Protists and the Origin of Eukaryotes   211

Chapter 27
Plants and the Colonization of Land   219

Chapter 28
Fungi   228

Chapter 29
Invertebrates and the Origin of Animal Diversity   233

Chapter 30
The Vertebrate Genealogy   244

Answer Section   254

Unit VI   Plants: Form and Function   267

Chapter 31
Plant Structure and Growth   269

Chapter 32
Transport in Plants   276

Chapter 33
Plant Nutrition   282

Chapter 34
Plant Reproduction and Development   286

Chapter 35
Control Systems in Plants   293

Answer Section   299

Unit VII   Animals: Form and Function   307

Chapter 36
Introduction to Animal Structure and Function   309

Chapter 37
Animal Nutrition   313

Chapter 38
Circulation and Gas Exchange   321

Chapter 39
The Body's Defenses   330

Chapter 40
Controlling the Internal Environment   338

Chapter 41
Chemical Signals in Animals   348

Chapter 42
Animal Reproduction   356

Chapter 43
Animal Development   363

Chapter 44
Nervous Systems   371

Chapter 45
Sensory and Motor Mechanisms   380

Answer Section   389

Unit VIII   Ecology   409

Chapter 46
Ecology: Distribution and Adaptations of Organisms   411

Chapter 47
Population Ecology   418

Chapter 48
Community Ecology   425

Chapter 49
Ecosystems   432

Chapter 50
Behavior   439

Answer Section   447

PREFACE

When I first thought about writing a *Student Study Guide* for Neil Campbell's *Biology*, Third Edition, I wanted to call it a structuring guide. The purpose of this book is to help you to structure and organize your developing knowledge of biology and to create your own personal understanding of the topics covered in the text. The four components of each chapter of this *Study Guide* are designed to help you in this task. The *Framework* identifies the overall picture; it provides a conceptual framework into which the chapter information fits. The *Chapter Summary* condenses the major concepts of each chapter. The *Structure Your Knowledge* section directs you to organize and relate these concepts by completing tables, labeling diagrams, answering essay questions, and constructing concept maps. In the *Test Your Knowledge* section, you are provided with a battery of objective questions to test your understanding. Suggested answers to the *Structure Your Knowledge* and *Test Your Knowledge* questions are provided in the *Answer Section* at the end of each unit.

What are the concept maps that often appear in this *Study Guide*? A concept map is a diagram that shows how ideas are organized and related. The *structure* of a concept map is a hierarchically organized cluster of concepts, enclosed in boxes and connected with labelled lines that explicitly show the relationships among the concepts. The *function* of a concept map is to help you structure your understanding of a topic and create meaning. The *value* of a concept map is in the thinking and evaluating required to create a map.

Developing a concept map for a group of concepts requires that you evaluate the relative importance of the concepts (which are most inclusive and important? which are less important and subordinate to other concepts?), arrange the concepts in a meaningful cluster, and draw connections between them that help you make their meanings explicit. (Additional information about concept mapping may be found in *Learning How to Learn*, Joseph D. Novak and D. Bob Gowin, Cambridge University Press, 1984.)

This book uses concept maps in several ways. A map of a chapter may be presented in the *Framework* section to show the organization of the key concepts in that chapter. More often, you will develop a concept map on a certain subset of ideas from the chapter. The answer section will present *suggested* concept maps. A concept map is an individual picture of your understanding at the time you make the map. As your understanding of an area develops, your concept map will evolve—sometimes becoming more complex and interrelated, sometimes becoming simplified and streamlined. Do not look to the answer section for the "right" concept map. *After* you have organized your own thoughts, look at the answer map to make sure you have included the key concepts (although you may have added more), to check that the connections you have made are reasonable, and perhaps to see another way to organize the information.

Biology is a visual science, and this third edition of the *Student Study Guide* includes diagrams and pictures taken from *Biology*, Third Edition, to help you see and understand the relationships among parts of structures and processes.

In the multiple choice questions presented in each chapter, you are asked to choose the best answer. Some answers may be partly correct; almost all choices have been written to test your ability to think and discriminate. Make sure you understand why the other choices are incorrect as well as why the given answer is correct.

Biology is a fascinating and broad subject. Campbell's *Biology* is filled with terminology and facts that are useful for understanding the major themes of modern biology. This *Student Study Guide* is intended to help you learn and recall information and, most importantly, to encourage and guide you as you develop your own understanding of and appreciation for biology.

Martha R. Taylor

INTRODUCTION: THEMES IN THE STUDY OF LIFE

FRAMEWORK

This chapter outlines nine themes that unify the study of biology and also describes the scientific construction of biological knowledge. A course in biology is neither a vocabulary course nor a classification exercise for the diverse forms of life. Biology is a collection of facts and concepts structured within theories and organizing principles. Recognizing the common themes within biology will help you to structure your knowledge of this fascinating and challenging study of life.

CHAPTER SUMMARY

Biology is the scientific study of life, an extension of our innate interest in life in its diverse forms. The scope of biology is immense, spanning from the submicroscopic to the complex web of ecosystems, from the present back through four billion years of evolutionary history. Recent advances in research methods, coupled with the large number of contemporary biologists working throughout the many subfields of biology, have led to an explosion of information. A beginning student can make sense of this expanding universe of knowledge by focusing on a few enduring themes that unify the study of biology.

A Hierarchy of Organization

Biological organization is based on a hierarchy of structural levels, with each level—from atoms, biological molecules, organelles, cells, tissues, organs, organ systems, organisms, populations, and communities, to ecosystems—building on the levels below it. The study of biology includes understanding these various levels of organization and the connections between them.

Emergent Properties

Interactions among components at each level of organization lead to the emergence of novel properties at the next level: the whole is greater than the sum of its parts. The concept of emergent properties of life may seem to, but does not support the doctrine of vitalism, which claims that life is driven by forces that defy explanation. Rather, the principles of physics and chemistry can be used to unravel and understand the properties of life that arise from the hierarchy of structural organization.

The characteristics or properties of life include order, reproduction, growth and development directed by heritable programs, energy utilization, responsiveness to the environment, homeostasis, and evolutionary adaptation.

The study of biology balances the pragmatic reductionist strategy, which breaks down complex systems to simple components, with holism, an approach in which the higher organizational levels of life are studied to comprehend the properties that arise from the interactions and integration of the system's parts.

The Cellular Basis of Life

The cell is the simplest structural level capable of performing all the activities of life; it is an organism's basic unit of structure and function. Hooke first described and named cells in 1665 when he observed a slice of cork with a simple microscope. Leeuwenhoek, a contemporary, developed lenses that permitted him to view the world of microscopic organisms. In 1839, Schleiden and Schwann concluded that all living things consist of cells. This cell theory now includes the idea that all cells come from other cells.

The recent advent of the electron microscope has revealed the complex structural organization of cells.

Two major types of cells are recognized. The simpler prokaryotic cell, found only in bacteria, lacks both a nucleus to enclose its DNA and most membrane-bound organelles. The eukaryotic cell, with its nucleus and numerous cytoplasmic compartments, is typical of all other living organisms.

Heritable Information

The biological "instructions" for the development and functioning of organisms are coded in the arrangement of the four nucleotides in DNA molecules and transmitted from parents to offspring in units of inheritance called genes. All forms of life use essentially the same genetic code to write and transcribe their heritable script.

A Feeling for Organisms

Barbara McClintock was able to discover an unexpected genetic mechanism, for which she received the Nobel Prize, because she was so familiar with corn, the organism she studied. This "feeling for the organism" provides the context and motivation for all levels of biological study.

The Correlation of Structure and Function

The form of a biological structure gives information about its functioning, and a study of function provides insight into structural organization. The principle that form fits function is illustrated at all levels of biological organization.

The Interaction of Organisms with Their Environment

Organisms affect and are affected by the physical and biological environments with which they interact. Within an ecosystem, nutrients cycle between the abiotic and biotic components, and energy flows from sunlight to photosynthetic organisms to the life forms that feed on plants and other organisms.

Unity in Diversity

About 1.5 million species, out of an estimated total of 5 to 30 million, have been identified and named. Taxonomy is the branch of biology that names organisms and classifies species into hierarchical groups. At the broadest unit of classification, life forms are organized into five kingdoms: Monera, Protista, Plantae, Fungi, and Animalia.

The kingdom Monera, which contains bacteria, is distinguished on the basis of the simple prokaryotic cell type. All other kingdoms have eukaryotic cells. Kingdom Protista contains mostly unicellular, or else simple multicellular forms. The other three kingdoms contain multicellular organisms characterized to a large extent by their mode of nutrition. Plants are photosynthetic; fungi are mostly decomposers that absorb their nutrients from dead organisms or organic wastes; and animals obtain their food by ingestion.

Within this diversity, living forms share a universal genetic code and commonalities in cell structure.

Evolution: The Core Theme of Biology

Evolutionary theory connects all of life by common ancestry. The history of living forms extends back over 3 billion years to the ancient prokaryotes. Each species is the tip of an evolutionary branch that connects closely related species with their common ancestors.

In *The Origin of Species*, published in 1859, Charles Darwin presented his case for evolution, or "descent with modification," that present forms evolved from a succession of ancestral forms guided by a mechanism called natural selection. Darwin synthesized the concept of natural selection from three generalized observations: (1) individuals vary in many heritable traits; (2) many more young are produced within a population than can be supported by the environment; and (3) individuals with traits best suited for an environment leave a larger proportion of offspring than do less fit individuals. This differential reproductive success results in the gradual accumulation of adaptations within a population.

According to Darwin, new species originate as unique traits and combinations of traits that arise randomly are selected for by differing environments. Descent with modification makes sense of both the unity and diversity of life; evolution is the core theme of biology.

Scientific Process: The Hypothetico-Deductive Method

Science, which is a way of knowing, emerges from our curiosity and drive to understand ourselves, the world, and the universe. Asking questions about nature and believing that those questions are answerable provide the basis of science.

The scientific process is based on the hypothetico-deductive method. A hypothesis is a tentative explanation for an observation or question. Deductive reasoning proceeds from the general to the specific, from a general hypothesis to specific predictions of results if the general premise is true. A hypothesis is usually tested by performing experiments to see whether predicted results occur.

There are five important aspects of hypotheses: hypotheses are possible causes or explanations; hypotheses reflect past experience with similar situations; having multiple hypotheses is good science; hypotheses must be testable using the hypothetico-deductive method; and hypotheses can be eliminated, but not proven.

The scientific process makes use of controlled experiments, in which subjects are divided into an experimental group and a control group. Both groups are treated alike except for the one variable that the experiment is trying to test.

The experiments of Reznick and Endler illustrate the hypothetico-deductive method. By maintaining control populations and following generations of guppies transferred from sites with pike-cichlids to sites with killifish for 11 years, they were able to conclude that natural selection due to differential predation was the most likely explanation for differences in life history characteristics between guppy populations.

Science is characterized as progressive and self-correcting. Scientists build on the work done by others, refining or refuting their ideas, both cooperating and competing with each other.

Facts, in the form of observations and experimental results, are prerequisites of science, but it is the new ways of organizing and relating those facts that advance science. Newton, Darwin, and Einstein are outstanding scientists because they synthesized theories that had great explanatory power. A theory is broader in scope than a hypothesis and is supported by a large body of evidence.

Science, Technology, and Society

Science and technology are interwoven as science uses the new instruments and techniques of technology to extend knowledge, and technology uses the information generated by science to guide the development of technological advances. The fields of molecular biology and genetic engineering illustrate this relationship. Technology contributes to our standard of living and has made it possible for the human population to expand rapidly, but at a price of severe environmental consequences, including deforestation, acid rain, nuclear accidents, toxic wastes, and species extinctions. Solutions to these problems will involve politics, economics, and cultural values, as well as science and technology.

Biology is a demanding science—partly because living systems are so complex and partly because biology incorporates concepts from chemistry, physics and math. This book presents a wealth of information. The basic themes of biology will help you understand, appreciate, and structure your growing knowledge of biology.

STRUCTURE YOUR KNOWLEDGE

1. This chapter presents nine unifying themes of biology. Briefly describe each of these:
    a. A hierarchy of organization
    b. Emergent properties
    c. The cellular basis of life
    d. Heritable information
    e. A feeling for organisms
    f. The correlation of structure and function
    g. The interaction of organisms with their environment
    h. Unity in diversity
    i. Evolution: the core theme of biology

2. Biology is the scientific study of life. How would you characterize a scientific approach to developing biological knowledge?

CHAPTER 1
INTRODUCTION: THEMES IN THE STUDY OF LIFE

Suggested Answers to Structure Your Knowledge

1. Unifying themes of biology:
   a. **A hierarchy of organization:** Living things exhibit a hierarchy of structural levels of organization, ranging from the organization of particles in atoms, through atoms in molecules, macromolecules, cellular organelles, cells, tissues, organs, organ systems, whole organisms, populations, and communities, to ecosystems.
   b. **Emergent properties:** Unique properties emerge at each structural level that were not evident at the simpler level, as a result of the interactions and organization of components.
   c. **The cellular basis of life:** Cells are the basic unit of structure and function of all living forms. All cells come from other cells. Prokaryotic cells, found in bacteria, and eukaryotic cells, found in all other organisms, are the two major types of cells distinguishable by their structural organization.
   d. **Heritable information:** The structures and processes of life are encoded in DNA, the universal language of inheritance.
   e. **A feeling for organisms:** The study of biology begins with a feeling for organisms—an attraction to the living world around us. Good scientists have a deep understanding and appreciation for the organisms with which they work. It is the organism that sets the context for studies that may range in focus from molecular to ecosystem biology.
   f. **The correlation of structure and function:** Within each level of organization is a correlation between structure and function—that is, between the physical arrangement of parts and the function they serve.
   g. **The interaction of organisms with their environment:** Organisms influence and are influenced by their environment. Ecosystems are characterized by the cycling of nutrients and the flow of energy.
   h. **Unity in diversity:** The diversity of life, as illustrated by the five-kingdom classification of organisms, is unified by the genetic code and the common characteristics of cells.
   i. **Evolution, the core theme of biology:** Evolution, the connection of all life forms through common ancestry with natural selection, the mechanism of accumulating adaptations to specific environments, is the core theme of biology.

2. The scientific study of life rests on the belief that natural phenomena have natural causes and that questions about life are answerable. Through the hypothetico-deductive method, scientists develop hypotheses, which are tentative answers to questions, and then deduce predictions. These can be eliminated as false, but not proven correct by the accumulation of evidence from experiments and observations. Scientists share evidence and conclusions, then repeat, reinforce, and refute each other's ideas. Theories are broad explanations of various phenomena that serve to organize biological knowledge, generate new questions, and direct future studies.

UNIT I

THE CHEMISTRY OF LIFE

"What the? ... This is lemonade! Where's my culture of amoebic dysentery?"

CHAPTER 2

ATOMS, MOLECULES, AND CHEMICAL BONDS

FRAMEWORK

This chapter considers the basic principles of chemistry that explain the behavior of atoms and molecules and that form the basis for our modern understanding of biology. Emergent properties are associated with each new level of structural organization as subatomic particles are organized into atoms and atoms into molecules. The following concept map sketches out the relationships among the key concepts included in this chapter.

CHAPTER SUMMARY

Understanding the structure and interaction of the atoms and molecules that constitute living organisms is essential to the study of the phenomenon we call life. A reductionist approach to the study of biology focuses on lower levels of organization. The principles of chemistry that govern how atoms and molecules behave lay the foundation for each successive structural level in the organization of life and for the emergent properties that accompany each level.

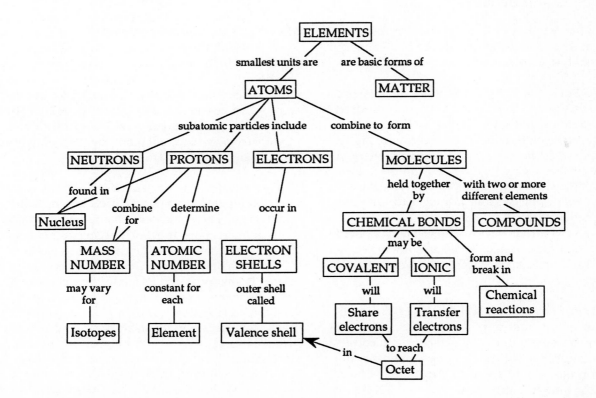

Matter: Elements and Compounds

Chemistry is the study of *matter*, anything that takes up space and has mass. Mass is a measure of the amount of matter within an object; weight is the measure of how strongly that mass is pulled by gravity. On earth, weight and mass may be considered synonymous.

The basic forms of matter are *elements*, substances that cannot be broken down to other types of matter by chemical reactions. Chemists have identified 92 naturally occurring elements and created about a dozen synthetic ones. Each element is symbolized by one or two letters.

A *compound* is made up of two or more elements combined in a fixed ratio. A compound usually has characteristics quite different from its constituent elements, an example of the emergence of novel properties in higher levels of organization.

Elements Essential to Life   About 25 elements are essential to life. Of these, carbon (C), oxygen (O), hydrogen (H), and nitrogen (N) make up 96% of living matter. The remaining 4% is composed of phosphorus (P), sulfur (S), calcium (Ca), and a few others. Some elements, like iron (Fe) and iodine (I), are required in very minute quantities and are called *trace elements*.

The Structure and Behavior of Atoms

An *atom* is the smallest unit of an element retaining the physical and chemical properties of that element. Each element has its own unique type of atom.

Subatomic Particles   Physicists have determined that atoms are composed of more than a hundred types of subatomic particles. Three stable subatomic particles are important to our understanding of atoms. Uncharged *neutrons* and positively charged *protons* are packed tightly together to form the *nucleus* of an atom. Negatively charged *electrons* orbit rapidly about the nucleus, electrically attracted to the positively charged nucleus.

Protons and neutrons have a similar mass of about 1.7×10^{-24} g or 1 dalton each. A *dalton* is the measurement unit for atomic mass. Electrons have negligible mass.

Atomic Number and Atomic Weight   The *atomic number* refers to the number of protons in the nucleus of an atom. Each element has a characteristic and consistent atomic number. Unless otherwise shown, the number of protons in an atom is equal to the number of electrons, and the atom has a neutral electrical charge. A subscript to the left of the symbol for an element indicates its atomic number; a superscript indicates mass number. The *mass number* of an element is equal to the number of protons and neutrons in its nucleus and approximates the mass of an atom of that element in daltons. The term *atomic weight* is often used to refer to the mass of an atom. The number of neutrons in an atom is equal to the difference between the mass number and the atomic number. An atom of phosphorus, $^{31}_{15}P$, contains 15 protons, 15 electrons, and 16 neutrons. The atomic weight of phosphorus is 31 daltons.

Isotopes   Although the number of protons is constant, the number of neutrons can vary within the atoms of an element, creating different *isotopes*. An element occurs in nature as a mixture of its isotopes. Some of the isotopes of elements are unstable or *radioactive*; their nuclei spontaneously decay, giving off particles and energy. The characteristic *half-life* for radioactive isotopes (the length of time it takes 50% of the radioactive atoms in a sample to decay) provides a basis for dating fossils. When an organism dies, radioactive isotopes from the environment are no longer taken into its body, and the radioactive isotopes in its body decay at their fixed rate. The half-life of carbon-14 (^{14}C) is about 5600 years. The ratio of carbon-14 to carbon-12 in a fossil is compared to the environmental ratio and indicates when the organism died.

Radioactive isotopes are important tools in biological research and medicine because they can be introduced into organisms and detected in minute quantities. Molecules can be labeled with radioactive isotopes and followed through the steps of a metabolic pathway or traced throughout an organism. Techniques using scintillation counters or autoradiography can determine the quantity and location of radioactively labeled molecules within a cell. Too great an exposure to radiation from decaying isotopes poses a significant hazard to life.

Energy Levels   The nucleus of an atom is extremely small compared to the large area in which the electrons orbit. The chemical behavior of an atom—the type of interactions it has with other atoms—is determined by the number and location of its electrons.

Energy is defined as the ability to do work. *Potential energy* is energy stored in matter as a consequence of the relative position of masses or charges. Matter naturally tends to move toward a lower level of potential energy, which is a more stable state, and requires the input of energy to return to a higher potential energy.

The negatively charged electrons are attracted to the positively charged nucleus; their potential energy increases as their distance from the nucleus increases.

Electrons can orbit in several different potential energy states, called *energy levels* or *electron shells*, surrounding the nucleus. The closer the shell to the nucleus, the lower the potential energy. Changes of the potential energy of electrons occur in steps of discrete amounts. To move to a shell farther from the nucleus, an electron must absorb energy, as happens when light energy excites an electron in the first stage of photosynthesis.

Electron Orbitals   The three-dimensional space or volume within which an electron is most likely to be found is called an *orbital*. No more than two electrons can occupy the same orbital. The first electron shell or energy level can contain two electrons in a single spherical orbital, called the 1s orbital. The second electron shell's four orbitals, each with the potential of containing two electrons, include a 2s spherical orbital and three dumbbell-shaped *p* orbitals located along the *x*, *y*, and *z* axes. Higher electron shells contain additional orbitals, but the outermost energy shell of an atom never contains more than eight electrons, located in its *s* and three *p* orbitals.

Electron Configuration and Chemical Properties
The chemical behavior of an atom is a function of its electron configuration—in particular, the number of *valence electrons* in its outermost energy shell, or *valence shell*. A valence shell of eight electrons is complete, resulting in an unreactive or inert atom. Atoms with incomplete valence shells are chemically reactive. The periodic table of elements is arranged in order of the sequential filling of electron orbitals. Atoms with the same number of electrons in their valence shell have similar chemical properties.

Chemical Bonds and Molecules

Atoms with incomplete valence shells can either share electrons with or completely transfer electrons to or from other atoms such that each atom is able to complete its valence shell. These interactions usually result in attractions, called *chemical bonds*, that hold the atoms together.

Covalent Bonds   When two atoms share a pair of valence electrons, a *covalent bond* is formed. Their valence shells overlap, and the electrons circulate about both nuclei. A *structural formula* indicates both the number and type of atoms and also the bonding within a molecule. Thus, H–H indicates a hydrogen molecule of two atoms of hydrogen held together by a covalent bond. O=O represents an oxygen molecule in which two pairs of valence electrons are shared between oxygen atoms, forming a *double covalent bond*.

A *molecular formula* (such as O_2) only indicates the kinds and numbers of atoms in a molecule. A molecule consists of two or more atoms held together by chemical bonds. A *compound*, such as H_2O, is a molecule formed from more than one kind of element.

The *valence* or bonding capacity of an atom is a measure of the number of covalent bonds it must form in order to complete its outer shell. The valences of the four most common elements of living matter are hydrogen 1, oxygen 2, nitrogen 3, and carbon 4.

The function of a molecule is often dependent on its shape or geometry. A methane molecule (CH_4) has a carbon atom in the center with four hydrogen atoms at the corners of a tetrahedron. A water molecule (H_2O) is approximately in the shape of a right angle.

Electronegativity is the attraction of an atom for shared electrons. If the atoms in a molecule have similar electronegativities, the electrons remain equally shared between the two nuclei, and the covalent bond is said to be *nonpolar*. If one element is more electronegative, it pulls the shared electrons closer to itself, creating a *polar covalent bond*. This unequal sharing of electrons results in a partial negative charge associated with the more electronegative atom and a partial positive charge associated with the atom from which the electrons are pulled.

Ionic Bonds   If two atoms are very different in their attraction for the shared electrons, the more electronegative atom may completely transfer an electron from another atom, forming an *ionic bond*. This transfer of a negatively charged electron results in the formation of charged atoms or molecules called *ions*. The atom that lost the electron is positively charged and called a *cation*. The atom that gained the electron is negatively charged and called an *anion*. The transfer of electrons allows atoms to achieve complete valence shells. The atoms are held together because of the attraction of their opposite charges. Ionic compounds, called salts, often exist as three-dimensional crystalline lattice arrangements held together by electrical attraction. The number of ions present in a salt crystal is not fixed, but the atoms are present in specific ratios (such as one-to-two Mg to Cl in magnesium chloride, $MgCl_2$).

Ion also refers to whole covalent molecules that are electrically charged. Ammonium (NH_4^+) is a positively charged ion; this covalently bonded molecule is missing one electron.

The sharing of electrons between atoms in the completion of their valence shells can be thought of as falling on a continuum from nonpolar covalent bonds in which electrons are equally shared, through polar covalent bonds, to ionic bonds in which electrons are shared so unequally as to be actually transferred from one atom to another.

Hydrogen Bonds   When a hydrogen atom is covalently bonded with an electronegative atom and thus has a partial positive charge, it can be attracted to another electronegative atom and form a *hydrogen bond*. Hydrogen bonds are responsible for many of the unusual properties of water.

Biological Importance of Weak Bonds   Hydrogen bonds, ionic bonds (both of which are weak in water), and other weak bonds form temporary interactions between molecules and are involved in many biological signals and processes. Weak bonds within large molecules such as proteins, help to create the three-dimensional shape and resulting activity of these molecules.

Chemical Reactions

Chemical reactions involve the making or breaking of chemical bonds in the transformation of matter into different forms. Matter is conserved in chemical reactions; the same number and kind of atoms are present in both *reactants* and *products*, although the rearrangement of electrons and atoms causes the properties of these molecules to be different. Thus molecules of hydrogen and oxygen can combine to form water. The reaction is written: $2H_2 + O_2 \longrightarrow 2H_2O$.

Most reactions are reversible—the products of the forward reaction can become reactants in the reverse reaction. The rate of a reaction is speeded by increasing the concentrations of reactants. As products accumulate, collisions resulting in the reverse reaction become more frequent. Eventually, *chemical equilibrium* may be reached when the forward and reverse reactions proceed at the same rate, and the concentrations of reactants and products no longer change. These relative concentrations will vary depending on the reaction; chemical equilibrium does not mean that reactants and products are equal in concentration.

Chemical Conditions on the Early Earth: Setting the Stage for the Origin and Evolution of Life

Most astronomers believe that all matter was once concentrated in a giant mass that blew apart with a "big bang" 10 to 20 billion years ago. Our sun formed about 5 billion years ago when most of the swirling matter in the cloud of dust that formed our solar system condensed in the center. Earth and the rest of the planets formed about 4.6 billion years ago as gravity attracted the remaining dust and ice to a few kernels.

Geologists believe that the initially cold Earth went through a molten period, when its components sorted into layers of different densities. Nickle and iron sank to the core, less dense material formed a mantle, and the least dense material solidified into a thin crust. The continents are borne on plates of crust that float on the mantle.

Scientists speculate that Earth's first atmosphere of hot hydrogen gas escaped and was replaced by gases from volcanoes and other vents. This early atmosphere probably consisted mostly of water vapor, carbon monoxide and carbon dioxide (CO, CO_2), nitrogen (N_2), methane (CH_4), and ammonia (NH_3). The first seas were formed by torrential rains. Lightning, volcanic activity, and UV radiation were quite intense. Chemical evolution in such an environment is believed to have given rise to the first cells.

STRUCTURE YOUR KNOWLEDGE

Take the time to write out or discuss your answers to the following questions. Then refer to the suggested answers at the end of this unit.

1. Describe an atom. Include the concepts of subatomic particles, mass, charge, and spatial arrangement.

2. Atoms can have various numbers associated with them. Define the following and show where each of them is placed relative to the symbol of an element, such as C: atomic number, mass number, atomic weight, valence. Which of these numbers is most related to the chemical behavior of an atom? Explain.

3. Explain what is meant by saying that there is a continuum from covalent bonds to ionic bonds in the range of ways that atoms share electrons.

TEST YOUR KNOWLEDGE

MULTIPLE CHOICE: *Choose the one best answer.*

1. Each element has its own characteristic atom in which
    a. the atomic weight is constant.
    b. the atomic number is constant.
    c. the mass number is constant.
    d. two of the above are correct.

2. Isotopes can be used in studies of metabolic pathways because
   a. their half-life allows a researcher to time an experiment.
   b. they are more reactive.
   c. the cell does not recognize the extra protons in the nucleus, so isotopes are readily used by the cell.
   d. their location or quantity can be experimentally determined because of their radioactivity.

3. At equilibrium in a reaction
   a. the forward and reverse reactions are occurring at the same rate.
   b. the reactants and products are in equal concentration.
   c. the forward reaction has gone further than the reverse reaction.
   d. both a and b are correct.

4. Oxygen has eight electrons. You would expect the arrangement of these electrons to be:
   a. eight in the second energy shell, creating an inert element.
   b. two in the first energy shell and six in the second, creating a valence of six.
   c. two in the first energy shell and six in the second, creating a valence of two.
   d. two in the first energy shell, four in the second, and two in the third, creating a valence of two.

5. A covalent bond between two atoms is likely to be polar if
   a. one of the atoms is much more electronegative than the other.
   b. the two atoms are equally electronegative.
   c. the two atoms are of the same element.
   d. the bond is part of a tetrahedrally shaped molecule.

6. Which of these classes of substances would be least soluble in the polar compound water?
   a. ionic compounds
   b. polar compounds
   c. nonpolar compounds
   d. salts

7. The most accurate structural formula for water is
   a. $H{\overset{O}{\diagup\diagdown}}H$
   b. H-O-H
   c. H_2O
   d. $H^+ O^= H^+$

8. A triple covalent bond would
   a. be very polar.
   b. involve the bonding of three atoms.
   c. produce a triangularly shaped molecule.

d. involve the sharing of six electrons.

9. It is difficult to speak of a molecule of the salt NaCl because
   a. each sodium ion is attracted to four chloride ions.
   b. salt occurs as a crystalline lattice of many sodium and chloride ions.
   c. the ratio of sodium and chlorine atoms may vary.
   d. the bonds in a salt crystal are weak and break easily.

10. A cation
    a. has gained an electron.
    b. can easily form hydrogen bonds.
    c. is nonpolar.
    d. has a positive charge.

The six elements most common in living organisms are:

$$^{12}_{6}C \quad ^{16}_{8}O \quad ^{1}_{1}H \quad ^{14}_{7}N \quad ^{32}_{16}S \quad ^{31}_{15}P$$

Use this information to answer questions 11 through 17.

11. How many electrons does phosphorus have in its valence shell?
    a. 15
    b. 5
    c. 7
    d. 8

12. What is the atomic weight of phosphorus?
    a. 15
    b. 16
    c. 31
    d. 46

13. A radioactive isotope of carbon has the mass number 14. How many neutrons does this isotope have?
    a. 6
    b. 8
    c. 12
    d. 14

14. What is the valence of nitrogen?
    a. 1
    b. 2
    c. 3
    d. 4

15. How many covalent bonds is a phosphorus atom most likely to form?
    a. 1
    b. 2
    c. 3
    d. 4

16. Based on electron configuration, which of these elements would have chemical behavior most like that of oxygen?
    a. C
    b. N
    c. P
    d. S

17. How many of these elements are found next to each other (side by side) on the periodic chart?
    a. one group of two
    b. two groups of two
    c. one group of two and one group of three
    d. all of them

18. Molecular shape is of biological importance because it
    a. can result in recognition and interaction of molecules.
    b. determines how fast reactions take place.
    c. relates to the potential energy stored in chemical bonds.
    d. determines the chemical properties of an atom.

19. What types of bonds are present in the following illustration of a water molecule interacting with an ammonia molecule?

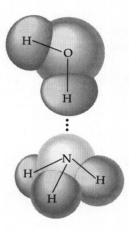

    a. hydrogen bonds between the oxygen and nitrogen atoms and their respective hydrogen atoms
    b. polar covalent bonds between the oxygen and nitrogen atoms and their respective hydrogen atoms and a hydrogen bond indicated by the dotted line
    c. an ionic bond indicated by the dotted line and nonpolar covalent bonds for the other indicated bonds
    d. a nonpolar covalent bond indicated by the dotted line and polar covalent bonds for the other indicated bonds

20. Which of the atoms shown in question 19 are electronegative?
    a. oxygen and nitrogen
    b. only oxygen
    c. hydrogen
    d. all atoms have equal electronegativity

CHAPTER 3

WATER AND THE
FITNESS OF THE ENVIRONMENT

FRAMEWORK

This chapter will introduce you to how emergent properties of water contribute to the biological fitness of the external and internal environment of living organisms.

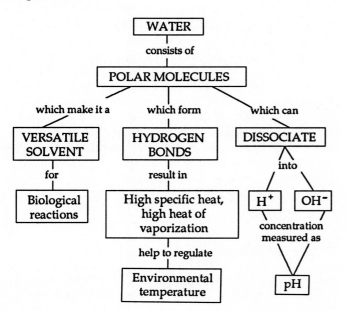

CHAPTER SUMMARY

Water makes up 70% to 95% of the cell content of living organisms and covers 75% of the earth's surface. It is the biological medium that makes life possible. Its unique properties make the external environment fit for living organisms and make the internal environment of organisms fit for the chemical and physical processes of life. The properties of water can be traced to the structure and interactions of its molecules.

Water Molecules and Hydrogen Bonding

Water molecules consist of two hydrogen atoms covalently bonded to an oxygen atom. The electronegative oxygen pulls the shared electrons away from the hydrogens, creating a polar covalent bond with a partial positive charge on each hydrogen and a partial negative charge associated with the oxygen. This asymmetric shape and the polarity of the bonds make water a *polar molecule*. The formation of hydrogen bonds, in which a hydrogen atom covalently bonded to an oxygen atom is also attracted to the negative oxygen atom of another molecule, creates a higher level of structural organization within water.

Some Extraordinary Properties of Water

Liquid Water Is Cohesive   Liquid water is unusually cohesive due to the constantly forming and reforming of hydrogen bonds that hold the molecules together. This *cohesion* creates a more structurally organized liquid and enables water to move against gravity in plants. The *adhesion* of water molecules to the walls of plant vessels also contributes to water transport. Hydrogen bonding between water molecules produces a high *surface tension* at the interface between water and air. Small insects are able to "walk on water" due to this surface tension.

Water Has a High Specific Heat   In a body of matter, *heat* is the total quantity of *kinetic energy*, the energy associated with the movement of atoms and molecules. *Temperature* measures the average kinetic energy of the molecules in a substance.

Temperature is measured by a *Celsius scale*. Water freezes at 0°C and boils at 100°C. Heat is measured by the *calorie* (cal). A calorie is the amount of heat energy it takes to raise 1 gram of water 1 degree Celsius. A *kilocalorie* (kcal, also designated as C) is 1000 calories, the amount of heat required to raise 1 kilogram of water 1 degree Celsius.

Specific heat is the amount of heat absorbed or lost when 1 gram of a substance changes temperature by 1 degree Celsius. Water's specific heat of 1 cal/gm/°C is unusually high compared with that of other common substances. This means that water must absorb or release a relatively large quantity of heat in order for its temperature to change. Heat must be absorbed to break hydrogen bonds before water molecules can move faster and temperature can rise, and conversely, heat is released when hydrogen bonds form. As the temperature of water drops slightly, many hydrogen bonds form and release a considerable amount of heat energy. The ability of large bodies of water to stabilize air temperature is due to the high specific heat of water. The high proportion of water in the environment and within organisms keeps temperature fluctuations within limits that permit life.

Water Has a High Heat of Vaporization   The transformation from a liquid to a gas is called vaporization or evaporation and happens when molecules with sufficient kinetic energy overcome their attraction to other molecules and escape into the air as gas. The addition of heat increases the rate of evaporation by increasing the kinetic energy of molecules. The *heat of vaporization* is the quantity of heat that must be absorbed for 1 gram of a liquid to be converted to a gas. Water has a high heat of vaporization (540 cal/gm) because a large amount of heat is needed to break the hydrogen bonds holding water molecules together. This property of water helps moderate the climate on earth. Solar heat is dissipated from tropical seas during evaporation and, as tropical air moves poleward, this heat is released as moisture condenses to form rain.

As a substance vaporizes, the liquid left behind loses the kinetic energy of the escaping molecules and cools down. *Evaporative cooling* helps to protect terrestrial organisms from overheating and contributes to the stability of temperatures in lakes and ponds.

Water Expands When It Freezes   As water cools below 4°C, it expands. By 0°C, each water molecule becomes hydrogen-bonded to four other molecules, creating a crystalline lattice that spaces the molecules apart. Ice is about 10% less dense than liquid water at 4°C, and therefore, it floats. The floating ice insulates the liquid water below, allowing it to be a warmer temperature than the air above it. The layer of ice allows aquatic organisms to live year round in the liquid water, which would not be possible if water froze solidly from the bottom upward in winter.

The formation and melting of ice make temperature fluctuations within the environment less abrupt as seasons change. Heat is released into the air when hydrogen bonds form as water solidifies into ice and snow. The melting of ice and snow absorbs heat as hydrogen bonds are broken, making spring warming more gradual.

Water Is a Versatile Solvent   A *solution* is a homogeneous mixture of two or more substances; the dissolving agent is called the *solvent*, and the substance that is dissolved is the *solute*. An *aqueous solution* is one in which water is the solvent. The positive and negative regions of water molecules are attracted to oppositely charged ions or partially charged regions of polar molecules. Thus, solute molecules become surrounded by water molecules and dissolve into solution. Ionic and polar substances are *hydrophilic*; they have an affinity for water due to electrical attractions and hydrogen bonding. Nonpolar and non-ionic compounds are *hydrophobic*; they will not mix with or dissolve in water.

Aqueous Solutions

Most of the chemical reactions of life take place in water. Biological chemistry requires an understanding of aqueous solutions.

Solute Concentration A *mole* is the amount of a substance that has a mass in grams numerically equivalent to its *molecular weight* (sum of the weight of all atoms in the molecule) in daltons. A mole of any substance has exactly the same number of molecules—6.02×10^{23}, called Avogadro's number. The *molarity* of a solution (abbreviated M) refers to the number of moles of a solute dissolved in 1 liter of solution.

Acids, Bases, and pH A water molecule can dissociate into a *hydrogen ion*, H^+ (which binds to another water molecule to form a *hydronium ion*, H_3O^+), and a *hydroxide ion* (OH^-). Although reversible and statistically rare, this dissociation into the highly reactive hydrogen and hydroxide ions has important biological consequences.

In pure water, the concentrations of H^+ and OH^- ions are the same; both are equal to 10^{-7} M. When acids or bases dissolve in water, the H^+ and OH^- balance shifts. An *acid* adds H^+ to a solution, whereas a *base* reduces H^+ in a solution by accepting hydrogen ions or by adding hydroxide ions (which then combine with H^+ and thus remove hydrogen ions). A strong acid or strong base is a substance that dissociates completely when mixed with water. A weak acid or base reversibly dissociates, releasing or binding H^+.

A solution with a higher concentration of H^+ than of OH^- is considered acidic. A basic solution has a higher concentration of OH^-. In a neutral solution, $[H^+]$ and $[OH^-]$ are both equal to 10^{-7} M. Brackets, [ ], indicate molar concentration. In any solution, the product of the $[H^+]$ and $[OH^-]$ is constant at 10^{-14} M. If the $[H^+]$ is higher, then the $[OH^-]$ is lower, due to the tendency of excess hydrogen ions to combine with the hydroxide ions in solution and form water. Likewise, an increase in $[OH^-]$ causes an equivalent decrease in $[H^+]$. If $[OH^-]$ is equal to 10^{-10}, then $[H^-]$ will equal 10^{-4}.

The hydrogen and hydroxide ion concentration can vary in different solutions by many orders of magnitude. The *pH scale* compresses the range of these concentrations because it is logarithmic. The pH of a solution is defined as the negative log (base 10) of the $[H^+]$: pH = $-$log $[H^+]$. For a neutral solution, $[H^+]$ is 10^{-7} M, and the pH equals 7. As the $[H^+]$ increases in an acidic solution, the pH value decreases. A pH value below 7 indicates an acidic solution; a value above 7 denotes a basic solution. The difference between each unit of the pH scale represents a tenfold difference in the concentration of $[H^+]$ and $[OH^-]$.

Most cells have an internal pH of 7. Even a slight shift in cellular pH can be harmful. *Buffers* within the cell minimize changes in concentration of hydrogen and hydroxide ions and thus maintain a constant pH. A buffer is a substance that accepts H^+ ions when they are in excess and donates H^+ ions to a solution when their concentration decreases. An acid and base in equilibrium with each other is typical of most buffering systems. The carbonic acid/bicarbonate buffering system is an important biological buffer. Carbonic acid acts as an acid to donate H^+ ions when pH starts to increase. When pH falls, the excess H^+ ions are accepted by the bicarbonate ion, which acts like a base.

$$H_2CO_3 \;\rightleftharpoons\; HCO_3^- + H^+$$

carbonic acid   bicarbonate
H^+ donor H^+ acceptor

Acid Rain: Upsetting the Fitness of the Environment

The environmental problem of acid rain emphasizes the sensitivity of life to pH. Acid rain, with a pH lower than the normal 5.6 pH of rain, is due to the reaction of water in the atmosphere with the sulfur oxides and nitrogen oxides released by the combustion of fossil fuels.

Acid rain has harmful effects on both terrestrial and freshwater ecosystems. Lowering the pH of the soil solution affects the solubility of minerals needed by plants. A lowered pH of lakes and ponds harms many species of fishes, amphibians, and aquatic invertebrates. The decline of European forests and the loss of fish populations from many of the lakes in the Adirondacks of New York are the results of acid rain. The reduction of acid rain depends on the development of industrial controls and antipollution devices.

STRUCTURE YOUR KNOWLEDGE

You need to be able to relate two major clusters of concepts to have a good understanding of water and its contribution to the fitness of the environment. Suggested concept maps are included at the end of this unit, but remember that your concept map should represent your own understanding. The value of this exercise is in your process of organizing these concepts for yourself.

1. The tendency of water to form hydrogen bonds between its polar molecules has a profound effect on temperature regulation within the environment. Select the key concepts that relate to this important property of water and create a concept map showing how the breaking and formation of hydrogen bonds are related to temperature.

2. To become proficient in the use of the concepts relating to pH, develop a concept map to organize your understanding of the following terms: pH, [H⁺], [OH⁻], acidic, basic, neutral, buffer, 1–14, acid-base pair. Remember to label connecting lines and add additional concepts as you need them.

TEST YOUR KNOWLEDGE

MULTIPLE CHOICE: *Choose the one best answer.*

1. Water contributes to the fitness of the environment because
   a. plants need water to grow.
   b. life evolved in water.
   c. the making and breaking of H bonds helps regulate temperature.
   d. the surface tension of water creates a niche for water organisms.

2. The polarity of the water molecule
   a. promotes the formation of hydrogen bonds.
   b. helps water to dissolve nonpolar solutes.
   c. lowers the heat of vaporization.
   d. does all of the above.

3. A low specific heat would mean that
   a. little heat must be absorbed or released to effect a temperature change.
   b. breaking hydrogen bonds releases a small amount of heat.
   c. breaking hydrogen bonds absorbs a small amount of heat.
   d. boiling temperature would probably be high.

4. Climates tend to be moderate by large bodies of water because
   a. a large amount of solar heat is absorbed by the gradual rise in temperature of the water.
   b. the gradual cooling of the water releases heat to the environment.
   c. the high specific heat of water helps to regulate air temperatures.
   d. of all of the above.

5. Temperature is a measure of
   a. specific heat.
   b. average kinetic energy of molecules.
   c. total kinetic energy of molecules.
   d. Celsius degrees.

6. Evaporative cooling is a result of
   a. a low heat of vaporization.
   b. a high heat of melting.

   c. a reduction in the average kinetic energy of the liquid remaining after molecules enter the gaseous state.
   d. sweating on a humid day.

7. Ice floats because
   a. air is trapped in the crystalline lattice.
   b. the formation of hydrogen bonds releases heat; warmer objects float.
   c. it has a larger surface area than liquid water.
   d. hydrogen bonding spaces the molecules farther apart, creating a less dense structure.

8. The molarity of a solution is equal to
   a. Avogadro's number of molecules in 1 liter of solvent.
   b. number of moles of a solute in 1 liter of solution.
   c. molecular weight of a solute in 1 liter of solution.
   d. number of solute particles in 1 liter of solvent.

9. A solution with a pH of 2, compared to a solution with pH 4,
   a. is twice as acidic.
   b. is 100 times more acidic.
   c. is 1000 times more acidic.
   d. has two times more [H⁺]

10. A buffer
   a. changes pH by a magnitude of 10.
   b. absorbs excess OH⁻.
   c. releases excess H⁺.
   d. is often a weak acid-base pair.

11. Which of the following is least soluble in water?
   a. polar compounds
   b. nonpolar compounds
   c. ionic compounds
   d. hydrophilic molecules

12. What accounts for the movement of water up xylem vessels in a plant?
   a. cohesion
   b. hydrogen bonding
   c. adhesion
   d. all of the above

13. What bonds must be broken for water to vaporize?
   a. polar covalent bonds
   b. nonpolar covalent bonds
   c. hydrogen bonds
   d. all of the above

14. How would you make a 0.1 M solution of acetic acid ($C_2H_4O_2$)? The mass numbers for these elements are C = 12, O = 16, H = 1.
   a. Mix 2 g carbon, 4 g hydrogen and 2 g oxygen in 1 liter of water.

    b. Mix 6 g of acetic acid with enough water to make 1 liter of solution.

    c. Mix 60 g of acetic acid with enough water to yield 1 liter of solution.

    d. Mix 0.1 mol of acetic acid with 1 mol of water.

15. How many molecules of acetic acid would be in the solution in question 14?

    a. 0.1

    b. 6

    c. 60

    d. 6×10^{22}

FILL IN THE BLANKS: *Complete the following table on pH.*

$[H^+]$	$[OH^-]$	pH	Acidic, basic, or neutral?
	10^{-11}	3	acidic
10^{-8}		8	
10^{-12}			
	10^{-5}		
		1	
	10^{-7}		

CARBON AND MOLECULAR DIVERSITY

FRAMEWORK

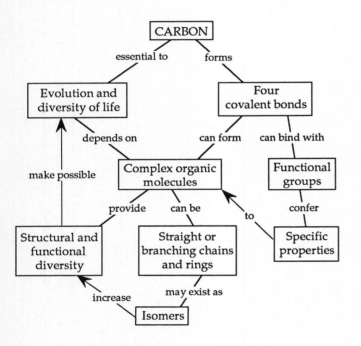

CHAPTER SUMMARY

The ability of carbon to form large and complex molecules is central to the evolution and diversity of life. The diverse functional properties of carbon-based molecules emerge from their molecular architecture.

The Foundations of Organic Chemistry

Organic chemistry is the study of carbon-containing molecules. Early organic chemists could not synthesize the complex molecules found in living organisms,

and therefore attributed the existence of life and the formation of these molecules to a life force independent of physical and chemical laws, a belief known as *vitalism.* Chemists began to synthesize organic compounds from inorganic substances in the 1800s. Recent experiments have demonstrated the spontaneous synthesis of organic compounds from inorganic compounds in an atmosphere similar to that hypothesized on primitive Earth. Mechanism, the philosophy underlying modern organic chemistry, holds that physical and chemical laws and explanations are sufficient to account for all natural phenomena, even the evolution of life.

The Versatility of Carbon in Molecular Architecture

Carbon has six electrons. To complete its valence shell, carbon forms four covalent bonds with other atoms. This tetravalence is at the center of carbon's ability to form large and complex molecules with characteristic three-dimensional shapes and properties. When carbon forms four single covalent bonds, it is shaped like a tetrahedron. The spatial arrangement of atoms can determine the function of a molecule in a cell.

Variation in Carbon Skeletons

The four covalent bonds of carbon can include single, double, or even triple bonds. Carbon atoms readily bond with each other, producing chains or rings of carbon atoms. These molecular backbones can vary in length, branching, placement of double bonds, and location of atoms of other elements. The simplest organic molecules are *hydrocarbons,* consisting of only carbon and hydrogen. Fossil fuels are composed of hydrocarbons. The nonpolar C—H bonds in hydrocarbon chains account for the hydrophobic behavior of fats.

Isomers Isomers are compounds with the same molecular formula but different structural arrangements and thus, different properties. *Structural isomers* differ in the arrangement of atoms and often in the location of double bonds. Ethyl alcohol and dimethyl ether (Figure 4.1a) have the same number and kinds of atoms but a different bonding sequence and very different properties.

Geometric isomers have the same sequence of covalently bonded atoms but differ in spatial arrangement due to the inflexibility of double bonds. Atoms freely rotate about a single covalent bond, but double bonds restrict that rotation and thus fix the spatial arrangement of a molecule. Maleic acid and fumaric acid (Figure 4.1b) are geometric isomers.

Stereoisomers are molecules that are mirror images of each other. An asymmetric carbon is one that is covalently bound to four different kinds of atoms or groups of atoms. Due to the tetrahedral shape of the asymmetric carbon, the four groups can be attached in spatial arrangements that are not superimposable on each other. Stereoisomers are left- and right-handed versions of each other and can differ greatly in their biological activity. Although *l*- and *d*-lactic acid look similar in a flat representation of their structures, they are not superimposable (Figure 4.1c). Usually only one form of a stereoisomer is biologically active because cells can differentiate between left- and right-handed versions.

Functional Groups

The properties of organic molecules are determined by groups of atoms, known as *functional groups*, which bond to the carbon skeleton and behave consistently from one carbon-based molecule to another. Because they are hydrophilic, these functional groups increase the solubility of organic compounds in water. Six functional groups are most important in the chemistry of life.

The *hydroxyl group* consists of an oxygen and hydrogen (–OH) covalently bonded to the carbon skeleton. The polarity of the hydroxyl group, due to the electronegative oxygen, increases a compound's solubility in water. Organic molecules with hydroxyl groups are called *alcohols*, and their names often end in *-ol*.

Carbonyl groups consist of a carbon double bonded to an oxygen (–C=O). If the carbonyl group is at the end of the carbon skeleton, the compound is called an *aldehyde*. If the carbonyl group is not at the end, the compound is called a *ketone*.

A *carboxyl group* consists of a carbon double bonded to an oxygen and also attached to a hydroxyl group (–COOH). Compounds with a carboxyl group are called *carboxylic acids* or organic acids because they tend to dissociate to release H^+. The two electronegative oxygens of this group pull the shared electrons away from the hydrogen.

An *amino group* consists of a nitrogen atom bonded to two hydrogens (–NH_2). Compounds with an amino group, called *amines*, can act as bases. The nitrogen, with its pair of unshared electrons, can attract a hydrogen ion, becoming –NH_3^+. Amino acids, the building blocks of proteins, contain both an amino and a carboxyl group (hence the name *amino acid*).

The *sulfhydryl group* consists of a sulfur atom bound to a hydrogen (–SH). The cross-linkings of sulfhydryl groups help to stabilize protein structure.

A *phosphate group* is bound to the carbon skeleton by an oxygen attached to a phosphorus atom bound to three other oxygen atoms (–$OPO_3^=$). An organic compound with a phosphate ion can store energy and transfer it to another molecule.

The Chemical Elements of Life: A Review

Carbon, oxygen, hydrogen, nitrogen, and smaller quantities of sulfur and phosphorus, all capable of forming strong covalent bonds, are combined into the

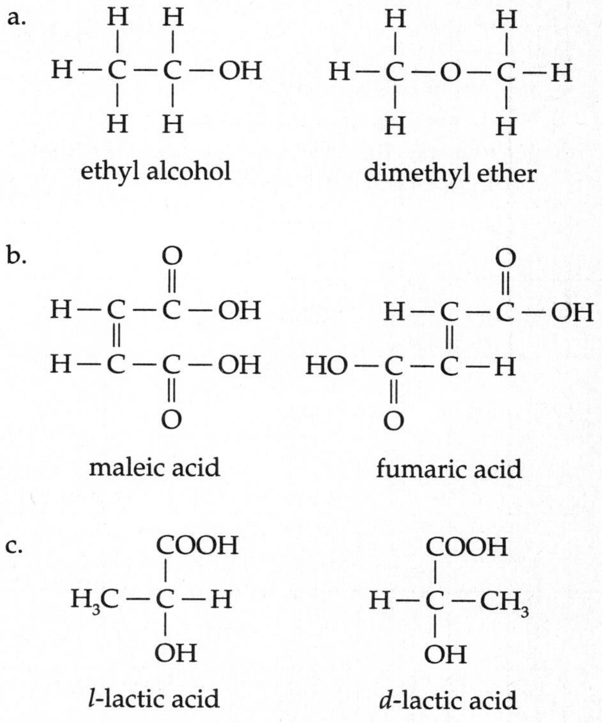

Figure 4.1 (a) Structural, (b) geometric, and (c) stereoisomers.

complex organic molecules of living matter. The versatility of carbon in forming four covalent bonds, linking readily with itself to produce chains and rings, and binding with other elements and functional groups, makes possible the incredible diversity of organic molecules.

STRUCTURE YOUR KNOWLEDGE

1. One characteristic of organic compounds is that they can exist in the form of isomers—molecules with the same molecular formula but different spatial arrangements of atoms and thus different properties. Construct a concept map that illustrates your understanding of the characteristics and significance of the three types of isomers.

2. Fill in the following table on the functional groups.

TEST YOUR KNOWLEDGE

MULTIPLE CHOICE: *Choose the one best answer.*

1. The tetravalence of carbon most directly results from
   a. its tetrahedral shape.
   b. its very slight electronegativity.
   c. its four electrons in the valence shell that can form four covalent bonds.
   d. its ability to form single, double, and triple bonds.

2. Hydrocarbons are not soluble in water because
   a. they are hydrophilic.
   b. the C–H bond is very nonpolar.
   c. they do not ionize.
   d. they store energy in the many C–H bonds along the carbon backbone.

3. Which functional group forms cross-links that stabilize protein structure?
   a. amino.
   b. carboxyl.
   c. phosphate.
   d. sulfhydryl.

4. Which of the following is not true of an asymmetric carbon atom?
   a. It is attached to four different atoms or groups.
   b. It can result in geometric isomers.
   c. It can result in stereoisomers.
   d. Its configuration is in the shape of a tetrahedron.

5. A reductionist approach to considering the structure and function of organic molecules would be based on
   a. mechanism.
   b. holism.
   c. determinism.
   d. vitalism.

6. The functional group that can cause an organic molecule to act as a base is
   a. –COOH.
   b. –OH.

Name of Group	Molecular Formula	Characteristics Conferred to Organic Compound
	–OH	
		Aldehyde or ketone, polar group
Carboxyl		
	–NH$_2$	
		Crosslinks stabilize protein structure
Phosphate		

c. –SH.
d. –NH₂.

7. The functional group that confers acidic properties on organic molecules is
   a. –COOH.
   b. –OH.
   c. –SH.
   d. –NH₂.

8. Which is *not* true about geometric isomers?
   a. They have different chemical properties.
   b. They have the same molecular formula.
   c. Their atoms and bonds are arranged in different sequences.
   d. The rigidity of the double carbon bond restricts movement.

9. The tetrahedral shape of the four covalent bonds of a carbon atom is due to
   a. the angles at which the shared electrons of the single covalent bonds are found.
   b. the difference in electronegativity of the bonds.
   c. the asymmetric nature of the carbon atom.
   d. geometric isomerism.

10. Which of the following is *not* true of amino acids?
    a. They can act as buffers.
    b. They are the building blocks of nucleic acids.
    c. They are often ionized in cells.
    d. They are named for their amino and carboxyl functional groups.

11. How many asymmetric carbons are there in the sugar galactose?
    a. 2
    b. 4
    c. 5
    d. 6

12. Which functional groups can act as acids?
    a. amino and sulfhydryl
    b. carbonyl and carboxyl
    c. carboxyl and phosphate
    d. alcohol and aldehyde

MATCHING: *Match the formulas to the terms below. Choices may be used more than once; more than one right choice may be available.*

1. ____&____ structural isomers
2. ____&____ geometric isomers
3. ____&____ stereoisomers
4. _____ carboxylic acid
5. _____ can make cross-link in protein
6. _____ alcohol
7. _____ aldehyde
8. _____ amino acid
9. _____ organic phosphate
10. _____ hydrocarbon
11. _____ amine
12. _____ ketone

CHAPTER 5

STRUCTURE AND FUNCTION OF MACROMOLECULES

FRAMEWORK

This chapter introduces the major classes of macromolecules that constitute living organisms. The central ideas of this chapter are that molecular function relates to molecular structure and that the diversity of molecular structure is the basis for the diversity of life. Combining a small number of monomers or subunits into unique sequences and three-dimensional structures creates a huge variety of macromolecules. The chart below briefly summarizes the major characteristics of the four classes of macromolecules.

CHAPTER SUMMARY

The four classes of large molecules found in living matter are carbohydrates, lipids, proteins, and nucleic acids. These giant molecules, called *macromolecules*, represent another level in the hierarchy of biological organization. The functions of these molecules are related to their complex and unique architectures.

Polymers

Polymers are large molecules formed from the linking together of many similar or identical small molecules, called *monomers*.

Polymers and Molecular Diversity Macromolecules are constructed from about 40 to 50 common monomers and a few rarer molecules. The limitless variety of polymers arises from the essentially infinite number of possibilities in the sequencing and arrangement of these basic subunits. The molecular basis for the diversity of life rests on the ordering of common small molecules into distinctive and unique macromolecules.

Making and Breaking Polymers The same chemical process forms macromolecules of all four classes. Monomers are joined by *polymerization,* which involves the removal of a molecule of water. In this condensation synthesis, one monomer loses a hydroxyl (OH) and the other contributes a hydrogen (H) to the water molecule, and a covalent bond between the monomers is formed. Energy is required to join

CLASS	MONOMERS	FUNCTIONS
Carbohydrates	Monosaccharides	Energy, raw materials, energy storage, structural compounds
Lipids	Glycerol, fatty acids	Energy storage, membranes, steroids, hormones
Proteins	Amino acids	Enzymes, structural compounds, movement, transport, hormones
Nucleic Acids	Nucleotides [pentose, nitrogenous base, phosphate group]	Heredity, code for amino acid sequence

monomers and the process is facilitated by enzymes.

Hydrolysis is the breaking of bonds between monomers through the addition of water molecules. A hydroxyl is joined to one monomer while a hydrogen is bonded with the other. Enzymes also control hydrolysis.

Carbohydrates

Carbohydrates include simple sugars, disaccharides and polysaccharides.

Monosaccharides Monosaccharides have the general formula of $(CH_2O)_n$. The number of these units forming a sugar varies from three to seven, with hexoses $(C_6H_{12}O_6)$, trioses, and pentoses found most commonly. A sugar can be an aldose or ketose, depending on the location of the carbonyl group (C=O). The other carbon atoms characteristically all bear hydroxyl groups (OH). Additional diversity among sugar molecules is provided by the spatial arrangement of parts around assymmetric carbons. The hexoses glucose and galactose, for instance, differ only in the arrangement around one asymmetric carbon. These small structural differences affect the recognition and interaction of molecules within cells.

In aqueous solutions, most monosaccharides are in a ring structure.

Glucose, a hexose, is broken down to yield energy in cellular respiration. Monosaccharides serve not only as the main fuel for cellular work, but also as the raw materials for synthesis of other small organic molecules, and as monomers that are synthesized into polysaccharides that serve as storage or structural molecules.

Disaccharides A *glycosidic linkage* is the bond formed between two monosaccharides. Maltose, a *disaccharide* formed from two glucose molecules, has a 1–4 linkage between the number one carbon of one glucose and the number four carbon of the other glucose. Sucrose, the common disaccharide we know as table sugar, is formed from the joining of a glucose and fructose molecule.

Polysaccharides Polysaccharides are macromolecules made from a few hundred to a few thousand monosaccharides. *Starch*, a storage molecule in plants, is a polymer made of glucose molecules joined by 1–4 linkages. Starch has a helical shape resulting from the angle of these bonds. Amylose is an unbranched, simple form of starch. Plants store sugar for later use in starch molecules. Most animals have enzymes to hydrolyze plant starch into glucose. Animals produce *glycogen*, a highly branched polymer of glucose, as their energy storage form.

Cellulose, the major component of plant cell walls, is the most abundant organic compound on earth. It differs from starch by the configuration of the ring form of glucose. In cellulose, the glucose monomers are in the beta (ß) configuration (the hydroxyl group is locked above the plane of the ring); in starch, there are 1–4 linkages between the alpha (α) glucose molecules. The resulting geometry of the glycosidic bonds is responsible for the different three-dimensional shapes and properties of these plant polysaccharides. Enzymes that digest starch are unable to hydrolyze the ß linkages of cellulose. Only a few organisms (some bacteria, microorganisms, and fungi) have enzymes that can digest cellulose. Cellulose is not digested by humans, but provides fiber to the diet.

In a plant cell wall, hydrogen bonds between hydroxyl groups hold parallel cellulose molecules together to form a microfibril. Several microfibrils are in turn intertwined to form a cellulose fibril, which may supercoil with other fibrils to form strong structural cables.

Chitin is a structural polysaccharide formed from modified amino sugar monomers and found in the exoskeleton of arthropods and the cell walls of many fungi.

Lipids

Fats, phospholipids, and steroids are a diverse assemblage of macromolecules, classed together as *lipids*, because they are all partially or wholly hydrophobic. They shun water due to the predominance of nonpolar hydrocarbon regions in their molecules.

Fats Fats are composed of fatty acids attached to the three-carbon alcohol, glycerol. A *fatty acid* consists of a long hydrocarbon tail (usually 16 or 18 carbons in length) with a carboxyl group at the "head" end.

Fatty acids are linked to glycerol by an ester linkage, a bond that forms between a hydroxyl and a carboxyl group. In a *triacylglycerol*, or fat, three fatty acids attach to one glycerol. Triglycerides is another name for fats.

Fatty acids with double bonds in their carbon skeletons are called *unsaturated*. The double bonds create a kink in the shape of the molecule and prevent the fat molecules from packing close together and becoming solidified at room temperature. *Saturated* fats have fatty acids with no double bonds in their carbon skeletons; each carbon atom is "saturated" with hydrogen. Most animal fats are saturated and solid at room temperature. Plant fats are generally unsaturated and are called oils. Diets rich in saturated fats have been linked to cardiovascular disease.

Fats are excellent energy storage molecules, containing twice the energy reserves as do carbohydrates such as starch. Adipose tissue, made of fat storage cells, also cushions organs and insulates the body.

Phosopholipids   *Phospholipids* consist of a glycerol linked to two fatty acids and a negatively charged phosphate group. Other small molecules may be attached to the phosphate group. The phosphate head of this molecule is hydrophilic and water soluble, whereas the two fatty acid chains are hydrophobic. This unique structure of phospholipids makes them ideal constituents of cell membranes. Arranged in a bilayer, the hydrophilic heads face toward the aqueous solutions inside and outside the cell, and the hydrophobic tails mingle in the center of the membrane.

Steroids   *Steroids* are a class of lipids distinguished by four connected carbon rings with various functional groups attached. *Cholesterol* is an important steroid that is a common component of animal cell membranes and a precursor for most other steroids, including many hormones.

Proteins

Proteins serve such diverse functions as structural support (fibers), storage of amino acids, transport (hemoglobin), internal coordination (hormones), movement (actin and myosin in muscle), defense (antibodies), reception of chemical signals, and enzymatic control of chemical reactions. This wide range of functions is made possible by the highly structured and unique three-dimensional shapes of these macromolecules.

Amino Acids   Amino acids serve not only as protein monomers; they may function also as individual molecules. Most amino acids are composed of an asymmetric carbon (called the α carbon) bonded to a carboxyl group, an amino group, a hydrogen, and a variable *side chain* called the R group. The R group confers the unique physical and chemical properties of each amino acid. Side chains may be nonpolar and hydrophobic, or polar or charged and thus hydrophilic. Acidic side chains contain carboxyl groups and are negatively charged; basic side chains contain an amino group (usually positively charged).

Polypeptide Chains   A *peptide bond* links the amino group of one amino acid with the carboxyl group of

another, yielding a polymer of amino acids called a *polypeptide chain*. The chain has a free carboxyl group at one end (the C-terminus) and a free amino group at the other (the N-terminus). The R groups of each amino acid extend out from this polypeptide backbone and participate in interactions that help create the three-dimensional structure of proteins. Polypeptides, each with its own unique sequence, vary in length from a few to a thousand or more amino acids.

Protein Conformation   Proteins have unique three-dimensional shapes, or *conformations*, created by the twisting or folding of one or more polypeptide chains. The unique conformation of a protein, which results from its sequence of amino acids, enables it to recognize and bind to other molecules.

Levels of Protein Structure   There are three structural levels in the conformation of a protein. A fourth level may be present when a protein consists of more than one polypeptide chain. *Primary structure* is the unique, genetically-coded sequence of amino acids within a protein. Even a slight deviation from the sequence of amino acids can severely affect a protein's function by altering the protein's conformation.

In the early 1950s, Sanger determined the primary structure of insulin through the laborious process of hydrolyzing the protein into small peptide chains, using chromatography to separate the small pieces, determining their sequences of amino acids, and then overlapping the sequences of small fragments created with different agents to reconstruct the whole polypeptide. Most of these steps are now automated, and the primary structures of hundreds of proteins have been determined.

Secondary structure involves the coiling or folding of the polypeptide backbone, stabilized by hydrogen bonds between the electronegative oxygen of one peptide bond and the weakly positive hydrogen attached to a nitrogen of another bond. An *alpha (α) helix* is a delicate coil produced by hydrogen bonding between every fourth peptide bond. Some fibrous proteins have alpha helices along most of their length. Globular proteins are more likely to have regions of alpha helix alternating with non-helical regions.

A *beta (ß) pleated sheet* is also held by repeated hydrogen bonds along the protein's backbone. This secondary structure forms when the polypeptide chain folds back and forth or when regions of the chain lie parallel to each other. Beta sheets are found in the dense core of many globular proteins and in some fibrous proteins.

Interactions between the various side chains of the constituent amino acids produce a protein's *tertiary* structure. *Hydrophobic interactions* between nonpolar side groups in the center of the molecule, hydrogen bonds, and ionic bonds between negatively and positively charged side chains produce a stable and unique shape to the protein. Strong covalent bonds, called *disulfide bridges*, may occur between the sulfhydryl side groups of cysteine monomers that have been brought close together by the folding of the polypeptide.

Quaternary structure occurs in proteins that are composed of more than one polypeptide chain. The individual polypeptide chains, called *subunits*, are held together in a precise structural arrangement. Connective tissue in animals gains its strength from collagen, a fibrous protein composed of three helical subunits supercoiled together. Hemoglobin is a globular protein consisting of four polypeptide chains.

What Determines Conformation?   The specific function of a protein is an emergent property of its three-dimensional architecture. Protein conformation is dependent upon the interactions among the amino acids making up the polypeptide chain and usually arises spontaneously as soon as the protein is synthesized in the cell. These interactions can be disrupted by changes in pH, salt concentration, temperature, or other aspects of the environment, and the protein may *denature*, losing its native conformation and thus its function. Any factor that disrupts the hydrogen bonding, hydrophobic interactions, ionic bonds, or disulfide bridges within a protein molecule will cause denaturation. Sometimes a denatured protein will reform to its three-dimensional conformation when returned to its normal environment.

The Protein-Folding Problem   The amino acid sequences of hundreds of proteins have been determined. Using the technique of X-ray crystallography, coupled with computer modeling and graphics, molecular biologists have established the three-dimensional shape of many of these molecules. But the rules of protein folding that could predict the conformation arising from a particular primary structure of a protein have been difficult to determine. Most proteins probably go through intermediate states on their way to their final form. As protein folding becomes better understood, scientists are learning to devise amino acid sequences for proteins that will be able to perform predetermined functions.

Nucleic Acids

Functions of Nucleic Acids: An Overview   Nucleic *acids* are macromolecules that carry and translate the code that directs all the cell's activities. *DNA, deoxyribonucleic acid*, is the genetic material that is inherited from one generation to the next and is reproduced in each cell of an organism. *RNA, ribonucleic acid*, reads the instructions coded in DNA and directs the synthesis of proteins, the ultimate enactors of the genetic program.

Nucleotides   Nucleic acids are polymers of *nucleotides*, monomers that consist of a pentose (five-carbon) sugar covalently bonded to a phosphate group and a nitrogenous base. There are two families of nitrogenous bases. *Pyrimidines*, including cytosine (C), thymine (T), and uracil (U), are characterized by six-membered rings of carbon and nitrogen atoms. Thymine is found only in DNA; uracil is found only in RNA. *Purines*, adenine (A) and guanine (G), add a five-membered ring to the pyrimidine ring. These compounds are called nitrogenous *bases* because the nitrogen tends to take up H^+. The pentose sugar is either *ribose*, in RNA, or *deoxyribose* (missing one hydroxyl group), in DNA. *Nucleosides* consist of a nitrogenous base joined to a sugar. Nucleotides have a phosphate group attached to the sugar.

Polynucleotides   Nucleotides are linked together into *polynucleotides* by *phosphodiester linkages*, which join the phosphate of one nucleotide with the sugar of the next. The nitrogenous bases extend from this repeating sugar-phosphate backbone. The unique sequence of bases in a gene codes for the specific amino acid sequence of a protein.

The Double Helix: An Introduction   DNA molecules consist of two polynucleotide chains spiraling around an imaginary axis in a *double helix*. In 1953, Watson and Crick first proposed this double helix arrangement, which consists of two sugar-phosphate backbones on the outside of the helix with their nitrogenous bases pairing and hydrogen bonding together in the inside. Adenine pairs with only thymine; guanine always pairs with cytosine. Thus, the sequences of nitrogenous bases on the two strands of DNA are complementary, predictable counterparts of each other. Because of this specific base-pairing property, DNA can replicate itself and precisely copy the genes of inheritance.

DNA and Proteins as Tape Measures of Evolution

Genes form the hereditary link between generations. Closely related members of the same species share more common DNA sequences. More closely related species also have a larger proportion of their DNA and proteins in common; this "molecular genealogy" provides evidence of evolutionary relationships.

STRUCTURE YOUR KNOWLEDGE

1. Create a concept map (or a brief table, chart, or summary, if you prefer) for each of the four families of macromolecules—carbohydrates, lipids, proteins, and nucleic acids. Include the monomers involved, the type of linkage, and the unique characteristics of the structure of the polymers. A few examples in each group would probably be helpful. Organize this information so that you can relate structure to function.

2. Describe the three structural levels in the conformation of a protein. In the following diagram of a portion of a polypeptide, label the types of interactions that are shown. What level of conformation do these interactions create?

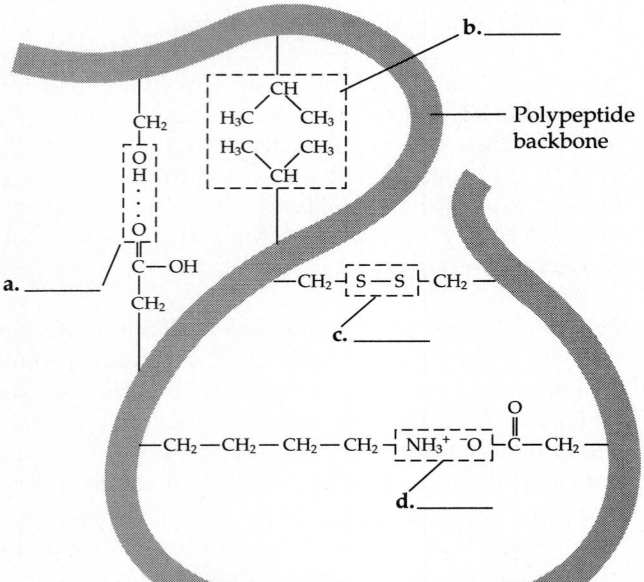

A. _____

B. _____

C. _____   D. _____

E. _____

F. _____

G. _____

TEST YOUR KNOWLEDGE

MATCHING OF FORMULAS: *Identify the type of compound shown by these formulas. Then match the chemical formulas with their description. Answers may be used more than once.*

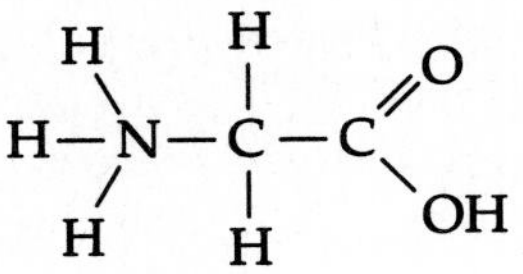

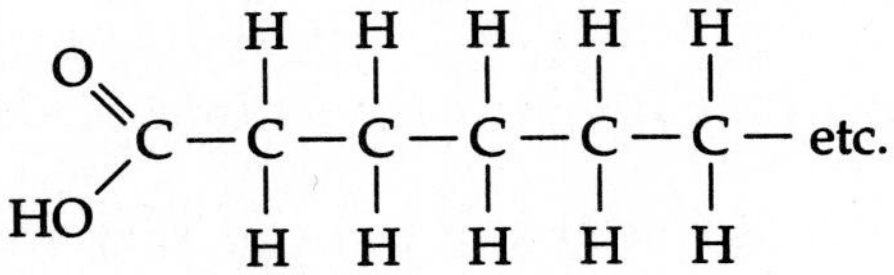

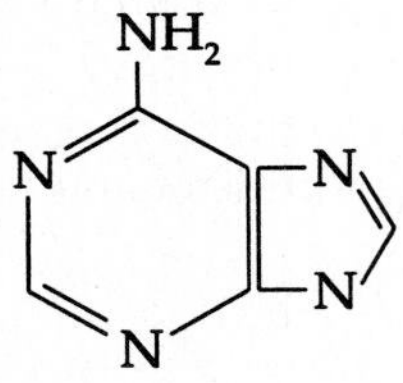

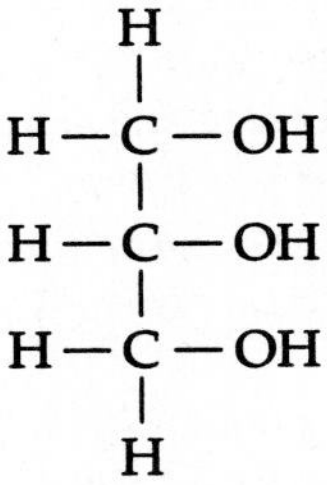

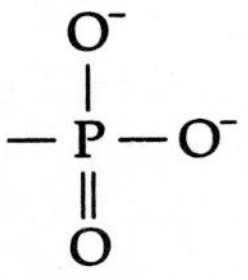

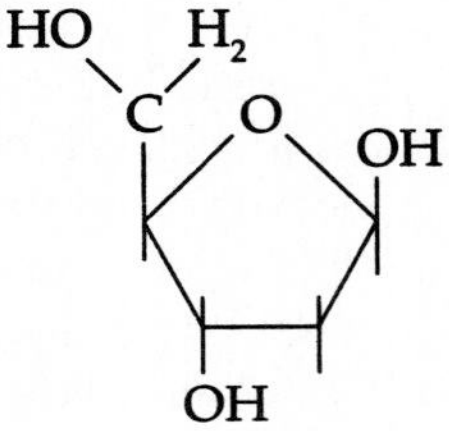

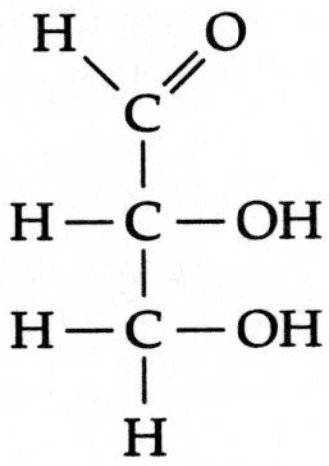

1. _____ molecules that would combine to form a fat

2. _____ molecule that would be attached to other monomers by a peptide bond

3. _____ molecules or groups that would combine to form a DNA nucleotide

4. _____ molecules that are carbohydrates

5. _____ molecule that is a purine

6. _____ monomer of a protein

7. _____ groups that would be joined by phosphodiester bonds

MATCHING: *Match the molecule with its class of macromolecules.*

1. _____ glycogen   **A.** carbohydrate

2. _____ cholesterol   **B.** lipid

3. _____ RNA   **C.** protein

4. _____ collagen   **D.** nucleic acid

5. _____ hemoglobin

6. _____ a gene

7. _____ triaclyglycerol

8. _____ enzyme

9. _____ cellulose

10. _____ chitin

MULTIPLE CHOICE: *Choose the one best answer.*

1. Polymerization is a process that
   a. creates bonds between amino acids in the formation of a peptide chain.
   b. involves the removal of a water molecule.
   c. links the phosphate of one nucleotide with the sugar of the next.
   d. involves all of the above.

2. Which of the following is *not* true of pentoses?
   a. They are found in nucleic acids.
   b. They can occur in a ring structure.
   c. They have the formula $C_5H_{12}O_5$.
   d. They have hydroxyl and carbonyl groups.

3. Disaccharides can differ from each other in all of the following ways *except*
   a. in the number of their monosaccharides.
   b. as stereoisomers.
   c. in the monomers involved.
   d. in the location of their glycosidic linkage.

4. Which of the following is *not* true of cellulose?
   a. It is the most abundant organic compound on earth.
   b. It differs from starch because of the configuration of glucose and the geometry of the glycosidic linkage.
   c. It is a highly branched polysaccharide.
   d. Few organisms have enzymes that hydrolyze its glycosidic linkages.

5. Plants store most of their energy as
   a. glucose.
   b. glycogen.
   c. starch.
   d. sucrose.

6. When a protein denatures, it
   a. loses its primary structure.
   b. loses its secondary and tertiary structure.
   c. becomes insoluble and precipitates.
   d. hydrolyzes into component amino acids.

7. The alpha helix of proteins is
   a. part of the tertiary structure and is stabilized by disulfide bridges.
   b. a double helix.
   c. stabilized by hydrogen bonds and commonly found in fibrous proteins.
   d. found in some regions of globular proteins and stabilized by hydrophobic interactions.

8. A fatty acid that has the formula $C_{16}H_{32}O_2$ is
   a. saturated.
   b. unsaturated.
   c. branched.
   d. hydrophilic.

9. Three molecules of the fatty acid in question 8 are joined to a molecule of glycerol ($C_3H_8O_3$). The resulting molecule has the formula
   a. $C_{48}H_{96}O_6$.
   b. $C_{51}H_{104}O_9$.
   c. $C_{51}H_{102}O_8$.
   d. $C_{51}H_{96}O_6$.

10. The molecule formed in question 9 is
    a. a triaclyglyceride.
    b. a lipid.
    c. a fat.
    d. all of the above.

11. Which of the following is hydrophobic?
    a. steroid
    b. chitin
    c. head end of phospholipid
    d. polynucleotide

12. Beta sheets are characterized by
    a. disulfide bridges between cysteine amino acids.
    b. back-and-forth folds of the polypeptide chain held together by hydrophobic interactions.
    c. folds stabilized by hydrogen bonds between segments of polypeptide chains.
    d. membrane sheets composed of phospholipids.

FILL IN THE BLANKS

1. The nitrogenous base absent in RNA is _____.

2. Cytosine always pairs with _____.

3. Adenine and guanine are _____.

4. A nitrogenous base joined to a pentose sugar is called a _____.

5. Adding a phosphate group to the molecule in question 4 yields a _____.

6. Proteins with more than one peptide chain have _____structure.

7. The conformation of a protein is determined by its _____.

8. The energy storage molecule of animals is _____.

9. Membranes are composed of a bilayer of _____.

10. The linkages between amino acids in a protein are called _____.

11. A technique used to determine the three-dimensional structure of macromolecules is _____.

12. The man who determined the amino acid sequence of insulin was _____.

CHAPTER 6

INTRODUCTION TO METABOLISM

FRAMEWORK

This chapter considers metabolism, the totality of the chemical reactions that take place in living organisms. Two key topics are emphasized: the energy transformations that underlie all chemical reactions and the role of enzymes in the "cold chemistry" of the cell.

The reactions in a cell either consume or release energy. The following illustration summarizes some of the components of the energy changes within a cell.

Enzymes are biological catalysts that lower the activation energy of a reaction and thus greatly speed up metabolic processes. An enzyme is a three-dimensional protein molecule with an active site specific for its substrate. Intricate control and feedback mechanisms produce the metabolic integration necessary for life.

CHAPTER SUMMARY

The Metabolic Map: An Overview

Metabolism includes the thousands of precisely coordinated, complex, efficient, and integrated chemical reactions in a cell. These reactions are ordered into metabolic pathways—sequenced and intricately branched routes controlled by enzymes. Through these pathways the cell creates and transforms the organic molecules that provide the material and energy for life. Metabolism is an emergent property arising from the organization and orderly interactions of molecules in cells.

Catabolic pathways release the energy stored in complex molecules through the breaking down or degra-

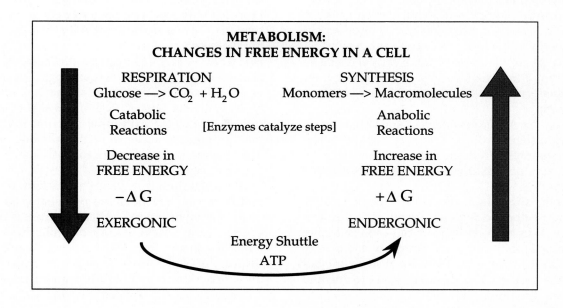

METABOLISM:
CHANGES IN FREE ENERGY IN A CELL

RESPIRATION	SYNTHESIS
Glucose —> CO_2 + H_2O	Monomers —> Macromolecules
Catabolic Reactions	Anabolic Reactions
[Enzymes catalyze steps]	
Decrease in FREE ENERGY	Increase in FREE ENERGY
$-\Delta G$	$+\Delta G$
EXERGONIC	ENDERGONIC

Energy Shuttle
ATP

dation of these molecules into simpler compounds. *Anabolic pathways* require energy to combine simpler molecules into more complicated ones. This energy is often supplied through the coupling of catabolic and anabolic pathways in a cell. Energy is involved in all metabolic processes. Thus, the study of cellular energy transformations, called *bioenergetics*, is essential to the understanding of metabolism.

Energy: Some Basic Principles

Energy has been defined as the capacity to do work, to move matter against an opposing force.

Forms of Energy   *Kinetic energy* is the energy of motion, of matter that is moving. This matter does its work by transferring its motion to other matter. Heat and light are forms of kinetic energy due to the motion of molecules and photons. *Potential energy* is the capacity of matter to do work as a consequence of its location or arrangement. Chemical energy is a form of potential energy stored in the bonds holding atoms together in molecules.

Energy Transformations   Energy can be converted from one form to another: kinetic energy into potential energy and vice versa. Plants change light energy to the chemical energy of sugar, and cells release the potential energy in sugar to drive cellular processes.

Two Laws of Thermodynamics   *Thermodynamics* is the study of energy transformations. The *first law of thermodynamics* states that energy can be neither created nor destroyed. Energy can be transferred between matter and transformed from one kind to another, but the total energy of the universe is constant. According to this principle of the *conservation of energy*, chemical reactions that either require or produce energy are merely transforming a set amount of energy into a different form.

The *second law of thermodynamics* states that every energy transformation or transfer results in an increasing disorder within the universe. *Entropy* is the term used as a quantitative measure of disorder or randomness.

A system, such as a cell, may become more ordered, but it does so with an attendant increase in the entropy of its surroundings. A cell can use highly ordered organic molecules as a source of the energy needed to create its own highly ordered structure, but it returns the simple molecules of carbon dioxide and water and heat to the environment.

In any energy transformation or transfer, some of the energy is converted to heat, a less-ordered kinetic energy. No energy transfer is 100% efficient. The quantity of energy in the universe may be constant, but its quality is not. Every process results in an increasing disorder in energy, often into the form of the random molecular motion of heat.

The Free Energy Concept: A Criterion for Spontaneous Change   A spontaneous process is a change that occurs without the input of external energy and that results in greater stability, reflected as an increase in the entropy of the universe. *Free energy* is a concept that allows us to consider whether a change in a system will be spontaneous. Free energy, a measure of the instability of a system, can be defined as the portion of a system's energy available to perform work when the system's temperature is uniform. Free energy (G) depends on the total energy of a system (H) and its entropy (S), such that $G = H - TS$. Absolute temperature (°Kelvin) is a factor because higher temperature increases random molecular motion, disrupting order and increasing entropy. The change in free energy during a reaction is represented by $\Delta G = \Delta H - T\Delta S$. For a reaction to be spontaneous, the free energy of the system must decrease ($-\Delta G$). The system must lose energy (decrease H) or become more disordered (increase S) or both. An unstable system is rich in free energy and has a tendency to change spontaneously to a more stable stage, potentially performing work in the process.

Chemical Energy and Life: A Closer Look   Using the reference of free energy, reactions can be classified as *exergonic* or *endergonic*. An exergonic ($-\Delta G$) reaction proceeds with a net release of free energy and is spontaneous. The magnitude of ΔG indicates the maximum amount of work the reaction can do. Endergonic reactions ($+\Delta G$) are nonspontaneous; they must absorb energy from the surroundings. The magnitude of ΔG indicates the minimum amount of work needed to drive the reaction. A total of 686 kcal of energy are released from the exergonic reaction that breaks a mole of glucose down to carbon dioxide and water. The reverse reaction in photosynthesis has a ΔG of +686 kcal/mol.

At equilibrium in a chemical reaction, the forward and backward reactions are proceeding at the same rate. At equilibrium, $\Delta G=0$ because there is no net free energy change. Moving toward equilibrium, the ΔG of a reaction is negative; when moving away from equilibrium, ΔG is positive. In order to drive a reaction away from its normal equilibrium, a cell must add free energy, usually by coupling the reaction with an exergonic reaction.

ATP and Cellular Work

A cell must perform several kinds of work: mechanical work involved in movement of the cell or parts of the cell, transport work in pumping molecules across membranes, and chemical work in driving endergonic reactions to synthesize cellular molecules. The immediate source of the energy to perform this work most often comes from *adenosine triphosphate*, or *ATP*.

Structure and Hydrolysis of ATP ATP is a nucleoside triphosphate, the purine base adenine bonded to the sugar ribose, which is connected to a chain of three phosphate groups. The triphosphate tail of ATP is unstable and the bonds between the phosphate groups can be broken by hydrolysis. Thus, ATP can be hydrolyzed to ADP (adenosine diphosphate) and an inorganic phosphate molecule, releasing 7.3 kilocalories of energy per mole of ATP. This quantity of free energy has been experimentally measured; the ΔG of the reaction in the cell is estimated to be closer to –10 to –12 kcal/mol.

Although the phosphate bonds in ATP are called high-energy bonds, they are actually weak bonds, easily hydrolyzed to yield more products and release energy.

How ATP Performs Work The transformation to a more stable state releases energy. In a cell, this energy can be used to transfer the phosphate group from ATP to another molecule, producing a *phosphorylated intermediate* that is more reactive. The phosphorylation of other molecules by ATP forms the basis for almost all cellular work.

Regeneration of ATP A cell regenerates ATP at a phenomenal rate. The formation of ATP from ADP and inorganic phosphate is endergonic, with a ΔG of +7.3 kcal/mol. Cellular respiration (the catabolic processing of glucose and other organic molecules) provides the energy for the regeneration of ATP. Plants can also produce ATP using light energy.

Metabolic Disequilibrium Respiration and other cellular chemical reactions are reversible and could reach equilibrium if the cell did not keep a steady supply of reactants and siphon off the products (as reactants for new processes or as waste products to be expelled). Due to the huge free energy difference, respiration continues to produce ATP as long as the cell can provide glucose and expel CO_2.

Enzymes

Thermodynamics can indicate what reactions are spontaneous but not how fast those reactions occur.

Enzymes are used by the cell to speed and regulate metabolic reactions so that life is possible. Enzymes are biological catalysts—agents that change the speed of a reaction but are unchanged by the reaction.

Enzymes Lower Activation Energy Chemical reactions rearrange atoms by breaking and forming chemical bonds. Energy must be absorbed to break bonds and is released when bonds form. *Activation energy*, or the *free energy of activation*, ($\Delta G^{\ddagger}$) is the energy that must be absorbed by reactants for their bonds to break. Reactants must reach the unstable *transition state*, in which bonds are more fragile and likely to break, and from which the reaction can proceed. This state can be reached by the addition of thermal energy from the surroundings, causing the reactant molecules to collide more often and more forcefully. Even in an exergonic reaction, in which ΔG is negative, energy must first be absorbed to reach the transition state.

The activation energy barrier is essential to life because it prevents the energy-rich macromolecules of the cell from decomposing spontaneously. For metabolism to proceed in a cell, however, $\Delta G^{\ddagger}$ must be reached. Heat, a normal source of activation energy in reactions, would be harmful to the cell and would also speed metabolic reactions indiscriminately. Enzymes are able to lower $\Delta G^{\ddagger}$ for specific reactions so that metabolism can proceed at cellular temperatures. Enzymes do not change ΔG for a reaction; they only speed up reactions that would otherwise occur very slowly. Enzymes are required for both exergonic and endergonic reactions.

Enzymes Are Specific Most enzymes are proteins, macromolecules with characteristic three-dimensional shapes. The specificity of an enzyme for the particular *substrate* on which it works is determined by its unique shape. The substrate is temporarily bound to its enzyme at the *active site*, a pocket or groove found on the surface of the enzyme molecule, that has a shape and charge arrangement complementary to the substrate molecule. When a substrate molecule enters the active site, the enzyme changes shape slightly, creating what is called an *induced fit* between substrate and active site, which enhances the ability of the enzyme to catalyze the chemical reaction.

The Catalytic Cycle of Enzymes The substrate is held in the active site by hydrogen or ionic bonds, creating an enzyme-substrate complex. The side chains (R groups) of some of the surrounding amino acids in the active site facilitate the conversion of substrate to product. The product then leaves the active site and the enzyme can bind with another substrate molecule.

The conversion is extremely fast; an enzyme can catalyze 1000 reactions or more per second.

Enzymes can catalyze reactions involving the joining of two reactants by providing active sites in which the substrates are bound closely together and properly oriented. An induced fit can stretch or bend critical bonds in the substrate molecule and make them easier to break. An active site may provide a microenvironment that is necessary for a particular reaction, such as a lower pH. Enzymes may also actually participate in a reaction by forming brief covalent bonds with the substrate.

The rate at which an enzyme molecule works partly depends on the concentration of its substrate. The speed of a reaction will increase with increasing substrate concentration up to the point at which all enzyme molecules are saturated with substrate molecules and working at full speed.

Factors Affecting Enzyme Activity   The activity of an enzyme depends on its three-dimensional shape, and this shape depends on environmental factors that affect the weak chemical bonds maintaining protein structure. The velocity of an enzyme-catalyzed reaction may increase with rising temperature up to the point at which increased thermal agitation begins to disrupt the hydrogen and ionic bonds that stabilize protein conformation. A change in pH may denature an enzyme by disrupting the hydrogen bonding of the molecule. Each enzyme has a temperature and pH optimum at which it is most active. Enzymes are sensitive to salt concentration because inorganic ions may interfere with ionic bonds within the enzyme molecule.

Cofactors are small molecules that bind with enzymes and are necessary for enzyme catalytic function. They may be inorganic, such as various metal atoms, or organic molecules called *coenzymes*. Most vitamins are coenzymes or precursors of coenzymes.

Enzyme inhibitors selectively disrupt the action of enzymes, either reversibly by binding with the enzyme with weak bonds or irreversibly by attaching with covalent bonds. *Competitive inhibitors* compete with the substrate for the active site of the enzyme. Increasing the concentration of substrate molecules may overcome this type of inhibition as long as the inhibitor does not bind too strongly to the active site. *Noncompetitive inhibitors* bind to a part of the enzyme separate from the active site and change the conformation of the enzyme, thus impeding enzyme action. Many pesticides are noncompetitive inhibitors of key enzymes and act as metabolic poisons. The cell itself uses selective inhibitors to control enzyme action and thus regulate metabolism.

The molecules that inhibit or activate enzyme activity may bind to an *allosteric site*, a receptor site not associated with the active site on the enzyme. Enzymes with allosteric sites are often complex molecules made of two or more polypeptide chains or subunits, each with its own active site. Allosteric sites are usually located where subunits join. The entire unit may oscillate between two confomational states, and the binding of an activator (or inhibitor) to the allosteric site stabilizes the catalytically active (or inactive) conformation. Allosteric enzymes may be critical regulators of metabolic pathways, such as an enzyme whose allosteric site fits both ATP and ADP and whose activity is thus regulated by the supply or need of ATP in the cell. Through a phenomenon called *cooperativity*, the induced fit binding of a substrate molecule to one polypeptide subunit can change the conformation such that the active sites of all subunits are more active.

The Control of Metabolism

Feedback Inhibition   *Feedback inhibition* commonly regulates metabolic pathways. The product of a pathway can act as an inhibitor of an enzyme early in the pathway and shut down the process when the cell has produced sufficient end product.

Structural Order and Metabolism   The complex internal structure of the cell serves to order metabolic pathways in space and time. Enzymes can be grouped into multienzyme complexes in which the enzymes regulating successive steps of a metabolic pathway are serially arranged. Specialized cellular compartments may contain high concentrations of the enzymes and substrates needed for a particular pathway. Or enzymes may be incorporated into the membranes of cellular compartments. Thus the structural organization of cells facilitates and controls metabolism.

Emergent Properties: A Reprise

This unit has illustrated how life is organized into a hierarchy of structural levels with emergent properties associated with each new level of order. The structure and interactions of atoms, molecules, monomers, and macromolecules have been linked to metabolism, the orderly chemistry characteristic of life.

STRUCTURE YOUR KNOWLEDGE

This chapter introduces many complex ideas concerning the thermodynamics of metabolism. Take the time to organize your understanding of small "chunks" of this information and then try to integrate these pieces into your picture of the energy transformations taking place within the cells of living organisms.

1. Create a simple concept map concerning energy: its definition, types, and the first two laws of thermodynamics.

2. Develop a concept map on free energy and ΔG. The value in this exercise is for you to wrestle with and organize these concepts for yourself. Do not turn to the suggested concept map until you have worked on your own understanding. Remember that the concept map in the answer section is only one way of structuring these ideas—have confidence in your own organization.

3. Now take a break from concept mapping and answer these questions about enzymes. (A map on enzymes would, of course, be helpful.)
   a. Why are enzymes so critical to the existence of life?
   b. What characteristics of proteins make them good catalysts?
   c. Briefly describe the process by which an enzyme catalyzes a reaction.
   d. How does a cell control enzyme action and regulate metabolism?
   e. In the graph of a reaction with and without an enzyme catalyst shown below, label the parts a through e.

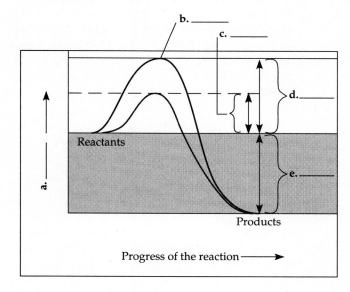

4. Label the three components (a through c) of the ATP molecule shown below. (d) Circle the region of instability in this molecule and show which

bonds are likely to break. (e) Explain why that part of the molecule is unstable and by what chemical mechanism the bond is broken. (f) Why is energy released?

TEST YOUR KNOWLEDGE

MULTIPLE CHOICE: *Choose the one best answer.*

1. Catabolic and anabolic pathways are often coupled in a cell because
   a. the intermediates of a catabolic pathway are used in the anabolic pathway.
   b. both pathways use the same enzymes.
   c. the free energy released from one pathway is used to drive the other.
   d. the activation energy of the catabolic pathway can be used in the anabolic pathway.

2. When glucose is converted to CO_2 and H_2O, changes in total energy, entropy and free energy are as follows:
   a. $-\Delta H, -\Delta S, -\Delta G$
   b. $-\Delta H, +\Delta S, -\Delta G$
   c. $-\Delta H, +\Delta S, +\Delta G$
   d. $+\Delta H, +\Delta S, +\Delta G$

3. When a protein forms from amino acids, the following changes apply:
   a. $+\Delta H, -\Delta S, +\Delta G$
   b. $+\Delta H, +\Delta S, -\Delta G$
   c. $-\Delta H, -\Delta S, +\Delta G$
   d. $-\Delta H, +\Delta S, +\Delta G$

4. Exergonic reactions
   a. release energy.
   b. are not spontaneous.
   c. usually result in a decrease in entropy.

d. all of the above are true.

5. A negative ΔG means that
   a. the quantity G of energy is available to do work.
   b. the reaction is spontaneous.
   c. the reactants have more free energy than the products.
   d. all of the above are true.

6. According to the first law of thermodynamics,
   a. for every action there is an equal and opposite reaction.
   b. every energy transfer results in an increase in disorder or entropy.
   c. the total amount of energy in the universe is conserved or constant.
   d. energy can be transferred or transformed, but disorder always increases.

7. Catabolic pathways
   a. combine molecules into more complex and energy-rich molecules.
   b. are usually coupled with anabolic pathways to which they supply energy in the form of ATP.
   c. involve endergonic reactions that break complex molecules into simpler ones.
   d. are spontaneous and, therefore, do not need enzyme catalysis.

8. Chemical energy
   a. is a form of potential energy.
   b. is stored in molecules because of the way atoms are bonded together..
   c. can be transferred from one molecule to another, but some of it will be transformed to heat energy.
   d. all of the above are true.

9. A spontaneous reaction
   a. proceeds rapidly.
   b. occurs without the addition of external energy (aside from ΔG‡).
   c. does not need to be catalyzed by enzymes.
   d. all of the above are true.

10. The formation of ATP from ADP and inorganic phosphate
    a. is an exergonic process.
    b. transfers the phosphate to another intermediate that becomes more reactive.
    c. produces an unstable, high-energy compound that can drive cellular work.
    d. has a ΔG of −7.3 kcal/mol.

11. At equilibrium,
    a. no enzymes are functioning.
    b. ΔG = O.
    c. the forward and backward reactions have stopped.
    d. all of the above are true.

12. An enzyme is capable of
    a. lowering the energy of activation of a reaction.
    b. changing the equilibrium of a reaction.
    c. increasing the change in free energy for a reaction.
    d. all of the above.

13. In cooperativity,
    a. a multienzyme complex contains all the enzymes needed for a metabolic pathway.
    b. a product of a pathway serves as a competitive inhibitor of an early enzyme in the pathway.
    c. a molecule bound to the active site of one subunit of an enzyme affects the active site of other subunits.
    d. the allosteric site is filled with an activator molecule.

14. Substrates are held in the active site of an enzyme by
    a. hydrogen and ionic bonds.
    b. the action of coenzymes and cofactors.
    c. the matching shape of the allosteric site.
    d. the lowering of the activation energy.

15. According to the induced-fit hypothesis,
    a. the binding of the substrate is an energy-requiring process.
    b. a competitive inhibitor can outcompete the substrate for the active site.
    c. the binding of the substrate changes the shape of the enzyme slightly and can stress or bend substrate bonds.
    d. the allosteric site creates a microenvironment ideal for the reaction.

FILL IN THE BLANKS

1. _____ is the totality of an organism's chemical processes.

2. _____ pathways require energy to combine molecules together.

3. _____ energy is the energy of motion.

4. _____ are more reactive molecules created by the transfer of a phosphate group from ATP.

5. _____ is the term for the measure of disorder or randomness.

6. _____ is the energy that must be absorbed by molecules to reach the transition state.

7. _____ inhibitors change the enzyme's conformation by binding to an allosteric site.

8. _____ are organic molecules that bind to enzymes and are necessary for their functioning.

9. _____ is a regulatory device in which the product of a pathway binds to an enzyme early in the pathway.

10. _____ enzymes change between two conformations depending on whether an activator or inhibitor is bound to them.

ANSWER SECTION

CHAPTER 2
ATOMS, MOLECULES, AND CHEMICAL BONDS

Suggested Answers to Structure Your Knowledge

1. An atom is the smallest unit of an element that maintains the physical and chemical properties of that element. Atoms are composed of subatomic particles, the largest and most stable of which are protons, neutrons, and electrons. Protons and neutrons are packed into the nucleus of an atom, whereas electrons travel at great speeds around the nucleus in orbitals within energy shells. Protons and neutrons both have a mass of 1 dalton; electrons have negligible mass. Neutrons do not carry a charge. Protons have a charge of +1; electrons have a charge of –1. Atoms that are not involved in chemical bonds are neutral because the number of electrons equals the number of protons. Ions are atoms or molecules that have gained or lost electrons, and their charge reflects the difference between the number of protons and electrons.

2. The atoms of each element have a characteristic number of protons in their nuclei, referred to as the *atomic number*. In a neutral atom, the atomic number also indicates the number of electrons.

   The *mass number* is an indication of the approximate mass of an atom and is equal to the number of protons and neutrons in the nucleus.

   The *atomic weight* refers to the atomic mass of an atom. It is equal to the mass number and is measured in the atomic mass unit of daltons. Protons and neutrons both have a mass of approximately 1 dalton.

   The *valence* is an indication of the bonding capacity of an atom. It is the number of covalent bonds that must be formed for an atom to complete its valence shell with eight electrons. The valence of an atom is most related to the chemical behavior of an atom because it is an indication of

the number of bonds the atom will make, or the number of electrons the atom must share in order to reach a filled valence shell.

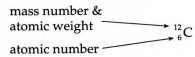

3. Ionic and nonpolar covalent bonds represent the two extremes along a continuum of electron sharing between atoms in a molecule. In ionic bonds the electrons are completely pulled away from one atom to the other, creating negatively and positively charged ions. In nonpolar covalent bonds the electrons are equally shared between two atoms. The middle ground of these two extremes is filled with polar covalent bonds in which a more electronegative atom pulls the shared electrons closer to it, producing a partial negative charge associated with that portion of the molecule and a partial positive charge associated with the atom from which the electrons are pulled. Because electrons are rapidly orbiting around the nuclei and constantly changing positions, a very polar bond may vacillate between being covalent and ionic.

Answers to Test Your Knowledge

Multiple Choice:

1. b	5. a	9. b	13. b	17. c
2. d	6. c	10. d	14. c	18. a
3. a	7. a	11. b	15. c	19. b
4. c	8. d	12. c	16. d	20. a

CHAPTER 3
WATER AND THE FITNESS OF THE ENVIRONMENT

Suggested Answers to Structure Your Knowledge

1.

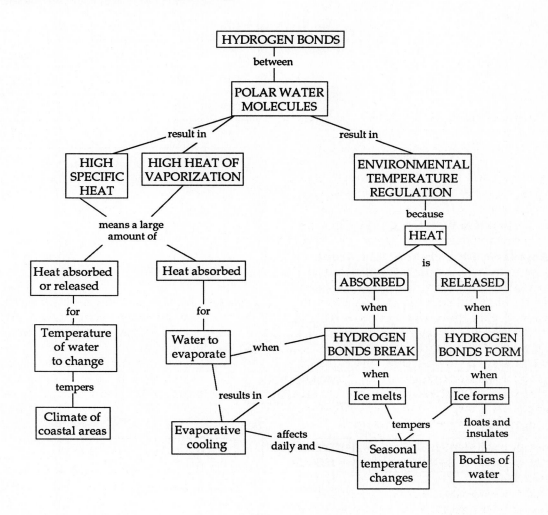

2.

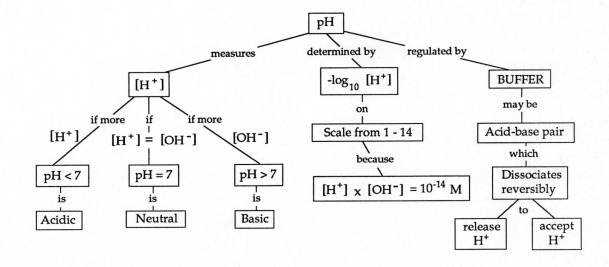

Answers to Test Your Knowledge

Multiple Choice:

1. c	4. d	7. d	10. d	13. c
2. a	5. b	8. b	11. b	14. b
3. a	6. c	9. b	12. d	15. d

Fill in the Blanks:

$[H^+]$	$[OH^-]$	pH	Acidic, basic, or neutral?
10^{-3}	10^{-11}	3	acidic
10^{-8}	10^{-6}	8	basic
10^{-12}	10^{-2}	12	basic
10^{-9}	10^{-5}	9	basic
10^{-1}	10^{-13}	1	acidic
10^{-7}	10^{-7}	7	neutral

CHAPTER 4
CARBON AND MOLECULAR DIVERSITY

Suggested Answers to Structure Your Knowledge

1.

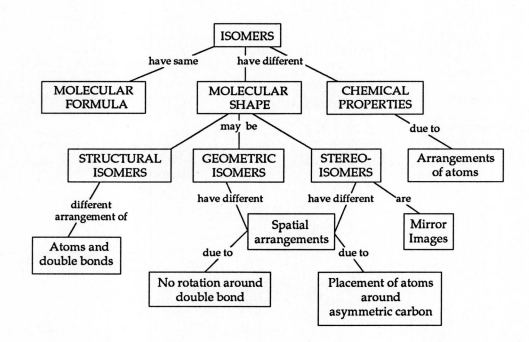

2.

Name of Group	Molecular Formula	Characteristics Conferred to Organic Compound
Hydroxyl	–OH	Polarity, solubility in water
Carbonyl	–C=O	Aldehyde or ketone, polar group
Carboxyl	–COOH	Acidic, can dissociate and release H^+
Amino	$-NH_2$	Basic, accepts H^+, becoming $-NH_3^+$
Sulfhydryl	–SH	Crosslinks stabilize protein structure
Phosphate	$-OPO_3^=$	Used in energy transfers

Answers to Test Your Knowledge

Multiple Choice:

1. c	**4.** b	**7.** a	**10.** b
2. b	**5.** a	**8.** c	**11.** b
3. d	**6.** d	**9.** a	**12.** c

Matching:

1. A, C	**4.** E	**7.** A, D, G	**10.** B, F
2. B, F	**5.** E	**8.** E	**11.** E
3. D, G	**6.** A, C, D, G	**9.** A, C	**12.** C

CHAPTER 5
STRUCTURE AND FUNCTION OF MACROMOLECULES

Suggested Answers to Structure Your Knowledge

1.

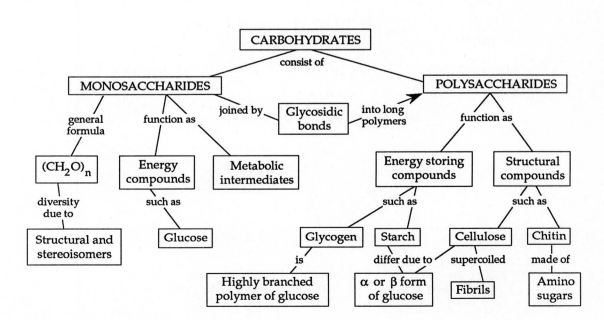

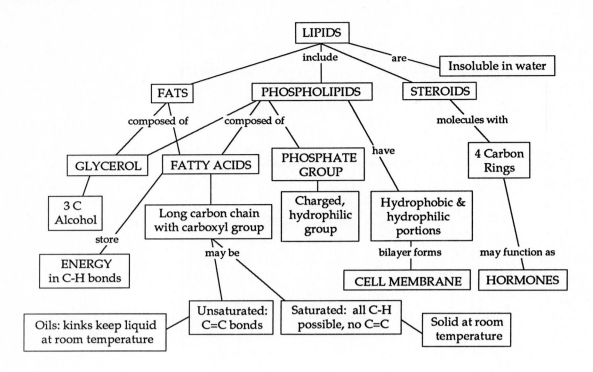

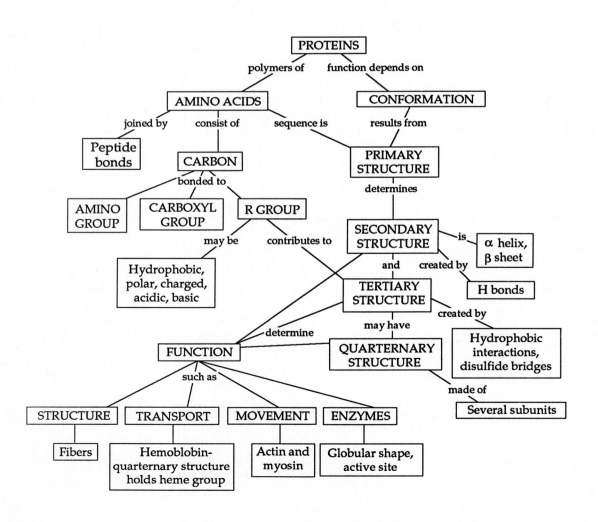

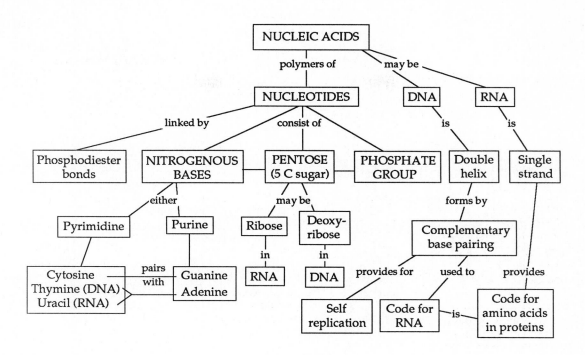

2. The primary structure of a protein is the specific, genetically-coded sequence of amino acids in a polypeptide chain. The secondary structure involves the coiling (alpha helix) or folding (beta pleated sheet) of the protein, stabilized by hydrogen bonds of the polypeptide backbone. The tertiary structure involves interactions between the side chains of amino acids and produces a characteristic three-dimensional shape for a protein. The four types of interactions that create the tertiary structure are listed below.

   a. hydrogen bond
   b. hydrophobic interaction
   c. disulfide bridge
   d. ionic bond

Answers to Test Your Knowledge

Matching of Formulas:

A. amino acid (glycine)
B. fatty acid
C. nitrogenous base (adenine)
D. glycerol
E. phosphate group

F. deoxyribose (pentose sugar)
G. sugar (triose)

1. D, B	**3.** C, E, F	**5.** C	**7.** E, F
2. A	**4.** F, G	**6.** A	

Matching:

1. A	**4.** C	**7.** B	**10.** A
2. B	**5.** C	**8.** C	
3. D	**6.** D	**9.** A	

Multiple Choice:

1. d	**4.** c	**7.** c	**10.** d
2. c	**5.** c	**8.** a	**11.** a
3. a	**6.** b	**9.** d	**12.** c

Fill in the Blanks:

1. thymine	**7.** primary structure or amino acid sequence
2. guanine	**8.** glycogen
3. purines	**9.** phospholipids
4. nucleoside	**10.** peptide bonds
5. nucleotide	**11.** X-ray chrystallography, computer modeling, graphics
6. quarternary	**12.** Frederick Sanger

CHAPTER 6
INTRODUCTION TO METABOLISM

Suggested Answers to Structure Your Knowledge

1.

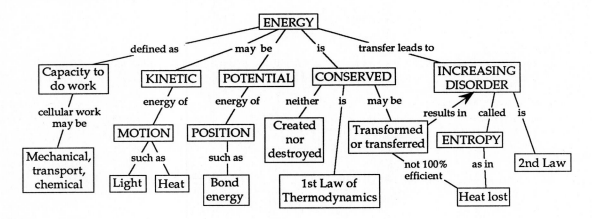

2.

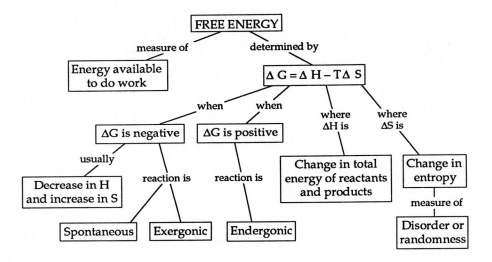

3. a. Enzymes are essential for the "cold chemistry" of life because they lower the free energy of activation of the reactions they catalyze and allow those reactions to occur extremely rapidly at a temperature conducive to life. Enzymes are also very specific, so that by regulating the enzymes it produces, the cell can regulate which of the myriad of possible chemical reactions take place at any given time.

b. The three-dimensional conformation of a protein permits it to have a characteristically shaped active site with specific chemical attributes. The shape and chemistry of the active site result in the specificity of an enzyme for a particular substrate.

c. A substrate binds to the active site on its enzyme with hydrogen and ionic bonds. The enzyme may change shape slightly, closing around the substrate. This induced fit strains the substrate's bonds or brings reactive groups in close proximity so that the transition state of the molecule(s) is reached and the reaction may proceed more easily. Once the reaction takes place, the substrate leaves the active site and another one enters. The speed of the reaction will increase with increasing substrate concentration until all enzyme molecules are saturated.

d. One way a cell can control enzyme action is through feedback inhibition, in which the product of a metabolic pathway then acts as an inhibitor of an enzyme operating earlier in that pathway. The cell can regulate which enzymes are produced and in what quantity through control of gene expression. The compartmentalization of the cell can order metabolic pathways and their substrates in space. Multienzyme complexes, where enzymes are sequentially arranged, may be located within cellular membranes.

e. **a.** free energy
   b. transition state
   c. $\Delta G^{\ddagger}$ (free energy of activation) with enzyme
   d. $\Delta G^{\ddagger}$ without enzyme
   e. ΔG of reaction

4. **a.** adenine

  b. ribose

  c. phosphates

  d. Circle the phosphate portion of molecule and indicate that terminal phosphate bond is most likely to break.

  e. The negatively charged phosphate groups are crowded together, and their mutual repulsion makes the bonds unstable. A hydrolysis reaction breaks the terminal phosphate bond and releases a molecule of inorganic phosphate.

  f. The products of hydrolysis (ADP and Pi) are more stable than ATP. The chemical change to a more stable state releases energy.

Answers to Test Your Knowledge

Multiple Choice:

1. c	**5.** d	**9.** b	**13.** c
2. b	**6.** c	**10.** c	**14.** a
3. a	**7.** b	**11.** b	**15.** c
4. a	**8.** d	**12.** a	

Fill in the Blanks:

1. metabolism	**6.** free energy of activation
2. anabolic	**7.** noncompetitive
3. kinetic	**8.** coenzymes
4. phosphorylated intermediate	**9.** feedback inhibition
5. entropy	**10.** allosteric

UNIT II

THE CELL

"He told you *that*? Well, he's pulling your flagellum, Nancy,"

CHAPTER 7

A TOUR OF THE CELL

FRAMEWORK

This chapter deals with the fundamental unit of life—
the cell. The complexities in the processes of life are
reflected in the complexities of the structure of the
cell. It is easy to become overwhelmed by the number
of new vocabulary terms for this array of cell
organelles and membranes. The following concept
map provides an organizational framework for the
wealth of detail found in a "tour of the cell."

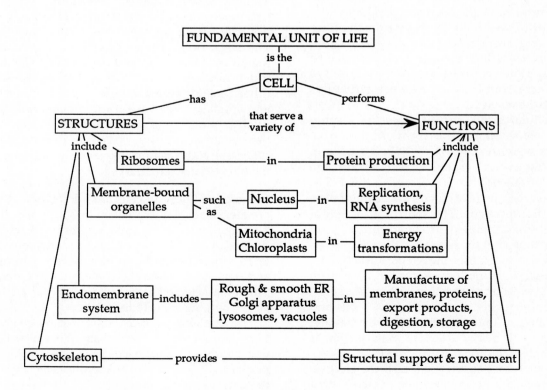

CHAPTER SUMMARY

The cell is the basic structural and functional unit of all living organisms. In the hierarchy of biological organization, the capacity for life emerges from the structural order of the cell. In this intricately integrated collection of subunits, structure is clearly correlated with function. The cell is able to sense and respond to its environment. Through the long history of evolution, environmental pressures have helped to shape the finely tuned structures and functions of this fundamental unit of life.

How Cells Are Studied

The growth of scientific knowledge and the development of new instruments and methods usually go hand-in-hand. The invention of the microscope in the 17th century led to the initial discovery and study of cells.

Microscopy The glass lenses of *light microscopes* refract (bend) the visible light passing through a specimen such that the projected image is *magnified*. Light microscopes can be built to magnify specimens 1500 times. *Resolving power* is a measure of the clarity of an image—determined by the minimum distance two points must be separated to be distinguished. The resolving power of the light microscope is limited by the wavelength of visible light, so that details finer than 0.2 μm (micrometers) cannot be resolved. Staining of specimens and using techniques such as dark field and phase contrast microscopy improve visibility by increasing contrast between structures large enough to be resolved.

Most subcellular structures, or *organelles,* are too small to be resolved by the light microscope. Although cells were discovered by Robert Hooke in 1665, their details were largely unknown until the development of the *electron microscope* in the 1950s. The electron microscope (EM) focuses a beam of electrons through the specimen. The short wavelength of electrons allows a resolution of about 0.2 nm (nanometer), a thousand times greater than that of the light microscope, and exposes the *cell ultrastructure.*

In a *transmission electron microscope (TEM)* a beam of electrons is passed through a thin section of a specimen, and electromagnets, acting as lenses, focus and magnify the image. Contrast is increased by staining preserved cells with heavy atoms of metals. TEM is used to study internal ultrastructure of cells.

In a *scanning electron microscope (SEM)* the electron beam scans the surface of a specimen coated with a thin gold film, exciting electrons from the specimen and collecting and focusing them onto a screen. The resulting image appears three-dimensional and shows the surface topography of the specimen.

Cytology is the study of cell structure. Modern cell biology integrates cytology with biochemistry to understand relationships between cellular structure and function.

Cell Fractionation *Cell fractionation* is a technique that separates major organelles of a cell so that their functions can be studied. Cells are homogenized or broken open by ultrasound or grinding. The resulting cellular soup is separated into component fractions by spinning in a *centrifuge,* a machine that spins test tubes at various speeds. *Ultracentrifuges* can spin at speeds of 80,000 rpm. First, the homogenate is spun slowly, and nuclei and large particles settle to form a pellet. The remaining supernatant is then centrifuged at increasing speeds, each time isolating smaller and smaller cellular components in the pellet. Each cellular fraction contains a large quantity of the same cellular components, thus permiting the isolated study of the metabolic functions of these organelles.

The Geography of the Cell: A Panoramic View

Prokaryotic and Eukaryotic Cells The kingdom Monera, which includes bacteria and cyanobacteria (blue-green algae), is characterized by having *prokaryotic cells*, which are cells with no nuclear membrane. The DNA of prokaryotic cells is concentrated in a region called the nucleoid. *Eukaryotic cells*, which are much more structurally complex, are found in the other four kingdoms of life: protists, plants, fungi, and animals. They have a true nucleus enclosed in a nuclear membrane and numerous organelles suspended in a semifluid medium called *cytosol*. The *cytoplasm* is the region between the nucleus and the membrane enclosing the cell.

Cell Size Almost all cells are microscopic; their sizes are measured in micrometers (μm; 1 mm = 1000 μm). The smallest known cells are certain bacteria with diameters of 0.1 μm—just large enough to pack in the DNA, ribosomes and enzymes necessary to reproduce and sustain life. Most bacterial cells range from 1 to 10 μm in diameter, whereas eukaryotic cells are ten times larger, ranging from 10 to 100 μm.

The small size of cells is dictated by geometry and the requirements of metabolism. Since area is proportional to the square of linear dimension while volume is proportional to its cube, eukaryotic cells, with 10 times the diameter of bacterial cells, have 100 times as much surface area but 1000 times greater volume. The *plasma membrane* surrounding every cell regulates exchange of sufficient oxygen, nutrients, and wastes to provide for the metabolism of the entire cell. The microscopic size of cells maintains sufficient surface area for exchange relative to the volume of the cell and limits the volume of cytoplasm that must be regulated by the nucleus.

The Importance of Compartmental Organization

The difference in the surface area:volume ratio in eukaryotic compared to prokaryotic cells is compensated in part by an extensive internal membrane system. These membranes compartmentalize the eukaryotic cell, providing local environments for specific metabolic functions, and participate in many metabolic processes through membrane-bound enzymes.

Membranes are composed of a bilayer of phospholipid molecules embedded with proteins. The specific molecular composition of a membrane varies according to its functions. Membranes play a key role in the organization and functioning of a cell's metabolism.

The Nucleus

The nucleus is surrounded by the *nuclear envelope,* a double membrane perforated by pores that are thought to regulate the movement of large macromolecules between the nucleus and the cytoplasm. The inner membrane is associated with a layer of protein that helps to maintain the shape of the nucleus and the organization of the genetic material.

Most of the cell's DNA is located in the nucleus, where, along with associated proteins, it is organized into *chromosomes.* Each eukaryotic species has a characteristic chromosomal number. Chromosomes are visible only when coiled and condensed in a dividing cell; otherwise they appear as a mass of stained material called *chromatin.*

The *nucleolus,* a round structure visible in the nondividing nucleus, consists of specialized regions of chromosomes, called *nucleolar organizers,* that have multiple copies of genes for ribosome synthesis. Ribosomal subunits are constructed in the nucleolus.

The genetic instructions for specific proteins are transcribed from DNA into *messenger RNA (mRNA)* in the nucleus. The messenger RNA passes into the cytoplasm, where it complexes with ribosomes which translate the message into the primary sequence of proteins.

Ribosomes

The prominence of nucleoli and the number of ribosomes are related to the rate of protein synthesis within a cell. *Ribosomes* are composed of a large and a small subunit that join together when they attach to a messenger RNA and begin protein synthesis. Most of the proteins produced by free ribosomes are used within the cytosol. Bound ribosomes, attached to the endoplasmic reticulum, usually make proteins that will be included within membranes or exported from the cell.

Subunits consist of protein and RNA and are not enclosed in a membrane. The difference in molecular composition between eukaryotic ribosomes and the smaller ribosomes of prokaryotes has medical consequences; some antibiotics specifically inhibit the prokaryotic ribosomes of bacteria without harming the activity of the patient's ribosomes.

The Endomembrane System

The *endomembrane system* of a cell consists of the nuclear envelope, endoplasmic reticulum, Golgi apparatus, lysosomes, vacuoles, and the plasma membrane. These membranes are all related either through direct contact or by the transfer of membrane segments by membrane-bound sacs called vesicles. Although related, the membranes differ in molecular composition and structure, depending on their functions.

Endoplasmic Reticulum   The *endoplasmic reticulum (ER)* is the most extensive portion of the endomembrane system. It is continuous with the outer nuclear membrane and encloses a network of interconnected tubules or compartments called *cisternae.* Ribosomes are attached to the cytoplasmic surface of *rough ER,* which manufactures membrane and secretory proteins. *Smooth ER* lacks ribosomes.

Smooth ER serves diverse functions, such as fat, phospholipid, steroid and sex hormone synthesis; storage and release of calcium ions for contraction of muscle cells; and detoxification of drugs and poisons. Barbituates, alcohol, and other drugs increase a liver cell's production of smooth ER and detoxification enzymes, thus leading to an increased tolerance (and thus reduced effectiveness) for these and other drugs.

Proteins intended for secretion are manufactured by membrane-bound ribosomes and then threaded into the cisternal space of the rough ER, where they fold into their conformation. Most secretory proteins are *glycoproteins.* A carbohydrate, usually an oligosaccharide (a molecule made of a "few" saccharides), is covalently bonded to these proteins by enzymes built into the ER membrane. The secretory proteins are transported from the rough ER in tiny membrane-bound *transport vesicles* budded off from a special region called *transitional ER.*

Rough ER manufactures membranes by inserting proteins formed by the attached ribosomes into the membrane. Both rough and smooth ER assemble phospholipids from precursors obtained from the cytosol with the aid of enzymes built into their membranes.

The Golgi Apparatus   Transport vesicles, pinched off from the ER, travel first to the *Golgi apparatus* where products of the ER are modified, stored, and directed to other locations.

The Golgi apparatus consists of a stack of flattened membrane sacs. Vesicles that bud from the ER join to a Golgi membrane at the *cis face*, adding to it their contents and membrane. Products that travel through the Golgi apparatus are usually modified or refined. Some macromolecules are manufactured within the Golgi. Vesicles pinch off from the other end of the Golgi apparatus, called the *trans face*. Golgi products may be tagged with different oligosaccharides or phosphate groups that help direct them to various parts of the cell.

Lysosomes   *Lysosomes* are membrane-enclosed sacs of hydrolytic enzymes used by the cell to digest macromolecules. The cell can maintain an acidic pH for these enzymes and also protect itself from unwanted digestion by containing hydrolytic enzymes within lysosomes.

Lysosomes function in a variety of circumstances. They digest food particles ingested by *phagocytosis* by cells such as protozoa and macrophages, and recycle the cell's own macromolecules by engulfing organelles or small bits of cytosol, a process known as autophagy. During development or metamorphosis, lysosomes may be programmed to destroy their cells to create a specific body form. Storage diseases are inherited defects in which a lysosomal enzyme is missing or defective, and lysosomes become packed with indigestible substances.

Vacuoles   Vacuoles are also membrane-enclosed sacs within the cell, distinguished by being larger than vesicles. *Food vacuoles* are formed as a result of phagocytosis. *Contractile vacuoles* pump excess water out of freshwater protozoa.

A large *central vacuole* is found in mature plant cells, surrounded by a membrane called the *tonoplast*. This vacuole stores organic compounds and inorganic ions for the cell. Hydrolytic enzymes are compartmentalized in the central vacuole and digest stored macromolecules and recycle organic compounds. Poisonous or unpalatable compounds that may protect the plant from predators, and dangerous metabolic byproducts may also be contained in the vacuole. Plant cells can increase in size with a minimal addition of new cytoplasm as their vacuoles absorb water and expand.

Relationships of Endomembranes: A Summary
Through the fusion of transport vesicles, or through direct continuity, the components of the endomembrane system are related. The nuclear envelope extends from the rough ER, which also connects to the smooth ER. Membranes pinched off from the ER move to the Golgi and on to lysosomes, vacuoles, and the plasma membrane. Though related, each type of membrane has structural modifications designed for its specific function. The structural compartmentalization of the endomembrane system organizes and facilitates the complexity of cellular function.

Peroxisomes (Microbodies)

Peroxisomes, also called microbodies, are compartments enclosed by a single membrane and filled with enzymes functioning in a variety of metabolic pathways, such as breaking down fats to smaller molecules or detoxifying alcohol and other poisons. Hydrogen peroxide (H_2O_2) is a toxic byproduct of these pathways, but peroxisomes contain an enzyme that converts it to water. Compartmentalizing these enzymes together is important to cellular function.

Specialized peroxisomes found in the tissues of germinating seeds contain enzymes that convert the oils stored in the seed to sugar for the developing seedling.

Peroxisomes, which do not bud from the endomembrane system, grow by incorporating proteins and lipids from the cytosol. They increase in number by dividing.

Energy Transducers: Mitochondria and Chloroplasts

Cellular respiration, the catabolic processing of fuels to produce ATP, occurs within the *mitochondria* of most eukaryotic cells. *Chloroplasts*, found only in plants and eukaryotic algae, produce organic compounds from carbon dioxide and water by absorbing solar energy. Mitochondria and chloroplasts are not considered part of the endomembrane system because their membrane proteins are made by ribosomes either free in the cytosol or contained within these organelles. They also contain a small amount of DNA that directs the synthesis of some of their proteins.

Mitochondria   The number of mitochondria found in a cell can range from hundreds to thousands, depending on the metabolic needs of the cell. Two membranes, each a phospholipid bilayer with unique embedded proteins, enclose a mitochondrion. A narrow intermembrane space exists between the smooth outer membrane and the convoluted inner membrane. The folds of the inner membrane, called *cristae*, create a large membrane surface area and enclose the *mitochondrial matrix*. Within this matrix are enzymes that control many of the metabolic steps of cellular respiration. Other important enzymes are built into the extensive inner membrane.

Chloroplasts *Plastids* are plant organelles that include amyloplasts, which store starch; chromoplasts, which contain pigments; and chloroplasts, which contain the green pigment chlorophyll and function in photosynthesis. All three types of plastids can develop in unspecialized cells from proplastids, depending on the location and the environment of the cell. Chloroplasts develop only when proplastids are exposed to light or when existing chloroplasts divide.

Chloroplasts, which are bounded by two membranes, are composed of a viscous fluid called the *stroma* and a membranous system of flattened sacs, called *thylakoids*, which enclose the thylakoid space. Thylakoids may be stacked together to form structures called *grana*.

The Cytoskeleton

The *cytoskeleton* is a network of fibers that function to give mechanical support; maintain or change cell shape; anchor or direct the movement of organelles and cytoplasm; and control movement of cilia, flagella, pseudopods, and even contraction of muscle cells. At least three types of fibers are involved in the cytoskeleton: *microtubules*, *microfilaments*, and *intermediate filaments*.

Microtubules Microtubules are hollow rods constructed of two kinds of globular proteins called α and ß tubulins. In addition to providing the major supporting framework of the cell, microtubules serve as tracks along which organelles move with the aide of specialized proteins called *motor molecules*. All eukaryotic cells have microtubules. In many cells they radiate out from an organizing mass called the *microtubule-organizing center* or *centrosome*. Other bundles of microtubules located near the plasma membrane contribute to cell shape.

The separation of chromosomes in a dividing cell is associated with microtubules assembled from the microtubule-organizing center. In animal cells a pair of centrioles, each composed of nine sets of triplet microtubules arranged in a ring, may help to organize microtubule assembly for cell division.

Cilia and *flagella* are extensions of eukaryotic cells, composed of and moved by microtubules. Cilia are numerous and short; flagella occur one or two to a cell and are longer. A flagellum moves with an undulating motion whereas a cilium generates force with a power stroke alternating with a recovery stroke (like an oar). Many protists use cilia or flagella to swim (or

actually crawl) through aqueous media. Sperm are propelled by flagella. Cilia or flagella attached to stationary cells of a tissue sweep fluid past the cell.

Both cilia and flagella are composed of two single microtubules surrounded by a ring of nine doublets of microtubules (a nearly universal "9 + 2" arrangement), all of which are enclosed in an extension of the plasma membrane. A *basal body*, structurally identical to a centriole, anchors the tubules of the cilium or flagellum to the cell. The sliding of the microtubule doublets past each other occurs as sidearms, composed of the protein *dynein*, alternately attach to adjacent doublets, pull down (as the conformation of dynein changes), release, and reattach. In conjunction with the radial spokes that anchor the doublets near the central microtubules, this action—driven by ATP—causes the bending of the flagella or cilia.

Microfilaments and Movement Microfilaments are solid rods consisting of a helix of two chains of molecules of the globular protein *actin*. In muscle cells, thousands of microfilaments of actin interdigitate with thicker filaments made of the protein *myosin*. The sliding of actin and myosin filaments past each other, driven by ATP-powered arms extending from the myosin, causes the shortening of the cell and thus the contraction of muscles.

Microfilaments are present in nearly all eukaryotic cells. They function in support, such as in the core of microvilli; in localized contractions, such as the pinching apart of animal cells when they divide and the extension and retraction of parts of moving cells; and in *cytoplasmic streaming*, which distributes materials through the cytoplasm of plant cells.

Intermediate Filaments Intermediate filaments are intermediate in size between microtubules and microfilaments. Unlike the other two fibers that are consistent in composition across all eukaryotic cells, intermediate fibers vary in protein makeup from one cell type to another. Intermediate fibers appear to be important in maintaining cell shape and anchoring certain organelles. The nucleus is often securely held in a web of intermediate fibers. These tension-bearing filaments may serve as the superstructure of the entire cytoskeleton.

The Cell Surface

Most cells synthesize and secrete some sort of covering over the plasma membrane.

Cell Walls A *cell wall* is a diagnostic feature of plant cells and is also found in prokaryotes, fungi, and some protists, but not in animals. Plant cell walls are composed of fibers of cellulose embedded in a matrix of polysaccharides and protein. The exact composition of the wall varies between species and even between cell types in the same plant.

The *primary cell wall* secreted by a young plant cell is relatively thin and flexible. Adjacent cells are connected by the *middle lamella*, a thin layer of polysaccharides called pectins that glue the cells together. When they stop growing, some cells secrete a thicker and stronger *secondary cell wall* between the plasma membrane and primary cell wall.

The Glycocalyx of Animal Cells Animal cells often secrete a *glycocalyx* made of sticky oligosaccharides that serves to strengthen the cell surface and glue cells together. The glycocalyx functions in cell–cell recognition.

Intercellular Junctions Intercellular junctions between membranes of neighboring cells improve communication and interaction between cells.

Plasmodesmata are channels in plant cell walls through which strands of cytoplasm connect bordering cells, and water and small solutes can move. The plasma membranes of adjacent cells are continuous through the channel, linking most cells of a plant into a living continuum.

There are three main types of intercellular junctions between animal cells. *Tight junctions* are fusions between adjacent cell membranes that create an impermeable seal across a layer of epithelial cells. *Desmosomes* are strong connections between adjacent cells, reinforced by intermediate filaments. *Gap junctions* allow for the exchange of materials between cells through protein-surrounded pores.

The Cell Is a Living Unit Greater Than the Sum of Its Parts

The compartmentalization and variety of organelles typical of cells exemplify the principle that structure correlates with function. The intricate functioning of a living cell emerges from the complex interactions of its many parts.

STRUCTURE YOUR KNOWLEDGE

1. **A.** This table lists the general functions performed by an animal cell. List the structures associated with each of these functions.

CELLULAR FUNCTIONS	ASSOCIATED ORGANELLES AND STRUCTURES
Cell division	
Information storage and transferal	
Energy conversions	
Manufacture membranes and products	
Digestion, recycling	
Conversion of H_2O_2 to water	
Structural integrity	
Movement	
Exchange with environment	
Cell–cell interaction	

B. This table lists structures that are unique to plant cells. Fill in the functions of these structures.

STRUCTURES UNIQUE TO PLANT CELLS	FUNCTIONS
Cell wall	
Central vacuole	
Chloroplast	
Amyloplast	
Plasmodesmata	

2. Is this a plant cell or animal cell? _____
Label the indicated structures.

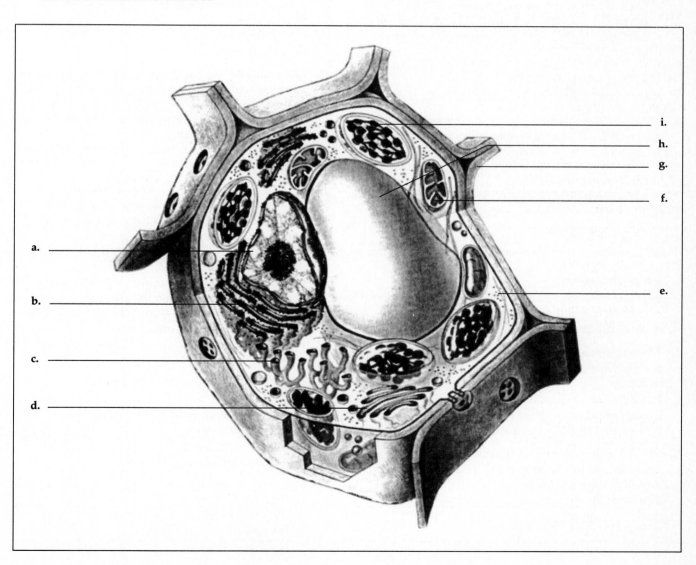

3. Create a diagram or flow chart to trace the development of a secretory product (such as a digestive enzyme) from the DNA code to its export from the cell.

TEST YOUR KNOWLEDGE

MULTIPLE CHOICE: *Choose the one best answer.*

1. Which of the following is/are *not* found in a prokaryotic cell?
   a. ribosomes
   b plasma membrane
   c. mitochondria
   d. a and c

2. Eukaryotic cells can be larger than prokaryotic cells because
   a. their plasma membrane is more expandable.
   b. their internal membrane system allows compartmentalization of functions and extra surface area for nutrient exchange and placement of enzymes.
   c. their DNA is localized in the nucleus whereas protein synthesis occurs in the cytoplasm.
   d. they have a better surface-area-to-volume ratio.

3. Which of the following is *not* a similarity among the nucleus, chloroplasts, and mitochondria?
   a. They all contain DNA.
   b. They all are bounded by a double phospholipid bilayer membrane.
   c. They can all divide to reproduce themselves.
   d. They all are derived from the endoplasmic reticulum system.

4. The pores in the nuclear envelope provide for the movement of
   a. proteins into the nucleus.
   b. ribosome subunits out of the nucleus.
   c. mRNA out of the nucleus.
   d. all of the above.

5. The nucleolus functions in
   a. synthesis and storage of components of ribosomes.
   b. formation of spindle fibers.
   c. condensation of chromosomes.
   d. direction of DNA synthesis.

6. The largest number of bound ribosomes most likely would be found in a cell
   a. with a high metabolic rate.
   b. that produces secretory products.
   c. with many cilia.
   d. that is actively dividing.

7. Which structure is *not* considered to be part of the endomembrane system?
   a. peroxisome
   b. smooth ER
   c. nuclear envelope
   d. lysosome

8. A plant cell grows larger primarily by
   a. increasing the number of vacuoles.
   b. synthesizing more cytoplasm.
   c. taking up water into its central vacuole.
   d. synthesizing more cellulose.

9. The innermost portion of a mature plant cell wall is the
   a. primary cell wall.
   b. secondary cell wall.
   c. middle lamella.
   d. glycocalyx.

10. The contractile elements of muscle cells are
    a. intermediate filaments.
    b. centrioles.
    c. microtubules.
    d. microfilaments.

11. Microtubules are components of all of the following *except*
    a. centrioles.
    b. the spindle apparatus for separating chromosomes in cell division.
    c. the pinching apart of the cytoplasm in animal cell division.
    d. flagella and cilia.

12. Of the following, which is probably the most common route for membrane flow in the endomembrane system?
    a. rough ER⟶Golgi⟶lysosomes⟶vesicles⟶plasma membrane
    b. rough ER⟶transitional ER⟶Golgi⟶vesicles⟶plasma membrane
    c. nuclear envelope⟶rough ER⟶Golgi⟶smooth ER⟶lysosomes
    d. rough ER⟶vesicles⟶Golgi⟶smooth ER⟶plasma membrane

13. Proteins to be used within the cytosol are generally synthesized
    a. by ribosomes bound to rough ER.
    b. by free ribosomes.
    c. by the nucleolus.
    d. within the Golgi apparatus.

14. Many of the proteins made by rough ER become part of
    a. the ribosomes.
    b. the cytosol.

c. membranes.
d. the cytoskeleton.

15. Plasmodesmata in plant cells are similar in function to
   a. desmosomes.
   b. tight junctions.
   c. gap junctions.
   d. glycocalyx.

16. Dynein is involved in
   a. the movement of microtubules.
   b. the contraction of microfilaments.
   c. the attachment of the microtrabecular lattice.
   d. the tensile strength of intermediate filaments.

17. The ultrastructure of a chloroplast could be best seen using
   a. transmission electron microscopy.
   b. scanning electron microscopy.
   c. phase contrast light microscopy.
   d. cell fractionation.

18. Resolving power of a microscope is a measure of
   a. the distance between two separate points.
   b. the sharpness or clarity of an image.
   c. the degree of magnification of an image.
   d. the depth of focus on a specimen's surface.

FILL IN THE BLANKS *with the appropriate cellular organelle or structure.*

1. _____ transport membranes and products to various locations

2. _____ infolding of mitochondrial membrane with attached enzymes

3. _____ sticky, supportive coat on animal cells

4. _____ small sacs with specific enzymes for particular metabolic pathway

5. _____ system of flattened sacs inside chloroplasts

6. _____ anchoring structure for cilia and flagella

7. _____ semifluid medium between nucleus and plasma membrane

8. _____ system of fibers that maintains cell shape, anchors organelles

9. _____ connection between animal cells that creates impermeable layer

10. _____ membrane surrounding central vacuole of plant cells

MATCHING: *Match the function and structure with the proper organelle or cellular component.*

FUNCTION	STRUCTURE	ORGANELLE
1. _____	_____	Chloroplast
2. _____	_____	Golgi apparatus
3. _____	_____	Lysosome
4. _____	_____	Mitochondrion
5. _____	_____	Nucleolus
6. _____	_____	Nucleus
7. _____	_____	Ribosome
8. _____	_____	Rough ER
9. _____	_____	Smooth ER
10. _____	_____	Vacuole

FUNCTION

A. produces RNA and ribosome subunits
B. manufactures proteins
C. contains chromosomes, produces mRNA
D. site of cellular respiration
E. site of photosynthesis
F. storage, cell elongation, found in center of plant cells
G. maintains cell shape
H. houses enzymes that synthesize steroids, detoxify drugs, store and release calcium ions
I. digests macromolecules
J. manufactures membranes and secretory products
K. processes products of rough ER

STRUCTURES

a. network of membranes with attached ribosomes
b. double membrane sac containing stacks of membrane sacs
c. large and small subunits, attach to mRNA
d. sac with hydrolytic enzymes, acid pH
e. round structure with multiple copies of genes for ribosomal RNA
f. single membrane-bound sac
g. surrounded by double membrane, inner layer of protein, with pores
h. large compartment formed from small vesicles
i. double membrane sac with cristae
j. stack of flattened sacs, cis, trans faces
k. membrane network without attached ribosomes

CHAPTER 8

MEMBRANE STRUCTURE AND FUNCTION

FRAMEWORK

This chapter presents the fluid mosaic model of membrane structure, relating the molecular structure of biological membranes to their function of regulating the passage of substances into the cell, out of the cell, and between intracellular compartments.

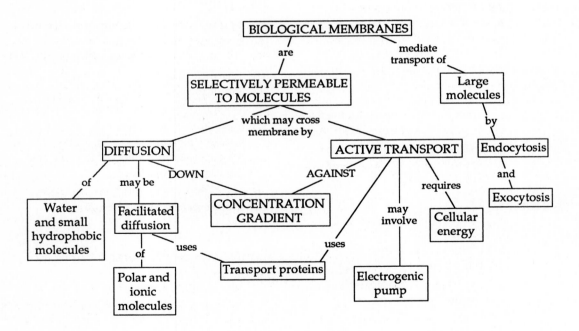

CHAPTER SUMMARY

The plasma membrane is the boundary of life, providing the cell with the potential to maintain a unique internal environment and to control the movement of materials into and out of the cell. Biological membranes are *selectively permeable*; their ability to discriminate in the chemical exchanges they allow is due to their unique structure.

Models of Membrane Structure

Two Generations of Membrane Models Over the years, scientists have created a series of models to explain membrane structure and function. The observation that lipid-soluble molecules were able to enter cells rapidly led to the conclusion that membranes were composed of lipids. In the early 1900s, membranes isolated from red blood cells were found to consist of lipids and proteins.

Membrane *phospholipids* are *amphipathic*, having both a hydrophobic tail and a hydrophilic head. In 1935, Davson and Danielli proposed a molecular model of membrane structure in which a double layer of phospholipids, with the hydrophobic hydrocarbon tails in the center and the hydrophilic heads facing the aqueous solution on both sides of the membrane, is covered with a coat of globular proteins. This sandwich model was consistent with the observed thickness of plasma membranes (adjusting the globular proteins to layers of protein in the pleated-sheet configuration) and with the apparent triple-layer staining results seen with electron microscopy in the 1950s. The Davson-Danielli model was widely accepted by the 1960s for all cellular membranes.

Variability later discovered in the thickness and other aspects of cellular membranes and the amphipathic nature of membrane proteins were inconsistent with the Davson-Danielli model. In 1972, Singer and Nicolson proposed their *fluid mosaic model* in which membrane proteins are embedded in the phospholipid bilayer with their hydrophilic regions extending out into the aqueous environment. The phospholipid bilayer is envisioned as fluid, with a mosaic of individually-inserted protein molecules shifting laterally within it.

This model is consistent with the known properties of membranes and is supported by evidence from freeze-fracture electron microscopy, which shows the interior of the bilayer with protein bumps dispersed in a smooth matrix. Freeze-fracture is a technique of freezing a specimen, fracturing it with a cold knife, etching the fractured surface to enhance contrast, and then coating the fractured surface with platinum and carbon. The resulting replica of the surface is examined in the electron microscope.

The fluid mosaic model is currently the most accepted and useful model for organizing existing knowledge and extending further research on membrane structure.

The Fluid Mosaic Model: A Closer Look   Membranes are held together primarily by hydrophobic attractions that allow the lipids and some of the proteins to drift laterally, keeping their hydrophobic regions in contact. The ability of membrane proteins of a hybrid mouse-human cell to intermingle illustrates the fluid nature of membranes. Some proteins do not drift due to their attachment to the cytoskeleton.

Phospholipids with unsaturated hydrocarbon tails maintain membrane fluidity at lower temperatures.

The steroid cholesterol, common in cell membranes of eukaryotes, restricts movement of phospholipids and thus reduces fluidity at warmer temperatures. Cholesterol also prevents the close packing of lipids and thus enhances fluidity at lower temperatures. Some cells may change the lipid composition of their membranes in response to changing temperature.

Each membrane has its own unique complement of membrane proteins, which determines most of the specific functions of that membrane. *Integral proteins* extend into the hydrocarbon region of the lipids. They may be unilateral, reaching partway through the membrane, or transmembrane, with two hydrophilic ends and a hydrophobic midsection. *Peripheral proteins* are attached to the surface of the membrane. Membranes are asymmetric; they have distinct inner and outer faces related to the directional orientation of their proteins, the composition of the lipid bilayers, and, in the plasma membrane, the attachment of carbohydrates to the exterior surface. The manufacture of the plasma membrane by ER (with the fusion of vesicles expanding the membrane) means that the exterior of the plasma membrane corresponds to the former interior of the ER.

Cell-cell recognition, the ability of a cell to determine if another cell is similar to it, is most likely based on recognition of membrane carbohydrates, which are usually branched oligosaccharides covalently bonded to lipids or, most often, to proteins. The glycolipids and glycoproteins in plasma membranes vary from species to species, individual to individual, and even among cell types.

The critical ability of membranes to regulate the passage of substances into and out of the cell and between cellular compartments is determined by their supramolecular structure, the unique architecture determined by the arrangement of many molecules into a higher level of organization.

Traffic of Small Molecules

The plasma membrane permits a regular exchange of nutrients, waste products, oxygen, and inorganic ions. This steady traffic of small molecules is regulated in a variety of ways.

Selective Permeability   Biological membranes are selectively permeable, which mean that the ease and rate at which small molecules pass through them differ. The hydrophobic center of the lipid bilayer impedes passage of ions and polar molecules. The rate at which hydrophobic molecules cross an artificial membrane is related to their lipid solubility and size. Small polar molecules, such as H_2O and CO_2, can cross a synthetic membrane rapidly, while ions and large polar molecules, such as glucose cannot permeate easily.

Transport proteins span the plasma membrane and provide mechanisms for movement of ions and moderately sized polar molecules. A uniport protein transports a single type of molecule; a symport transports two different molecules at the same time in the same direction; and an antiport transports two molecules in opposite directions. These proteins are highly specific for particular molecules, providing either a hydrophilic channel or physical binding and transport.

Diffusion and Passive Transport   *Diffusion* is the movement of a substance down its *concentration gradient* due to random molecular motion. This spontaneous process decreases free energy and increases entropy (disorder, randomness) in a system. Much of the exchange across membranes occurs by diffusion. The cell does not expend energy when substances diffuse across membranes down their concentration gradient; therefore, the process is called *passive transport*.

Osmosis: A Special Case of Passive Transport
Osmosis is the diffusion of water across a selectively permeable membrane. Water will move from a *hypoosmotic* solution across a membrane into a *hyperosmotic* solution—in other words, from a region with a lesser concentration of solutes into a region with a greater concentration of solutes. *Isosmotic* solutions have equal solute concentrations, and there is no net movement of water across a membrane separating them. Water diffuses down its own concentration gradient, which is affected by the binding of water molecules to solute particles. The attraction between water and solute molecules lowers the proportion of unbound water that is free to cross the membrane. The direction of osmosis is determined by the difference in total solute concentration, regardless of the kinds of solute molecules involved.

Osmotic pressure, which can be determined by an instrument called an osmometer, is a measure of the tendency for a solution to take up water when separated from pure water by a selectively permeable membrane. This pressure is proportional to *osmotic concentration*; it increases as the concentration of solute particles increases. Water moves from a solution with a lower osmotic pressure (fewer solute particles—hypoosmotic) to a solution with a higher osmotic pressure (higher osmotic concentration—hyperosmotic).

An animal cell placed in a hyperosmotic environment will lose water and crenate (shrivel). If placed in a hypoosmotic environment, the cell will gain water, swell, and possibly lyse (burst). Cells without rigid walls must either live in an isosmotic environment, such as salt water or isosmotic body fluids, or have

iso-osmotic

adaptations for *osmoregulation*, the control of water balance. Protists in fresh water environments may have membranes that are less permeable to water and contractile vacuoles that expel excess water as their osmoregulatory mechanism.

The cell walls of plants, fungi, prokaryotes, and some protists play a role in water balance within hypoosmotic environments. Water moving into the cell builds pressure as the cell swells against its cell wall, resulting in a dynamic equilibrium in the movement of water and creating a *turgid* cell. Turgid cells provide mechanical support for nonwoody plants. Plant cells in an isosmotic surrounding are *flaccid*. In a hyperosmotic medium, a plant cell undergoes *plasmolysis*, the pulling away of the plasma membrane from the cell wall as the cell shrivels.

Facilitated Diffusion   *Facilitated diffusion* involves the diffusion of polar molecules and ions across a membrane with the aid of transport proteins. Transport proteins are similar to enzymes in several ways: they are specialized for the solute they transport, presumably with a specific binding site for that solute; they can reach a maximum rate of transport when all transport proteins are saturated by a high concentration of the solute; and they can be inhibited by molecules that resemble their normal solute. A proposed model to explain the mechanism used by transport proteins suggests that the conformation of the protein changes when it binds to the solute and this structural change serves to translocate the binding site and the attached solute from one side of the membrane to the other. Facilitated diffusion is considered passive transport because, although it speeds diffusion, it cannot change the direction of movement of a solute down its concentration gradient.

Active Transport   *Active transport*, requiring the expenditure of energy by the cell, is essential for a cell to be able to maintain internal concentrations of small molecules that differ from the concentration in the environment. Transport proteins that engage in active transport use metabolic energy, usually in the form of ATP, to pump molecules against their concentration gradient. The terminal phosphate group of ATP may be transferred to the protein, inducing it to change its conformation and translocate the bound solute across the membrane. The *sodium-potassium pump*, also called sodium-potassium ATPase because it acts as an enzyme to hydrolyze ATP, allows the cell to exchange Na^+ and K^+ across animal cell membranes, creating a greater concentration of potassium ions and a lesser concentration of sodium ions within the cell.

The Special Case of Ion Transport   Cells have a *membrane potential*, a voltage across their plasma membrane created by the electrical potential energy resulting from the separation of opposite charges. The cytoplasm of a cell is negatively charged compared to extracellular fluid. The membrane potential favors the diffusion of cations (positively charged ions) into the cell and anions out of the cell. Diffusion of an ion is affected by membrane potential and its own concentration gradient; thus an ion diffuses down its *electrochemical gradient*.

Why is cytoplasm usually negatively charged, such that a membrane potential exists? At cellular pH, most proteins and other macromolecules are negatively charged. *Electrogenic pumps*, membrane proteins that generate voltage across a membrane by active transport of ions, contribute most to the charge differences. The sodium-potassium pump, the major electrogenic pump in animal cells, moves three Na$^+$ ions out of the cell for every two K$^+$ ions it pumps in. A proton pump that transports H$^+$ out of the cell generates membrane potential in plants, fungi, and bacteria. Proton pumps are also important in the membranes of mitochondria and chloroplasts.

Cotransport Cotransport is a mechanism through which the active transport of one solute is indirectly driven by an ATP-powered pump that transports another substance against its gradient. Energy is stored by concentrating a substance on one side of the membrane. As that substance diffuses back down its concentration gradient crossing the membrane through specific transport proteins, other solutes may be cotransported against their concentration gradients. The proton pump of plant cells drives the active transport of amino acids and sugars by coupling them with hydrogen ions so that both cross the membrane via symport proteins.

Traffic of Large Molecules: Endocytosis and Exocytosis

Large molecules, such as proteins and polysaccharides, are transported across the plasma membrane through the processes of exocytosis and endocytosis. In *exocytosis*, the cell secretes macromolecules by the fusion of vesicles with the plasma membrane. In *endocytosis*, a region of the plasma membrane sinks inward and pinches off to form a vesicle containing material that had been outside the cell. *Phagocytosis* is a form of endocytosis in which pseudopodia wrap around a food particle and create a vacuole. The vacuole then fuses with a lysosome containing hydrolytic enzymes. In *pinocytosis*, droplets of extracellular fluid are taken into the cell in small vesicles. *Receptor-mediated endocytosis* allows a cell to acquire specific substances that bind with receptor sites clustered in *coated pits* on the cell surface. A layer of the fibrous protein

clathrin on the cytoplasmic side of the membrane reinforces the coated pits. *Ligands*, molecules that bind specifically to receptor sites, are carried into the cell when the coated pit buds off to form a *coated vesicle*.

Animal cells use receptor-mediated endocytosis to take in cholesterol for the synthesis of membranes and other steroids. Familial hypercholesteremia is an inherited disease in which receptor sites are missing, and cholesterol accumulates in the blood, leading to early atherosclerosis.

STRUCTURE YOUR KNOWLEDGE

1. Explain the relationship between the structure of membranes, as depicted by the fluid mosaic model, and their permeability to (a) nonpolar hydrophobic molecules and (b) polar or ionic molecules.

2. Create a concept map to illustrate your understanding of osmosis.

3. The following diagrams illustrates two examples of transport proteins.

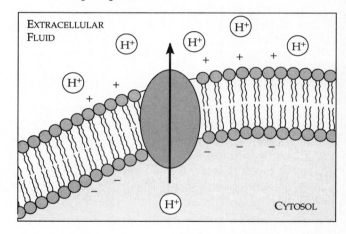

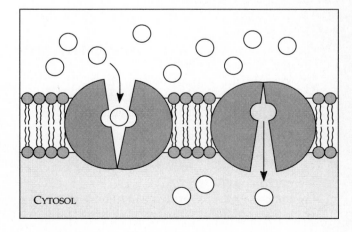

a. Label the diagram that represents facilitated diffusion.

How can you tell?

Does the cell expend energy in this transport? Why or why not?

What types of solute molecules may be moved by this type of transport?

b. Label the diagram that shows active transport.

How can you tell?

Does the cell expend energy in this transport? Why or why not?

TEST YOUR KNOWLEDGE

MULTIPLE CHOICE: *Choose the one best answer.*

1. Glycoproteins and glycolipids are important for
   a. facilitated diffusion.
   b. active transport.
   c. cell-cell recognition.
   d. cotransport.

2. Amphipathic molecules
   a. have hydrophilic and hydrophobic regions.
   b. include phospholipids and integral membrane proteins.
   c. may be held to one another by hydrophobic interactions.
   d. include all of the above.

3. Which of the following is *not* true about osmosis?
   a. It increases free energy in a system.
   b. Water moves from a hypoosmotic to a hyperosmotic solution.
   c. Osmotic pressure increases with increasing osmotic concentration.
   d. It increases the entropy in a system.

4. One of the problems with the Davson-Danielli sandwich model for membranes was that
   a. it did not explain the easy permeability by hydrophobic molecules.
   b. it did not account for the actual thickness of the plasma membrane.
   c. it could not explain where the hydrophobic regions of the membrane proteins would align.
   d. it could not account for the membrane potential.

5. Evidence for the fluid mosaic model of membrane structure came from
   a. the freeze-fracture technique of electron microscopy.
   b. the movement of proteins in hybrid cells.
   c. the amphipathic nature of membrane proteins.

d. all of the above.

6. Membrane proteins that function in active transport are likely to be
   a. peripheral proteins.
   b. integral proteins.
   c. ATPases.
   d. both b and c.

7. Ions diffuse across membranes down their
   a. electrochemical gradients.
   b. electrogenic gradients.
   c. electrical gradients.
   d. concentration gradients.

8. The fluidity of membranes in cold weather may be maintained by
   a. increasing the number of phospholipids with saturated hydrocarbon tails.
   b. activating an H^+ pump.
   c. increasing the concentration of cholesterol in the membrane.
   d. increasing the proportion of integral proteins.

9. Water, a polar molecule, can easily pass through membranes because it
   a. is small enough to pass between the lipids of the membrane.
   b. always has a strong concentration gradient.
   c. creates enough pressure to force its way through the membrane.
   d. has specialized carrier molecules that facilitate its diffusion.

10. A plant cell placed in a hypoosmotic environment will
    a. plasmolyze.
    b. crenate.
    c. become turgid.
    d. become flaccid.

11. Which of the following is *not* true of the carrier molecules involved in facilitated diffusion?
    a. They increase the speed of transport across a membrane.
    b. They can concentrate solute molecules on one side of the membrane.
    c. They have specific binding sites for the molecules they transport.
    d. They may undergo a conformational change upon binding of solute.

12. The membrane potential of a cell may be created by
    a. the proton pump.
    b. charge differences on either side of the membrane.
    c. the sodium-potassium pump.
    d. all of the above.

13. Cotransport may involve
    a. active transport of two solutes through a symport.
    b. transport of one solute against its concentration gradient in tandem with another that is diffusing down its concentration gradient.
    c. ion diffusion against the electrochemical gradient created by an electrogenic pump.
    d. receptor-mediated endocytosis.

14. Exocytosis involves all of the following *except*
    a. clathrin coated pits.
    b. the fusion of a vesicle, usually from the ER or Golgi apparatus, with the plasma membrane.
    c. a mechanism to transport carbohydrates to the outside of plant cells during the formation of cell walls.
    d. a mechanism to rejuvenate the plasma membrane.

15. The proton pump in plant cells is the functional equivalent of an animal cell's
    a. cotransport mechanism.
    b. sodium-potassium pump.
    c. contractile vacuole for osmoregulation.
    d. receptor-mediated endocytosis of cholesterol.

16. Cell walls provide an adaptive advantage for certain organisms by
    a. providing structural support.
    b. maintaining cell turgidity.
    c. resisting excessive water influx.
    d. all of the above.

17. Pinocytosis may involve
    a. the fusion of a newly formed food vacuole with a lysosome.
    b. receptor-mediated endocytosis and the formation of coated vesicles.
    c. the pinching in of the plasma membrane around droplets of external fluid.
    d. pseudopod extension as vesicles move along the cytoskeleton and fuse with the plasma membrane.

18. A freshwater *Paramecium* is placed into salt water. Which of the following events would occur?
    a. an increase in the action of its contractile vacuole
    b. swelling of the cell until it becomes turgid
    c. swelling of the cell until it lyses
    d. shriveling or crenation of the cell

19. Watering a house plant with too concentrated a solution of fertilizer can result in wilting because
    a. the uptake of ions into plant cells increases their osmotic pressure.
    b. the soil solution becomes hyperosmotic, causing the cells to lose water.
    c. the plant will grow faster than it can transport and maintain proper water balance.
    d. diffusion down the electrochemical gradient will cause a disruption of membrane potential and accompanying loss of water.

20. A single layer of phospholipid molecules coats a beaker of water. Which part of the molecules will face the air?
    a. the phosphate group
    b. the hydrocarbon tails
    c. both head and tail because the molecules will lie sideways
    d. the phospholipids would dissolve in the water and not form a membrane coat

21. An explanation for the movement of water from a hypoosmotic solution through a semipermeable membrane to a hyperosmotic solution is that
    a. the hypoosmotic solution has a greater osmotic pressure, driving water through the membrane.
    b. solute molecules move down their concentration gradient and water that is bound to the solute moves along with it.
    c. solute molecules bind water in the hyperosmotic solution, leaving more water molecules free to move in the hypoosmotic solution.
    d. the free energy of the hyperosmotic side is much higher and water moves in to lower that free energy.

Use the U-tube setup to answer questions 22 through 24.

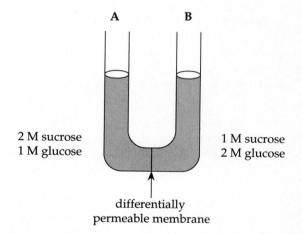

The solutions in the two arms of this U-tube are separated by a membrane that is permeable to water and glucose but not to sucrose. Side A is filled with a solution of 2 M sucrose and 1 M glucose. Side B is filled with 1 M sucrose and 2 M glucose.

22. Initially, the solution in side A, with respect to that in side B, is
    a. hypoosmotic.
    b. hyperosmotic.
    c. isosmotic.
    d. saturated.

23. After the system reaches equilibrium, what changes are observed?
    a. The water level is higher in side A than in side B.
    b. The water level is higher in side B than in side A.

c. The molarity of glucose is higher in side A than in side B.
d. The molarity of sucrose has increased in side A.

24. During the period before equilibrium is reached, which molecule(s) will show net movement through the membrane?
    a. water
    b. glucose
    c. water and glucose
    d. water and sucrose

CHAPTER 9

CELLULAR RESPIRATION: HARVESTING CHEMICAL ENERGY

FRAMEWORK

This chapter details the intricate processes through which cells harvest energy from organic molecules. The catabolic pathways of glycolysis and respiration capture the chemical energy in glucose and other fuels and store it in ATP. Glycolysis, occurring in the cytosol, produces a small amount of ATP and pyruvate and NADH, both of which may then enter the mitochondria for respiration. A mitochondrion consists of a matrix in which the enzymes of the Krebs cycle are localized, a highly folded inner membrane in which enzymes and the molecules of the electron transport chain are embedded, and an intermembrane space between the two membranes to temporarily house H⁺ pumped across the inner membrane during the redox reactions of the electron transport chain. Through a chemiosmotic mechanism, a proton motive force drives oxidative phosphorylation as protons move back through ATP synthases located in the membrane. The following diagram illustrates the key components of these integrated processes.

CHAPTER SUMMARY

Cells require energy to perform the work associated with living: synthesis of macromolecules, pumping of substances across membranes, movement, reproduction, and maintenance of their complex, ordered structure. For most cells, this energy ultimately comes from the sun through the organic molecules produced by photosynthetic organisms. Cells obtain the chemical energy stored in their food by systematically degrading, with the help of enzymes, energy-rich molecules into simpler, energy-poor waste products.

Fermentation, which occurs without oxygen, is the partial degradation of sugars to release energy. *Cellular respiration*, the most common catabolic pathway, uses oxygen to break down glucose (or other energy-rich organic compounds) and obtain energy in the usable form of ATP. This exergonic process has a free energy change of −686 kcal/mol glucose. The common equation for cellular respiration is as follows:

$$C_6H_{12}O_6 + 6O_2 \longrightarrow 6CO_2 + 6H_2O + \text{Energy (ATP + heat)}$$

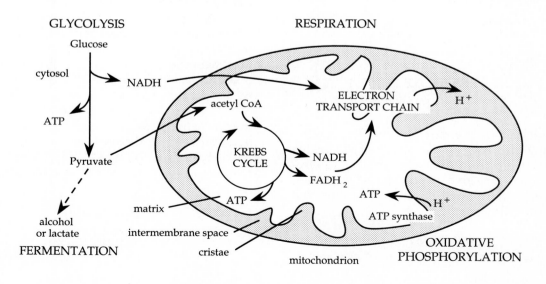

The waste products of respiration, CO_2 and water, are the raw materials chloroplasts use in photosynthesis to produce glucose and oxygen. The chemical elements essential to life are recycled. Energy, however, is not. It is converted from solar energy to potential energy temporarily stored in chemical compounds, harnessed by cells to do work, and finally released in the highly entropic form of heat energy.

ATP and Cellular Work: A Review

ATP, adenosine triphosphate, is very reactive because of the instability of the bonds between its three negatively charged phosphate groups. When enzymes shift the terminal phosphate group to another molecule, this phosphorylated molecule is energized and can perform work. The cell regenerates its supply of ATP by phosphorylating ADP, using energy from respiration for ATP synthesis.

Chemical Background: Respiration as an Oxidation-Reduction Process

The transfer of electrons in oxidation and reduction reactions releases energy stored in food molecules which drives the synthesis of ATP.

Introduction to Redox Reactions   Oxidation-reduction or *redox reactions* involve the partial or complete transfer of one or more electrons from one reactant to another. The substance that loses electrons is *oxidized* and acts as a *reducing agent* to the substance that gains electrons and is thus *reduced*. This reduced substance acts as an *oxidizing agent* by accepting electrons from the substance that is oxidized. These relationships can be illustrated as follows:

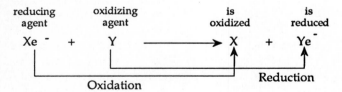

Oxygen, which is highly electronegative, attracts electrons strongly and is one of the most powerful oxidizing agents. As electrons shift toward a more electronegative atom, they give up potential energy. Thus, chemical energy is released in a redox reaction that passes electrons to oxygen.

Respiration: A Stepwise Redox Reaction   In respiration, the energy from the oxidation of glucose and the reduction of oxygen is captured in ATP. Organic molecules with an abundance of hydrogen are rich in electrons that are shared equally with carbon. Respiration is a redox process that transfers hydrogen to oxygen, where these electrons are held closer to oxygen, thus releasing their potential energy.

The oxidation of glucose as hydrogen is transferred to oxygen does not occur all at once, but rather in a series of discrete steps, each catalyzed by a specific enzyme. At certain steps, two hydrogen atoms are removed by enzymes called dehydrogenases and passed to an oxidizing agent, NAD^+ (nicotinamide adenine dinucleotide). Actually, two electrons and one proton are passed to NAD^+ and a proton (H^+) is released into the environment. NAD^+ is called an electron acceptor; its reduced form is NADH. Electrons retain most of their potential energy when they are transferred to NAD^+.

Unlike the explosive release of energy when H_2 and O_2 combine to form water, energy from respiration is slowly released in a series of small steps in an *electron transport chain* until the electrons from hydrogen atoms that were removed from organic molecules are captured by oxygen.

Overview of Cellular Respiration

Respiration occurs in three stages: *glycolysis*, the *Krebs cycle*, and the *electron transport chain and oxidative phosphorylation*. Glycolysis, occurring in the cytosol, breaks glucose into two molecules of pyruvate. The Krebs cycle, located in the mitochondrial matrix, converts a derivative of pyruvate into carbon dioxide. Some of the steps of glycolysis and the Krebs cycle transfer electrons to NAD^+, reducing it to NADH. NADH passes electrons to the electron transport chain, from which they eventually combine with hydrogen ions and oxygen to form water. The energy released at each step of the electron transport chain is used to synthesize ATP by oxidative phosphorylation. The inner mitochondrial membrane, separating the matrix from the intermembrane space, is the location of the electron transport chain and oxidative phosphorylation.

More than 30 molecules of ATP are generated for each molecule of glucose oxidized to carbon dioxide and water. Most of the ATP produced by respiration comes from oxidative phosphorylation. About 10% is produced by *substrate-level phosphorylation* during glycolysis and the Krebs cycle. In this type of ATP synthesis, an enzyme transfers a phosphate group from a substrate to ADP.

Glycolysis: A Closer Look

Glycolysis, the breakdown of glucose to pyruvate, occurs in the cytosol and consists of ten steps, each catalyzed by a specific enzyme. There are two major segments to the process: an energy-investment stage and an energy-yielding phase. In the first five steps, two molecules of ATP are consumed as glucose is split into two phosphorylated three-carbon molecules of glyceraldehyde phosphate. The energy yielding-phase begins as an enzyme catalyzes the oxidation and phosphorylation of glyceraldehyde phosphate, forming NADH and diphosphoglycerate. One phosphate is then transferred to ATP by substrate-level phosphorylation. In the next two steps, the substrate is rearranged so that the remaining phosphate bond becomes unstable, leading to the final step which yields another ATP and pyruvate. The conversion of the two molecules of glyceraldehyde phosphate to two molecules of pyruvate produces two NADH molecules and four ATPs by substrate-level phosphorylation. For each molecule of glucose, glycolysis yields a net gain of twos ATPs, since two ATPs are used to prime the process.

The Krebs Cycle: A Closer Look

The two molecules of pyruvate produced by glycolysis still hold more than three-fourths of the chemical energy stored in glucose. If oxygen is present, pyruvate enters the mitochondria to be oxidized and release the remaining energy during the Krebs cycle.

Before the Krebs cycle begins, a series of steps occur within a multienzyme complex in the mitochondria: a carboxyl group is removed from the 3-car-

bon pyruvate and released into the cytosol as CO_2; the remaining 2-carbon group is oxidized to form acetate with the accompanying reduction of NAD^+ to NADH; and coenzyme A is attached to the acetate by an unstable bond, forming *acetyl CoA*.

The Krebs cycle has eight steps, each catalyzed by a specific enzyme in the mitochondrial matrix. The acetyl fragment of acetyl CoA is added to oxaloacetate to form citrate, which is progressively decomposed back to oxaloacetate. For each turn of the Krebs cycle, two carbons enter in the reduced form from acetyl CoA; two carbons exit completely oxidized as CO_2; three molecules of NAD^+ are reduced to NADH; one molecule of reduced $FADH_2$ is formed; and one ATP molecule is made by substrate-level phosphorylation. The Krebs cycle is summarized in figure 9.1. There are two turns of the Krebs cycle for each glucose molecule oxidized.

The Electron Transport Chain and Oxidative Phosphorylation: A Closer Look

Glycolysis and the Krebs cycle produce only 4 ATPs by substrate-level phosphorylation per glucose molecule. Most of the energy extracted from food is stored in the molecules of NADH and $FADH_2$ and released by the electron transport chain to power oxidative phosphorylation.

Pathway of Electron Transport Most components of the chain are proteins with tightly bound, non-protein prosthetic groups. These groups shift between reduced and oxidized states as they accept and donate electrons. NADH passes its high-energy electrons to a flavoprotein (FMN), which becomes oxidized as it passes the electrons to the next electron carrier, an iron-sulfur protein (Fe.S). This molecule passes electrons to ubiquinone (CoQ), which passes the electrons on to a series of molecules called *cytochromes*, proteins with an iron-containing *heme group*. The last cytochrome of the chain passes its electrons to oxygen, which picks up a pair of H^+ from the aqueous medium and forms water.

$FADH_2$, the other reduced electron acceptor of the Krebs cycle, adds its electrons to the chain at a lower energy level; thus one-third less energy for ATP synthesis is derived from $FADH_2$ than from NADH.

Chemiosmosis: Coupling Electron Flow to ATP Synthesis Located in the inner mitochondrial membrane are many copies of an enzyme called *ATP synthase* that uses the energy of a proton (hydrogen ion)

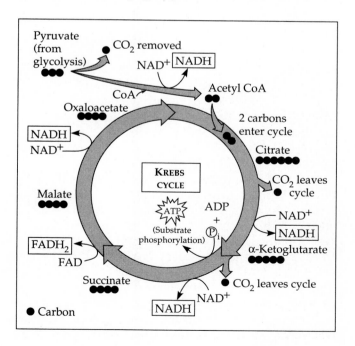

gradient to drive ATP synthesis. The electron transport chain pumps H^+ from the matrix into the intermembrane space. The H^+ leak back across the membrane through the ATP synthase complexes that span the membrane. Through this mechanism, called chemiosmosis, the exergonic passage of protons is coupled with the phosphorylation of ATP.

The electron carrier molecules are organized into three complexes, each of which spans the membrane. Electrons are transferred between the complexes by CoQ and cytochrome c, carriers that are mobile in the membrane. Some members of the transport chain accept and release protons along with electrons, and these protons are pumped across the membrane at three points along the electron transport chain. The proton gradient stores potential energy, referred to as the *proton-motive force*. This force drives H^+ through ATP synthase complexes, synthesizing ATP.

Respiratory poisons provide evidence for the chemiosmotic model of oxidative phosphorylation. Some poisons, such as cyanide, block the passage of electrons along the electron transport chain. No protons are pumped and no ATP is made. Another group of poisons inhibits ATP synthase, resulting in a higher proton gradient than usual because protons are not moving through the enzyme. Poisons called uncouplers make the membrane leaky to H^+, destroying the proton gradient so that no ATP is made, and the energy released by electron transport is transformed into heat. Special modification of the mitochondrial membrane allows certain tissues, such as brown fat in bats, to produce metabolic heat from the electron transport chain by preventing the formation of the proton gradient.

Chloroplasts use chemiosmosis to make ATP during photosynthesis, using light energy to generate the proton gradient that powers the ATP synthases. Bacteria generate proton gradients across their plasma membranes and use the proton-motive force to generate ATP, pump molecules across the membrane, and even move flagella. P. Mitchell received the Nobel Prize for his formulation of the chemiosmotic theory in his work with bacteria.

The Chemiosmotic Model Incorporates Many Biological Themes   Energy conversion is the function of mitochondria and an excellent example of an emergent property from the interaction of molecules within an intact organelle. The structure of the membranes, with molecules precisely embedded in them, and the intermembrane space and matrix of the mitochondria correlates with the functions of electron transport, H^+ pumping, and the chemiosmotic mechanism of oxidative phosphorylation. The similarity of chemiosmosis in mitochondria, chloroplasts, and bacteria points to their evolutionary connection.

The ATP Ledger for Respiration   The tally for ATP formed from the oxidation of a molecule of glucose to six molecules of carbon dioxide includes the substrate-level phosphorylation of ATP in glycolysis (4 ATP) and the Krebs cycle (2 ATP), and the ATP generated by chemiosmotic mechanisms in the electron transport chain (32 ATP—approximately 3 for each of the 8 NADH produced within the mitochondrion and 2 for each of the 2 $FADH_2$ from Krebs and for the 2 NADH from glycolysis that are passed to $FADH_2$ to enter the mitochondrion), minus the 2 ATP required initially to phosphorylate glucose. The total is 36 ATP. In prokaryotes, the yield is 38 because NADH from glycolysis does not have to pass its electrons across a mitochondrial membrane.

The yield of ATP per molecule of glucose oxidized is only an estimate because electron transport and ATP synthesis are loosely linked by the proton gradient. The use of the gradient to transport solutes across the mitochondrial membrane, and differences in the permeability of the membrane to H^+ may affect the yield of ATP from respiration. The efficiency of respiration in its energy conversion is approximately 63%: complete oxidation of a mole of glucose has a $\Delta G = -686$ kcal/mole, while the amount of energy stored in 36 molecules of ATP is 432 kcal. The rest of the energy is released as heat and used to maintain body temperature, with excess heat dissipated through cooling mechanisms.

Fermentation: The Anaerobic Alternative

Glycolysis produces 2 ATP per glucose molecule, under both *aerobic* and *anaerobic* conditions. Without oxygen, glycolysis is part of fermentation—the anaerobic conversion of organic nutrients to a waste product that regenerates NAD^+, the oxidizing agent for glycolysis. This process captures a small percentage of the energy stored in glucose.

In *alcohol fermentation*, pyruvate is converted into a two-carbon acetaldehyde and CO_2 is released. Acetaldehyde is then reduced by NADH to form ethanol (ethyl alcohol). The alcohol fermentation of yeasts, a fungus, is used in brewing. In *lactic acid fermentation*, pyruvate is reduced by NADH to form lactate and recycle NAD^+; no CO_2 is released. The dairy industry uses the lactic acid fermentation of certain fungi and bacteria to make cheese and yogurt. Muscle cells make ATP by lactic acid fermentation in vigorous exercise when energy demand is high and oxygen supply is low.

Comparison of Aerobic and Anaerobic Catabolism

In the three major catabolic schemes—aerobic respiration, fermentation, anaerobic respiration—glucose or other fuels are oxidized and their high-energy electrons are passed to NAD^+. The ultimate acceptor of these electrons differs for the three schemes. In *aerobic respiration*, the final electron acceptor is oxygen. Organisms that rely on aerobic respiration for energy are called *strict aerobes*; they must have oxygen to survive. *Strict anaerobes*, found in a few groups of bacteria, are poisoned by oxygen. The final electron acceptor in *anaerobic respiration* is a substance such as sulfate or nitrate. Both anaerobic and aerobic respiration are considered to be cellular respiration; they involve catabolic pathways that use electron transport chains to make ATP.

Fermentation produces ATP by substrate-level phosphorylation without the use of an electron transport chain. The final electron acceptor is pyruvate or one of its derivatives. Some strict anaerobes rely entirely on fermentation; they do not have electron transport chains. *Facultative anaerobes*, such as yeasts and some bacteria, can make ATP by fermentation or respiration, depending upon whether oxygen is available. When oxygen is available, pyruvate is converted to acetyl CoA, which is oxidized in the Krebs cycle. Without oxygen, pyruvate serves as the final electron acceptor.

Evolutionary Significance of Glycolysis

Glycolysis is common to fermentation and respiration. This most widespread of all metabolic processes probably evolved as the method of ATP formation by ancient prokaryotes before oxygen was available in the atmosphere. The cytoplasmic location of glycolysis, not using any of the membrane-bound organelles of eukaryotic cells, is also evidence of its antiquity.

Catabolism of Other Molecules

Fats, proteins, and carbohydrates can all be used by cellular respiration to make ATP. Starch and glycogen are hydrolyzed to glucose. Digestion of disaccharides provides glucose and other monosaccharides that are converted to glucose. Proteins can be used for fuel after they are digested into their constituent amino acids, which are then deaminated by the removal of their amino groups. Depending on its structure, an amino acid can enter into respiration at several sites (pyruvate, acetyl CoA, or an intermediate of the Krebs cycle). The digestion of fats yields glycerol, which is converted to an intermediate of glycolysis, and fatty acids, which are broken down by a process called *beta oxidation* to two-carbon fragments that enter the Krebs cycle as acetyl CoA. A gram of fat produces more than twice as much ATP through respiration as a gram of carbohydrate.

Biosynthesis

Not all the organic molecules of food are oxidized as fuel to make ATP; many provide the carbon skeletons the cell needs for biosynthesis. Some monomers from digestion, such as amino acids, can be directly incorporated into the cell's macromolecules. Many of the intermediates of glycolysis and the Krebs cycle serve as precursors for the cell's anabolic pathways. The molecules of carbohydrates, fats and proteins can all be interconverted through the intermediates of metabolism to provide for the cell's needs. Metabolism is a very intricate and adaptable network.

Control of Respiration

Metabolic pathways are strictly controlled under the basic principles of supply and demand. Through feedback inhibition, the end product of a pathway inhibits the enzyme that initiates the pathway. Respiration is controlled by the supply of ATP in the cell. The allosteric enzyme that catalyzes an early step of glycolysis, phosphofructokinase, is inhibited by ATP and activated by ADP. Phosphofructokinase is also inhibited by citrate transported from the mitochondria into the cytosol, thus synchronizing the rates of glycolysis and the Krebs cycle. Other enzymes located at key intersections help to maintain metabolic balance. The processes through which cells harvest energy and raw materials from food molecules are intricately connected and controlled.

STRUCTURE YOUR KNOWLEDGE

1. In the following diagram, fill in the major molecules that enter and exit glycolysis, the Krebs cycle, and oxidative phosphorylation. Show the numbers of each of these molecules as they lead to the generation of approximately 36 ATP for each molecule of glucose processed.

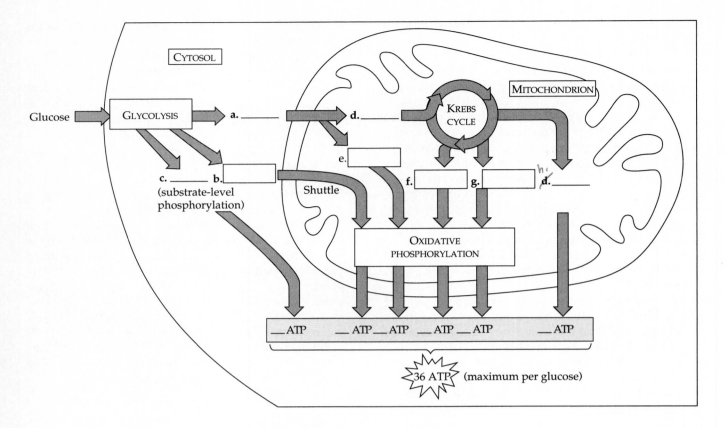

2. Create a concept map to organize your understanding of chemiosmosis and oxidative phosphorylation.

TEST YOUR KNOWLEDGE

MULTIPLE CHOICE: *Choose the one best answer.*

1. Which of the following is *not* true of a substance that is phosphorylated?
   a. It has an unstable phosphate bond.
   b. Its reactivity has been increased; it is primed to do work.
   c. It has entered the oxidative phosphorylation process.
   d. It has been reduced.

2. Chemiosmosis involves
   a. the diffusion of water down a concentration gradient.
   b. a proton-motive force that then drives ATP formation.
   c. a proton gradient that drives the redox reactions in the cristae.
   d. an ATP synthase that pumps H^+ across the mitochondrial membrane.

3. In a redox reaction,
   a. the substance that is reduced loses energy.
   b. the substance that is oxidized loses energy.
   c. the oxidizing agent donates electrons.
   d. the reducing agent is the most electronegative.

4. Which of the following is *not* true of oxidative phosphorylation?
   a. It uses oxygen as the ultimate electron donor.
   b. It involves the redox reactions of the electron transport chain.
   c. It involves an ATP synthase located in the inner mitochondrial membrane.
   d. It produces three ATP for every NADH that is oxidized.

5. Substrate-level phosphorylation
   a. involves the shifting of a phosphate group from ATP to a substrate.
   b. can use NADH or $FADH_2$.
   c. accounts for all the ATP formed by anaerobic respiration.
   d. takes place both in the cytosol and in the mitochondrial matrix.

6. Facultative anaerobes
   a. are poisoned by oxygen.
   b. make ATP by fermentation or aerobic respiration, depending on availability of oxygen.

c. must use some molecule other than oxygen as the final electron acceptor.

d. include plants that can produce their own oxygen but undergo anaerobic respiration at night.

7. Glycolysis is not as energy-productive as respiration because
a. NAD^+ is regenerated by alcohol or lactate production, without the high-energy electrons passing through the electron transport chain.
b. it is the pathway common to fermentation and respiration.
c. it does not take place in a specialized membrane-bound organelle.
d. pyruvate is more reduced than CO_2; it still contains much of the energy from glucose.

8. The electron carrier molecules CoQ and cytochrome c
a. are reduced as they pass electrons on to the next molecule.
b. contain heme prosthetic groups.
c. are mobile carriers that transfer electrons between the electron carrier complexes.
d. transport H^+ into the mitochondrial matrix, establishing the proton-motive force.

9. When pyruvate is converted to acetyl CoA
a. CO_2 and ATP are released.
b. a multienzyme complex removes a carboxyl group and attaches a coenzyme.
c. one turn of the Krebs cycle is completed.
d. NAD^+ is regenerated so that glycolysis can continue.

10. How many molecules of CO_2 are generated for each molecule of acetyl CoA introduced into the Krebs cycle?
a. 2
b. 3
c. 4
d. 6

11. Support for the chemiosmotic theory comes from
a. the production of only two ATP from the oxidation of $FADH_2$
b. the difference in the membrane potential between the inner mitochondrial membrane and the plasma membrane.
c. the study of poisons called uncouplers that prevent the formation of the proton gradient and result in no synthesis of ATP.
d. the presence of the electron transport molecules on the inside of the cristae where they can bind to H^+.

12. Which of the following reactions is incorrectly paired with its location?

a. ATP synthesis—inner membrane of the mitochondrion
b. fermentation—cell cytoplasm
c. glycolysis—cell cytoplasm
d. Krebs cycle—cristae of mitochondrion

13. In the reaction $C_6H_{12}O_6 + 6 O_2 \longrightarrow 6 CO_2 + 6 H_2O$,
a. oxygen becomes reduced.
b. glucose becomes reduced.
c. oxygen becomes oxidized.
d. water is a reducing agent.

14. What do muscle cells in oxygen deprivation gain from the conversion of pyruvate?
a. ATP and lactate
b. NAD^+ and lactate
c. CO_2 and lactate
d. ATP and alcohol

15. Glucose, made from six radioactively labeled carbon atoms, is fed to yeast cells in the absence of oxygen. How many molecules of radioactive alcohol (C_2H_5OH) are formed from each molecule of glucose?
a. 0
b. 1
c. 2
d. 3

16. Which of the following produces the most ATP per gram?
a. glucose, because it is the starting place for glycolysis
b. glycogen or starch, because they are polymers of glucose
c. fats, because they are highly reduced compounds
d. proteins, because of the energy stored in their tertiary structure

17. Fats and proteins can be used as fuel in the cell because
a. they are converted to glucose by enzymes.
b. they are converted to intermediates of glycolysis or the Krebs cycle.
c. they can pass through the mitochondrial membrane to enter the Krebs cycle.
d. they contain unstable phosphate bonds.

18. Which is *not* true of the enzyme phosphofructokinase? It is
a. an allosteric enzyme.
b. inhibited by citrate.
c. the pacemaker of glycolysis and respiration.
d. inhibited by ADP.

19. We can produce and store fat in our bodies, even if out diet is fat-free because

a. carbohydrates such as breads and potatoes are fattening.

b. a genetic predisposition may interfere with feedback inhibition.

c. excess carbohydrates and proteins can be converted to fat through intermediates of glycolysis and the Krebs cycle.

d. fats contain more energy than proteins or carbohydrates.

20. The fact that chemiosmosis is found in mitochondria, chloroplasts, and bacteria indicates that

a. it was the first metabolic pathway to evolve.

b. mitochondria and chloroplasts may have evolved from bacteria.

c. electron transport chains probably evolved three separate times.

d. oxidative phosphorylation is more productive than substrate-level phosphorylation.

CHAPTER 10

PHOTOSYNTHESIS

FRAMEWORK

This chapter describes how photosynthesis captures light energy from the sun and stores it in organic molecules. Chlorophyll and other pigments are assembled into photosystems in the thylakoid membranes of chloroplasts. In the light reactions, energized electrons reduce $NADP^+$ and energized electrons are passed down a redox electron transfer chain, transporting H^+ across the thylakoid membrane and leading to the chemiosmotic synthesis of ATP in photophosphorylation. Water is split to replace the electrons of chlorophyll trapped in NADPH, and oxygen is evolved. In the Calvin cycle, which takes place in the stroma, CO_2 is fixed to the five-carbon RuBP, which splits and is reduced to glyceraldehyde phosphate, an energy-rich sugar. NADPH and ATP for the Calvin cycle were produced during the light reactions. The following diagram summarizes these processes.

CHAPTER SUMMARY

In photosynthesis, plants convert the light energy of the sun into chemical energy stored in organic molecules made from carbon dioxide and water. Organisms obtain the organic molecules they require for energy and carbon skeletons by *autotrophic* or *heterotrophic nutrition*. Autotrophs "feed themselves" in the sense that they make their own organic molecules from inorganic raw materials. Plants, algae, and some protists and prokaryotes are *photoautotrophs*, which use light as the source of energy for the manufacture of organic molecules. Some bacteria are *chemoautotrophs*, organisms that produce their organic compounds using energy obtained from oxidizing inorganic substances.

Heterotrophs obtain their organic compounds from other organisms. They may eat plants or animals or decompose organic litter, but almost all are ultimately dependent on plants for food and oxygen.

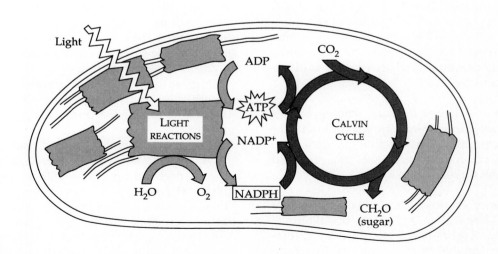

Chloroplasts: Sites of Photosynthesis

Chloroplasts, found mainly in the *mesophyll* tissue of the leaf, contain *chlorophyll*, the green pigment that absorbs the light energy that drives photosynthesis. A chloroplast consists of a double membrane surrounding a dense fluid called the stroma and an elaborate thylakoid membrane system enclosing the thylakoid space. Thylakoid sacs may be layered to form grana. Chlorophyll is contained in the thylakoid membrane, and the conversion of light energy into chemical energy occurs in the thylakoids. The production of carbohydrate from this chemical energy and CO_2 occurs in the stroma. CO_2 enters and O_2 leaves the leaf through *stomata*. Veins carry water from the roots to leaves and distribute sugar to nonphotosynthetic tissue.

Photosynthetic prokaryotes lack chloroplasts, but chlorophyll is built into their plasma membranes or other membranes.

How Plants Make Food: An Overview

The process of photosynthesis is represented as follows: 6 CO_2 + 12 H_2O + light energy $\longrightarrow$ $C_6H_{12}O_6$ + 6 O_2 + 6 H_2O. This equation shows that water is both consumed and created during photosynthesis. If only the net consumption of water is considered, the equation for photosynthesis is the reverse of respiration. Both processes occur in plant cells.

The Splitting of Water Using evidence from bacteria that use hydrogen sulfide (H_2S) rather than water for photosynthesis and create sulfur instead of oxygen as a byproduct, C.B. van Niel hypothesized in the 1930s that all photosynthetic organisms need a hydrogen source and that plants split water as their hydrogen source, releasing oxygen. Twenty years later scientists confirmed this hypothesis by using a heavy isotope of oxygen (^{18}O) to label the CO_2 and water used in photosynthesis. Labeled O_2 was produced only when water, rather than carbon dioxide, was the radioactively labeled molecule.

Photosynthesis as a Redox Process Photosynthesis is a redox process like respiration, but differs in the direction of electron flow. The electrons increase their potential energy when they travel from water to sugar and light provides this energy.

The Two Stages of Photosynthesis Solar energy is converted into chemical energy in the thylakoids in the *light reactions*. Light energy absorbed by chlorophyll drives the transfer of electrons and hydrogen from water to the electron acceptor *NADP$^+$* (nicoti-namide adenine dinucleotide phosphate), which is reduced to NADPH. Water is split and releases oxygen during this process. ATP is formed during the light reactions by *photophosphorylation*.

In the *Calvin cycle*, carbon is incorporated into organic compounds by a process known as *carbon fixation*, and these compounds are then reduced to form carbohydrate. NADPH and ATP, from the light reactions, supply the reducing power and chemical energy needed for the Calvin cycle. These "dark reactions," which take place in the stroma, reduce CO_2 to carbohydrate.

How the Light Reactions Capture Solar Energy: A Closer Look

The Nature of Sunlight In the giant thermonuclear reactor of the sun, fusion reactions convert mass to energy as hydrogen atoms combine to form helium. This *electromagnetic energy*, also called radiation, travels as rhythmic wave disturbances of electrical and magnetic fields. The distance between the crests of the electromagnetic waves, their *wavelength*, ranges from less than a nanometer to more than a kilometer. Within this *electromagnetic spectrum*, the narrow band of radiation with wavelengths of 400 to 700 nm is called *visible light*, because humans can see it as various colors.

Light has wave-like properties but also behaves as if it consists of discrete particles called quanta or *photons*, which have a fixed quantity of energy. The amount of energy in a photon of light is inversely related to its wavelength. Thus a photon of red light, with a long wavelength, has less energy than a photon of violet light.

Photosynthetic Pigments Light may be reflected, transmitted, or absorbed when it strikes matter. Substances that absorb visible light are called *pigments*. A *spectrophotometer* measures the ability of a pigment to absorb various wavelengths of light. The *absorption spectrum* of *chlorophyll a* shows that it absorbs blue and red light best. Chlorophyll is green because it reflects or transmits green wavelengths of light.

An *action spectrum* shows the relative rates of photosynthesis under different wavelengths of light. The action spectrum for photosynthesis does not exactly match the absorption spectrum of chlorophyll *a* because other pigments in the chloroplasts absorb light of different wavelengths. These other pigments transfer energy to chlorophyll *a*, which is the only pigment that can directly channel energy into the light reactions. These accessory pigments include chlorophyll *b* and the *carotenoids*, yellow and orange pigments that extend the spectrum of light that can power photosynthesis.

The Photooxidation of Chlorophyll   When a pigment molecule absorbs energy from a photon, one of the molecule's electrons is elevated to an orbital where it has more potential energy and the pigment is said to be in an *excited state*. Only photons whose energy is equal to the difference in energy between the excited state and *ground state*, when the electron is in its normal orbital, are absorbed by a molecule.

The excited state is unstable. Generally the potential energy of the electron, created from the light energy of the photon, is released as heat as the electron drops back to its ground-state orbital. Some pigments also emit lower-energy photons of light, called *fluorescence*, as their electrons return to ground state.

Chlorophyll removed from chloroplasts will fluoresce red light upon illumination. However, when chlorophyll bound in a thylakoid membrane absorbs light, a redox reaction occurs: the chlorophyll is photooxidized and a molecule located nearby, called a *primary electron acceptor*, becomes reduced.

The Two Photosystems   Chlorophyll *a* and the accessory pigments are clustered together in the thylakoid membrane in assemblies of several hundred pigment molecules. One pair of chlorophyll *a* molecules can donate its excited electrons to the primary electron acceptor because they are located in the *reaction center* of the pigment assembly. Other pigment molecules in the assembly act as light-gathering antennae to absorb photons and pass the energy from molecule to molecule until it reaches the reaction center. This light-harvesting unit of the thylakoid membrane, consisting of the antenna complex, reaction-center chlorophyll *a*, and its primary electron acceptor is called a *photosystem*.

There are two **types** of photosystems in the thylakoid membrane. The pair of chlorophyll molecules at the reaction center of *photosystem I* are called *P700*, after the wavelength of light (700 nm) they absorb best. At the reaction center of *photosystem II* are chlorophyll *a* molecules called *P680*, absorbing light best at 680 nm. The differences between these two molecules and between these molecules and the rest of the chlorophyll *a* molecules in the photosystem, are the specific proteins to which they are bound in the reaction center of the thylakoid membrane and their proximity to their primary electron acceptors. Structure, as always, relates to function.

Cyclic Electron Flow   Of the two possible routes of electron flow in the light reactions, *cyclic electron flow*

is the simpler, involving only photosystem I and generating only ATP. Electrons are excited to a higher energy level by the absorption of photons of light and are passed by a series of redox reactions along an electron transport chain until they return to P700, again at their ground state energy level. These electrons are passed from the primary electron acceptor to a series of compounds that includes ferrodoxin (Fd), plastoquinone (Pq), a complex of two cytochromes, and plastocyanin (Pc). The electron transport chain pumps hydrogen ions across the thylakoid membrane, storing energy in the form of a proton-motive force of an H^+ gradient. This energy is tapped by ATP synthase enzymes in the thylakoid membrane using a chemiosmotic mechanism similar to that found in mitochondria. The synthesis of ATP in chloroplasts is called photophosphorylation because it is driven by light energy. *Cyclic photophosphorylation* refers to ATP formation during cyclic electron flow. Neither NADPH nor O_2 are generated by cyclic electron flow.

Noncyclic Electron Flow   In *noncyclic electron flow*, electrons pass continuously from water to $NADP^+$. Both photosystems are involved in this process. Electrons excited from P700 of photosystem I are not returned to P700, but are trapped in their high-energy state in NADPH. These electrons are captured by the primary acceptor, passed to Fd, and then to $NADP^+$ which they reduce to NADPH. The oxidized P700 now becomes a potent oxidizing agent. It becomes reduced by electrons activated by photosystem II. When P680 absorbs light, electrons are trapped by the primary electron acceptor of photosystem II, then passed through the electron transport chain used in cyclic electron flow (from Pq to cytochrome complex to Pc), and ultimately to P700. The energy released as the electrons lose their potential energy in the transport chain pumps H^+ across the thylakoid membrane, creating the proton-motive force to drive ATP synthesis in the process called *noncyclic photophosphorylation*.

Electrons are restored to oxidized P680 by a water-splitting enzyme that removes two electrons from water. The oxygen atom released when water is split immediately combines with another oxygen to form O_2, the byproduct of photosynthesis. The net effect of noncyclic electron flow is to move electrons from their low potential-energy state in water to their high potential-energy state in NADPH. ATP is also generated and O_2 released.

NADPH and ATP are produced in roughly equal quantities by noncyclic electron flow. However, the sugar-producing stage of photosynthesis, the Calvin cycle, requires more ATP than NADPH. The additional ATP is supplied by cyclic photophosphorylation.

Review: Comparison of Chemiosmosis in Chloro-plasts and Mitochondria Chemiosmosis in mitochondria and chloroplasts is very similar. The key difference is that in respiration chemical energy from food is converted into ATP, whereas in chloroplasts light energy is converted to chemical energy stored in ATP.

In the chloroplast, the components of the electron transport chain pump protons from the stroma into the thylakoid compartment. As hydrogen ions diffuse back through ATP synthase in the thylakoid membrane, ATP is formed on the stroma side, where it is available for the Calvin cycle. In the light, the proton gradient across the thylakoid membrane is as great as 3 pH units (pH of 5 in thylakoid compartment and pH 8 in stroma). The gradient is eliminated in the absence of light. Jagendorf provided convincing evidence for the chemiosmotic theory in the 1960s when he induced thylakoids to make ATP in the dark by artificially creating a pH gradient across the membrane.

Several laboratories are working on a tentative model for the organization of components of the photosystems in the thylakoid membrane. As it is now understood, electrons are removed from water within the thylakoid compartment and crisscross the membrane, passing through photosystem II and photosystem I to NADPH. When the mobile electron carrier plastoquinone transfers electrons to the cytochrome complex, protons are carried into the thylakoid compartment. The proton gradient is increased by the addition of protons from split water in the thylakoid compartment and the removal of hydrogen in the stroma during the reduction of $NADP^+$. Both ATP and NADPH are synthesized on the side of the membrane facing the stroma.

How the Calvin Cycle Makes Sugar: A Closer Look

In the Calvin cycle, the ATP and NADPH formed by the light reactions drive the reduction of CO_2 to sugar. *Glyceraldehyde 3-phosphate* is the three-carbon sugar formed by three turns of the cycle in the following steps. First, CO_2 is fixed by being added to a five-carbon sugar, ribulose bisphosphate (RuBP). This reaction is catalyzed by the enzyme RuBP carboxylase (rubisco), probably the most abundant protein on earth. The unstable six-carbon intermediate that is formed splits immediately into two molecules of 3-phosphoglycerate, which are then phosphorylated by ATP to form two molecules of 1,3-diphosphoglycerate. Two electrons from NADPH reduce this compound to create the high energy sugar glyceraldehyde

phosphate, the same sugar that is formed when glucose is split in glycolysis.

The cycle must turn three times, reducing three CO_2 molecules, to create one molecule of glyceraldehyde phosphate and regenerate the three molecules of the five-carbon RuBP used in the cycle. The rearrangement of five molecules of glyceraldehyde phosphate into the three molecules of RuBP requires three more ATPs. A total of nine ATPs and six molecules of NADPH are consumed in the formation of one glyceraldehyde phosphate. This product can be synthesized into a variety of carbohydrates and other organic compounds. The process of photosynthesis involves the integration of the light reactions and Calvin cycle within the thylakoid membrane and stroma of intact chloroplasts.

Photorespiration: An Evolutionary Relic?

Rubisco, the enzyme that initiates the Calvin cycle, can accept O_2 in place of CO_2 when a high concentration of O_2 builds up in the air spaces in the leaf. When oxygen is added to RuBP, the five-carbon product splits into a three-carbon molecule that remains in the cycle and a two-carbon compound, glycolate, which leaves the chloroplast, enters a peroxisome, and is broken down to release CO_2. This seemingly wasteful process that removes fixed carbon from the Calvin cycle and produces no ATP is called *photorespiration*. It is most common on hot, dry, bright days when leaves close their stomata to conserve water and oxygen, formed from the light reactions, collects in the leaf.

One hypothesis holds that photorespiration is an evolutionary relic from the time when there was little O_2 in the atmosphere and the ability of rubisco to distinguish between O_2 and CO_2 was not critical. Now, in our oxygen-rich atmosphere, photorespiration can result in the loss of as much of 50% of the carbon fixed in the Calvin cycle and, therefore, represents a considerable agricultural liability.

C_4 Plants

C_3 *plants* incorporate CO_2 directly into the Calvin cycle by way of rubisco, making them susceptible to photorespiration. C_4 *plants* have evolved a different mode of carbon fixation that serves to minimize photorespiration. CO_2 is first added to phosphoenol-pyruvate (PEP) with the aid of an enzyme (*PEP carboxylase*) that has no affinity for O_2. The four-carbon oxaloacetate formed by this carbon fixation in the *mesophyll cells* of the leaf is usually converted to malate and transported to *bundle-shealth cells* tightly packed around the veins of the leaf. This four carbon compound is broken down to release CO_2, which is then in high enough concentrations that rubisco will accept CO_2 and initiate the Calvin cycle. C_4 plants are able to photosynthesize efficiently in hot, arid regions using C_4 photosynthesis.

CAM Plants

CAM plants illustrate another evolutionary adaptation to very dry climates. These plants (mostly cacti, many bromeliads, and several others) close their stomata during the day to prevent water loss, but open them at night to take up CO_2 and incorporate it into a variety of organic acids. These carbon-containing compounds are stored during the night in vacuoles of mesophyll cells and then broken down to release CO_2 for photosynthesis during daylight. Unlike the C_4 pathway, the *CAM* pathway (crassulacean acid metabolism, named after the plant family in which it was first discovered) does not structurally separate carbon fixation from the Calvin cycle; instead, the two processes are separated in time.

The Fate of Photosynthetic Products

Carbohydrate is transported through the plant in the form of sucrose, a disaccharide. About 50% of the organic material produced by photosynthesis is used as fuel for cellular respiration in the mitochondria of plant cells; the rest is used for carbon skeletons for synthesis of organic molecules (mainly cellulose), stored as starch, or lost through photorespiration. The essential process of photosynthesis produces about 160 billion metric tons of carbohydrate per year.

STRUCTURE YOUR KNOWLEDGE

1. In a diagrammatic form, outline the key events of photosynthesis. Trace the flow of electrons through photosystem II and I, the transfer of ATP and NADPH formed by the light reactions into the Calvin cycle, and the major steps in the production of glyceraldehyde phosphate. Note where these reactions occur in the chloroplast.

2. What is the difference between an absorption and an action spectrum? Describe the following figure and explain its significance.

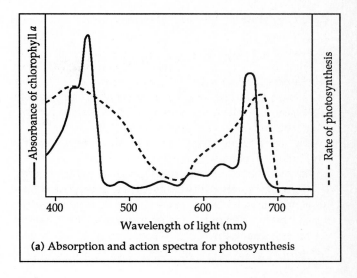

(a) Absorption and action spectra for photosynthesis

3. Create a concept map to help you develop your understanding of the chemiosmotic synthesis of ATP in photophosphorylation.

4. The chemiosmotic synthesis of ATP is similar in mitochondria and chloroplasts. In the following diagram, label the mitochondrion and chloroplast, and then fill in the locations and structures on the corresponding sides of the figure.

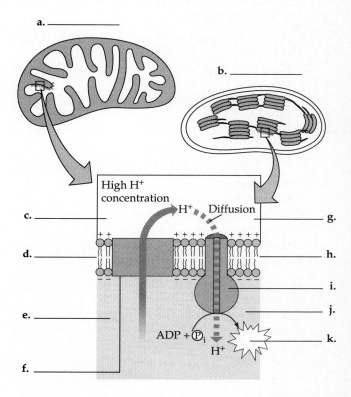

TEST YOUR KNOWLEDGE

MULTIPLE CHOICE: *Choose the one best answer.*

1. Chemoautotrophs
   a. use hydrogen sulfide as their hydrogen source for the photosynthesis of their organic compounds.
   b. "feed themselves" by obtaining energy in the chemical bonds of organic molecules.
   c. oxidize inorganic compounds to obtain energy to drive the synthesis of their organic compounds.
   d. live as decomposers of inorganic chemicals in organic litter.

2. Which of the following processes and locations is mismatched?
   a. light reactions—grana
   b. electron transport chain—thylakoid membrane
   c. Calvin cycle—stroma
   d. ATP synthase—double membrane surrounding chloroplast

3. Photosynthesis is a redox process in which
   a. CO_2 is reduced and water is oxidized.
   b. $NADP^+$ is reduced and RuBP is oxidized.
   c. CO_2, $NADP^+$, and water are reduced.
   d. O_2 acts as an oxidizing agent and water acts as a reducing agent.

4. Blue light has more energy than red light. Which of the following statements is true?
   a. Blue light has a longer wavelength than red light.
   b. Red light has a longer wavelength than blue light.
   c. Blue light contains more photons than red light.
   d. Blue light has a broader electromagnetic spectrum than red light.

5. A spectrophotometer measures
   a. the absorption spectrum of a substance.
   b. the action spectrum of a substance.
   c. the amount of energy in a photon.
   d. the wavelength of visible light.

6. Accessory pigments within chloroplasts are responsible for
   a. driving the splitting of water molecules.
   b. absorbing photons of different wavelengths of light and passing that energy to reaction center molecules.
   c. extending the absorption spectrum of chlorophyll *a*.
   d. both b and c.

7. Which of the following is *not* true of the reaction center molecule?
   a. It is always a chlorophyll *a* molecule.
   b. It is named for the wavelength of light it absorbs best.
   c. It is bound to specific proteins in the thylakoid membrane and located close to its primary electron acceptor.
   d. It accepts energized electrons from the antenna complex.

8. Noncyclic electron flow in the chloroplast results in the production of
   a. ATP.
   b. ATP and NADPH.
   c. ATP, NADPH, and O_2.
   d. Glyceraldehyde 3-phosphate.

9. The chlorophyll known as P680 is reduced by electrons from
   a. photosystem I.
   b. photosystem II.
   c. water.
   d. NADPH.

10. CAM plants avoid photorespiration by
    a. fixing CO_2 into organic acids during the night.
    b. fixing CO_2 into four-carbon compounds in bundle-sheath cells.
    c. fixing CO_2 into four-carbon compounds in the mesophyll.
    d. using an enzyme that has a higher affinity for CO_2 than does rubisco.

11. Electrons that flow through the two photosystems have their lowest potential energy in
    a. P700.
    b. P680.
    c. NADPH.
    d. water.

12. Chloroplasts can make carbohydrate in the dark if provided with
    a. ATP and NADPH and CO_2.
    b. an artificially induced proton gradient.
    c. organic acids or four-carbon compounds such as malate.
    d. a source of hydrogen.

13. In the chemiosmotic synthesis of ATP, H^+ diffuses through the ATP synthase
    a. from the stroma into the thylakoid compartment.
    b. from the thylakoid compartment into the stroma.
    c. from the cytoplasm into the matrix.
    d. from the cytoplasm into the stroma.

14. How many "turns" of the Calvin cycle are required to produce one molecule of glucose?
    a. 1
    b. 2
    c. 3
    d. 6

15. In green plants, most of the ATP for synthesis of proteins, cytoplasmic streaming, and other cellular activities comes directly from
    a. Photosystem I.
    b. the Calvin cycle.
    c. photophosphorylation.
    d. oxidative phosphorylation.

16. The six molecules of glyceraldehyde phosphate formed from three turns of the Calvin cycle are converted into
    a. three molecules of glucose.
    b. three molecules of RuBP and one glyceraldehyde phosphate.
    c. one molecule of glucose and four molecules of 3-phosphoglycerate.
    d. one glyceraldehyde phosphate and three four-carbon intermediates.

17. NADPH is used in the Calvin cycle to reduce
    a. an unstable six-carbon intermediate.
    b. RuBP.
    c. 1,3-diphosphoglycerate.
    d. glyceraldehyde phosphate.

18–25. Indicate if the following events occur during
    a. respiration
    b. photosynthesis
    c. both respiration and photosynthesis
    d. neither respiration nor photosynthesis

18. _____ Chemiosmotic synthesis of ATP

19. _____ Reduction of oxygen

20. _____ Reduction of CO_2

21. _____ Reduction of NAD^+

22. _____ Oxidation of water

23. _____ Oxidation of $NADP^+$

24. _____ Oxidative phosphorylation

25. _____ Electron flow along a cytochrome chain

CHAPTER 11

REPRODUCTION OF CELLS

FRAMEWORK

This chapter details the process through which cells create new cells with identical genetic material. The following concept map summarizes cell division.

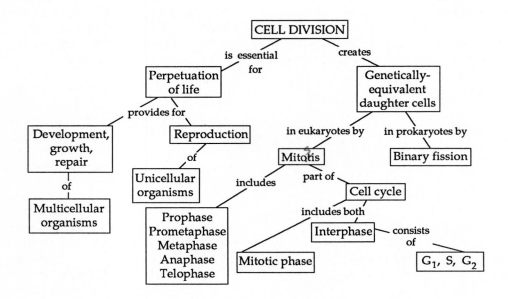

CHAPTER SUMMARY

Cell division, the process through which a cell reproduces itself, is the basis for the perpetuation of all life. Cell division creates duplicate offspring in unicellular organisms and provides for growth, development and repair in multicellular organisms. The process of recreating a structure as intricate as a cell necessitates the exact duplication and equal division of the DNA containing the cell's genetic program (the *genome*).

Bacterial Reproduction

Prokaryotes, the bacteria, reproduce themselves by a process known as *binary fission*. First, two copies of the single circular DNA molecule are produced and attached to the plasma membrane. Growth of the membrane separates the duplicate chromosomes, the membrane pinches in between the chromosomes, and a cell wall develops—creating two identical daughter cells.

An Introduction to Eukaryotic Chromosomes

Copying and dividing the tens of thousands of genes of a eukaryotic cell involves a precise process. Grouping genes on chromosomes simplifies their management and distribution in cell division. Each species has a characteristic number of chromosomes in each somatic cell; reproductive cells (egg and sperm) have half that number of chromosomes.

Each chromosome is a very long DNA molecule with associated proteins that help to structure the chromosome and control the activity of the genes. This DNA-protein complex, called *chromatin*, is organized into a coiled and folded fiber. Prior to cell division, a cell copies its genome by duplicating each chromosome. Replicated chromosomes consist of two identical *sister chromatids*, joined together at a specialized region called a *centromere*. The two sister chromatids separate during *mitosis* (the division of the nucleus) and then the cytoplasm divides during *cytokinesis*, producing two separate, genetically-equivalent daughter cells.

A type of cell division called *meiosis* produces daughter cells that have half the number of chromosomes of the parent cell. With the fertilization of egg and sperm, which are formed by meiosis, the chromosome number is restored in the somatic cells of the new offspring.

The Cell Cycle: An Overview

The *cell cycle* includes the *M phase* (mitotic phase), in which the chromosomes and cytoplasm are divided, and *interphase*, during which most of the cell's growth, metabolic activities and chromosome replication occur. Interphase, usually lasting 90% of the time of the cell cycle, includes the G_1 phase, the *S phase*, and the G_2 phase. Most organelles and cell components are produced continuously throughout the G_1, S, and G_2 phase; DNA synthesis occurs only during the S phase.

Mechanics of Cell Division: A Closer Look

Although mitosis is a dynamic continuum of chromosome separation followed by cytoplasm division, it is conventionally described in five stages: *prophase*, *prometaphase, metaphase, anaphase,* and *telophase*. Figure 11.1 illustrates interphase and the stages of mitosis.

Just outside the nucleus in late interphase are two centrosomes, centers for the organization of microtubules. In animal cells a pair of centrioles is associat-ed with each centrosome. The chromosomes, which replicated during the S phase, are in the form of indistinguishable chromatin fibers.

During prophase, the nucleoli disappear and the chromatin fibers coil and fold into visible chromosomes consisting of sister chromatids joined at the centromere. The centrosomes move to opposite poles of the cell and the microtubules of the spindle assemble.

In prometaphase, the nuclear membrane disintegrates. Bundles of microtubules, called spindle fibers, extend from the poles. Kinetochore microtubules attach to the kinetochore, a specialized structure in the centromere region of each chromatid.

In metaphase the centromeres of all the chromosomes become aligned on the metaphase plate, with the kinetochores of sister chromatids facing opposite poles.

Anaphase involves the physical separation of the sister chromatids; these split at the centromeres and the now-individual chromosomes move toward opposite poles. The poles of the cell move further apart.

By telophase, equivalent sets of chromosomes have gathered at the two poles of the cell. Nuclear envelopes form from fragments of the parent nuclear envelope and contributions from the endomembrane system. Nucleoli reappear, the chromatin fibers unravel, and cytokinesis occurs.

Structure and Function of the Mitotic Spindle The *mitotic spindle* consists of fibers made of microtubules and associated proteins. Spindle fibers are bundles of several microtubules. The organizing center of the spindle is the centrosome. Microtubules elongate or shorten by the addition or removal of tubulin proteins at the + end, the end away from the centrosome. A pair of centrioles is centered in the centrosome of animal cells, but is not required for normal spindle operation.

The centrosome duplicates during interphase. The two centrosomes move from near the nucleus toward the poles of the cell during prophase and prometaphase, as spindle microtubules elongate from them. By the end of prometaphase, some of the spindle microtubules attach to the kinetochore of each chromatid. Alternate tugging on the chromosome by opposite microtubules moves the chromosome to the equator.

In metaphase, the chromosomes are aligned at the metaphase plate, with about 15-35 microtubules, called kinetochore microtubules, attached to each kinetochore. Nonkinetochore microtubules extend out from the centrosome; some of these overlap at the midline with microtubules from the opposite pole.

The centromere of each chromosome divides in anaphase, and the sister chromatids (now considered chromosomes) separate and move toward the poles. The mechanism of this movement is still under investigation, but researchers have found the motor protein dynein in kinetochores. The chromosome may be walked along the kinetochore microtubules as they shorten by depolymerizing at their kinetochore end.

The extension of the spindle poles away from each other as the cell elongates is probably due to the sliding of nonkinetochore microtubules past each other where they overlap at the equator. This sliding movement is thought to involve ATP-driven molecules similar to the dynein cross-bridges responsible for movement of flagella.

Mechanisms of Cytokinesis During telophase, nuclei reform and cytokinesis divides the elongated cell. *Cleavage* is the process that separates the two daughter cells in animals. A *cleavage furrow* on the cell surface forms as a *contractile ring* of microfilaments, made of the protein actin, begins to contract on the cytoplasmic side of the membrane. The cleavage furrow deepens until the dividing cell is pinched in two.

In plant cells, a *cell plate* forms at the site of the old metaphase plate from the fusion of membrane vesicles derived from the Golgi apparatus. The two sides of the membrane thus formed join with the plasma membrane, creating two daughter cells each with its own continuous plasma membrane. A new cell wall develops between these membranes.

The process of mitosis used in plants and animals makes it possible to perpetuate the large genomes of eukaryotes. Mitosis is far more complex than binary fission, the mechanism of cell reproduction found in bacteria. Intermediate mechanisms of cell division, present in several groups of unicellular algae, may provide models of how this evolutionary progression occurred.

Control of Cell Division

Normal growth, development, and maintenance depend on proper control of the timing and rate of cell division. Control mechanisms have been studied through the use of *cell culture*. In this technique, cells isolated from plants or animals are grown in the laboratory in glass containers supplied with a nutrient solution called a growth medium.

Researchers using cell culture have identified factors that stimulate or inhibit cell division. Certain nutrients and regulatory substances called *growth factors* are often essential for normal cell division. Mammalian fibroblast cells have receptors on their plasma membranes for "platelet-derived growth factor" (PDGF). This growth factor stimulates cell division when released from blood platelets at the site on an injury.

Density-dependent inhibition of cell division is related to diminishing supplies of essential nutrients or growth factors or overcrowding. Most normal vertebrate cells will stop dividing when they come in contact with other cells or when they lose contact with their substrates, either a culture vessel or the extracellular matrix of a tissue.

An important control period in cell division seems to occur during the G_1 phase of the cell cycle. Internal or external cues determine whether the cell will switch into a nondividing state (called the G_0 *phase*) or continue past a so-called *restriction point* into the S, G_2, and M phases. Cell size, determined by the cytoplasm:genome ratio, is an important cue to passing the restriction point.

A threshold concentration of a complex of proteins know as *MPF*, or maturation promoting factor, is necessary for a cell to move from G_2 to mitosis. A high concentration of MPF allows the cell to progress through mitosis and a decrease in MPF accompanies the transition to G_1 of the next interphase. MPF is a *protein kinase*, an enzyme that activates other proteins, including enzymes, through phosphorylation by ATP. Target proteins may also be protein kinases, leading to a cascade of activation steps. MPF appears to be the master switch that controls correct functioning of the mitotic sequence.

Active MPF consists of two associated proteins: *cdc2* and *cyclin*. The "cell-division cycle" cdc2 protein is the kinase part of the complex, but it is active only when attached to cyclin. The concentration of MPF varies throughout the cell cycle, as the availability of cyclin fluctuates. Cyclin is produced at a constant rate throughout the cycle, but an enzyme activated by MPF destroys cyclin. Thus as the MPF concentration rises during mitosis, cyclin is destroyed and active MPF levels fall at the end of mitosis.

Control of the cell cycle occurs during G_1 when nutrients, growth factors, and cell density help determine whether the cell will pass the restriction point or become a nondividing G_0 cell. The cell proceeds through the restriction point upon reaching a threshold size. Chromosomes duplicate in S phase and the cell continues to grow in G_2. A threshold concentration of MPF moves the cell from interphase to mitosis, and MPF controls the steps of mitosis.

Abnormal Cell Division: Cancer Cells

Cancer cells escape from the body's normal control mechanisms. If they do stop dividing, they seem to do so at random points in the cell cycle. Cancer cells do not exhibit density-dependent inhibition when grown in cell culture. Moreover, most cancer cells in cell culture appear to be "immortal;" they continue to divide indefinitely instead of stopping after the typical 20 to 50 divisions of normal mammalian cells.

When a normal cell is *transformed* or converted to a cancer cell, the body's immune system usually destroys it. If it proliferates to form a *tumor*, a mass of cancer cells develops within a tissue. *Benign tumors* remain at their original site and can be removed by surgery. *Malignant tumors* cause cancer as they spread to other parts of the body. Malignant tumor cells may have abnormal metabolism and unusual numbers of chromosomes. They lose their attachments to other cells and can spread to surrounding tissues or enter the blood and lymph system and *metastasize*. Radiation and chemicals are used to treat tumors that metastasize. Much remains to be learned about the cell division processes of both normal and cancerous cells.

STRUCTURE YOUR KNOWLEDGE

1. Describe the life of one chromosome as it proceeds through an entire cell cycle.

2. Draw a sketch of a mitotic spindle that includes the components of the spindle and lists the functions of each part.

3. Explain the significance of the restriction point and MPF concentrations to the cell cycle.

TEST YOUR KNOWLEDGE

FILL IN THE BLANK: *Identify the appropriate stage of the cell cycle.*

1. _____ most cells that have finished dividing move into this phase

2. _____ sister chromatids separate and chromosomes move apart

3. _____ mitotic spindle begins to form

4. _____ cell plate forms or cleavage furrow pinches cells apart

5. _____ chromosomes duplicate

6. _____ chromosomes line up at equatorial plane

7. _____ phase after DNA replication

8. _____ chromosomes become visible

9. _____ kinetochore-microtubule interactions move chromosomes to midline

10. _____ restriction point occurs in this phase

MULTIPLE CHOICE: *Choose the one best answer.*

1. One of the major differences in the cell division of prokaryotic cells compared to eukaryotic cells is that
   a. cytokinesis does not occur.
   b. genes are not duplicated on chromosomes.
   c. the duplicated chromosomes are attached to the nuclear membrane and are separated from each other as the membrane grows.
   d. the chromosomes do not separate along a mitotic spindle.

2. Centrosomes are
   a. the regions at which two sister chromatids attach.
   b. associated with centrioles in plant cells.
   c. microtubule organizing centers.
   d. centered at the metaphase plate during spindle formation.

3. The longest part of the cell cycle is
   a. interphase.
   b. G_1 phase.
   c. G_2 phase.
   d. mitosis.

4. Cytokinesis may involve
   a. the separation of sister chromatids.
   b. the contraction of the contractile ring of microfilaments.
   c. depolymerization of kinetochore fibers.
   d. a protein kinase that phosphorylates other enzymes.

5. Humans have 46 chromosomes. That number of chromosomes will be found in
   a. cells in anaphase.
   b. the egg and sperm cells.
   c. the somatic cells.
   d. all the cells of the body.

6. Nonkinetochore fibers
   a. are found only in animal cells.
   b. attach along the chromosomes, not to the kinetochore region of the centromere.
   c. seem to be involved in moving the chromosomes toward the poles.
   d. seem to be involved in pushing the poles of the cell apart.

7. Which of the following would not be exhibited by cancer cells?
   a. changing levels of MPF concentration
   b. passage through the restriction point
   c. density-dependent inhibition
   d. metastasis

8. Which of the following is *not* true of a cell plate?
   a. It forms at the site of the metaphase plate.
   b. It results from the fusion of microtubules.
   c. It fuses with the plasma membrane.
   d. A cell wall is laid down between its membranes.

9. A cell that passes the restriction point will most likely
   a. undergo chromosome duplication.
   b. have just completed cytokinesis.
   c. continue to divide only if it is a cancer cell.
   d. show a drop in MPF concentration.

10. Sister chromatids
    a. have one-half the amount of genetic material as does the original chromosome.
    b. start to move along kinetochore fibers toward opposite poles during telophase.
    c. each have their own kinetochore.
    d. are formed during prophase.

11. The rhythmic changes in cyclin concentration in a cell cycle are due to
    a. its increased production once the restriction point is passed.
    b. the cascade of increased production once its enzyme is phosphorylated by cdc2.
    c. its destruction by an enzyme phosphorylated by MPF.
    d. the changing ratio of cytoplasm to genome.

12. The growth hormone PDGF
    a. causes cell division of many cells in cell culture.
    b. is released from platelets at the site of an injury.
    c. binds to a specific plasma membrane receptor on fibroblast cells and stimulates cell division.
    d. does all of the above.

**CHAPTER 7
TOUR OF THE CELL**

Suggested Answers to Structure Your Knowledge

1. A. Functions and structures of animal cells.

CELLULAR FUNCTIONS	ASSOCIATED ORGANELLES AND STRUCTURES
Cell division	Nucleus, chromosomes, centrioles, microtubules (spindle), micro-filaments (actin-myosin aggregates pinch apart cell)
Information storage and transferal	Nucleus, chromosomes, DNA—>RNA—>enzymes and other proteins
Energy conversions	Mitochondria
Manufacture membranes and products	Ribosomes, rough and smooth ER, Golgi apparatus, vesicles
Digestion, recycling	Lysosomes, food vacuoles
Conversion of H_2O_2 to water	Peroxisomes
Structural integrity	Cytoskeleton: microtubules, microfilaments, intermediate filaments; glycocalyx
Movement	Cilia and flagella (microtubules), microfilaments (actin)
Exchange with environment	Plasma membrane
Cell–cell interaction	Desmosomes, tight and gap junctions, glycocalyx

1. B. Structures and functions unique to plant cells

STRUCTURES UNIQUE TO PLANT CELLS	FUNCTIONS
Cell wall	Structural support, middle lamella glues cells together
Central vacuole	Storage, digestion, growth of cell by water absorption
Chloroplast	Photosynthesis, production of carbohydrates
Amyloplast	Starch storage
Plasmodesmata	Cytoplasmic connection between cells

2. Diagram is of a plant cell.
   A. Nucleus
   B. Rough endoplasmic reticulum
   C. Smooth endoplamsic reticulum
   D. Golgi apparatus
   E. Ribosomes
   F. Mitochondrion
   G. Cell wall
   H. Central vacuole
   I. Chloroplast

3.

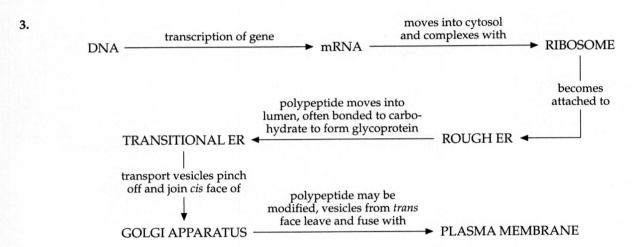

Answers to Test Your Knowledge

Multiple Choice:

1. c	**4.** d	**7.** a	**10.** d	**13.** b	**16.** a
2. b	**5.** a	**8.** c	**11.** c	**14.** c	**17.** a
3. d	**6.** b	**9.** b	**12.** b	**15.** c	**18.** b

Fill in the Blanks:

1. transport vesicles
2. cristae
3. glycocalyx
4. peroxisomes
5. grana
6. basal body
7. cytosol
8. cytoskeleton
9. tight junction
10. tonoplast

Matching:

1. E b	5. A e	9. H k
2. K j	6. C g	10. F h
3. I d	7. B c	
4. D i	8. J a	

CHAPTER 8
MEMBRANE STRUCTURE AND FUNCTION

Suggested Answers to Structure Your Knowledge

1. The fluid mosaic model depicts a membrane composed of a phospholipid bilayer with embedded and surface proteins. Small hydrophobic molecules easily dissolve in the hydrophobic interior of the membrane and pass through by diffusion. The rate of permeability is determined by the size and relative lipid solubility of the molecule. Small polar or ionic compounds cannot pass easily through the hydrophobic center of the membrane. They must cross by facilitated diffusion using specific transport proteins either through channels in the protein or by a conformational change in the protein, caused by the binding of the solute, which shifts the solute from one side of the membrane to the other.

Large
Small polar – H_2O, CO_2
can pass through

2.

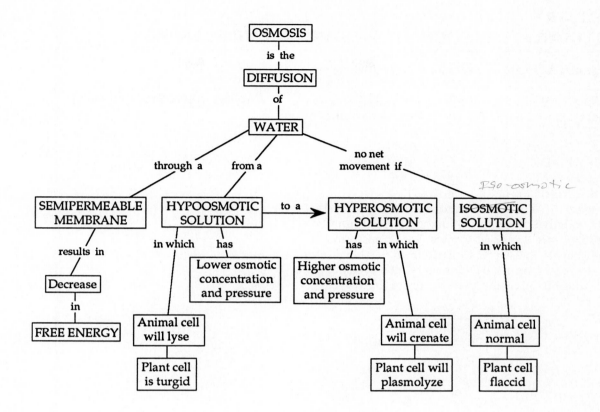

3. a. The diagram on the right represents facilitated diffusion because the solute is moving down its concentration gradient as it crosses the membrane. The cell does not expend energy in this transport. Polar molecules and ions, which could not pass through the hydrophobic center of the membrane, may be moved by facilitated diffusion.

b. The diagram on the left represents active transport because the hydrogen ions are clearly moving against their electrochemical gradient. Energy from the cell must be expended to drive this electrogenic pump.

Answers to Test Your Knowledge

Multiple Choice:

1. c	**5.** d	**9.** a	**13.** b	**17.** c	**21.** c
2. d	**6.** d	**10.** c	**14.** a	**18.** d	**22.** c
3. a	**7.** a	**11.** b	**15.** b	**19.** b	**23.** a
4. c	**8.** c	**12.** d	**16.** d	**20.** b	**24.** c

Explanation for answer to #23. This problem involves both osmosis and diffusion. Although the solution is initially isosmotic, glucose will diffuse down its concentration gradient until it reaches dynamic equilibrium with a 1.5 M concentration on both sides of the tube. As this is occuring, osmotic pressure is increasing on side A and water is crossing the membrane into side A until the solution again becomes isosmotic.

CHAPTER 9
CELLULAR RESPIRATION: HARVESTING CHEMICAL ENERGY

Suggested Answers to Structure Your Knowledge

1. Summary of molecules involved in the oxidation of one molecule of glucose to yield a maximum of 36 ATPs.

a. 2 Pyruvate	**e.** 2 NADH
b. 2 NADH	**f.** 6 NADH
c. 2 net ATP	**g.** 2 FADH$_2$
d. 2 Acetyl CoA	**h.** 2 ATP

Numbers of ATPs across bottom: 2 from glycolysis, 4 from oxidative phosphorylation of 2 NADH from glycolysis, 6 from 2 NADH formed from transformation of 2 pyruvate to 2 acetyl CoA, 18 from 6 NADH from Krebs cycle, 4 from FADH$_2$ from Krebs cycle, and 2 from substrate phosphorylation during Krebs cycle.

Answers to Test Your Knowledge

1. c	**5.** d	**9.** b	**13.** a	**17.** b
2. b	**6.** b	**10.** a	**14.** b	**18.** d
3. b	**7.** d	**11.** c	**15.** c	**19.** c
4. a	**8.** c	**12.** d	**16.** c	**20.** b

2.

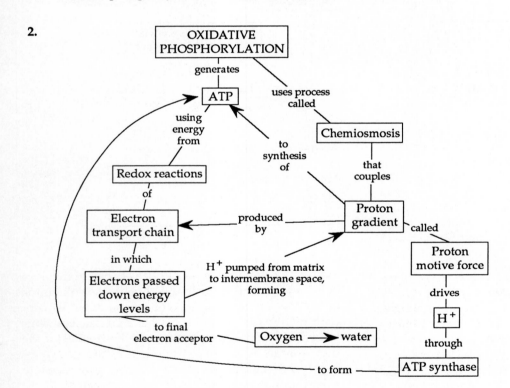

CHAPTER 10
PHOTOSYNTHESIS

Suggested Answers to Structure Your Knowledge

1.

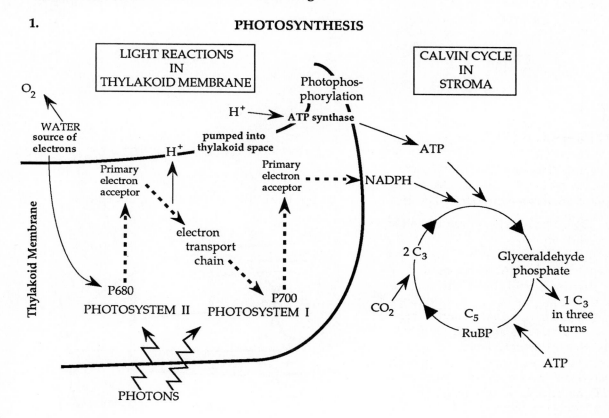

PHOTOSYNTHESIS

2. An absorption spectrum shows the wavelengths of light that are best absorbed by a particular pigment. An action spectrum presents the effectiveness or rate of a reaction at various wavelengths. The solid line shows the absorbance of chlorophyll *a* for the various wavelengths of visible light. The dotted line indicates the rate of photosynthesis at those wavelengths. This graph shows that the absorption spectrum of chlorophyll *a* and the action spectrum for photosynthesis are not identical. Some wavelengths of light, particularly in the yellow and orange range, result in a higher rate of photosynthesis than would be indicated by the absorption of those wavelengths just by chlorophyll *a*. These differences are accounted for by accessory pigments that absorb light energy from different wavelengths and pass that energy on to the reaction center molecules.

3.

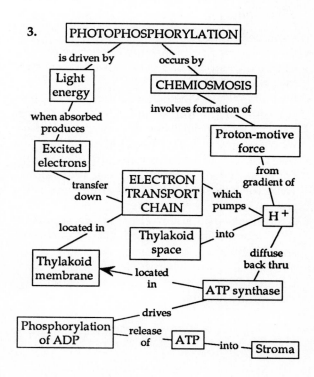

4. a. mitochondrion
   b. chloroplast
   c. intermembrane
      space
   d. inner membrane
   e. matrix

   f. electron transport chain
   g. thylakoid space
   h. thylakoid membrane
   i. ATP synthase
   j. stroma
   k. ATP

Answers to Test Your Knowledge

Multiple Choice:

1. c	**6.** b	**11.** d	**16.** b	**21.** a
2. d	**7.** d	**12.** a	**17.** c	**22.** b
3. a	**8.** c	**13.** b	**18.** c	**23.** d
4. b	**9.** c	**14.** d	**19.** a	**24.** a
5. a	**10.** a	**15.** d	**20.** b	**25.** c

CHAPTER 11
REPRODUCTION OF CELLS

Suggested Answers to Structure Your Knowledge

1. Journey of one chromosome through cell cycle:

Interphase: 90% of cell cycle; growth, metabolism, DNA replication.
- G_1 Phase: the chromosome, consisting of chromatin fiber made of DNA and associated proteins, is diffuse throughout the nucleus. Proteins, stabilizing the coiling and folding of the fiber, may contribute to the control of gene activity. RNA molecules are being transcribed from genes that are switched on.
- S Phase—synthesis of DNA: the chromosome is replicated; two exact copies, called sister chromatids, are produced and held together at a region called the centromere.
- G_2 Phase: growth and metabolic activities of cell continue.

Mitosis: cell division
- Prophase—the chromosome, consisting of two sister chromatids, becomes tightly coiled and folded.
- Prometaphase—kinetochore fibers from opposite ends of the mitotic spindle attach to the kinetochores of the sister chromatids; the chromosome moves toward midline.
- Metaphase—the centromere of the chromosome is aligned at the metaphase plate along with the centromeres of the other chromosomes.
- Anaphase—the sister chromatids separate (now considered to be individual chromosomes) and move to opposite poles.
- Telophase—chromatin fiber of chromosome uncoils and is surrounded by reforming nuclear membrane.

2. Structure and function of a mitotic spindle:

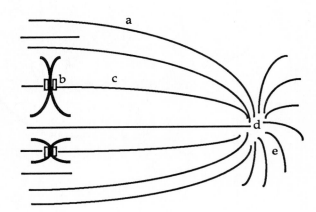

  a. Nonkinetochore Fiber: bundles of microtubules that push poles apart by sliding past fibers from the opposite pole
  b. Kinetochore: region of centromere where kinetochore fiber attaches
  c. Kinetochore Fiber: move chromosomes to metaphase plate and separate chromosomes as the microtubules disassemble
  d. Centrosome: region of organization of microtubules of mitotic spindle (associated with centrioles in animal cells)
  e. Aster: radiating spindle fibers found in animal cells.

3. Significance of the restriction point and MPF concentrations to the cell cycle:

Both the restriction point and MPF are involved in regulating the cell cycle. Whether or not a cell continues to divide depends on the nutrient level, presence of appropriate growth hormone, and lack of interference by other density-dependent inhibition factors. When conditions are favorable, the cell may pass the restriction point during the latter part of the G_1 phase. Once the cell reaches a sufficient size, indicated by its cytoplasm:genome ratio, the chromosomes duplicate and the cell irreversibly moves toward cell division. The cellular concentration of MPF, however, determines when mitosis begins. This protein kinase is responsible for phosphorylating the key enzymes and other protein kinases that orchestrate the complex events of mitosis. The level of MPF rises in a cell as increasing quantities of the protein cyclin complex with the protein cdc2, creating active MPF. One of the effects of MPF, however, is to activate an enzyme that destroys cyclin. Thus the level of active MPF declines in the cell as the process of mitosis comes to an end.

Answers to Test Your Knowledge

Fill in the Blank:

1. G_0
2. anaphase
3. prophase
4. telophase
5. S phase
6. metaphase
7. G_2
8. prophase
9. prometaphase
10. G_1 phase

Multiple Choice:

1. d
2. c
3. a
4. b
5. c
6. d
7. c
8. b
9. a
10. c
11. c
12. d

UNIT III

THE GENE

Embarrassing moments at gene parties

CHAPTER **12**

MEIOSIS AND SEXUAL LIFE CYCLES

FRAMEWORK

This chapter introduces the process of meiosis and the concept of genetic variation as resulting from crossing over, independent assortment, and random fertilization. The sexual life cycles of animals, plants, algae, fungi, and protists all include meiosis and fertilization, but vary in the timing of these events.

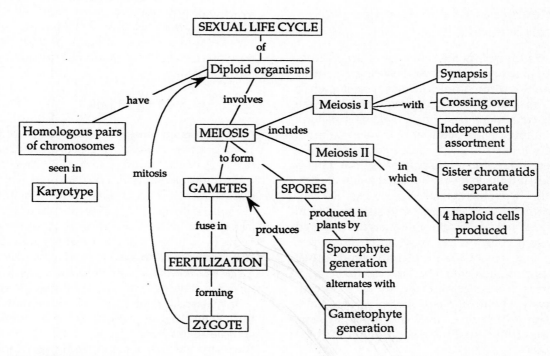

CHAPTER SUMMARY

Organisms reproduce their own kind. *Genetics* is the scientific study of the transmission of traits from parents to offspring, or *heredity*, and the *variation*, or differences, between and within generations.

Genes, DNA, and Chromosomes: A Brief Orientation

The inheritance of traits from parents to offspring involves the transmission of discrete units of information, known as *genes*. Specific sequences of the four

93

nucleotides that comprise DNA, the substance of genes, translate into instructions for synthesizing proteins, such as enzymes that then guide the development of inherited traits.

Precise copies of an organism's genes are packaged into sperm or ova. Upon fertilization, genes from two parents combine. The DNA of a eukaryotic cell is packaged into a species-specific number of chromosomes, each of which contains hundreds or thousands of genes. A gene's *locus* is its location on a chromosome.

Sexual and Asexual Reproduction: A Comparison

In *asexual reproduction*, a single parent passes all its genes on to its offspring, which are its exact genetic copies. A *clone* is a group of genetically identical offspring of an asexually reproducing individual.

In *sexual reproduction*, an individual receives a unique combination of genes inherited from two parents. The genetic variation created by sexual reproduction is a result of chromosome behavior during the sexual life cycle.

Introduction to Sexual Life Cycles: The Human Example

An organism's *life cycle* is the sequence of stages from conception to production of its own offspring. In *somatic cells*, there are two chromosomes of each type, known as *homologous chromosomes* or homologues, which have the same length, centromere position, and staining pattern. A gene controlling a particular trait is found at the same locus on each chromosome of a homologous pair.

A *karyotype* is an ordered display of an individual's chromosomes. It is made by cutting the individual chromosomes out of a photograph taken of white blood cells that were stimulated to undergo mitosis, arrested in metaphase, and stained. The chromosomes are arranged into homologous pairs by size, shape, and banding pattern. Karyotyping may be used to identify chromosomal abnormalities associated with inherited disorders.

Sex chromosomes determine the sex of a person: females have two homologous *X* chromosomes; males have nonhomologous *X* and *Y* chromosomes. Chromosomes other than the sex chromosomes are called *autosomes*.

Somatic cells contain a set of chromosomes from each parent; humans have a total of 46 chromosomes. *Gametes*, ova and sperm, contain a single set chromosomes. The *haploid* number (n) of chromosomes for humans is 23. Fusion of sperm and ovum into a *zygote*, called *fertilization*, or *syngamy*, combines the paternal and maternal set of genes. The *diploid* zygote then divides by mitosis to produce the somatic cells of the body, all of which contain the diploid number ($2n$) of 46 chromosomes.

Meiosis is a special type of cell division that halves the chromosome number in gametes to compensate for the doubling that occurs at fertilization. An alternation between diploid and haploid conditions, involving the processes of meiosis and fertilization, is characteristic of the life cycle of all sexually reproducing organisms.

The Variety of Sexual Life Cycles

In most animals, meiosis occurs in the formation of gametes, which are the only haploid cells in the life cycle. In many fungi and some protists, the only diploid stage is the zygote. Meiosis occurs right after the gametes fuse, and mitosis then creates a multicellular haploid organism. Gametes are produced by mitosis in these organisms.

Plants and some species of algae (protists) have a type of life cycle called *alternation of generations* that includes both diploid and haploid multicellular stages. The multicellular diploid *sporophyte* stage produces haploid *spores* by meiosis. These spores undergo mitosis and develop into a multicellular haploid plant, the *gametophyte*, which produces gametes by mitosis. Gametes fuse to form a diploid zygote that develops into the next sporophyte generation.

Meiosis: A Closer Look

As in mitosis, chromosome replication precedes meiosis. There are two consecutive cell divisions: *meiosis I* and *meiosis II*, which result in four haploid daughter cells.

In interphase I, each chromosome replicates, producing two genetically identical chromatids attached at their centromeres. During prophase I, which lasts more than 90% of the time required for meiosis, homologous chromosomes *synapse*, producing a *tetrad*, and segments of nonsister chromatids may criss-cross, forming *chiasmata* that help hold the homologues together as they move toward the metaphase plate.

In metaphase I, the synapsed chromosomes line up on the metaphase plate, with their centromeres attached to spindle fibers from opposite poles. The homologous pairs separate in anaphase I, with one homologue moving toward each pole. In telophase I, a haploid set of chromosomes, each composed of two chromatids, reaches each pole. Cytokinesis usually occurs during telophase I and may be followed by a brief period of interkinesis, when nuclei reform, or the two haploid cells may proceed directly into meiosis II. There is no replication of genetic material prior to this second division.

Meiosis II looks like a regular mitotic division, in which chromosomes line up individually on the metaphase plate, and sister chromatids separate and move apart in anaphase II. At the end of telophase II, there are four haploid daughter cells.

Comparison of Mitosis and Meiosis

Mitosis produces daughter cells that are genetically identical to the parent cell. Meiosis produces haploid cells that differ genetically from their parent cell and from each other. The three unique events that produce this result occur during meiosis I:

In prophase I, when homologous chromosomes synapse, genetic material is rearranged by crossing over, which is visible during this stage by the appearance of X-shaped regions called chiasmata. Neither synapsis nor crossing over occurs in mitosis. In metaphase I, chromosomes line up in pairs, not as individuals, on the metaphase plate. During anaphase I, the homologous pairs separate and one homologue goes to each pole. Centromeres do not divide and sister chromatids remain together in Meiosis I.

Meiosis II is identical to mitosis in that sister chromatids separate. It is not preceded by a replication and results in the formation of four haploid cells, each with half the number of chromosomes as the parent cell.

Sexual Sources of Genetic Variation

Chromosome activity during meiosis and fertilization is responsible for most of the variation found from one generation to the next.

Independent Assortment of Chromosomes The first meiotic division results in an assortment of maternal and paternal chromosomes in the two daughter cells. Each homologous pair lines up independently at the metaphase plate; the orientation of the maternal and paternal chromosomes is random. The number of possible combinations of maternal and paternal chromosomes is 2^n, where n is the haploid number. For humans, the number of possible chromosome combinations is 2^{23}—8 million distinct gametes that an individual can produce.

Crossing Over In prophase I, segments of nonsister chromatids synapse and *cross over*, resulting in new genetic combinations of maternal and paternal genes on the same chromosome. The exact mechanism of synapsis is not known, but it involves a *synaptonemal complex*, a protein structure that helps to bring homologous chromosomes into precise alignment.

Random Fertilization The random nature of fertilization adds to the genetic variability established in meiosis. A given pair of parents can produce a zygote with any of 64 trillion (8 million x 8 million) diploid combinations.

Genetic Variation and Evolution

In Darwin's theory of natural selection, heritable variations present in a population result in adaptation as the individuals best-suited to an environment produce the most offspring. If the environment changes, different heritable variations may prevail. The process of sexual reproduction, which includes independent assortment of chromosomes, crossing over and random fertilization, is a source of this variation. Mutations, which are rare changes in DNA, are the ultimate source of genetic diversity in populations.

STRUCTURE YOUR KNOWLEDGE

1. Create a diagram of an animal sexual life cycle, including the major events and processes.

2. Describe the key events of the following stages of meiosis:

a. Interphase I
b. Prophase I
c. Metaphase I
d. Anaphase I
e. Metaphase II
f. Anaphase II

3. Label the following stages of meiosis and place them in the proper order.

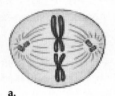 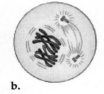  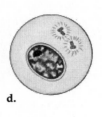  

a.   b.   c.   d.   e.   f.

a.   b.   c.
d.   e.   f.

4. Create a concept map that compares and contrasts mitosis and meiosis.

TEST YOUR KNOWLEDGE

MULTIPLE CHOICE: *Choose the one best answer.*

1. The restoration of the diploid chromosome number after halving in meiosis is due to
   a. synapsis.
   b. fertilization.
   c. mitosis.
   d. DNA replication.

2. A karyotype is a
   a. genotype of an individual.
   b. unique combination of chromosomes found in a gamete.
   c. blood type determination of an individual.
   d. picture display of an individual's chromosomes.

3. Autosomes are
   a. sex chromosomes.
   b. chromosomes that occur singly.
   c. chromosomal abnormalities that result in genetic defects.
   d. none of the above.

4. A synaptonemal complex would be found during
   a. prophase I of meiosis.
   b. fertilization or fusion of gametes.
   c. metaphase II of meiosis.
   d. prophase of mitosis.

5. During the first meiotic division (meiosis I)
   a. homologous chromosomes separate.
   b. the chromosome number becomes haploid.
   c. crossing over between nonsister chromatids occurs.
   d. all of the above occur.

6. A cell with a diploid number of 6 could produce gametes with how many different combinations of chromosomes?
   a. 6
   b. 8
   c. 12
   d. 64

7. The DNA content of a cell is measured in the G_2 phase. After meiosis I, the DNA content of one of the two cells produced would be
   a. equal to that of the G_2 cell.
   b. twice that of the G_2 cell.
   c. one-half that of the G_2 cell.
   d. one-fourth that of the G_2 cell.

8. In many fungi and some protists,
   a. the zygote is the only haploid stage.
   b. gametes are formed by meiosis.
   c. the adult organism is haploid.
   d. the gametophyte generation produces gametes by mitosis.

9. In the alternation of generations found in plants,
   a. the sporophyte generation produces spores by meiosis.
   b. the gametophyte generation produces gametes by mitosis.
   c. the zygote will develop into a sporophyte generation by mitosis.
   d. all of the above are correct.

10. Which of the following is *not* a source of genetic variation in sexually reproducing organisms?
    a. crossing over
    b. replication of DNA during S phase.
    c. independent assortment of chromosomes
    d. random fertilization of gametes

11. Meiosis II is similar to mitosis because
    a. sister chromatids separate.
    b. homologous chromosomes separate.
    c. DNA replication precedes the division.
    d. they both take the same amount of time.

12. Homologous chromosomes
    a. have identical genes.
    b. have genes for the same traits at the same loci.
    c. are found in gametes.
    d. are all of the above.

13. Asexual reproduction of a diploid organism would
    a. be impossible
    b. involve meiosis.
    c. produce identical offspring.
    d. show variation among sibling offspring.

14. In a sexually reproducing species with a diploid number of 8, how many different combinations of paternal and maternal chromosomes would be possible in the offspring?
    a. 8
    b. 16
    c. 64
    d. 256

15. The calculation of offspring in question #14 includes variation resulting from
    a. crossing over.
    b. random fertilization.
    c. independent assortment of chromosomes.
    d. both b and c.

16. Sexual reproduction involves the alternation of haploid and diploid stages of life, but only plants (and some algal protists) are said to have a life cycle with an alternation of generations. What sets them apart?
    a. Plants have an alternation of multicellular diploid life forms.
    b. A multicellular haploid stage alternates with a multicellular diploid stage.
    c. A diploid sexual flower develops on the haploid asexual plant.
    d. Each generation, haploid and diploid, can complete a life cycle without the other.

CHAPTER 13

MENDEL AND THE GENE IDEA

FRAMEWORK

Through his work with garden peas, Mendel developed the fundamental principles of inheritance and the laws of segregation and independent assortment. This chapter describes the basic monohybrid and dihybrid crosses that Mendel performed to establish that inheritance involves particulate genetic factors (genes) that segregate independently in the formation of gametes and recombine to form offspring. The laws of probability can be applied to predict the outcome of genetic crosses.

The phenotypic expression of genotype may be affected by such factors as incomplete dominance, multiple alleles, pleiotropy, epistasis, and polygenic inheritance, as well as the environment. Genetic screening and counseling for recessively inherited genetic disorders use Mendelian principles to analyze human pedigrees.

CHAPTER SUMMARY

What is the genetic basis for the variation evident in the individuals of a population and how is this variation transmitted from parents to offspring? The genetic material of two parents is not blended nor irreversibly mixed in offspring, but is passed on to future generations as discrete heritable units or genes. In the 1860s, Gregor Mendel developed the fundamental principles that established this "particulate model" of inheritance.

Mendel's Model: A Case Study in the Scientific Process

Mendel's Experimental Approach   Mendel worked with garden peas, a good choice of study organism

because they are easy to grow, their fertilization is easily controlled, and their offspring can be quantified.

Mendel studied seven *characters*, or heritable features, that occurred in two alternative forms. He used *true-breeding* varieties of pea plants, which means that self-fertilizing parents always produce offspring with the parental form of the character. To follow the transmission of these well-defined traits, Mendel performed *hybridizations* in which he cross-pollinated plants that were true breeding for alternate forms of the same characteristic (a *monohybrid* cross), and then allowed the next generation to self-pollinate. The true-breeding parental plants are the *P generation* (parental); the results of the first cross are the F_1 *generation* (first filial); and the next generation, from the self-cross of the F_1, is known as the F_2 *generation*.

Mendel's Law of Segregation   Mendel found that the F_1 offspring did not show a blending of the parental characteristics. Instead, only one of the parental forms of the trait was found in the hybrid offspring. In the F_2 generation, however, the missing parental form reappeared in the ratio of 3:1—three offspring with the parental trait shown by the F_1 to one offspring with the reappearing trait.

Mendel's explanation for this phenomenon contains four parts: (1) there are alternate forms, now called *alleles*, for the heritable factors called genes; (2) an organism has two alleles for each inherited trait, one received from each parent; (3) when two different alleles occur together, one of them, called the *dominant allele*, may be completely expressed while the other, the *recessive allele*, has no observable effect on the organism's appearance; and (4) gene pairs separate (segregate) during the formation of gametes, so an egg or sperm only carries one allele for each inherited trait. This explanation is consistent with the behavior of chromosomes in diploid organisms, in which homologous pairs of chromosomes separate in gamete formation, and the paired condition is reestablished by the random fusion of gametes at fertilization.

Mendel's segregation hypothesis explains the 3:1 ratio observed in the F_2 plants. During the segregation of allele pairs in the formation of F_1 gametes, half the gametes receive one allele while the other half receive the alternate allele. Random fertilization of gametes results in one-fourth of the plants having two dominant alleles, half having one dominant and one recessive allele, and one-fourth receiving two recessive alleles. Since the dominant allele is the one expressed in the half of the offspring that are hybrid, the ratio of plants showing the dominant to recessive trait is 3:1. A Punnett square can be used to predict the results of simple genetic crosses.

Dominant alleles are often symbolized by a capital letter, recessive alleles by a small letter. An organism that has a pair of identical alleles for a character is said to be *homozygous*. If the organism has two different alleles, it is said to be *heterozygous* for that character. Homozygotes are true breeding; heterozygotes are not, since they produce gametes with one or the other allele that can combine to produce homozygous dominant, heterozygous, and homozygous recessive offspring, in the genotypic ratio 1:2:1. *Phenotype* is an organism's expressed traits; *genotype* is its genetic makeup.

A *testcross* breeds an organism of unknown genotype with a recessive homozygote to determine the genotype of an organism expressing the dominant phenotype. Since the recessive homozygote can contribute only recessive alleles, if there is a 1:1 phenotypic ratio of offspring, the unknown parent must have been heterozygote.

Inheritance as a Game of Chance General rules of probability apply to the segregation of alleles and the reconstitution of pairs at fertilization. The probability of an event occurring is the ratio of the times that event could occur over all the possible events; for example, the probability of drawing an ace from a deck of cards is 4/52.

The outcome of independent events is not affected by previous events. However, the *rule of multiplication* states that the probability that a certain combination of independent events will occur simultaneously is equal to the product of the separate probabilities of the independent events. The probability of a particular genotype being formed by fertilization is equal to the product of the probabilities of forming each type of gamete needed to produce that genotype. The probability of getting a homozygous recessive offspring from a cross of heterozygotes is equal to the probability of getting one gamete with the recessive allele (1/2) multiplied by the probability of the other gamete having the recessive allele ($1/2 \times 1/2 = 1/4$).

If the genotype can be formed in more than one way, then the *rule of addition* states that its probability is equal to the sum of the separate probabilities of the different ways the event can occur. For example, a heterozygote offspring can occur if the egg contains the dominant allele and the sperm the recessive (1/4 probability) or vice versa (1/4). Therefore, the heterozygote offspring would be the predicted result from a hybrid cross half of the time.

These rules of probability account for the 1:2:1 ratio of genotypes seen in the offspring of a hybrid cross. Large samples usually will conform more closely to predicted distributions.

Mendel's Law of Independent Assortment Mendel deduced the law of segregation from working with monohybrid crosses, crosses involving parents differing in only a single trait. He used *dihybrid crosses*, involving parents differing in two traits, to determine whether the two traits were transmitted independently of each other or together, in the same combination in which they were inherited from the parent plants.

If the two pairs of genes segregate independently, then gametes from an F_1 hybrid generation (AaBb) should contain four combinations of genes in equal quantities (AB, Ab, aB, ab). The random fertilization of these four classes of gametes should result in 16 (4 x 4) gamete combinations that produce four phenotypic categories in a ratio of 9:3:3:1 (nine offspring showing both dominant traits, three showing one dominant trait and one recessive, three showing the opposite dominant and recessive traits, and one showing both recessive traits). Mendel obtained these ratios when he categorized the F_2 progeny of dihybrid crosses, providing evidence for his *law of independent assortment*. This principle states that genes assort independently from each other when alleles segregate in the formation of gametes.

Fairly complex genetics problems can be solved by applying the rules of probability to segregation and independent assortment. The probability of a particular genotype arising from a cross can be determined by considering each gene involved as a separate monohybrid cross and then multiplying the probabilities of all the independent events involved in the final genotype.

Extending Mendelian Genetics

Incomplete Dominance Some characteristics show *incomplete dominance*, in that the F_1 hybrids have a phenotype intermediate between that of the parents.

Neither allele masks the expression of the other, and the phenotype of heterozygotes is distinguishable from the two homozygous conditions. Heterozygote carnations are pink because they have only half as much red pigment as do red homozygotes.

So What Is a Dominant Allele?

In a case of an allele showing complete dominance, the phenotype of the heterozygote is indistinguishable from that of the homozygous dominant. At the other end of the continuum, alleles that exhibit *codominance* are both separately expressed in the phenotype, as in the case of the *MN* blood cell type in which both types of surface proteins are present. Intermediate phenotypes are characteristic of alleles showing incomplete dominance. In the case of pink carnations, however, what appears macroscopically to be incomplete dominance is actually codominance at the microscopic level—both red and colorless plastids are produced in the pink flowers.

Although it is sometimes stated that a dominant allele "masks" the expression of a recessive allele, it is important to remember that the two alleles in a cell do not interact at all. The mechanism of action of an allele depends on the sequence of its DNA and the resulting protein (usually an enzyme) for which it codes.

The third major point about dominance relates to the relative abundance of alleles. Whether or not an allele is dominant or recessive has no relation to how common it is in a population.

Multiple Alleles Most genes exist in more than two allelic forms. The gene that determines human blood groups has three alleles. The alleles I^A and I^B are codominant with each other; each codes for a carbohydrate found on the surface of red blood cells. The I^AI^B genotype results in type AB blood. The allele *i* is recessive to I^A and I^B. Individuals with the *ii* genotype have type O blood. The I^AI^A and I^Ai genotypes result in type A blood, I^BI^B and I^Bi in type B blood. Blood type is critical in transfusions because, if the carbohydrate coating on the donor's blood cells is foreign to the recipient, the recipient's antibodies will cause agglutination, or clumping, of blood cells within the blood vessels.

Pleiotropy Pleiotropy is the characteristic of a single gene having multiple phenotypic effects in an individual. Many hereditary diseases, with complex sets of symptoms, are caused by a single gene.

Epistasis In *epistasis*, a gene at one locus may affect the expression of another gene. F_2 ratios that differ from the typical 9:3:3:1 often indicate epistasis. In the case of mouse coat color, presence of the dominant allele B results in a black mouse as long as another allele, C, is present to allow pigment to be deposited in the hair. In the F_2 of a heterozygous cross (BbCc), offspring occur in the ratio of 9 black:3 brown:4 white due to the epistatic effect of a cc genotype.

Polygenic Inheritance Quantitative characters, such as height or skin color, vary along a continuum in a population. Such variation is usually due to *polygenic inheritance*, where two or more genes have an additive effect on a single phenotypic trait. Each dominant allele contributes one "unit" to the expressed trait. A polygenic trait may result in a normal distribution (forming a bell-shaped curve) of the trait within a population.

Nature Versus Nurture: Environmental Impact on Phenotype The phenotype of an individual is the result of complex interactions between its genotype and the environment. Genotypes have a phenotypic range called a *norm of reaction* within which the environment influences phenotypic expression. Polygenic characters are often *multifactorial*, meaning that a combination of genetic and environmental factors influence phenotype.

Integrating a Mendelian View of Heredity and Variation The phenotypic expression of most genes is influenced by other genes and by the environment. An organism's overall phenotype is a product of its overall genotype and its environmental history. But the theory of particulate inheritance and Mendel's principles of segregation and independent assortment form the basis for modern genetics.

Mendelian Inheritance in Humans

Human Pedigrees A family *pedigree* is a family tree with the history of a particular trait shown across the generations. By convention, circles represent females, squares are used for males, and solid symbols indicate individuals that express the phenotype in question. Parents are joined by a horizontal line and offspring are listed below parents from left to right in order of birth. The genotype of individuals in the pedigree can often be deduced by following the patterns of inheritance.

For genetic disorders inherited as simple Mendelian traits, geneticists, physicians, and genetic counselors use pedigree analysis to make statistical predictions concerning the probability that a trait may be inherited.

Recessively Inherited Disorders Only homozygous-recessive individuals express the phenotype for the 1,000 or so genetic disorders that are inherited as simple recessive traits. *Carriers* of the disorder are heterozygotes who are phenotypically normal but may transmit the recessive allele to their offspring. Each mating between heterozygotes has a 1/4 chance of producing an offspring with the homozygous-recessive disorder. A phenotypically normal child produced by this mating has a 2/3 chance of being a heterozygote and, thus, a carrier.

Genetic disorders are usually unevenly distributed among racial or cultural groups due to the different genetic history of each group. Cystic fibrosis is the most common lethal genetic disease in the United States; it is found more frequently in Caucasians than in other races. This recessive allele results in defective chloride pumps in cell membranes. Accumulating chloride within cells alters osmotic balance, leading to the build-up of thickened mucus in various organs and a predisposition to pneumonia and other infections.

Tay-Sachs disease is a lethal disorder in which the brain cells are unable to metabolize a type of lipid that then accumulates and damages the brain, resulting in an early death. There is a disproportionately high incidence of this disease among Ashkenazi Jews, Jewish people whose ancestors came from Central Europe.

Sickle cell anemia is the most common inherited disease among African-Americans. Due to a single amino acid substitution in the hemoglobin protein, the molecules tend to crystallize, causing red blood cells to deform into a sickle shape and triggering blood clotting and other symptoms. Heterozygous individuals are said to have sickle cell trait, but are usually healthy. The resistance to malaria that accompanies the sickle cell trait may explain why this lethal recessive allele remains in relatively high frequency in areas where malaria is common.

The likelihood of two mating individuals carrying the same rare deleterious allele increases when the individuals have common ancestors. *Consanguineous* matings, between siblings or close relatives, are more likely to produce offspring homozygous for various traits, including harmful or lethal traits. Matings between "blood" relatives are indicated on pedigrees by double lines.

There is some debate over the extent to which human consanguinity increases the risk of inherited diseases. Among zoo animals and domesticated animals, inbreeding has resulted in a higher incidence of harmful recessive traits, but has enabled some endangered species to avoid extinction.

Dominantly Inherited Disorders A few human disorders are due to dominant genes. In achondroplasia,

dwarfism is due to a single copy of a mutant allele. Homozygosity for this dominant allele is lethal.

Dominant lethal alleles are more rare than recessive lethals because the harmful allele cannot be masked in the heterozygote. Most dominant lethal alleles are the result of mutations and kill the developing organism before it can reproduce and pass on the new form of the gene. A late-acting lethal dominant allele, however, can be passed on if the symptoms do not develop until after an organism is old enough to have reproduced. Huntington's disease is a degenerative disease of the nervous system that does not develop until later in life. Children of a person who develops Huntington's have a 50% chance of having inherited the dominant lethal allele. Medical researchers have recently developed a method to detect the lethal gene without waiting for the symptoms to begin.

Genetic Screening and Counseling The risk of a genetic disorder being transmitted to offspring can sometimes be determined through genetic counseling and testing before a child is conceived or in the early stages of pregnancy.

The probability of a child having a genetic defect can be determined by considering the family history of the disease. If two prospective parents both have siblings who had the disorder, both sets of prospective grandparents must have been carriers. The parents each have a 2/3 chance that they are heterozygote carriers. The probability that both parents are carriers is 2/3 x 2/3; the chance that two heterozygotes will have a recessive homozygous child is 1/4. The overall chance that a child will inherit the disease (4/9 x 1/4—following the rule of multiplication) is 1/9. Should this couple have a baby that has the disease, this establishes that they are both carriers, and the chance that a subsequent child would have the disease is 1/4.

Determining whether parents are carriers is important for determining the risk of a genetic disorder. Biochemical tests that permit carrier recognition have been developed for several heritable disorders.

Amniocentesis is a procedure done between the fourteenth and sixteenth week of pregnancy by extracting a small amount of amniotic fluid from the sac surrounding the fetus. The fluid is analyzed biochemically, and fetal cells present in the fluid are cultured for several weeks, then karyotyped to check for certain chromosomal defects.

Chorionic villi sampling is a technique in which a small amount of the fetal tissue is suctioned from the placenta. These rapidly growing cells can be karyotyped immediately. This procedure can be performed at only 8 to 10 weeks of pregnancy.

Ultrasound is a simple, noninvasive procedure that

can reveal major abnormalities. It uses sound waves to produce an image of the fetus. Fetoscopy, the insertion of a needle-thin viewing scope and light into the uterus, allows the fetus to be checked for anatomical problems.

Some genetic disorders can be detected at birth. The recessively inherited disorder phenylketonuria (PKU) is caused by the lack of an enzyme needed to break down the amino acid phenylalanine. An accumulation of phenylalanine and its by-product causes severe mental retardation. Routine screening is now used to detect the disease in newborns; if detected, the diet of the child is adjusted and the child develops normally.

Multifactorial Disorders Many diseases have genetic and environmental components. These multifactorial disorders include heart disease, diabetes, cancer, and others.

STRUCTURE YOUR KNOWLEDGE

1. Relate Mendel's two laws of inheritance to the behavior of chromosomes in meiosis that you studied in Chapter 12.

2. Tall red-flowered plants are crossed with short white-flowered plants. The resulting F_1 generation consists of all tall pink-flowered plants. Assuming that height and flower color are each determined by a single gene locus, predict the results of an F_1 cross of *TtRr* plants. Fill in the F_1 gametes and F_2 genotypes in the following Punnett square and the phenotypes and predicted ratios of the F_2 generation.

3. Mendel worked with characters that exhibited two alternate forms: smooth or wrinkled, green or yellow, tall or short. In his F_1 generation, the dominant allele was always expressed, with the recessive trait reappearing in the F_2 offspring. But not all allele pairs operate by complete dominance; some show incomplete dominance or codominance. Taking into account the mechanisms by which genotype becomes expressed as phenotype, explain how this range in dominance can occur.

TEST YOUR KNOWLEDGE

MATCHING: *Match the definition with the correct term.*

1. _____ codominance
2. _____ homozygous
3. _____ heterozygous
4. _____ phenotype
5. _____ polygenic trait
6. _____ pleiotropy

A. true breeding variety
B. cross between two hybrids
C. cross that involves two different gene pairs
D. an allele that is not expressed in the phenotype
E. the physical characteristics of an individual
F. genotype with two different alleles for same locus

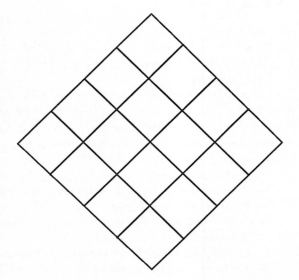

PHENOTYPE	PREDICTED RATIO
_____	_____
_____	_____
_____	_____
_____	_____
_____	_____
_____	_____

7. _____ epistasis

8. _____ testcross

9. _____ dihybrid cross

10. _____ incomplete dominance

G. genotype with multiple alleles for same locus

H. one gene influences the expression of another gene

I. both alleles are expressed in heterozygote

J. single gene with multiple phenotypic effects

K. heterozygote is intermediate between homozygous phenotypes

L. two or more genes with additive effect on phenotype

M. cross with homozygous recessive to determine genotype of unknown

MULTIPLE CHOICE: *Choose the one best answer.*

1. The genotypic ratio resulting from a cross
   a. will be equal to the phenotypic ratio.
   b. can be predicted with a Punnett square.
   c. will not be typical if the gene involved has multiple alleles.
   d. will indicate which allele is dominant and which is recessive.

2. The phenotypic ratio of a dihybrid cross
   a. will be equal to the genotypic ratio.
   b. will vary from what is expected if the gene involved is pleiotropic.
   c. will vary from what is expected if the gene involved has multiple alleles.
   d. will include all the different phenotypic combinations of gametes.

3. Recessive alleles
   a. are not active in a cell.
   b. are expressed in the phenotype when present as both alleles of a gene pair.
   c. are always the least frequently found alleles in a population.
   d. often code for mutations.

4. A multifactorial disease
   a. can usually be traced to consanguineous matings.
   b. is caused by recessively inherited lethal genes.
   c. has both genetic and environmental causes.
   d. has a collection of symptoms traceable to a pleiotropic gene.

5. The F_2 generation
   a. has a phenotypic ratio of 3:1.
   b. is the result of the self-fertilization or crossing of F_1 individuals.
   c. can be used to determine the genotype of individuals with the dominant phenotype.
   d. will show up in a dihybrid cross.

6. According to Mendel's law of segregation,
   a. there is a 50% probability that a gamete will get a dominant allele.
   b. gene pairs segregate independently of other genes in gamete formation.
   c. allele pairs separate in gamete formation.
   d. the laws of probability determine gamete formation.

7. A 1:1 phenotypic ratio in a testcross indicates that
   a. the alleles are dominant.
   b. one parent must have been homozygous recessive.
   c. the dominant phenotype parent was a heterozygote.
   d. the alleles segregated independently.

8. According to Mendel's law of independent assortment,
   a. an individual heterozygous for three genes should produce six different gametic combinations of alleles.
   b. allele pairs segregate in gamete formation.
   c. the F_2 phenotypic ratio of a dihybrid cross of two true breeding varieties should be 9:3:3:1, not 3:1.
   d. all of the above are correct.

9. Quantitative traits
   a. are found in large numbers of offspring.
   b. may exist in a normal distribution within a population.
   c. may be the result of pleiotropic genes.
   d. may be the result of varying degrees of dominance of a gene.

10. The probability that a particular genotype may result from a cross
    a. can be determined as the product of the probabilities of the formation of the gametes needed to produce the genotype.
    b. can be determined from the genotypic ratio for the cross.
    c. will depend upon the genotypes of the parents.
    d. is related to all of the above.

11. Carriers of a disease
    a. are indicated by solid symbols on a family pedigree.
    b. are heterozygotes for the gene that can cause the disease.
    c. will produce children with the disease.
    d. usually are involved in consanguineous matings.

12. A lethal recessive allele is more likely to be maintained in a population
    a. if it is somehow beneficial in the heterozygous condition.
    b. if it is not expressed until late in life.
    c. if it can be identified by genetic screening.
    d. both a and b are correct.

13. Dominant lethal alleles
    a. are expressed late in life.
    b. are lethal in both the homozygous and heterozygous condition.
    c. are more likely to occur in consanguineous matings.
    d. are responsible for cystic fibrosis, Tay-Sachs disease and sickle cell anemia.

14. With amniocentesis,
    a. a couple is tested to see if they are carriers of a genetic disorder.
    b. a fetus is observed for anatomical problems.
    c. fetal cells of the placenta can be karyotyped.
    d. fetal cells from the protective fluid can be cultured and then karyotyped.

15. If both parents are carriers of a lethal recessive gene, the probability that their child will inherit and express the disorder is
    a. 2/3 x 2/3 x 1/4, or 1/9.
    b. 1/2.
    c. 1/4.
    d. 1/2 x 1/2 x 1/4, or 1/16.

16. Feather color in budgies (Fig. 13.9) is determined by two different genes that affect the pigmentation of the outer feather and its core. $Y\_B\_$ is green; $yyB\_$ is blue; Y_bb is yellow; and $yybb$ is white. A green budgie is crossed to another green budgie ($Y\_B\_$ x $Y\_B\_$). You do not know the complete genotype of either bird. Which of the following results is *not* possible?
    a. all green offspring
    b. all blue offspring
    c. all white offspring
    d. all of the above are possible, but with different probabilities

17. After obtaining two heads from two tosses of a coin, the probability of tossing the coin and obtaining a head is
    a. 1/2.
    b. 1/4.
    c. 1/8.
    d. 1/16.

18. The probability of tossing three coins simultaneously and obtaining three heads is
    a. 1/2.
    b. 1/4.
    c. 1/8.
    d. 1/16.

GENETICS PROBLEMS

1. Blood typing has often been used as evidence in paternity cases, when the blood type of the mother and child may indicate that a man alleged to be the father could not possibly have fathered the child. For the following mother and child combinations, indicate which blood groups of potential fathers would be exonerated.

Blood group of mother	Blood group of child	Man is exonerated if he belongs to blood group(s)
AB	A	_____
O	B	_____
A	AB	_____
O	O	_____
B	A	_____

2. Polydactyly (extra fingers and toes) is due to a dominant gene. A father is polydactyl, the mother has the normal phenotype, and they have had one normal child. What is the genotype of the father? Of the mother? What is the probability that a second child will have the normal number of digits?

3. For the following genotypes, indicate what proportion of the gametes have the indicated genes.

Parental genotype	Type of gamete	Proportion expected
AABb	AB	_____
AaBb	ab	_____
AABbcc	ABc	_____
AaBbCc	ABc	_____

4. For the following crosses, indicate the probability of obtaining the indicated genotype in an off-

spring. Remember it is easiest to treat each gene separately as a monohybrid cross and then combine the probabilities.

Cross	Offspring	Probability
AAbb x *AaBb*	*AAbb*	_____
AaBB x *AaBb*	*aaBB*	_____
AABBcc x *aabbCC*	*AaBbCc*	_____
AaBbCc x *AaBbcc*	*aabbcc*	_____

5. In dogs, black (*B*) is dominant to chestnut (*b*), and solid color (*S*) is dominant to spotted (*s*). What are the genotypes of the parents that would produce a cross with 3/8 black solid, 3/8 black spotted, 1/8 chestnut solid, and 1/8 chestnut spotted puppies? (Hint: first determine what genotypes the offspring must have before you deal with the fractions.)

6. The height of spike weed is a result of polygenic inheritance involving three genes, each of which can contribute 5 cm to the plant. The base height of the weed is 10 cm, and the tallest plant can reach 40 cm. If a tall plant (*AABBCC*) is crossed with a base-height plant (*aabbcc*), what is the height of the F_1 plants? How many phenotypic classes will there be in the F_2?

7. In guinea pigs, the gene for production of melanin is epistatic to the gene for the deposition of melanin. The dominant allele *M* causes melanin to be produced; mm individuals cannot produce the pigment. The dominant allele *B* causes the deposition of a lot of pigment and produces a black guinea pig, whereas only a small amount of pigment is laid down in bb animals, producing a light-brown color. Without an *M* allele, no pigment is produced so the allele *B* has no affect and the guinea pig is white. A homozygous black guinea pig is crossed with a homozygous recessive white: *MMBB* x *mmbb*. Give the phenotypes of the F_1 and F_2 generations.

8. When hairless hamsters are mated with normal-haired hamsters, about one-half the offspring are hairless and one-half are normal. When hairless hamsters are crossed with each other, the ratio of normal-haired to hairless is 1:2. How do you account for the results of the first cross? How would you explain the unusual ratio obtained in the second cross?

9. If two medium-tailed pigs were mated and the liter produced included three stub-tailed piglets, six medium-tailed, and four long-tailed piglets, what would be the simplest explanation of these results?

10. The ability to taste phenylthiocarbamide (PTC) is controlled in humans by a single dominant allele (*T*). A woman nontaster married a man taster and they had three children, two boy tasters and a girl nontaster. All the grandparents were tasters. Create a pedigree for this family for this trait. (Solid symbols should signify nontasters.) Where possible, indicate whether tasters are *TT* or *Tt*.

11. Fur color in rabbits is determined by a single gene locus for which there are four alleles. Four phenotypes are possible: black, Chinchilla (gray color caused by white hairs with black tips), Himalayan (white with black patches on extremities), and white. The black allele (*C*) is dominant over all other alleles, the Chinchilla allele (*C^{ch}*) is dominant over Himalayan (*C^h*), and the white allele (*c*) is recessive to all others. A black rabbit is crossed with a Himalayan, and the F_1 consisted of a ratio of 2 black: 2 Chinchilla. Can you determine the genotypes of the parents?

A second cross was done between a black rabbit and a Chinchilla. The F_1 contained a ratio of 2 black: 1 Chinchilla: 1 Himalayan. Can you determine the genotypes of the parents of this cross?

12. A dominant allele *B* produces bristles in fruit flies when it is present in the heterozygote. When it is homozygous (*BB*), it is lethal. Homozygous recessive (*bb*) flies are nonbristled. Another gene *S* acts to suppress the action of *B*, but it is also lethal when homozygous (*SS*). The ss genotype has no effect on bristles. Two nonbristled flies in which a B allele is being suppressed are crossed. What is the phenotypic ratio of the F_1? If the bristled flies from the F_1 are backcrossed to parental flies, what phenotypic ratio would be predicted for the offspring?

13. Albinism (lack of skin pigmentation) is caused by a recessive allele. Consider the following human pedigree for this trait (solid symbols represent individuals who are albinos). From your knowledge of Mendelian inheritance, determine the probable genotypes of the parents in generation I, the mates in generation II, and son 4 in generation III. Can you determine the genotype of son 3 in generation II? Why or why not? (Let *AA* and *Aa* represent normal pigmentation and *aa* be the albino genotype.)

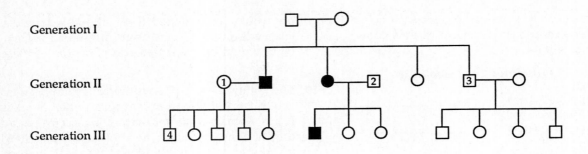

Generation I
    Genotype of father _____ of mother _____

Generation II
    Genotype of mate 1 _____ of mate 2 _____

Generation III
    Genotype of son 4 _____

Can you predict genotype of son 3? Explain.

CHAPTER 14

THE CHROMOSOMAL BASIS OF INHERITANCE

FRAMEWORK

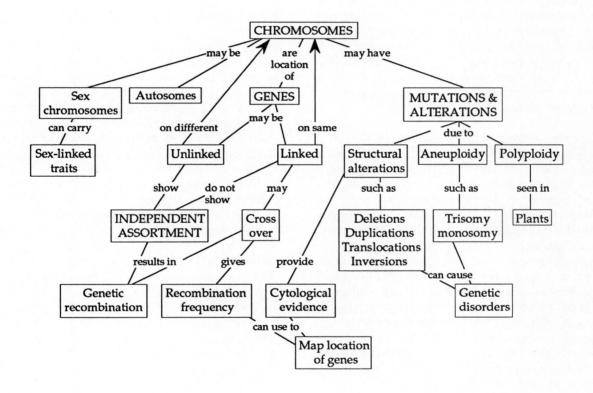

CHAPTER SUMMARY

The Chromosome Theory of Inheritance

By 1900, three botanists had independently reproduced the plant breeding results that Mendel had explained 35 years previously. Mendel's principles, combined with cytological evidence of the processes of mitosis and meiosis gathered in the late 1800s, led to the *chromosome theory of inheritance* that Mendelian "heritable factors" are located on chromosomes which undergo segregation and independent assortment in the process of gamete formation.

Morgan and the Drosophila School Morgan, working with the fruit fly, *Drosophila melangaster*, was the first to associate a specific gene with a specific chromosome. Fruit flies are excellent organisms for genetic studies because they are prolific breeders and have only four pairs of chromosomes, which are easily distinguishable with a microscope. The sex chromosomes occur as *XX* in females and *XY* in male flies.

The normal phenotype for a character, which is found most commonly in nature, is called the *wild type*, whereas alternative traits, assumed to have arisen as mutations in a wild-type allele, are called *mutant phenotypes*. The notation commonly used by geneticists employs a small letter to signify a recessive mutant allele and the small letter with a superscripted plus sign for the wild-type allele. A capital letter is used to signify dominant mutant alleles.

Morgan discovered a mutant white-eyed male fly that he mated with a wild-type red-eyed female. The F_1 were all red-eyed and the F_2 showed the 3:1 phenotypic ratio typical of a simple dominant trait. In the F_2, however, all female flies were red-eyed, whereas half of the males were red-eyed and half were white-eyed.

Morgan deduced that the gene for eye color was *sex-linked*, occurring only on the *X* sex chromosome. Males only have one *X* and their phenotype is determined by the eye-color allele they inherit from their mother. This association of a specific gene with a chromosome provided evidence for the chromosome theory of inheritance.

Linked Genes   Genes that are located on the same chromosome are *linked*. Because chromosomes are inherited as a unit, the law of independent assortment does not apply to linked genes. Traits are usually inherited together if their genes are located on the same chromosome, and offspring, therefore, will usually have the same phenotypic combination for those traits as the parents.

Morgan performed a testcross of heterozygote wild-type flies with flies homozygous recessive for black bodies and vestigial wings, and found that the offspring were not in the predicted 1:1:1:1 phenotypic classes. Rather, most of the offspring were the same phenotypes as the parents—either wild-type (gray, normal wings) or double mutant (black, vestigial). Morgan deduced that these traits were inherited together because their genes were located on the same chromosome.

The Chromosomal Basis of Recombination

The production of offspring with new combinations of traits inherited from two parents is called *genetic recombination*.

Recombination of Unlinked Genes: Independent Assortment   In a dihybrid cross between a heterozygote (*AaBb*) and a homozygous recessive (*aabb*), one-half of the offspring will be *parental types* and have

phenotypes like one or the other parent, and one-half of the offspring, called *recombinants*, will have combinations of the two traits that are unlike the parents (*Aabb* or *aaBb*). This 50% frequency of recombination is observed when two genes are located on different chromosomes and results from the random alignment of homologous chromosomes at metaphase I and the resulting independent assortment of alleles.

Recombination of Linked Genes: Crossing Over   Linked genes do not assort independently, and one would not expect to see recombination of parental traits in the offspring. However, there is recombination in linked genes that is due to crossing over, the reciprocal trade between non-sister chromatids of synapsed homologous chromosomes during prophase of Meiosis I.

Genetic Maps Based on Crossover Data

Morgan's group was the first to work out methods for mapping genes on particular chromosomes. Sturtevant suggested that recombination frequencies reflect the relative distance between genes. If genes are located farther apart on the chromosome, there is a greater probability that a crossover event will occur between them. Sturtevant used recombination data to locate genes on a chromosomal map. He defined one map unit (or centimorgan) as equal to a 1% recombination frequency.

The sequence of genes on a chromosome can be determined by finding the recombination frequency between different pairs of genes, which is calculated as the proportion of recombinant offspring out of the total number of offspring. For example, if a and b have 12% recombination and are 12 map units apart, b and c are 3 map units apart, and c and a are 9 units apart, then the sequence of genes must be a-c-b.

Linkage cannot be determined if genes are 50 or more map units apart because they would exceed the maximum of 50% frequency of recombination typical of unlinked genes. Distant genes on the same chromosome may be mapped by adding the recombination frequencies determined for intermediate genes.

Sturtevant and his co-workers clustered the genes for the various known mutations of *Drosophila* into four groups of linked genes. Given that there are four sets of chromosomes in *Drosophila*, these results provided additional evidence for the hypothesis that genes are located on chromosomes.

Crossover data provide only relative, not precise, locations of genes on chromosomes. The frequency of crossing-over may vary along the length of a chromosome. Due to multiple crossovers between genes located further apart, recombination data tend to underestimate actual map unit distance. *Cytological mapping* pinpoints the exact locations of gene loci, often using a technique that associates a mutant phenotype with a visible chromosomal defect or other feature.

Sex Chromosomes and Sex-Linked Inheritance: A Closer Look

The Chromosomal Basis of Gender in Humans Gender is a phenotypic character usually determined by the presence or absence of special chromosomes. Females, who are *XX*, produce ova that each contain an *X* chromosome. Males, who are *XY*, produce two kinds of gametes, each sperm has either an *X* or a *Y* chromosome. Whether of not the gonads of an embryo develop into testes or ovaries depends on the presence or absence of the gene *Sry*, found on the *Y* chromosome.

Bees and ants have no sex chromosomes. Males develop from unfertilized eggs and are haploid. Females develop from fertilized eggs and are diploid.

Sex-Linked Disorders in Humans Sex chromosomes may carry genes for traits that are not related to gender. In humans, most *X*-linked genes (coding for what are usually called sex-linked traits) do not have corresponding loci on the *Y* chromosome. Males, therefore, inherit their sex-linked alleles only from their mothers; daughters inherit sex-linked alleles from both parents.

Recessive sex-linked traits are seen more often in males, since they are hemizygous for sex-linked genes. Females need to inherit recessive alleles on both their X chromosomes to express a recessive trait. *Duchenne's muscular dystrophy* is a sex-linked disorder resulting from the lack of a key muscle protein. *Hemophilia* is a sex-linked recessive trait characterized by excessive bleeding due to the malfunction of a blood-clotting factor. This disorder has passed through the royal families of several European kingdoms due to intermarriage among the nobility. Color-blindness is also a sex-linked trait.

X-inactivation in Females Only one of the X chromosomes is fully active in most mammalian female somatic cells. The other X chromosome is inactive and contracted into a Barr body located inside the nuclear membrane. Lyon demonstrated that the selection of which X chromosome is inactivated is a random event occurring independently in embryonic cells. A female who is heterozygous for a sex-linked trait is a mosaic and will express one allele in approximately half her cells and the alternate in the other cells.

Chromosomal Alterations

Alterations of Chromosome Number Nondisjunction occurs when a pair of homologous chromosomes does not separate properly in meiosis I or sister chromatids do not separate in meiosis II. As a result, a gamete receives either two or none of that chromosome. A zygote formed with one of these aberrant gametes has a chromosomal alteration known as *aneuploidy*, a non-typical number of chromosomes. The zygote will be either trisomic for a chromosome (chromosome number is $2N + 1$) or monosomic ($2N - 1$ chromosome number). Aneuploid organisms usually have a set of symptoms caused by the abnormal dosage of genes on the extra or missing chromosome. A mitotic nondisjunction early in embryonic development is likely to have a substantial impact on the organism.

Polyploidy is a chromosomal alteration in which an organism has more than two complete chromosome sets, as in triploid ($3N$) or tetraploid ($4N$) organisms. Polyploidy is common in the plant kingdom and has played an important role in the evolution of plants. In animals, complete polyploids are rare. More common are mosaic polyploids, in which patches of tetraploid tissue grow from a cell in which sister chromatids did not separate in mitosis.

Alterations of Chromosome Structure Chromosome breakage can result in chromosome fragments that are lost, called *deletion*; that join to the homologous chromosome, called *duplication*; that join a nonhomologous chromosome, called *translocation*; or that rejoin the original chromosome in the reverse orientation, called *inversion*. Errors in crossing over can result in a deletion and duplication in non-sister chromatids, caused by nonequal exchange of chromatids.

A homozygous deletion is usually lethal because most genes are necessary for an organism's existence. Duplications and translocations also are typically harmful. Even though all the genes are present in proper quantities in inversions and translocations, the phenotype may be altered due to the *position effects* of neighboring genes on the expression of the relocated genes.

Chromosomal Alterations in Human Disease The frequency of aneuploid zygotes may be fairly high in humans, but development is usually so disrupted that the embryos spontaneously abort. Some genetic diseases, expressed as "syndromes" of characteristic traits, are the result of aneuploidy.

Down syndrome, affecting one out of every 700 children born, is the most common serious birth defect in the United States. Trisomy of chromosome 21 results in characteristic facial features, short stature, heart defects, and mental retardation. The incidence of Down syndrome increases for older mothers, perhaps because older women are less likely to spontaneously abort trisomic embryos. Pregnant women over 35 may have amniocentesis to check for trisomy 21.

Patau syndrome, caused by trisomy of chromosome 13, and Edwards syndrome, caused by a trisomy of chromosome 18, produce such serious defects that the victims of these syndromes rarely live more than a year.

Most sex-chromosome aneuploidies upset the genetic balance less than do autosomal aneuploidies, perhaps because so few genes are located on the Y chromosome and extra X chromosomes are inactivated as Barr bodies. XXY males exhibit *Klinefelter syndrome*, a condition in which the individual has abnormally small testes, is sterile, may have feminine body contours, and is usually of normal intelligence. Additional X or Y chromosomes (XXYY, XXXY, etc.) usually result in individuals with Klinefelter syndrome who also are mentally retarded.

Males with a single extra Y chromosome may be somewhat taller than average males, but they do not exhibit any well-defined syndrome. Trisomy X results in *metafemales* with limited fertility. Monosomy X individuals (XO) exhibit *Turner syndrome* and are phenotypically female, sterile individuals with short stature and usually normal intelligence.

Structural alterations of chromosomes, such as deletions or translocations, may be associated with specific human disorders. The cri du chat syndrome is caused by a deletion in chromosome 5; chronic myelogenous leukemia is a cancer that is associated with a reciprocal chromosomal translocation; and some individuals with Down syndrome have an extra part of a third chromosome 21 attached to another chromosome.

Parental Imprinting of Genes

Some traits, including some genetic disorders, seem to depend on which parent supplied the alleles for the trait. A child with a deletion of a segment of chromosome 15 will exhibit Prader-Willi syndrome if the defective chromosome came from the father or Angelman syndrome if the chromosome came from the mother. When both copies of chromosome 15 come from the mother, in a chromosomal alteration called uniparental disomy, the offspring develops Prader-Willi syndrome. These phenomena provide evidence for *genomic imprinting*, a hypothesis that certain genes have different effects depending on which parent they were inherited from. The resulting offspring then re-inprints these genes according to gender.

Genomic inprinting may help to explain *fragile-X-syndrome*, a mental retardation more prevalent in males than females and seemingly linked to whether the fragile-X chromosome was inherited from the mother.

Extranuclear Inheritance

Exceptions to Mendelian inheritance are found in the case of extranuclear genes located in cytoplasmic organelles, such as mitochondria and plant plastids that are transmitted to offspring in the cytoplasm of the ovum. In 1909, Correns observed the inheritance of variegated leaf coloration from the maternal plant. Maternal inheritance of mitochondrial genes in mammals is due to the large cytoplasmic contribution of the ovum.

STRUCTURE YOUR KNOWLEDGE

1. Morgan discovered and bred a white-eyed male fruit fly. Explain how the hereditary pattern of this mutant allele gave evidence for the chromosome theory of inheritance.

2. Two of the genes Mendel used to support his law of independent assortment were actually located on the same chromosome. Explain why genes located more than 50 map units apart behave as though they are not linked. How can one determine whether these genes are linked and what the relative distance is between them?

3. You have found a new mutant phenotype in fruit flies that you suspect is recessive and sex-linked. What is the best cross you could make to confirm your predictions?

4. Fill in the following table of human disorders or syndromes related to chromosomal abnormalities. What explanation can you give for the adverse phenotypic effects associated with these chromosomal alterations?

TEST YOUR KNOWLEDGE

MULTIPLE CHOICE: *Choose the one best answer.*

1. The chromosomal theory of inheritance states that
   a. genes are located on chromosomes.
   b. chromosomes and their associated genes undergo segregation during meiosis.
   c. chromosomes and their associated genes undergo independent assortment in gamete formation.
   d. all of the above are correct.

2. A wild type is
   a. the phenotype found most commonly in nature.

NAME OF DISORDER OR SYNDROME	CHROMOSOMAL ALTERATION INVOLVED	SYMPTOMS OR ASSOCIATED TRAITS
	Trisomy 21	
		Sterility, small testes, feminine body contours, normal intelligence or mental retardation
Turner Syndrome		
	Deletion in Chromosome 5	Mental retardation, small head, unusual cry
CML	reciprocal translocation of chromosome 22 & 9	

  b. the dominant allele.
  c. designated by a small letter if it is recessive or a capital letter if it is dominant.
  d. your basic party animal.

3. Sex-linked traits
  a. are carried on an autosome but expressed only in males.
  b. are coded for by genes located on a sex chromosome.
  c. are found in only one or the other sex, depending on the sex-determination system of the species.
  d. are always inherited from the mother in mammals and fruit flies.

4. Genetic and cytological maps for the same chromosome
  a. are both based on mutant phenotypes and recombination data.
  b. may have different sequences of genes.
  c. have both the same sequence of genes and intergenic distances.
  d. have the same sequence of genes but different intergenic distances.

5. Genetic recombination
  a. results in recombinant offspring.
  b. occurs in the fertilization process.
  c. occurs by independent assortment and crossing over in the formation of gametes.
  d. involves all of the above.

6. A 1:1:1:1 ratio of offspring from a dihybrid testcross indicates that
  a. the genes are linked.
  b. the genes are not linked.
  c. crossing over has occurred.
  d. the genes are 25 map units apart.

7. Two genes will probably assort independently if they are
  a. on nonhomologous chromosomes.
  b. on nonsister chromatids.
  c. far apart on homologous chromosomes.
  d. either a or c.

8. The genetic event that results in metafemales (*XXX*) is probably
  a. nondisjunction.
  b. uniparental disomy.
  c. deletion of the *Sry* gene.
  d. any of the above.

9. A son inherits color-blindness from his
  a. mother.
  b. father.
  c. mother only if she is color-blind.
  d. father only if he is color-blind.

10. Genomic imprinting
  a. explains cases where the gender of the parent from whom an allele is inherited affects the expression of that allele.
  b. is greatest in females because of the larger maternal contribution of cytoplasm.
  c. may explain the transmission of Duchenne's muscular dystrophy.
  d. applies to all of the above.

11. Nondisjunction
  a. occurs when homologous chromosomes do not separate properly in meiosis I.
  b. occurs when sister chromatids do not separate in meiosis II.
  c. can result in both trisomic and monosomic offspring.
  d. may involve any of the above.

12. In translocation,
    a. a fragment of a chromosome joins to its homologue.
    b. a fragment of a chromosome joins to a nonhomologous chromosome.
    c. the breakage and refusion of a chromosome does not affect phenotype.
    d. the offspring shows aneuploidy.

13. A triploid individual
    a. is the result of a duplication.
    b. has a $2N + 1$ chromosome number.
    c. has a $3N$ chromosome number.
    d. results from cells in which chromosomes did not separate during mitosis.

14. Extra dosages of genes
    a. usually have deleterious effects.
    b. can be caused by duplications.
    c. are found in Down, Patau, and Edwards syndromes.
    d. all of the above.

15. Sex-chromosome aneuploidies upset genetic balance less than do autosomal aneuploidies because
    a. extra X chromosomes are inactivated as Barr bodies.
    b. the Y chromosome only determines sex; it carries no genes.
    c. of genomic imprinting.
    d. of all the above.

16. Which of the following is *not* true of maternal inheritance?
    a. It is the result of the larger contribution of egg cytoplasm.
    b. It gave evidence for the existence of cytoplasmic genes.
    c. Mitochondrial, ribosomal, and plant plastid genes come from the mother.
    d. It does not follow Mendelian principles of inheritance.

17. Which of the following chromosomal mutations does not alter genic balance, but may alter phenotype because of position effects?
    a. deletion
    b. inversion
    c. duplication
    d. aneuploidy

18. The work of Mary Lyon showed that
    a. females are a genetic mosaic due to random nonseparation of chromatids during mitosis.
    b. males inherit their sex-linked traits from their mothers.
    c. the selection of which X chromosomes form Barr bodies during early embryonic development is random.
    d. maternal inheritance is a result of the large amount of cytoplasm in the ovum.

19. In birds, sex is determined by a ZW chromosome scheme. Males are ZZ and females are ZW. A sex-linked lethal recessive gene occurs in pigeons. What would be the sex ratio in the offspring of a cross between a male heterozygous for the lethal gene and a normal female?
    a. 2:1 male to female
    b. 1:2 male to female
    c. 1:1 male to female
    d. 4:3 male to female

GENETICS PROBLEMS

1. Two normal color-sighted individuals produce the following children and grandchildren. Fill in the probable genotype of the indicated individuals in this pedigree. Squares are males, circles are females, and solid symbols represent color-blindness. Choose an appropriate notation for the genotypes.

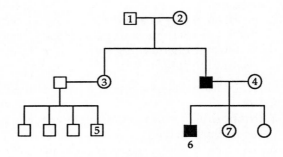

GENOTYPES

1. _____   5. _____

2. _____   6. _____

3. _____   7. _____

4. _____

2. The following recombination frequencies were found. Determine the order of these genes on the chromosome.

a—c	10%	b—c	4%	c—d	20%
a—d	30%	b—d	16%		
a—e	6%	b—e	20%		

3. In guinea pigs, black (B) is dominant to brown (b), and solid color (S) is dominant to spotted (s). A heterozygous black, solid-colored pig is mated

with a brown, spotted pig. The total offspring for several litters are

black solid	16
black spotted	5
brown solid	5
brown spotted	14

Are these genes linked or nonlinked? If they are linked, how many map units are they apart?

4. A woman is a carrier for a sex-linked lethal gene that causes spontaneous abortions. She has nine children. How many of these children do you expect to be boys?

5. A recessive sex-linked gene in fruit flies produces vermillion eye color (vv females and vY males have vermillion eyes). An autosomal recessive produces brown eye color ($bwbw$). Flies that are homozygous recessive for both vermillion and brown genes have white eyes. Determine the outcome of a cross between a wild type female (v^+v^+ bw^+bw^+) and a white-eyed male ($vYbwbw$). What would be the F_1 of a reciprocal cross between a white-eyed female and a red-eyed male (v^+Y bw^+bw^+)? Determine the phenotypic ratio of the F_2 for the first of these crosses.

6. A dominant sex-linked gene B produces white bars on black chickens, as seen in the Barred Plymouth Rock breed. A clutch of chicks has equal numbers of black and barred chicks. If only the females are found to be black, what were the genotypes of the parents? If males and females are evenly represented in the black and barred chicks, what were the genotypes of the parents? (Remember how sex is determined in birds.)

CHAPTER 15

THE MOLECULAR BASIS OF INHERITANCE

FRAMEWORK

This chapter outlines the key evidence that was gathered to establish DNA as the molecular basis of inheritance. Watson and Crick's double helix, with its rungs of specifically paired nitrogenous bases and twisting side ropes of phosphate and sugar groups, provided the three-dimensional model that explained DNA's ability to encode a great variety of information and produce exact copies of itself through semiconservative replication. The replication of DNA is an extremely fast and accurate process involving many enzymes and proteins.

CHAPTER SUMMARY

Deoxyribonucleic acid, DNA, is the genetic material, the substance of genes, the basis of heredity. Nucleic acids' unique ability to direct their own replication allows for the precise copying and transmission of DNA to all the cells in the body and from one generation to the next. DNA encodes the blueprints that direct and control the biochemical, anatomical, physiological, and behavioral traits of organisms.

The Search for the Genetic Material: A Case Study of the Scientific Process

The role of DNA in heredity was first established through work with microorganisms—bacteria and viruses. By the 1940s, chromosomes were known to carry hereditary information and to consist of proteins and DNA. Most scientists believed that the proteins carried the genetic program because of their known specificity and heterogeneity. The simple, rep-

etitious nucleic acids did not seem capable of coding the wealth of information needed for heredity.

Evidence That DNA Can Transform Bacteria The work of Griffith in 1928 provided the first evidence that the genetic material was some sort of heat-stable chemical. Griffith worked with two strains of *Streptococcus pneumoniae*—a smooth strain (S) that synthesized a polysaccharide capsule and a rough strain (R) that did not form a coat. Only live S strain cells caused pneumonia when injected in mice. However, when Griffith injected a mixture of heat-killed S cells and live R cells, the mice died. Moreover, live S cells could be isolated from the blood of these mice. R cells had somehow acquired the ability to make polysaccharide coats from dead S cells. These bacteria had incorporated external genetic material in a process now called *transformation*.

Avery worked for a decade to identify the transforming agent by purifying chemicals from heat-killed S cells. In 1944, he and his colleagues McCarty and MacLeod announced that DNA was the molecule that transferred genetic information. Their claim, however, was met with skepticism.

Evidence That Viral DNA Can Program Cells Viruses consist of little more than DNA, or sometimes RNA, contained in a protein coat. They reproduce by infecting another cell and commandeering that cell's metabolic machinery. Bacteriophages, or *phage*, are viruses that infect bacteria. In 1952, Hershey and Chase showed that DNA was the genetic material of a phage known as T2 which infects the bacterium *Escherichia coli (E. coli)*.

Hershey and Chase devised an experiment using radioactive isotopes to determine whether the phage's DNA or protein was transferred to the bacteria. One batch of T2 was grown with radioactive sulfur that became incorporated into protein; another was grown with radioactive phosphorus that labelled the DNA.

The labelled T2 cells were allowed to infect separate samples of *E. coli*. The cultures were blended to separate phage protein coats from *E. coli* and centrifuged to isolate the heavier bacterial cells from the lighter viral particles. In the samples with the labeled proteins, radioactivity was found in the supernatant, indicating that the phage protein did not enter the bacterial cells. In the samples with the labeled DNA, most of the radioactivity was found to have entered the bacterial cell fraction. When these *E. coli* cells were returned to culture, they released phages containing radioactive phosphorus. Hershey and Chase concluded that viral DNA is injected into the host cell whereas most of the proteins remain outside.

Additional Evidence That DNA Is the Genetic Material of Cells Circumstantial evidence that DNA was the genetic material came from the observation that prior to mitosis, a eukaryotic cell doubles its DNA content and that diploid cells have twice as much DNA as haploid gametes of the same organism.

Chargaff, in 1947, reported that DNA composition is species-specific; he found that each species he studied had a different ratio of nitrogenous bases. Chargaff also determined that the number of adenines and thymines were approximately equal and the number of guanines and cytosines were also equal in the DNA from all the organisms he studied. The A=T and G=C properties of DNA, known as Chargaff's Rules, were explained once the double helix was discovered.

Discovery of the Double Helix

By the early 1950s, the arrangement of covalent bonds in a nucleic acid polymer was established, and the race was on to determine the three-dimensional structure of DNA. Pauling in California and Wilkins and Franklin in London were working on the problem, but Watson and Crick were the first to solve the DNA puzzle.

Crick was studying protein structure using a technique called X-ray crystallography. An X-ray beam passed through a crystal can expose photographic film to produce a pattern of spots that a crystallographer can interpret into information about three-dimensional atomic structure. Watson saw an X-ray photo produced by Franklin that indicated that the basic shape of DNA was a helix. He and Crick deduced that the helix had a width of 2 nanometers with its purine and pyrimidine bases stacked 0.34 nanometers apart. This width suggested that the helix consisted of two strands, thus the term *double helix*.

Watson and Crick constructed wire models to build a double helix that would conform to the X-ray measurements and the known chemistry of DNA. They finally arrived at a model that paired the nitrogenous bases on the inside of the helix with the sugar-phosphate chains on the outside. The helix makes one full turn every 3.4 nanometers; thus ten layers of nucleotide pairs, stacked 0.34 nm apart, are present in each turn of the helix.

At first Watson assumed that each of the four bases paired with itself. To produce the molecule's uniform 2 nm width, however, a purine must pair with a pyrimidine. The molecular arrangement of the side groups of the bases permit two hydrogen bonds to form between adenine and thymine and three hydrogen bonds between guanine and cytosine. This complementary pairing explains Chargaff's rules.

In April 1953, Watson and Crick published a paper in *Nature* reporting the double helix as the molecular model for DNA.

DNA Replication: The Basic Concept

Watson and Crick noted that the base-pairing rule of DNA sets up a mechanism for its replication. Each side of the double helix is an exact complement to the other. When the two sides of a DNA molecule separate in the replication process, each strand serves as a template for rebuilding a double-stranded molecule. New nucleotides are joined to the exposed nucleotides sticking out from the sugar-phosphate backbone according to the specific base-pairing rules.

Their *semiconservative model* of gene replication predicts that the two daughter DNA molecules each have one old strand from the parent DNA and one newly formed strand. In contrast, a conservative model predicts that the parent strand remains intact and the duplicated molecule is totally new, whereas a dispersive model predicts that all four strands of the two DNA molecules are a mixture of parent and new DNA.

Meselson and Stahl tested these models by growing *E. coli* in a medium with ^{15}N, a heavy isotope that the bacteria incorporated into their nitrogenous bases. When centrifuged, this DNA could be separated by density from nonlabeled DNA. Cells with labeled DNA were transferred to a medium with the lighter isotope, ^{14}N. After one generation of bacterial growth, the DNA extracted from the culture was all of intermediate density; it contained hybrids of one parental heavier DNA strand and one newly-formed lighter DNA strand. A second replication produced both light and hybrid DNA, confirming the semiconservative model.

A Closer Look at DNA Replication

DNA replication is extremely rapid and accurate. The 6 billion base pairs found in your 46 chromosomes can be precisely copied in just a few hours. More than a dozen enzymes and other proteins are involved in this intricate process.

Getting Started: Origins of Replication   Replication begins at special sites, called origins of replication, where proteins that initiate replication bind to a specific sequence of nucleotides and separate the two strands. Replication spreads in both directions from these replication "bubbles." Bacterial and viral DNA molecules have only one replication origin, but large eukaryotic DNA molecules have hundreds to thousands of origin sites. Each end of a replication bubble where new DNA is being formed is called a replication fork because of its Y shape.

Elongating a New DNA Strand   Enzymes called DNA polymerases connect nucleotides to the growing end of the new DNA strand. A nucleoside triphosphate lines up with its complementary base on the template strand; the hydrolysis of its two tail phosphate groups provides the energy for polymerization.

The two strands of a DNA molecule are antiparallel; their sugar-phosphate backbones run in opposite directions. The deoxyribose sugar of each nucleotide is connected to its own phosphate group at its 5' carbon and connects to the phosphate group of the adjacent nucleotide by its 3' carbon. Thus a strand of DNA has polarity with a 5' end, where the final nucleotide's phosphate group attaches to the 5' carbon of its deoxyribose, and a 3' end, where the 3' carbon of the end nucleotide is not attached to a neighboring nucleotide. DNA polymerase adds nucleotides to the 3' end of a growing strand; DNA is replicated in a 5'$\longrightarrow$3' direction.

Because the DNA strands run in an antiparallel direction, the simultaneous synthesis of both strands presents a problem. The *leading strand* is the new 5'$\longrightarrow$3' strand being formed as DNA polymerase moves along the template as the replication fork progresses. The *lagging strand* is created as a series of short segments, called Okazaki fragments, that are formed in the 5'$\longrightarrow$3' direction away from the replication fork. An enzyme called *DNA ligase* joins the fragments. It is thought that both parent strands of DNA thread through a polymerase dimer in the replication fork, with the template for the lagging strand forming a loop before passing through the active site of the polymerase in the correct direction.

DNA polymerase cannot initiate synthesis of a DNA strand, but can only add nucleotides to an existing chain. A *primer* of RNA is needed to start the chain. An enzyme called *primase* pairs about ten RNA nucleotides to a short portion of the DNA strand to form the primer. Each fragment on the lagging strand requires an RNA primer. A continuous strand of DNA is produced after an enzyme replaces the RNA primer with DNA nucleotides and ligase joins the fragments.

Many other proteins are involved in DNA replication: Enzymes called *helicases* unwind the helix and separate the parent strands, and *single-strand binding proteins* keep the separated strands apart.

Proofreading   Initial pairing errors in nucleotide placement may occur as often as 1 per 10,000 bases. The amazing accuracy of DNA replication is achieved by proofreading and correcting pairing errors. In bacteria, DNA polymerase checks each new nucleotide against its template and backs up and replaces incorrect nucleotides. It is not known whether DNA polymerase or other enzymes perform this proofreading function in eukaryotic cells.

DNA Repair

DNA molecules may be altered by reactive chemicals, radioactive emissions, X-rays and ultraviolet (UV) light. These changes, or mutations, may be corrected by many types of DNA repair enzymes. In *excision repair*, the damaged strand is cut out by a repair enzyme and the gap is correctly filled through the action of a DNA polymerase and DNA ligase. This type of repair happens frequently in skin cells, where ultraviolet rays of sunlight cause adjacent thymine bases to link covalently, forming thymine dimers.

STRUCTURE YOUR KNOWLEDGE

1. Fill in the following chart about these key investigators and the evidence they provided about the structure and function of DNA.

INVESTIGATOR	ORGANISMS USED	TECHNIQUES, EXPERIMENTS	CONCLUSIONS
Griffith			
Avery			
Hershey & Chase			
Chargaff			
Franklin			
Watson & Crick			
Messelson & Stahl			

2. In this diagram showing the replication of DNA, label the following items: leading and lagging strands; Okazaki fragment, where DNA polymerase, DNA ligase, helicase, primase, and single-strand binding proteins operate; RNA primer; replication fork; 5' and 3' ends of parental DNA.

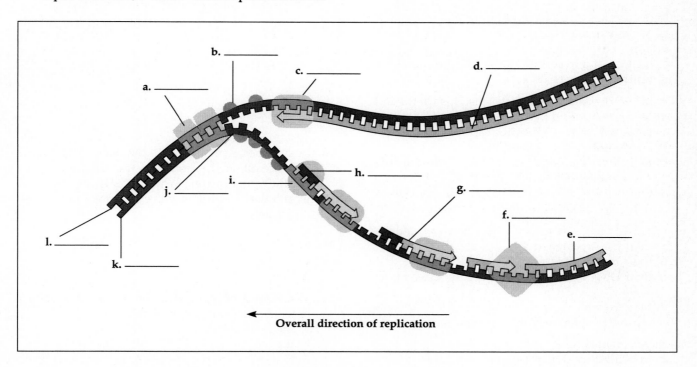

Overall direction of replication

TEST YOUR KNOWLEDGE

MULTIPLE CHOICE: *Choose the one best answer.*

1. One of the reasons most scientists believed proteins were the carriers of genetic information was that
   a. proteins were more heat stable than nucleic acids.
   b. the protein content of duplicating cells always doubled prior to division.
   c. proteins were much more complex molecules than nucleic acids and were known to be highly specific.
   d. early experimental evidence pointed to proteins as the hereditary material.

2. Transformation involves
   a. the transfer of genetic material, often from one bacterial strain to another.
   b. the creation of a strand of RNA from a DNA molecule.
   c. the infection of bacterial cells by phage.
   d. the type of semiconservative replication shown by DNA.

3. According to Chargaff's Rules,
   a. A=C and T=G.
   b. A=T and G=C.
   c. DNA replication must be semiconservative.
   d. each species has a different ratio of nitrogenous bases.

4. In her work with DNA, Franklin used
   a. X-ray crystallography.
   b. paper chromatography.
   c. heavy isotopes of sulfur and phosphorus.
   d. centrifugation and ^{15}N.

5. In his work with pneumonia-causing bacteria and mice, Griffith found that
   a. DNA was the transforming agent.
   b. the R and S strains mated.
   c. heat-killed S cells could cause pneumonia when mixed with heat-killed R cells.
   d. that some heat-stable chemical was transferred to R cells to transform them into S cells.

6. When T2 phage are grown with radioactive sulfur,
   a. their DNA is tagged.
   b. their proteins are tagged.
   c. their DNA is found to be of medium density in a centrifuge tube.
   d. they transfer their radioactivity to *E. coli* chromosomes when they infect the bacteria.

7. Meselson and Stahl
   a. provided evidence for the semiconservative model of DNA replication.
   b. were able to separate phage protein coats from *E. coli* by using a blender.
   c. found that DNA labelled with ^{15}N was of intermediate density.
   d. grew *E. coli* on labelled phosphorus and sulfur.

8. Watson and Crick concluded that each base could not pair with itself because
   a. there would not be room for the helix to make a full turn every 3.4 nanometers.
   b. the width of 2 nm would not permit two purines to pair together.
   c. the bases could not be stacked 0.34 nm apart.
   d. identical bases could not hydrogen bond together.

Use these centrifuge tubes showing density bands of DNA to answer questions 9 and 10.

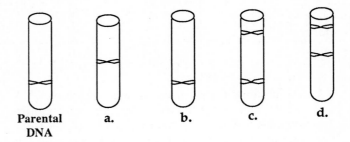

Parental   a.   b.   c.   d.
DNA

9. *E. coli* grown on a medium of heavy nitrogen (^{15}N) are transferred to a ^{14}N medium for one generation of growth. DNA extracted from these cells is centrifuged. What density distribution of DNA would you predict for this experiment from the possible DNA density bands illustrated above?
   a.
   b.
   c.
   d.

10. In an experiment similar to that in question 9, the cells are allowed to grow on the ^{14}N medium for another generation. What density distribution would you predict for this experiment from those illustrated above?
    a.
    b.
    c.
    d.

11. The joining of nucleotides in the polymerization of DNA requires energy from
    a. the hydrolysis of GTP.
    b. the hydrolysis of ATP.
    c. the loss of a pyrophosphate from incoming nucleoside triphosphates.
    d. hydrogen bonding.

12. Continuous elongation of a new DNA molecule along one strand of DNA
    a. occurs on the leading strand.
    b. occurs because DNA ligase can only elongate in the 5'——→3' direction.
    c. makes a single Okazaki fragment.
    d. all of the above.

13. Helicases
    a. unwind DNA strands prior to replication.
    b. rewind the newly formed DNA molecules.
    c. provide excision repair of damaged DNA.
    d. separate DNA strands and hold them apart for replication.

14. DNA ligase is needed to
    a. join Okazaki fragments on the leading strand.
    b. join the 5' end of an Okazaki fragment to the 3' end of the elongating DNA strand.
    c. join the 3' end of repaired segments of DNA to the 5' end of the gap.
    d. do all of the above.

15. Which of the following statements about DNA polymerase is incorrect?
    a. It is found only in eukaryotes.
    b. It is able to proofread and correct for errors in its base pairing in bacteria.
    c. It is unable to join free nucleotides unless an RNA primer is present.
    d. It only works in the 5'——→3' direction.

16. Thymine dimers, covalent links between adjacent thymine bases in DNA, may be induced by UV light. When they occur, they are repaired by
    a. excision enzymes that cut out the damaged region.
    b. DNA polymerase.
    c. ligase.
    d. all of the above.

Use the following diagram to answer questions 17 through 20.

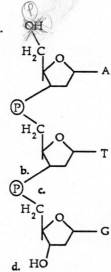

17. Which letter indicates the 5' end of this single DNA strand?
    a.
    b.
    c.
    d.

18. At which letter would the next nucleotide be added?
    a.
    b.
    c.
    d.

19. Which letter indicates a phosphodiester bond formed by DNA polymerase?
    a.
    b.
    c.
    d.

20. The base sequence of the DNA strand made from this template would be (from top to bottom)
    a. A T C
    b. C G A
    c. T A C
    d. U A C

CHAPTER 16

FROM GENE TO PROTEIN

FRAMEWORK

This chapter deals with the pathway from DNA to RNA to proteins. The diagram on p. 121 shows the important steps in transcription, RNA processing, and translation.

CHAPTER SUMMARY

The hereditary information of DNA is contained in specific sequences of nucleotides. These sequences are translated into proteins, which are responsible for the characteristics of an organism and thus form the link between genotype and phenotype.

Evidence That Genes Specify Enzymes

In 1909, the British physician Garrod first suggested that genes determine phenotype through the action of enzymes that control chemical processes in the cell. Garrod reasoned that inherited diseases were attributable to an inability to make certain enzymes.

How Genes Control Metabolism In the 1930s, Beadle and Ephrussi speculated that the mutations causing various eye colors in *Drosophila* resulted from a nonfunctioning enzyme at some point in the metabolic pathway leading to pigment formation. A short time later, Beadle and Tatum, working with mutants of a bread mold, *Neurospora crassa*, demonstrated the relationship between genes and enzymes.

Beadle and Tatum studied several nutritional mutants, called *auxotrophs*, that could not grow on the *minimal medium* of agar mixed with inorganic salts, sucrose, and biotin that sufficed for wild-type

Neurospora. By growing an auxotroph on *complete growth medium* and then transferring bits of this fungus to various combinations of minimal medium and added nutrients, Beadle and Tatum were able to identify the specific defective metabolic pathway for each auxotroph. Three classes of *Neurospora* mutants that were unable to synthesize arginine were tested by providing each of them with the different precursors of the pathway. Beadle and Tatum reasoned that the metabolic pathway of each class was blocked at a different enzymatic step. Assuming that each mutant was defective in a single gene, Beadle and Tatum formulated the "one gene—one enzyme" hypothesis that the function of a gene is to control production of a specific enzyme.

One Gene—One Polypeptide Molecular biologists revised this hypothesis to "one gene—one protein," because not all proteins are enzymes. Because many proteins consist of more than one polypeptide chain, each of which is codified by its own gene, the axiom has been further revised to "one gene—one polypeptide."

An Overview of Protein Synthesis

RNA is the link between a gene and the protein for which it codes. All three molecules are composed of specific sequences of monomers: nucleotides in DNA and RNA and amino acids in proteins.

Transcription is the transfer of information from DNA to *messenger RNA*, using the "language" of nucleic acids. *Translation* transfers information from RNA to a polypeptide, changing from the language of nucleotides to that of amino acids.

In prokaryotes, which lack a nucleus, transcription of DNA to mRNA and translation of mRNA by ribosomes into protein occur almost simultaneously. In eukaryotes, the mRNA must exit the nucleus before

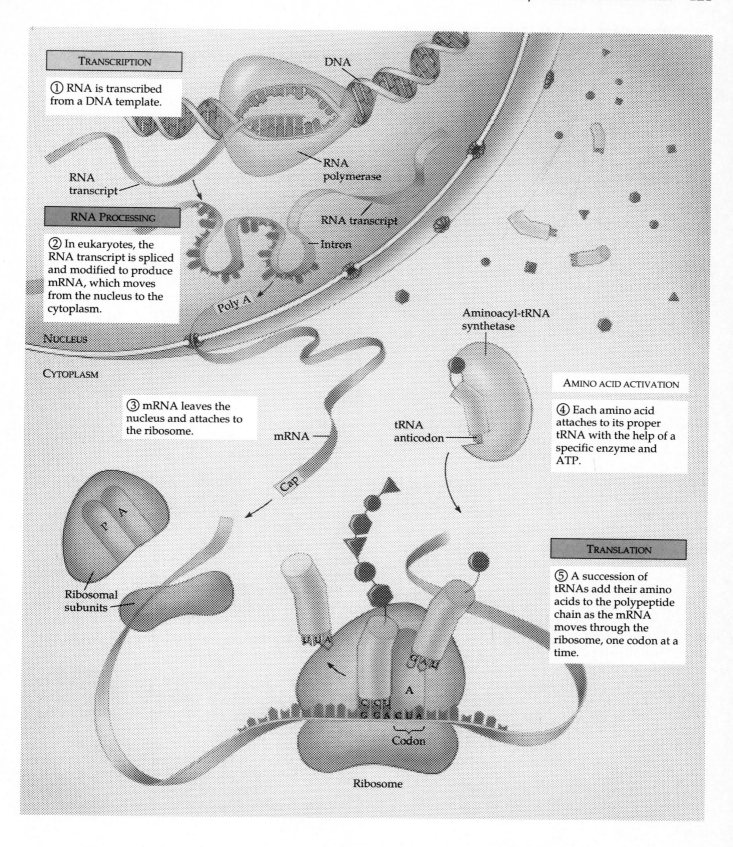

TRANSCRIPTION

① RNA is transcribed from a DNA template.

DNA

RNA polymerase

RNA transcript

RNA transcript

RNA PROCESSING

② In eukaryotes, the RNA transcript is spliced and modified to produce mRNA, which moves from the nucleus to the cytoplasm.

Intron

Poly A

NUCLEUS

CYTOPLASM

Aminoacyl-tRNA synthetase

AMINO ACID ACTIVATION

④ Each amino acid attaches to its proper tRNA with the help of a specific enzyme and ATP.

③ mRNA leaves the nucleus and attaches to the ribosome.

mRNA

Cap

tRNA anticodon

Ribosomal subunits

TRANSLATION

⑤ A succession of tRNAs add their amino acids to the polypeptide chain as the mRNA moves through the ribosome, one codon at a time.

Codon

Ribosome

translation can occur. *RNA processing*, the modification of mRNA within the nucleus, occurs only in eukaryotes. The basic sequence of protein synthesis in all cells is DNA——→RNA——→protein.

The Genetic Code

How can four nucleotides be translated into 20 different amino acids? Groups of nucleotides could form a code, but a two-base code would provide only 16 (4^2) unique arrangements, not enough to specify all the amino acids. A triplet code would provide 64 (4^3) possible codons, which is more than enough. The translation of nucleotides into amino acids does involve a *triplet code*, a sequence of three nucleotides, called a *codon*, to specify each amino acid.

The coding strand of a gene is the strand of the DNA molecule whose codons are transcribed into mRNA. The same strand of a long DNA molecule can be the coding strand for one gene and the noncoding strand for another. The mRNA is complementary to the DNA template since its bases follow the same base pairing rules, with the exception that uracil substitutes for thymine in RNA. Thus the codon ATG is transcribed in mRNA as UAC. Each mRNA codon then specifies one of the 20 amino acids to be sequenced in the polypeptide chain.

The genetic code was "cracked" in the early 1960s by a series of experiments that determined the amino acid translation of each of the codons of nucleic acids. Nirenberg synthesized artificial mRNA by linking uracil RNA nucleotides. Adding this "poly-U" to an in vitro system of a test tube containing all the biochemical ingredients necessary for protein synthesis, Nirenberg obtained a polypeptide containing the single amino acid phenylalanine. The codons AAA, GGG, and CCC were deciphered in the same manner.

Using more elaborate techniques to decode the triplets with mixed bases, scientists had deciphered all 64 triplets by the mid-1960s. Sixty-one of the triplets code for amino acids. The three remaining mRNA codons function as stop signals. AUG both codes for methionine and functions as the start signal for translation; thus all polypeptides are synthesized with methionine as their first amino acid.

The code is often redundant, meaning that more than one codon may specify a single amino acid. This redundancy, called "degeneracy," often occurs as differences in the third base of the triplet. The code is never ambiguous; no codon specifies two different amino acids.

In general, a particular nucleotide sequence on DNA is read in only one reading frame, starting at a start triplet and reading each triplet sequentially.

Evolutionary Significance of a Common Genetic Language

The genetic code of codons and their corresponding amino acids is the same for most organisms. A bacterial cell can translate the genetic messages of human cells. Recently, exceptions to the constancy of the genetic code have been found in several single-celled ciliates and in the DNA of mitochondria and chloroplasts. The near universality of a common genetic language, however, lends compelling evidence to the antiquity of the code and the evolutionary connection of all living organisms.

A Closer Look at Transcription

Transcription begins when enzymes called RNA *polymerases* separate the two strands of DNA and link RNA nucleotides that base-pair along the coding strand to the 3' end of the growing RNA polymer. A *transcription unit* is the entire sequence of DNA, including the initiation and termination sites, that is transcribed into one mRNA molecule. In eukaryotes, this unit represents a single gene; in prokaryotes, a transcription unit may include a few related genes.

Bacteria have one type of RNA polymerase. Eukaryotes have three types; the one that synthesizes mRNA is called RNA polymerase II.

Binding of RNA Polymerase and Initiation of Transcription   *Transcription factors* are proteins that aid RNA polymerase II in locating and binding to a promoter region. The promoter region includes the initiation site and the *TATA box*, a stretch of DNA rich in thymine and adenine bases which is upstream from the initiation site and which is recognized by a transcription factor. RNA polymerase must bind to the complex of promoter region and transcription factor for transcription to begin. In prokaryotes, RNA polymerase includes a "sigma" subunit that aids in promoter recognition and binding.

Elongation of the RNA Strand   RNA polymerase II untwists and separates the double helix, exposing DNA nucleotides for base pairing with RNA nucleotides, which are added to the 3' end of the growing polymer. As the RNA is made, it peels away from the DNA template and the DNA double helix reforms. Several RNA molecules may be transcribed simultaneously from a single gene through the action of multiple molecules of RNA polymerase, allowing a cell to produce large quantities of a particular protein.

Termination of Transcription   Transcription ends when RNA polymerase recognizes a termination site on the coding strand, often the sequence AATAAA. Other proteins probably aid RNA polymerase in this recognition and the subsequent release of the newly made RNA strand. In prokaryotes, this mRNA can be translated immediately. In eukaryotes, RNA is processed before it leaves the nucleus as mRNA.

A Closer Look at Translation

Transfer RNA (tRNA) carries amino acids from elsewhere in the cytoplasm to ribosomes, where they are added to the growing end of a polypeptide chain in the sequence indicated by the codons on mRNA. Transfer RNA molecules are specific for the amino acid they carry. They each have a specific base triplet, called an *anticodon*, that binds to a complementary codon on mRNA, thus assuring that amino acids are arranged in the sequence prescribed by the transcription from DNA.

Transfer RNA   As with other RNAs, transfer RNA is transcribed in the nucleus of a eukaryote and moves into the cytoplasm where it can be used repeatedly. These single-stranded, short RNA molecules are arranged into a clover-leaf shape by hydrogen bonding between complementary base sequences and then folded into a three-dimensional, roughly L-shaped structure. At one end of the L, a protruding loop holds a specialized base triplet called the anticodon. The anticodon base pairs with its complementary codon on mRNA. At the other, 3′ end of the tRNA molecule is the attachment site for its amino acid.

Sixty-one codons for amino acids can be read from mRNA, but there are only about 45 different tRNA molecules. A phenomenon known as *wobble* enables the third nucleotide of some tRNA anticodons to pair with more than one kind of base in the codon. Thus one tRNA can recognize more than one mRNA codon, all of which code for the same amino acid carried by that tRNA. For example, the modified base inosine (I) is in the third position on several tRNAs and can pair with U, C, or A.

Aminoacyl-tRNA Synthetases   Aminoacyl-tRNA synthetases are a family of enzymes that bind one type of amino acid with its appropriate tRNA molecule to create an amino acid-tRNA complex, also known as aminoacyl-tRNA. Each amino acid has a specific enzyme to which it binds along with an ATP molecule that releases two phosphates and joins to the amino acid as AMP. The appropriate tRNA then displaces the AMP and covalently bonds to the amino acid. The amino acid–tRNA complex is released from the enzyme and can carry the amino acid to a ribosome.

Ribosomes   Ribosomes facilitate the specific coupling of tRNA anticodons with mRNA codons during protein synthesis. They consist of two subunits, each composed of proteins and a specialized form of RNA called *ribosomal RNA (rRNA)*. Ribosomes have a binding site for mRNA, a *P site* that holds the tRNA carrying the growing polypeptide chain, and an *A site* that binds to the tRNA carrying the next amino acid. The transfer of this amino acid to the carboxyl end of the growing polypeptide chain is catalyzed by one of the ribosomal proteins.

Building a Polypeptide   The three stages of protein synthesis—chain initiation, chain elongation, and chain termination—all require proteins (mostly enzymes). The first two stages also require energy, which is provided by GTP (guanosine triphosphate).

The initiation stage begins as the mRNA and an initiator tRNA bind to the small subunit of the ribosome. The tRNA, carrying methionine, attaches to the start codon AUG on the mRNA. With the aid of proteins called *initiation factors* and the expenditure of one GTP, the large subunit of the ribosome attaches to the small one. The initiator tRNA fits into the P site of the now-functional ribosome, and the A site is vacant, ready for the next tRNA.

The addition of amino acids in the elongation stage involves several proteins called *elongation factors* and occurs in a three-step cycle: codon recognition, peptide bond formation, and translocation. In the codon recognition step, an elongation factor brings the correct tRNA into the A binding site of the ribosome. With energy from the hydrolysis of a phosphate bond of GTP, the mRNA codon in the A site forms a hydrogen bond with the tRNA anticodon. In the second step, an enzyme called *peptidyl transferase* catalyzes the formation of a peptide bond between the polypeptide held in the P site and the amino acid in the A site. The polypeptide is released from the tRNA that was holding it and is now held by the amino acid in the A site. The tRNA carrying the growing polypeptide is now translocated to the P site, a process requiring energy from the hydrolysis of a GTP molecule. The next mRNA codon moves into the A site as the mRNA ratchets through the ribosome in the 5′⟶3′ direction.

Termination occurs when a *termination codon*—UAA, UAG, or UGA—reaches the A site of the ribosome and binds with a protein called *release factor*. Peptidyl transferase now attaches a water molecule to the polypeptide chain, freeing the completed polypeptide from the tRNA in the P site. The ribosome then separates into its small and large subunits.

Polyribosomes   A mRNA may be translated simultaneously by several ribosomes in clusters called *polyribosomes*.

From Polypeptide to Functional Protein   During and following translation, a polypeptide folds spontaneously into its secondary and tertiary structure. The protein may need to undergo other changes before it can begin its function in the cell. Amino acids may be chemically modified; one or more amino acids at the beginning of the chain may be enzymatically removed; segments of the polypeptide may be excised; the chain may be cleaved into several pieces; or several polypeptides may associate into a quaternary structure.

Protein Targeting

All ribosomes are identical, whether they are free ribosomes, which synthesize cytosol proteins, or ER-bound ribosomes, which make membrane and secretory proteins. All protein synthesis begins in the cytoplasm. If a protein is destined to be a secretory protein, its polypeptide chain will begin with a signal sequence of amino acids that enables the ribosome to attach to a receptor site on the ER membrane. As the growing polypeptide threads into the cisternal space, the signal sequence is enzymatically removed. A ribosome remains free in the cytosol if the mRNA does not code for an ER signal sequence. Other signal sequences are thought to direct newly made proteins to specific sites such as mitochondria or chloroplasts.

Comparing Protein Synthesis in Prokaryotes and Eukaryotes: A Review

The basic processes of transcription and translation are the same in bacteria and eukaryotes, although the RNA polymerases and ribosomes differ. In bacteria, these processes occur almost simultaneously, whereas in eukaryotes, transcription is physically separated from translation by the nuclear envelope. Also in eukaryotes, RNA processing may occur before RNA leaves the nucleus.

RNA Processing in Eukaryotes

Alteration of the Ends of mRNA   During RNA processing, mRNA is modified by the addition of a cap of modified guanine nucleotides at the 5' end and a string of adenine nucleotides, called a *poly-A tail*, at the 3' end. The cap serves as a recognition site for the small ribosomal subunit, and both the cap and poly-A tail protect the ends of the mRNA from hydrolytic enzymes.

RNA Splicing   Long segments of noncoding base sequences, known as *introns* or intervening sequences,

have been found within the boundaries of eukaryotic genes. The remaining coding regions are called *exons*, since they are expressed in protein synthesis. An entire transcript is made of the gene, forming an oversized RNA called *heterogeneous nuclear RNA (hnRNA)*. The introns are removed and the exons joined before the RNA leaves the nucleus. This *RNA splicing* also occurs in the production of tRNA and rRNA.

Signals for RNA splicing are sets of a few nucleotides at either end of each intron. Small nuclear ribonucleoproteins (snRNPs), composed of proteins and small nuclear RNA (snRNA), are components of a molecular complex called a *spliceosome* that is involved in RNA splicing. The spliceosome snips an intron out of the RNA transcript and connects the adjoining exons. Several other schemes of RNA processing have been identified for tRNA and rRNA. Some splicing occurs with the intron RNA acting as an enzyme and catalyzing the process itself.

Ribozymes, RNA molecules that act as enzymes, make obsolete the statement "all enzymes are proteins." Ribosomal RNA has enzymatic functions during translation and intron RNA may assist in RNA splicing. RNA is a versatile molecule with many different shapes and functions.

Several hypotheses attempt to explain the functions of introns. One is that they are somehow involved in regulating gene activity or in the flow of mRNA into the cytoplasm. Another hypothesis is that they facilitate recombination between exons to create a diversity of proteins.

The splicing of the RNA transcript may allow different cells to use the same gene to synthesize different proteins. Exons may code for polypeptide *domains*, functional segments of a protein, such as binding sites. The arrangement of these domains by splicing can yield different mRNAs.

Mutations and Their Effects on Proteins

Mutations, changes in a cell's genetic information, may involve large portions of a chromosome or, as in a *point mutation*, affect just one or a few nucleotides. If the mutation is in a gamete, it may be passed on to offspring. Mutations may be reflected in the proteins translated from a gene. A deleterious effect on phenotype is described as a genetic disorder or hereditary disease.

Types of Mutations   Point mutations within a gene are either base-pair substitutions or base-pair insertions or deletions.

A *base-pair substitutions* replaces one nucleotide and its complementary partner with another pair of nucleotides. Due to the redundancy of the genetic code, some base-pair substitutions do not affect gene translation. If the third nucleotide of a codon is changed, it may still code for the same amino acid.

A substitution may result in the insertion of a different amino acid without altering the character of the protein if the new amino acid has similar properties or is not located in an area crucial to that protein's function.

A base-pair substitution that results in a different amino acid in a critical portion of a protein, such as the active site of an enzyme, may significantly affect protein function. Most often this type of mutation is harmful to the cell.

An incorrectly coded amino acid is called a *missense mutation*. *Nonsense mutations* occur when the point mutation changes an amino acid codon into a stop codon, prematurely halting the translation of the polypeptide chain and probably creating a nonfunctional protein.

Base-pair *insertions* or *deletions* that are not in multiples of three nucleotides alter the reading frame. All nucleotides downstream from the mutation will be improperly grouped into codons, creating extensive missense and usually prematurely ending in nonsense. These *frameshift mutations* almost always produce nonfunctional proteins.

Mutagenesis   *Mutagenesis*, the generation of mutations, can occur in a number of ways. *Spontaneous mutations* include base-pair substitutions, insertions,

and deletions that occur during DNA replication or repair. Physical agents, such as X-rays and UV light, and various chemical agents that cause mutations are called *mutagens*. Chemical mutagens include base analogues that substitute for normal bases in DNA synthesis and result in mispairing and base-pair substitutions. The Ames test measures the mutagenic strength of compounds by testing their ability to cause mutations in bacteria.

What Is a Gene

Our definition of a gene has evolved from Mendel's heritable factors, to Morgan's loci along chromosomes, to Beadle and Tatum's one gene–one polypeptide axiom. Research continually refines our understanding of the structural and functional aspects of genes, which now include introns and promoter regions. The best working definition of a gene is that it is a region of DNA that is required to produce an RNA molecule.

STRUCTURE YOUR KNOWLEDGE

1. Fill in the following chart that summarizes the processes of transcription and translation in eukaryotic cells.

	TRANSCRIPTION	TRANSLATION
Template		
Location		
Molecules involved		
Enzymes needed		
Control— start & stop		
Product		
Processing involved		
Energy source		

2. What is the genetic code? Explain redundancy and the wobble hypothesis. What is meant by saying that the genetic code is almost universal?

3. Prepare a concept map showing the types and consequences of mutations.

TEST YOUR KNOWLEDGE

MULTIPLE CHOICE: *Choose the one best answer.*

1. Beadle and Tatum's study of *Neurospora* showed that
   a. auxotrophs could not grow on minimal medium.
   b. mutants had defective enzymes in their metabolic pathways.
   c. the occurrence of different mutants with different defects in their metabolic pathway supported a one gene–one enzyme hypothesis.
   d. all of the above.

2. Transcription involves the transfer of information from
   a. DNA to RNA.
   b. RNA to DNA.
   c. mRNA to an amino acid sequence.
   d. DNA to an amino acid sequence.

3. Which of the following is *not* true of an anti-codon?
   a. It consists of three nucleotides.
   b. It is the basic unit of the genetic code.
   c. It extends from one end of a tRNA molecule.
   d. It may pair with more than one codon, especially if it has the base inosine in its third position.

4. RNA polymerase
   a. is the protein responsible for the production of ribonucleotides.
   b. is the enzyme that creates hydrogen bonds between nucleotides on the DNA strand and their complementary RNA nucleotides.
   c. in the enzyme that transcribes exons, but does not transcribe introns.
   d. begins transcription at a promoter sequence and moves along the coding strand of DNA in a 5'$\longrightarrow$3' direction.

5. Transfer RNA
   a. forms hydrogen bonds with the anticodon in the A site of a ribosome.
   b. binds to its specific amino acid in the active site of an aminoacyl–tRNA synthetase.
   c. uses GTP as the energy source to bind its amino acid.
   d. does all of the above.

6. Translocation involves
   a. the hydrolysis of a GTP molecule.
   b. the movement of the tRNA in the A site to the P site.
   c. the movement of the mRNA strand one triplet length in the A site.
   d. all of the above.

7. Changes in a polypeptide following translation may involve
   a. the addition of sugars or lipids to certain amino acids.
   b. the action of enzymes to add amino acids at the beginning of the chain.
   c. the removal of poly-A from the end of the chain.
   d. all of the above.

8. A single mRNA may produce several proteins at a time by
   a. the action of several ribosomes in a cluster called a polyribosome.
   b. several RNA polymerase molecules working sequentially.
   c. association with rough ER.
   d. containing several promoter regions.

9. A major difference in the protein synthesis of prokaryotes and eukaryotes is that
   a. transcription and translation of a gene can occur simultaneously in prokaryotes.
   b. RNA processing occurs in the cytoplasm of prokaryotes.
   c. polyribosomes are found only in eukaryotes.
   d. signal sequences guide mRNA out of the nucleus in eukaryotes.

10. In RNA processing,
   a. exons are excised before the mRNA is translated.
   b. assemblies of protein and snRNPs, called spliceosomes, may catalyze splicing.
   c. the RNA transcript that leaves the nucleus may be much longer than the original transcript.
   d. large quantities of rRNA are assembled into ribosomes.

11. Base-pair substitutions may have little effect on the resulting protein for all of the following reasons *except* which one?
   a. The redundancy of the code may result in no change in translation.
   b. As long as the substitution is three nucleotides, the reading frame is not altered.

c. The missense mutation may not occur in a critical part of the protein.

d. The new amino acid may have similar properties to the replaced one.

12. A ribozyme is
  a. an exception to the one gene–one RNA molecule axiom.
  b. an enzyme that adds the 5′ cap and poly-A tail to mRNA.
  c. an example of rearrangement of protein domains caused by RNA splicing.
  d. an RNA molecule that functions as an enzyme.

13. A signal sequence
  a. is most likely to be found on proteins produced by bacterial cells.
  b. directs an mRNA molecule into the cisternal space of ER.
  c. is a sign to help bind the small ribosomal unit at the initiation codon.
  d. would be the first 20 or so amino acids of a protein destined for secretion from the cell.

14. A base deletion early in the coding sequence of a gene may result in a protein that is
  a. functionally disrupted by the mutation.
  b. prematurely terminated.
  c. largely missense coding due to a frameshift mutation.
  d. all of the above.

15. The bonds between the anticodon of a tRNA molecule and the complementary codon of mRNA are
  a. formed by the input of energy from GTP.
  b. catalyzed by peptidyl transferase.
  c. hydrogen bonds.
  d. formed by the input of energy from ATP.

16. A prokaryotic gene 600 nucleotides long can code for a polypeptide chain of about how many amino acids?
  a. 200
  b. 300
  c. 600
  d. 1800

17. Peptidyl transferase
  a. translocates the tRNA holding the polypeptide chain from the A to the P site.
  b. hydrolyzes a GTP to pair the codon and anticodon in the A site.
  c. catalyzes the formation of a peptide bond between a polypeptide and the amino acid in the A site.
  d. binds the initiator tRNA to the start codon in the P site.

18. How many amino acids could be coded for if there were five instead of four different nucleotide bases and codons consisted of doublets (two nucleotides) instead of triplets?
  a. 10
  b. 25
  c. 32
  d. 125

FILL IN THE BLANKS: *Complete the table of DNA codons, mRNA codons, anticodons, and corresponding amino acids. Use the following portion of the genetic code to help you answer these questions.*

CODON	AMINO ACID	CODON	AMINO ACID
AUG	Methionine	GGG	Glysine
AAG	Lysine	GCA	Alanine
CCA	Proline	UGU	Cysteine
GUC	Valine	UAG	Stop

DNA codon			TTC	
mRNA codon				UAG
Anticodon		CAG		
Amino acid	methionine			

CHAPTER 17

MICROBIAL MODELS: THE GENETICS OF VIRUSES AND BACTERIA

FRAMEWORK

The concept maps below outline the major genetic characteristics of viruses and bacteria.

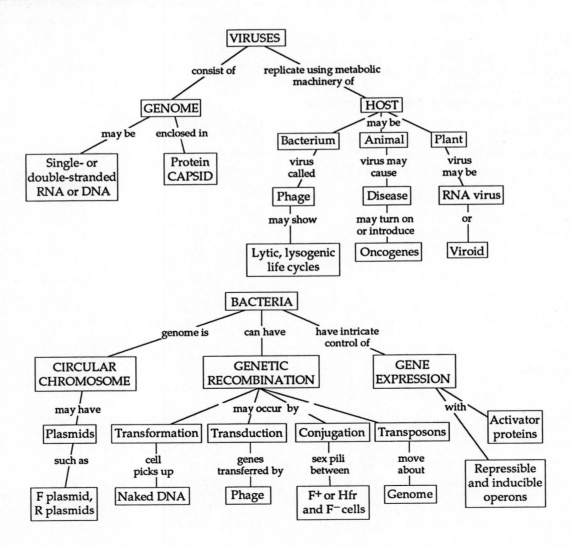

CHAPTER SUMMARY

Molecular biology originated in the laboratories of microbiologists. The study of viruses and bacteria has provided information about the molecular genetics of all organisms, an appreciation for the special genetic features of microbes, an understanding of how viruses and bacteria cause disease, and new powerful techniques of manipulating genes that have had an immense impact on basic research and biotechnology.

The Discovery of Viruses

The search for the cause of tobacco mosaic disease led to the discovery of viruses. Mayer, in 1883, found that, although he could not find any microbe, he could transmit the disease by spraying sap from an infected plant onto a healthy plant. He concluded that the disease was caused by an unusually small bacteria. Ivanowsky found that infected sap passed through a filter designed to remove bacteria still transmitted the disease. The disease could not be caused by a bacterial toxin that passed through the filter, because the infectious agent in the filtered sap was able to reproduce in the plant and then infect other plants. Unlike bacteria, the infectious agent could not be cultivated on nutrient media, nor was it inactivated by alcohol. In 1935, Stanley crystallized the infectious particle, now known as tobacco mosaic virus (TMV). Since that time, many viruses have been seen with the electron microscope.

Viral Structure and Replication: An Overview

A virus is an infectious particle that consists simply of nucleic acid enclosed in a protein shell.

Viral Genomes Viral genomes may be single- or double-stranded DNA, or single- or double-stranded RNA. As few as four or as many as several hundred genes may be contained on a single linear or circular nucleic acid molecule.

Capsids and Envelopes The *capsid*, or protein shell is built from a large number of often identical protein subunits (capsomeres) and may be rod-shaped (helical), polyhedral, or more complex in shape. Membranous *envelopes*, derived from membranes of the host cell but also including viral proteins and glycoproteins, may cloak the capsids of viruses found in animals.

Complex capsids are found among *bacteriophages*, the viruses that infect bacteria. The seven phages that infect the bacterium *Escherichia coli* were the first to be studied. The three phages—T2, T4, and T6—have a similar capsid structure consisting of an icosahedral (20-sided) head and a protein tailpiece with tail fibers for attaching to a bacterium.

How Viruses Replicate: Viral Infection Viruses are obligate intracellular parasites that lack metabolic enzymes and other equipment needed to express their genes and reproduce. Each virus type has a limited *host range* of cells that it can infect. This specificity is due to proteins on the outside of the virus that recognize only certain receptor molecules on the host cell surface.

Once the viral genome enters the host cell, the cell's enzymes, nucleotides, amino acids, ribosomes, ATP and other resources are used to make copies of the viral genome and capsid proteins. DNA viruses use host DNA polymerases to copy their genome, whereas RNA viruses bring their own enzymes for replicating their RNA genome.

After replication, viral nucleic acid and capsid proteins spontaneously assemble to form new virus particles within the host cell, a process called *self-assembly*. Hundreds or thousands of newly formed viruses are released, often destroying the host cell in the process.

Bacterial Viruses

Bacteriophages are the best understood of viruses. Double-stranded DNA viruses can reproduce by the lytic and lysogenic cycles.

The Lytic Cycle A replication cycle of a virus that culminates in death of the host cell is known as a *lytic cycle*. *Virulent* phages cause their host cells to lyse and release newly-produced phages.

The T4 phage uses its tail fibers to stick to a receptor site on the surface of an *E. coli* cell. The sheath of the tail contracts using energy from ATP stored in the tailpiece, and thrusts its viral DNA into the cell, leaving the empty capsid attached to the outside of the cell. The *E. coli* cell begins to transcribe and translate phage genes, one of which codes for an enzyme that chops up host cell DNA. Nucleotides from the degraded bacterial DNA are used to create viral DNA. Capsid proteins are made and assembled into phage tails, tail fibers, and polyhedral heads. The viral components assemble into 100 to 200 phage particles that are released after a lysozyme is manufactured that digests the bacterial cell wall.

The lytic cycle takes only 20 to 30 minutes at 37° C, during which time the T4 population increases more than a 100-fold. Clear holes in a growing bacterial culture indicate the lysis of bacterial cells by successive generations of a single phage particle.

Bacteria defend against viral infection by mutations that change their receptor sites and by producing *restriction enzymes* that chop up viral DNA once it enters the cell. Bacterial hosts and their viral parasites evolve in response to each other's new defenses and offenses.

The Lysogenic Cycle *Temperate phages* are viruses that reproduce without killing their host. In this *lysogenic cycle*, viral DNA inserts into the bacterial chromosome and is replicated when the bacteria reproduces.

Some viruses, including the phage lambda (λ) can enter either a lytic cycle or a lysogenic cycle. In the lysogenic cycle, most of the genes of the inserted phage genome, known as a *prophage*, are inactive. A *repressor protein* gene keeps the other genes switched off. Reproduction of the host cell replicates the prophage along with the bacterial DNA. The prophage can become virulent if it is excised from the bacterial chromosome, usually in response to environmental stimuli, and starts the lytic cycle.

Expression of prophage genes may cause a change in the bacterial phenotype. Several disease-causing bacteria, including those causing diptheria and botulism, would be harmless except for the expression of prophage genes that code for toxins.

Animal Viruses

Diverse Reproductive Cycles of Animal Viruses A membranous envelope surrounding the capsid is present in several groups of animal viruses. Glycoproteins extending from the viral membrane attach to receptor sites on the host cell plasma membrane and the two membranes fuse, transporting the capsid into the cell. The viral genome replicates using host cell ribosomes for protein synthesis. In some viruses, glycoproteins synthesized by ER are deposited in patches in the plasma membrane where new viruses bud off within an envelope derived from the host's plasma membrane.

The envelopes of herpesviruses come from the host nuclear membrane. The herpesvirus' double-stranded DNA can integrate into the cell's genome as a *provirus*, similar to a bacterial prophage, and result in recurring herpes infections in times of stress.

Viruses are classified based on the relationship of their genome to the messenger RNA that is translated into viral proteins. All classes of viral genomes are found in animal viruses. In the complicated reproductive cycle of *retroviruses*, the viral RNA genome is transcribed into DNA by a viral enzyme, *reverse transcriptase*. This viral DNA is then integrated into a chromosome, where it is transcribed by the host cell into viral RNA, which acts both as new viral genome and template for viral proteins. *HIV (human immunodeficiency virus)* is a retrovirus that causes *AIDS (acquired immunodeficiency syndrome)*.

Viral Diseases in Animals The symptoms of a viral infection may be caused by toxins produced by infected cells, toxic components of the viruses themselves, cells killed or damaged by the virus, or the body's defense mechanisms fighting the infection. The ability of infected tissue to regenerate or repair itself may determine the severity of the damage from a viral infection.

Vaccines are harmless derivatives of pathogens that induce the immune system to react more vigorously against the actual disease agent. Jenner experimented in 1796 with using the cowpox virus to vaccinate against smallpox. Vaccinations have greatly reduced the incidence of many viral diseases.

Unlike bacteria, viruses use the host's cellular machinery to replicate, and few drugs have been found to treat or cure viral infections. Some antiviral drugs effective against a number of human viruses are analogues of purine nucleosides and thus interfere with viral nucleic acid synthesis.

Viruses and Cancer Viruses that can cause cancer in animals are called *tumor viruses*. When tumor viruses infect cells growing in tissue culture, the cells are transformed—they assume rounded shapes and lose the contact inhibition regulation of growth. Viral nucleic acid becomes permanently integrated into host cell DNA. There is strong evidence linking viruses, such as the hepatitis B, Epstein-Barr, papiloma viruses and the retrovirus HTLV-I, to certain types of human cancer.

Oncogenes are genes responsible for triggering cancerous transformation in cells. Surprisingly, these genes have been found not only in tumor viruses but also within the genomes of normal cells of many species. These genes generally code for cellular growth factors or the receptor proteins for growth factors. Some tumor viruses lack oncogenes but transform cells simply by turning on the cells' oncogenes. Carcinogens may also act by turning on cellular oncogenes.

Plant Viruses and Viroids

Most plant viruses are RNA viruses. Plant viral diseases may spread through vertical transmission from a parent plant or through horizontal transmission from an external source. Plant injuries increase susceptibility to viral infections, and insects can act as carriers or vectors of viruses.

Viral particles spread easily through the plasmodesmata, the cytoplasmic connections between plant cells. Reducing the spread of disease and breeding resistant varieties are the best preventions for plant viral infections.

Viroids are very small molecules of naked RNA that can disrupt the metabolism of a plant cell and severely stunt plant growth. Viroid RNA has been found to be very similar to sequences of self-excising introns in some eukaryotic rRNA genes. It is likely that viroids disrupt the regulatory control of cellular genes.

The Evolutionary Origin of Viruses

Viruses inhabit a gray area between life and non-life—they are inert molecules containing a genetic program that can be expressed only within living cells.

The genomes of viruses are more similar to those of their host cells than to the genomes of viruses infecting other hosts. Viruses may have evolved from fragments of cellular nucleic acids that moved from one cell to another and eventually evolved special packaging. Sources of viral genomes may have been *plasmids*, self-replicating circles of DNA found in bacteria and yeast, and *transposons*, segments of DNA that can change location on a chromosome.

Bacterial Genomes: Replication and Mutation

Bacteria are a valuable organisms for the study of molecular genetics. Their single, circular chromosome is simpler in structure than eukaryotic chromosomes, which have more associated proteins and a thousand times more DNA. The bacterial double-stranded DNA molecule carries about 3000 genes and is tightly packed into a region of the cell called the *nucleoid*. Many bacteria also have plasmids, small rings of DNA with a few genes.

Replication of the bacterial chromosome proceeds bidirectionally from a single origin prior to binary fission. Most bacteria in a colony are genetically identical. Mutations, although statistically rare (one in every 1×10^{-7} cell divisions) produce a great deal of genetic diversity because bacteria can reproduce as rapidly as once every 20 minutes.

Genetic Recombination and Gene Transfer in Bacteria

Genetic recombination, the combining of genetic material from two individuals, also adds to the genetic diversity of bacterial populations. Evidence that genetic recombination occurs in bacteria is provided by growing two mutant stains of *E. coli*, each of which is unable to produce a particular amino acid, together on a complete culture medium. When later transferred to minimal media, numerous colonies grow that are now able to synthesize both amino acids. The cells must have combined genes from both strains. Genetic recombination in bacteria involves the mechanisms of transformation, transduction, and conjugation rather than the eukaryotic mechanisms of meiosis and fertilization.

Transformation   Some bacteria can take up segments of naked DNA in a process called *transformation*. When the foreign DNA is integrated into the bacterial chromosome by an exchange similar to crossing-over, a new combination of genes is produced. Many bacteria have surface proteins specialized for uptake of DNA. *E. coli* can be artificially induced to take up foreign DNA, a procedure important to biotechnology.

Transduction   Bacteriophages can transfer genes from one bacterium to another by *generalized transduction*, when a random piece of host DNA is accidentally packaged within a phage capsid and introduced into a new bacterium, or by *specialized transduction*, when bacterial genes adjacent to a prophage insertion site are excised with the prophage from the bacterial chromosome. By either method, recombination occurs when the newly introduced bacterial genetic material replaces the homologous region of a bacterial chromosome.

Conjugation and Plasmids   In *conjugation*, two cells temporarily join by an appendage called sex pili and transfer DNA from one to the other. The ability to form pili and donate DNA requires a plasmid called the *F plasmid*.

Plasmids replicate independently, either in concert with the bacterial chromosome or on their own schedule. Some plasmids, called *episomes*, can reversibly incorporate into the cell's chromosome. Temperate viruses are considered to be episomes. Viruses may have evolved from episomes. Plasmid genes are not required for reproduction and survival under normal conditions, but may confer advantages to bacteria in stressful environments.

Bacterial cells containing the F plasmid are called F+ (for fertility). The F plasmid replicates in synchrony with the bacterial chromosome, and F+ cells pass the trait to daughter cells. The plasmid also replicates before conjugation and is transferred to the recipient cell, changing it from an F− "female" cell to an F+ "male" cell.

The F plasmid is an episome. Cells in which it is inserted into the bacterial chromosome are called Hfr cells, for "high frequency of recombination." When these cells undergo conjugation, the F plasmid transfers an attached copy of the bacterial chromosome to the recipient cell. Movement may disrupt the mating, resulting in a partial transfer of genes. Recombination between the new DNA and the recipient cell's chromosome produces new genetic combinations in this bacterium and its offspring.

Hfr bacteria of a given strain always transfer chromosomal genes in the same order, beginning with the F plasmid. The loci of genes can be determined by experiments which interrupt mating at different time intervals. The order of genes can be determined by genetic analysis of the recombinants cultured from the different time samples.

R plasmids carry genes that code for antibiotic-destroying enzymes. R plasmids can be transferred to non-resistant cells during conjugation, creating the medical problem of antibiotic-resistant pathogens.

Transposons   Transposable genetic elements, or transposons, are mobile segments of DNA that can change location in a genome or replicate, thus remaining in its original position and also inserting in a new location. Unlike other forms of genetic recombination where alleles exchange between homologous regions, transposons can move genes to totally new areas.

Insertion sequences are the simplest transposons, consisting of only a transposase gene and inverted repeats, which serve as recognition sites for transposase. Transposase is the enzyme responsible for cutting and ligating DNA required for transposition. Other enzymes are also required, such as DNA polymerase that forms direct repeats at both ends of a transposon in its new site.

Insertion sequences cause mutations which often inactivate a gene. Even though insertion sequences transpose rarely in *E. coli*, they are a significant source of genetic variation due to the rapid proliferation of bacteria.

No known benefit for insertion sequences has been identified. They may be related to viruses, such as phage Mu, which causes a high mutation rate by transposing repeatedly to random insertion sites. Insertion sequences transpose less frequently than such temperate phages, however, and lack the genes for capsid proteins that would facilitate an extracellular stage.

Complex transposons contain other genes between two insertion sequences. They have been shown to confer selective advantage by moving beneficial genes, such as those for antibiotic resistance, about in the bacterial genome.

Wandering genes were first described by McClintock from her corn breeding experiments in the 1940s and 50s. Transposons have since been found to be important components of eukaryotic genomes.

Control of Gene Expression in Prokaryotes

Genetic variation resulting from mutation and recombination makes natural selection possible as populations of bacteria adapt to environmental conditions. An individual cell, however, is able to respond to environmental conditions by controlling enzyme activity through feedback inhibition or enzyme synthesis through regulation of gene expression. Regulated genes can be switched on and off as metabolic needs change. The operon model for gene regulation was first described by Jacob and Monod in 1961.

Operons: The Basic Concept   Operons are clusters of genes on a bacterial chromosome that are controlled by a single promoter and operator. The *structural genes* (genes that code for polypeptides) within an operon may code for the different enzymes of a single metabolic pathway. The *trp* (tryptophan) operon of *E. coli* contains five structural genes for the enzymes of the tryptophan pathway. Because a single promoter region, where RNA polymerase binds, controls all the genes of an operon, a large mRNA molecule will be produced with the transcript of all the enzymes for the pathway.

An *operator* is a segment of DNA between the promoter region and the structural genes that controls the access of RNA polymerase to the structural genes. Operators are normally "on." A *repressor* is a protein that binds to a specific operator, blocking attachment of RNA polymerase and thus turning the operator "off."

Regulatory genes code for repressor proteins and usually produce them at a slow rate. The activity of the repressor protein is determined by the presence or absence of a metabolite (a precursor, intermediate, or end product of a metabolic pathway). In the *trp* operon, tryptophan is a *corepressor* that binds to the repressor protein, changing its conformation into its active state, which has a high affinity for the operator and switches off the *trp* operon. Thus, rising tryptophan levels in the cell signal the shutdown of the *trp* operon. Should the tryptophan concentration fall, repressor proteins are no longer bound with tryptophan, the operator is no longer repressed, RNA polymerase attaches to the promoter, and mRNA for tryptophan synthesis is produced.

Repressible vs. Inducible Enzymes: Two Types of Negative Regulation The synthesis of *repressible enzymes*, such as the enzymes of the tryptophan pathway, is inhibited by a metabolite. The synthesis of *inducible enzymes* is stimulated by a metabolite. The *lac* operon, controlling lactose metabolism in *E. coli*, is an inducible operon that contains three structural genes. One of these genes codes for ß-galactosidase, an enzyme that cleaves lactose. Unlike the *trp* repressor, which is inactive until bonded with its corepressor, the *lac* repressor is innately active, binding to the *lac* operator and switching off the operon. Allolactose, an isomer of lactose, acts as an *inducer* that binds to and inactivates the repressor protein, so that the operon can be transcribed.

Repressible enzymes are usually found in anabolic pathways, where the pathway's end product serves to switch off enzyme synthesis and prevent overproduction of metabolic products. Genes for repressible enzymes are switched on until the repressor is activated. Inducible enzymes are found in catabolic pathways, where the presence of nutrient molecules stimulates the production of enzymes necessary for their breakdown. Genes for inducible enzymes are switched off until a metabolite inactivates the repressor. Both types of regulation are examples of negative control: they involve a repressor protein that, when active, binds with the operator and blocks transcription.

CAP: An Example of Positive Gene Regulation With a regulatory system that uses positive control, an activator molecule interacts directly with the genome to turn on transcription.

E. coli preferentially use glucose as their source of energy and carbon skeletons. Most of the enzymes for glucose catabolism are coded for by *constitutive genes*, genes that are continuously transcribed. Should glucose levels fall, transcription of operons for other catabolic pathways can be increased through the action of *catabolite activating protein (CAP)*. Cyclic AMP (cAMP), derived from ATP, accumulates in the cell when glucose is missing and binds with CAP. The cAMP-CAP complex attaches to the promoter region and stimulates transcription by increasing the promoter's ability to associate with RNA polymerase. A low level of cAMP in the cell signifies that sufficient glucose is present for an energy source; the cAMP-CAP complex disassociates; the activating protein releases from the promoter; and transcription of the operon slows down.

The regulation of the *lac* operon includes negative control by the repressor protein that is inactivated by the presence of lactose, and positive control by CAP that functions when complexed with cAMP. *E. coli* is able to control its gene expression and conserve its RNA and protein synthesis depending upon the presence of glucose and the availability of catabolites (such as lactose) when glucose supply is limited.

STRUCTURE YOUR KNOWLEDGE

1. Create a concept map or diagram that describes the lytic and lysogenic cycles of bacteriophage.

2. In the following diagram of an operon for repressible enzymes located on a stretch of a bacterial chromosome, identify components a through i.

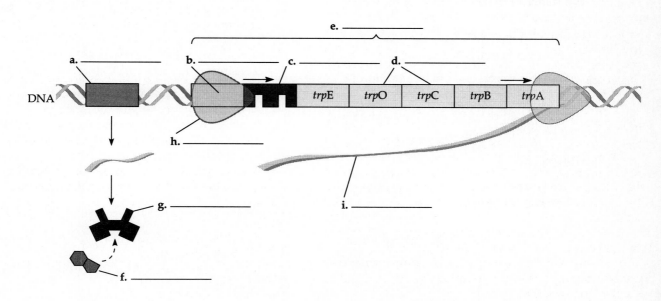

3. Develop a concept map to illustrate your understanding of the mechanisms by which bacteria regulate their gene expression in response to varying metabolic needs. Distinguish repressible and inducible operons, which are both examples of negative control, and activator proteins, which illustrate positive control of gene expression.

TEST YOUR KNOWLEDGE

MATCHING: *Match the term with its description.*

1. _____ viral enzyme that uses RNA as a template for DNA synthesis
2. _____ plasmid with genes for antibiotic resistance
3. _____ bacterial enzyme that chops up foreign DNA
4. _____ can replicate independently in cell or incorporate into bacterial chromosome
5. _____ transfer of genes through bridge joining two bacteria
6. _____ enzyme that cuts and ligates DNA at inverted repeats
7. _____ uptake of naked DNA by bacterial cell
8. _____ virus that reproduces by lytic cycle
9. _____ injection of host cell DNA incorporated into phage capsid into a new cell
10. _____ virus that reproduces by lysogenic cycle

A. conjugation

B. episome

C. reverse transcriptase

D. lysozyme

E. virulent virus

F. F plasmid

G. R plasmid

H. restriction enzyme

I. temperate virus

J. transposon

K. transduction

L. transformation

M. transposase

MULTIPLE CHOICE: *Choose the one best answer.*

1. The study of the genetics of viruses and bacteria has done all the following *except*
   a. provide information on the molecular biology of all organisms.
   b. illuminate the sexual reproductive cycles of viruses.
   c. develop new techniques for manipulating genes.
   d. develop an understanding of the causes of cancer.

2. Beijerinck concluded that the cause of tobacco mosaic disease was not a filterable toxin because
   a. the infectious agent could not be cultivated on nutrient media.
   b. a plant sprayed with filtered sap would develop the disease.
   c. the infectious agent could be crystallized.
   d. the infectious agent reproduced in a plant infected with filtered sap.

3. Viral genomes may be any of the following *except*
   a. several molecules of single-stranded DNA.
   b. double-stranded RNA.
   c. a circular DNA molecule.
   d. a linear single-stranded RNA molecule.

4. Retroviruses have a gene for reverse transcriptase that
   a. uses viral RNA as a template for making complementary RNA strands.
   b. uses viral RNA as a template for DNA synthesis.
   c. destroys the host cell RNA.
   d. translates RNA into proteins.

5. Virus particles are formed from capsid proteins and nucleic acid molecules
   a. spontaneously by self-assembly.
   b. at the direction of viral enzymes.
   c. by using host cell enzymes.
   d. using energy from ATP stored in the tailpiece.

6. The study of lambda, a phage of *E. coli*
   a. led to the discovery of the *lac* operon.
   b. showed that DNA was the genetic material.
   c. led to the discovery of the lytic and lysogenic replication cycles.
   d. revealed the structure of an icosahedral head and tail fibers attached to the tailpiece.

7. Vertical transmission of a plant viral disease may involve
   a. the movement of viral particles through the plasmodesmata.
   b. the inheritance of an infection from a parent plant.

c. bacteriophage as vectors.

d. insects as vectors carrying viral particles between plants.

8. Bacteria defend against viral infection through the action of
   a. antibiotics that they produce.
   b. restriction enzymes that chop up foreign DNA.
   c. their R plasmids.
   d. reverse transcriptase.

9. Drugs that are effective in treating viral infections
   a. induce the body to produce antibodies.
   b. inhibit the action of viral ribosomes.
   c. interfere with the synthesis of viral nucleic acid.
   d. change the cell-recognition sites on the host cell.

10. Which of the following is *not* true of tumor viruses
    a. They can integrate viral nucleic acid into the host cell genome.
    b. They transform cells growing in tissue culture into rounded cells that lose their contact inhibition.
    c. They may not always contain oncogenes but may turn on the host cell's oncogenes.
    d. They are transferred by bacteriophage.

11. The herpesvirus
    a. acts as a provirus when its DNA becomes incorporated into the host cell's genome.
    b. is a retrovirus that uses reverse transcriptase to transcribe DNA from its RNA genome.
    c. has an envelope derived from the host cell's plasma membrane.
    d. has been linked to HIV, the virus that causes AIDS.

12. The F plasmid of bacteria
    a. carries genes for sex pili production.
    b. may be transferred and convert a cell from F⁻ to F⁺.
    c. is an episome.
    d. all of the above

13. Tiny molecules of naked RNA are
    a. retroviruses.
    b. transposons.
    c. viroids.
    d. episomes.

14. Regulatory genes are genes that
    a. code for repressor proteins.
    b. are usually constitutive.
    c. are not contained in the operon they control.
    d. all of the above

15. Inducible enzymes
    a. are usually involved in anabolic pathways.

b. are produced when a metabolite inactivates the repressor protein.
c. are produced when an activator molecule enhances the attachment of RNA polymerase with the promoter.
d. all of the above

16. In *E. coli*, tryptophan switches off the *trp* operon by
    a. binding to the promoter.
    b. binding to the operator.
    c. binding to the repressor and increasing the latter's affinity for the operator.
    d. inactivating the repressor protein.

17. A mutation that renders the regulatory gene of a repressible operon inactive would result in
    a. continuous transcription of the structural genes.
    b. inhibition of transcription of the structural genes.
    c. irreversible binding of the repressor to the promoter.
    d. no difference in transcription rate when an activator protein was present.

18. The role of the specific metabolite in an inducible operon is to
    a. bind to the promoter region and increase the affinity of RNA polymerase for the promoter.
    b. bind to the operator region and block the attachment of RNA polymerase to the promoter.
    c. bind to the repressor protein and inactivate it.
    d. increase the production of inactive repressor proteins.

MATCHING: *Match these components of the lac operon with their functions:*

1. _____ β-galactosidase
2. _____ cAMP-CAP complex
3. _____ lactose
4. _____ operator
5. _____ promoter
6. _____ regulator gene
7. _____ repressor
8. _____ structural gene

A. is inactivated when attached to lactose
B. codes for synthesis of repressor
C. hydrolyzes lactose
D. stimulates gene expression
E. repressor attaches here
F. RNA polymerase attaches here
G. acts as inducer that inactivates repressor
H. codes for an enzyme

CHAPTER 18

GENOME ORGANIZATION AND EXPRESSION IN EUKARYOTES

FRAMEWORK

The lengthy nucleotide sequences of eukaryotic chromosomes are packed tightly, first into nucleosomes, which are a complex of histone proteins and the DNA helix; then these are coiled into 30 nm chromatin fibers, which are looped and condensed into chromatin. A typical gene may include an enhancer, a promoter region, and exons, which code for protein or RNA products, separated by large noncoding sequences called introns. The genomes of an organism's cells may vary due to gene amplification or loss and rearrangements of DNA sequences.

Control of gene expression occurs at various places in the steps from transcription to post-transcription to translation. Chemical signals, such as steroid hormones, are important in gene regulation. Cancer develops when genes that normally control cell development and division or suppress tumors transform into oncogenes.

CHAPTER SUMMARY

Eukaryotic cells in multicellular organisms control the expression of their genes in response to their external and internal environments and also to direct the *cellular differentiation* necessary to create specialized cells. The sheer size of a typical eukaryotic genome, along with the more complex structures of chromosomes, adds to the challenge of gene replication and expression.

In both prokaryotes and eukaryotes, gene activity is regulated by DNA-binding proteins. In eukaryotes, the greater complexity of chromosome structure, gene organization, and cellular structure provides more avenues for control of gene activity than in prokaryotes.

Genome Organization at the Microscopic Level

Each chromosome consists of a single, extremely long DNA molecule complexed with a large amount of protein. During interphase, chromatin is extended and intertwined, whereas discrete, condensed chromosomes appear during mitosis. The DNA in the chromosomes is compacted by multiple levels of folding.

Nucleosomes, or "Beads on a String" *Histones* are small, positively charged proteins that bind tightly to the negatively charged DNA to make up chromatin. Partially unfolded chromatin appears as a string of beads, each bead a *nucleosome* consisting of the DNA helix wound around a protein core of four pairs of different histone molecules. A fifth histone, H1, may attach to the outside of the "bead." Nucleosomes may limit the access of transcription proteins to DNA.

Higher Levels of DNA Packing The 30 nm chromatin fiber, or 30 nm solenoid, is a tightly coiled cylinder of nucleosomes organized with the aid of histone H1. The looped domain is a loop of the 30 nm chromatin fiber. Looped domains may also coil and fold, further compacting the chromatin in the metaphase chromosome. A framework of nonhistone proteins may anchor the loops.

Interphase chromatin consists of nucleosomes, 30 nm fiber, and looped domains that may be attached to the nuclear envelope by protein scaffolding. Chromatin visible with the light microscope during interphase is called *heterochromatin*, a highly compacted DNA that is not actively transcribed. One of the two X chromosomes in female mammalian somatic cells is present as a highly compacted Barr body, composed of heterochromatin. The more open form of interphase chromatin is called *euchromatin*.

Genome Organization at the Molecular Level

Unlike the prokaryote genome, in which most of the DNA contains uninterrupted codes for proteins (or RNA), the eukaryote genome has multiple copies of some DNA sequences and contains mostly noncoding DNA, long stretches of which interrupt coding sequences.

Repetitive Sequences Highly repetitive, short sequences of nucleotides make up 10% to 25% of the DNA in multicellular eukaryotes. Called *satellite DNA* because its base composition is sufficiently different from the rest of the cell's DNA to be able to isolate it by ultracentrifugation, this DNA is located primarily at the centromeres and appears to serve a structural rather than a genetic role in the cell.

Multigene Families Most genes are present in single copies. *Multigene families* are collections (often clustered but occasionally scattered in the genome) of similar or identical genes that probably evolved from a single ancestral gene. Identical genes usually code for RNA, not protein. The identical genes coding for the major rRNA molecules are arranged one after another, hundreds to thousands of times, forming huge tandem arrays that enable cells that are actively synthesizing proteins to produce the millions of ribosomes they need.

Examples of multigene families of nonidentical genes are the two families of genes that code for the α and ß polypeptide chains of hemoglobin. Different versions of each chain are clustered together and are expressed at the appropriate time during development.

Families of identical genes most likely arose by repeated gene duplication, a phenomenon that seems to occur due to mistakes in DNA replication and recombination. Nonidentical gene families probably arose from mutations in duplicated genes. *Pseudogenes*, sequences of DNA similar to real genes but lacking the signals for gene expression, may be present within gene families. The globin pseudogenes, which lack introns and have poly-A tails, may have been transcribed from RNA by reverse transcription and moved about the genome by transposition.

Organization of a Typical Eukaryotic Gene: A Review A typical protein-encoding gene consists of a promoter sequence, where RNA polymerase attaches, and a sequence of introns interspersed among the coding exons. After transcription, RNA processing removes the introns, and adds the guanosine cap at the 5' end and a poly-A tail at the 3' end. *Enhancers* are non-coding sequences, which may be located at some distance from the promoter, that influence transcription.

Arrangement of Coordinately Controlled Genes Unlike prokaryotic genes, which are often organized into operons that are controlled by the same regulatory sites and transcribed together, eukaryotic genes coding for enzymes in the same metabolic pathway are often scattered throughout the chromosomes and separately transcribed. The integrated control of these scattered genes may involve a specific nucleotide sequence associated with each related gene that serves as a recognition signal for a regulatory protein, which then transcribes or represses all the genes in synchrony.

Genome Plasticity

Chemical changes, as well as physical changes, in an organism's genome influence gene expression.

Gene Amplification and Selective Gene Loss Multiple copies of rRNA genes are present in the genomes of most eukaryotic cells. In an amphibian ovum, a million or more extra copies of the rRNA genes are synthesized and contained as extrachromosomal circles of DNA, enabling the developing egg to make the huge numbers of ribosomes needed for protein synthesis after it becomes fertilized. Selective synthesis of extra DNA is called *gene amplification*.

In certain insects, genes may be selectively lost and whole or parts of chromosomes may be eliminated from some cells early in development.

DNA Methylation In a process called *DNA methylation*, methyl groups ($-CH_3$) are added to DNA bases (usually cytosine) after DNA synthesis. Genes are more heavily methylated in cells in which they are not expressed; drugs that inhibit methylation can induce gene reactivation, even in Barr bodies. DNA methylation may be a long-term control mechanism of gene expression.

Rearrangements in the Genome Transposons may affect gene expression if they move into the middle of a coding sequence of a gene or insert into a sequence that regulates transcription. A transposon carrying a gene may activate that gene if it inserts downstream from a promoter.

The two mating types found in yeasts are determined by the presence of either the alpha (α) or the *a* allele at the mating-type locus. Yeast cells can change their mating type by a *cassette mechanism* which switches in a copy of the other allele. Unexpressed, "silent" copies of both alleles are located elsewhere on the same chromosome.

Immunoglobulin genes permanently rearrange DNA segments during cellular differentiation. B-lymphocytes are white blood cells that produce highly-specific proteins, known as antibodies or *immunoglobulins*, that recognize and bind to viruses, bacteria, and other invading molecules. Each differentiated B-lymphocyte and its descendants produce one specific antibody. During differentiation, the antibody gene is pieced together from a number of different, widely separated, DNA sequences. Much of the variation in antibodies arises from different combinations of variable and constant regions of the four polypeptides that form the complete antibody molecule. The formation of millions of different kinds of antibody genes shows how a chemical rearrangement of DNA yields somatic cells with distinctive genomes.

Control of Gene Expression

Transcriptional Control of Gene Expression Both prokaryotic and eukaryotic cells have transcription factors that bind to the promoter region and function in concert with RNA polymerase to begin transcription. In addition, eukaryotic genomes have transcription factors that stimulate transcription of specific genes when they bind to enhancer sites, which may be thousands of nucleotides away from the promoter and coding sequence. A hairpin loop in the DNA may bring the transcription factor that is attached to the enhancer into contact with the transcription factors and RNA polymerase at the promoter. The hundred or so transcription factors that have been identified fall into three classes of proteins based on the structure of their functional region.

Post-Transcriptional Control of Gene Expression Gene expression, measured by the amount and types of proteins that are made, can be blocked or stimulated at any post-transcriptional step:

In the nucleus, RNA processing involves the removal of the coded intron DNA segments and the splicing together of exons, as well as the addition of a 5' cap and a poly-A tail. The exit of mRNA from the nucleus, probably through a nuclear pore, may also present an opportunity for controlling gene expression, but little is known about regulation at this step of the process.

Prokaryotic mRNA molecules degrade after only a few minutes of activity; prokaryotic organisms, therefore, can vary patterns of protein synthesis rapidly in response to environmental changes. Eukaryotic mRNA, in contrast, can last hours or even weeks; the length of time before they are degraded by cellular enzymes relates to the quantity of protein synthesis they can direct.

The translation of certain eukaryotic mRNA can be delayed until initiated by a control signal. A great deal of mRNA is synthesized and stored in egg cells, waiting to be translated during the period of active protein synthesis following fertilization.

Protein initiation factors may be necessary for translation to begin within eukaryotic cells; therefore they serve as control factors for gene expression. Following translation, polypeptides are often cleaved or chemical groups added to yield an active protein. Selective degradation of proteins also serves as a control mechanisms in the cell.

The Role of Small Molecules in Regulating Eukaryotic Gene Expression

As in prokaryotes, small organic molecules combine with regulatory proteins to influence transcription in eukaryotes. Steroid hormones affect gene transcription in target cells.

Chromosome Puffs: Evidence for the Regulatory Role of Steroid Hormones in Insects Chromosome puffs appear along the polytene chromosomes in the salivary glands of *Drosophila* larvae. These DNA loops may make the DNA more accessible to RNA polymerase. Autoradiography has shown that the puffs correspond to regions of active RNA synthesis. The location of the puffs shifts, especially during critical times in development, indicating that genes are turned on and off selectively. These changes in puff patterns can be induced by ecdysone, the steroid hormone that initiates molting, showing that gene regulation is responsive to chemical signals.

Action of Steroid Hormones in Vertebrates A steroid hormone, which is lipid soluble, diffuses across the plasma membrane and enters the nucleus, where it binds with a receptor protein and releases the inhibitory protein that was associated with the receptor. The now-activated receptor protein can bind to an enhancer region and initiate transcription of a steroid-responsive gene. The steroid serves as a chemical signal to switch on specific genes.

Five key points summarize the control of gene expression in eukaryotes: Different cell types of a multicellular organism express different genes; physical and chemical rearrangements of the genome make certain genes available or unavailable for expression; regulation of gene expression may occur at each step from gene to functional protein; control of transcription is especially important, with the binding of regulatory proteins to enhancer sequences activating transcription of specific genes; and some of these DNA-binding proteins are responsive to hormones and other chemical cues.

Gene Expression and Cancer

Cancer may be caused by physical agents or by chemicals called *carcinogens,* all of which cause mutations in certain genes. *Oncogenes,* or cancer-causing genes, were first found in certain RNA viruses and later recognized in the genomes of humans and other animals. These cellular *proto-oncogenes* code for proteins that control cell growth, division, and adhesion. When converted to oncogenes, they cause uncontrolled growth or changes in cell adhesion.

A gene that has an essential function in a normal cell may become an oncogene if it is present in more copies than normal (amplification); undergoes transposition or chromosomal translocation (both of which may separate it from its normal control regions); or has a mutated nucleotide sequence that creates a more active or resilient protein. Mutations in *tumor-suppressor genes* can contribute to the onset of cancer if a normal protein that prevents uncontrolled cell growth is no longer made.

More than one somatic mutation may be needed to produce a cancerous cell. In the case of colorectal cancer, three DNA changes need to occur: the activation of a cellular oncogene and the inactivation of two tumor-suppressor genes.

Virus-induced or virus-associated cancers are thought to account for 15% of human cancer cases. The virus might add an oncogene or affect a proto-oncogene or tumor-suppressor gene, but other environmentally-induced DNA changes are probably needed for cancer to develop.

STRUCTURE YOUR KNOWLEDGE

1. Eukaryotic chromosomes have a much more complex physical organization than does a prokaryotic chromosome. Label the following diagrams of the levels of chromatin packing.

2. In addition to the physical complexity of eukaryotic chromosomes, you have learned that they include much more than a simple sequence of

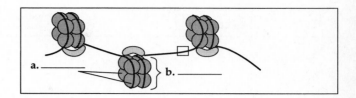

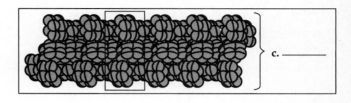

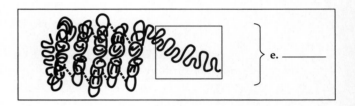

coding genes. Describe the molecular organization (types of nucleotide sequences) typical of a eukaryotic chromosome.

3. Fill in the following table to help you organize the major mechanisms that can regulate the expression of eukaryotic genes.

TEST YOUR KNOWLEDGE

MULTIPLE CHOICE: *Choose the one best answer.*

1. The control of gene expression is more complex in eukaryotic cells because
   a. chromosomes are contained in a nucleus.
   b. gene expression differentiates specialized cells and responds to the environment.
   c. the chromosomes are linear and more numerous.
   d. operons are controlled by more than one promoter region.

LEVEL OF CONTROL	TYPE OF CONTROL	EXAMPLES
Availability of genes		
Transcriptional control		
Post-transcriptional control		
Translational control		
Post-translational control		

2. Histones are
   a. small, positively charged proteins that bind tightly to DNA.
   b. small bodies in the nucleus involved in rRNA synthesis.
   c. basic units of DNA packing consisting of DNA wound around a protein core.
   d. repeating arrays of six nucleosomes organized around an H1 molecule.

3. Heterochromatin
   a. has a higher degree of packing than does euchromatin.
   b. is visible with the light microscope during interphase.
   c. is not actively involved in transcription.
   d. is all of the above.

4. DNA methylation of cytosine residues
   a. can be induced by drugs that reactivate genes.
   b. may contribute to the formation of the inactive Barr bodies found in female somatic cells.
   c. produces the promoter regions that specifically bind RNA polymerase.
   d. all of the above.

5. Chromosome puffs along the polytene chromosomes of *Drosophila*
   a. appear at specific sites during developmental stages and give visual evidence of selective gene activation.
   b. are looped domains, or folds in the chromatin fiber, containing genes that are expressed in a coordinated fashion.
   c. are regions of DNA methylation causing highly compacted and inactive areas on the chromosome.
   d. contain collections of multigene families.

6. Which of the following is *not* true of enhancers?
   a. They may be located thousands of nucleotides upstream from the genes they affect.
   b. When bound with transcription factors, they interact with the promoter region and other transcription factors to increase the activity of a gene.
   c. They may complex with steroid-activated receptor proteins and thus selectively activate specific genes at appropriate stages in development.
   d. They are located within the promoter region and, when complexed with a steroid or other small molecule, they release an inhibitory protein and thus make DNA more accessible to RNA polymerase.

7. Which of the following is *not* an example of post-transcriptional control of gene expression?
   a. mRNA stored in the cytoplasm needing a control signal to initiate translation
   b. the length of time mRNA lasts before it is degraded
   c. rRNA genes amplified in tandem arrays
   d. RNA processing before mRNA exits from the nucleus

8. Pseudogenes are
   a. tandem arrays of rRNA genes that enable actively synthesizing cells to create enough ribosomes.

b. highly repetitive short sequences found around the centromeres.

c. genes that can become oncogenes when induced by carcinogens.

d. sequences of DNA that are similar to real genes but lack signals for gene expression.

9. Which of the following is *not* true? Proto-oncogenes
   a. are genes that control cell growth and differentiation.
   b. may be activated by carcinogens.
   c. are introduced into the cell by viruses.
   d. are noncancerous.

10. A gene can develop into an oncogene when it
    a. is present in more copies than normal.
    b. undergoes a transposition or translocation that removes it from its control region.
    c. develops a mutation that creates a more active or resistant protein.
    d. does any of the above.

11. Gene amplification involves
    a. the production of loud genes.
    b. the enlargement of chromosomal areas undergoing transcription.
    c. the synthesis of extra copies of genes to meet particular metabolic needs.
    d. the transformation of genes into oncogenes.

12. Which of the following is *not* an example of the plasticity of the genome that may result from chemical changes in chromosomes?
    a. the compaction of chromosomes into heterochromatin
    b. the cassette mechanism for altering mating types in yeast
    c. selective gene loss, especially in certain insects
    d. the rearrangements of immunoglobulin genes in differentiating B lymphocytes

CHAPTER 19

DNA TECHNOLOGY

FRAMEWORK

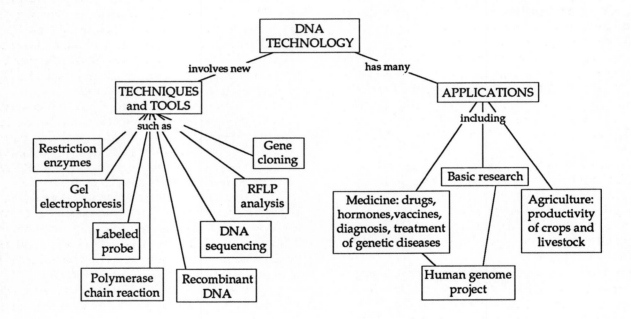

CHAPTER SUMMARY

Genetic engineering, the manipulation of genetic material for practical purposes, has begun an industrial revolution in *biotechnology*. This use of living organisms to manufacture desirable products dates back centuries, from the use of microorganisms to make cheese and wine and the selective breeding of plants and animals. But advances in *recombinant DNA* technology, along with new methods for manipulating and analyzing DNA, have resulted in hundreds of new products and major advances in basic knowledge of the organization and regulation of the eukaryotic genome. Genes

from different organisms and species are now routinely combined to form recombinant DNA and inserted into living cells in which these genes are expressed and their products studied or harvested.

Basic Strategies of Gene Manipulation and Analysis

Before 1975, the technology for altering genes involved finding natural mutants or creating new mutations with mutagenic radiation or chemicals and then propagating the genes through breeding. Selection of the desired mutants involved methods that would permit only the desired organism to survive and grow. Bacteria can transfer genes from one bacterial strain to another by the processes of transformation, conjugation, and transduction by phage. Geneticists learned to use these processes for the detailed molecular study of prokaryotic and phage genes in the 1950s and '60s. The study of eukaryotic genes has been made possible with recombinant DNA techniques.

Recombinant DNA and Gene Cloning Recombinant DNA was first made *in vitro* (outside of cells) by inserting DNA into bacterial plasmids. The plasmids are put back into bacterial cells where they replicate as the bacteria reproduce, thus "cloning" the gene of interest.

Restriction enzymes protect bacteria from foreign DNA by cutting up nonbacterial DNA in a process called *restriction*. Most restriction enzymes recognize short nucleotide sequences and cut at specific points within them. The cell protects its own DNA from restriction by methylating nucleotide bases within recognition sequences. Recognition sequences are symmetrical sequences of four to eight nucleotides running in opposite directions on the two strands, such as GAATTC on one strand and CTTAAG on the other. The restriction enzyme usually cuts phosphodiester bonds in a staggered way, between the same adjacent nucleotides on both strands, leaving "sticky ends" of short single-stranded sequences on both sides of the cut.

DNA from different sources can be recombined in the laboratory when the strands are cut by the same restriction enzyme and the complementary bases on the resulting sticky ends of the *restriction fragments* pair by hydrogen bonding and are then joined together by *DNA ligase*, which catalyzes the formation of phosphodiester bonds.

Restriction fragments can be separated on the basis of their size and electrical charge by *gel electrophoresis*. Due to the negative charge of their phosphate groups, nucleic acids migrate through the electric field produced in a thin slab of gel toward the positive electrode at a rate inversely proportional to their size. The band patterns of restriction fragments produced in the gel can be used to identify specific viral, plasmid, and chromosomal DNA. Individual fragments can be isolated from the gel and still retain their biological activity.

The plasmid method of gene cloning involves treating plasmids with a restriction enzyme that cuts the DNA ring and disrupts a gene whose activity is easily determined. The clipped plasmids are mixed with foreign DNA that has been treated with the same restriction enzyme and has complementary sticky ends. The sticky ends hydrogen-bond, and DNA ligase seals the recombinant molecules. The plasmids are introduced into bacterial cells by transformation and reproduce as

the bacteria develop clones of cells. The resulting replication of the inserted foreign genes is called *gene cloning*. Clones carrying the recombinant DNA are identifiable because they have lost the characteristic conferred by the disrupted gene and gained characteristics of the foreign DNA. Resistance to a particular antibiotic is often used as an easily identified characteristic.

Plasmids serve as *cloning vectors* by introducing recombinant DNA into bacterial cells by transformation. Bacteriophage can also serve as vectors. Recombinant phage DNA is created using restriction enzymes and DNA ligase to insert restriction fragments of foreign DNA into the viral genome. Inserted into a bacterial cell, the phage DNA replicates itself to form new phage particles, simultaneously cloning the inserted genes. These phages can then infect other bacterial cells and transmit the foreign DNA.

Recombinant DNA can also be cloned in eukaryotic cells such as yeast and animals cells growing in culture. Naked DNA can be taken up from the surrounding medium and become incorporated into chromosomal DNA or replicate itself as a plasmid. Yeast cells have plasmids, and recombinant plasmids have been engineered that combine yeast and bacterial DNA and can replicate in either type of cell.

Other means for introducing DNA into eukaryotic cells include *electroporation*, in which an electric pulse briefly opens holes in the plasma membrane through which DNA can enter. DNA can be injected into cells using microscopically thin needles or fired into plant cells on microscopic metal particles.

The genes used for cloning can be isolated from an organism by cutting its DNA into thousands of pieces with restriction enzymes and inserting them into plasmids or viral DNA. The collection of the huge number of DNA segments from a genome, derived from this shotgun approach, is called a *genomic library*.

Creating artificial, *complementary DNA (cDNA)* eliminates the problem caused by large introns that can make a eukaryotic gene too large to clone. Using reverse transcriptase from retroviruses, mRNA—which has no introns—is used as a template to produce cDNA. This cDNA will code for mRNA that can be more easily translated by bacterial cells, which lack RNA processing enzymes. Bacterial transcription also requires the attachment of a promoter and other transcription signals to the cDNA. A partial genomic library can be produced from the mRNA molecules found in a cell and contains only the genes that are expressed (transcribed) within the cell.

Finding a bacterial clone that contains the gene of interest within a huge genomic library is a difficult challenge. If the gene is translated into protein, the presence of the protein can be determined by its activity or structure (using antibodies). The gene itself can be detected using a radioactively labeled nucleic acid sequence called a *probe*, which has complementary sequences to segments of the gene. Once the desired clone is identified, it can be grown in culture and the gene of interest isolated in large quantities.

Large-scale, commercial production of gene products has been accomplished using bacteria to express eukaryotic genes, despite differences between prokaryotic and eukaryotic mechanisms for transcription and translation. Difficulties have been overcome by using plasmid vectors that produce many gene copies per cell; changing the gene's promoter to a highly active one; and attaching the eukaryotic gene to a bacterial gene produced normally in large quantities. When bacterial cells are engineered to secrete the protein product, the task of purification is simplified.

Yeasts present an excellent vehicle for recombinant DNA technology because of their ease of culture; ability to take up DNA by transformation, since they have plasmids; and capacity to express eukaryotic genes, since they are eukaryotic cells. Cultured animal and plant cells must be used in some genetic research and in commercial applications that require synthesis of complex products such as antibodies.

DNA Synthesis and Sequencing   Gene sequencing is made possible with restriction enzymes, which cut long DNA sequences into fragments, and gel electrophoresis, which can separate fragments that differ by only one nucleotide.

In the Sanger method of DNA sequencing, four samples of a single-stranded restriction fragment are incubated with radioactively labeled primer, DNA polymerase, the four nucleotides, and one of four modified nucleotides (dideoxy forms) that block further synthesis when they are incorporated into a growing DNA strand. The result is four sets of radioactive strands of varying lengths that can be separated by gel electrophoresis. The nucleotide sequence can be read from the sequence of bands produced in the gel.

Thousands of DNA sequences are being collected in computer data banks where they can be analyzed for genetic control elements and similarities among genes of different organisms, and automatically translated into amino acid sequences.

Amplifying DNA by the Polymerase Chain Reaction (PCR)   Polymerase chain reaction (PCR) is a technique developed in 1985 that has revolutionized molecular biology by being able to rapidly produce millions of copies of a section of DNA in vitro. A solution of DNA containing the region of interest is heated to separate the strands, incubated with specially synthesized primers that bind upstream from the target sequence, supplied

with the four nucleotides and DNA polymerase, and then heated again to repeat the process.

RFLP Analysis   *Restriction fragment length polymorphisms, RFLPs,* provide valuable markers for mapping chromosomes, diagnosing genetic diseases, and solving violent crimes. Restriction fragment length refers to the length of DNA fragments, shown by the pattern of bands produced by gel electrophoresis, that result when DNA is cut by a specific restriction enzyme. Fragment length variations or polymorphisms occur as a result of differences in base sequences that may delete or add *restriction sites* to homologous chromosomes. When these naturally occurring variations in DNA occur frequently in a population, geneticists can use them as reference points along a chromosome. Comparing patterns of inheritance of genetic diseases and RFLP markers has led to the mapping of several disease genes.

In RFLP analysis, DNA extracted from white blood cells is mixed with a restriction enzyme, and the resulting restriction fragments are separated by gel electrophoresis. Using a technique called *Southern blotting*, the DNA on the gel is denatured by heating and the single-stranded fragments are blotted onto a filter. The bands of interest are identified by using a radioactively labeled probe. Autoradiography reveals the fragments that have base-paired with the probe.

Applications of DNA Technology

Use of DNA Technology in Basic Research   New techniques in DNA research have facilitated the study of eukaryotic gene structure and function on a molecular level and contributed to research advances in almost all fields of biology.

Genes can be produced in large amounts and used as labeled probes to locate similar DNA segments in the same or other genomes, providing information on evolutionary relationships. Complementary DNA probes allow location and study of the natural form of the gene, including its regulatory and noncoding sequences.

DNA probes can map genes on eukaryotic chromosomes with an *in situ hybridization* technique. Labeled DNA probes base-pair with intact chromosomes on a microscope slide, and autoradiography and chromosome staining show the location of the gene.

Many molecules controlling cell metabolism and development are produced in such small quantities that they cannot be purified and characterized by normal biochemical techniques. The production of cDNA from mRNA molecules and its cloning in bacteria provides large enough quantities of these proteins to investigate their structures and functions.

The Human Genome Project There are four complementary components to the current effort to map the human genome: genetic (linkage) mapping, physical mapping, nucleotide sequencing, and analysis of genomes of other species. Locating genetic markers, including RFLPs, throughout the chromosomes will facilitate mapping other genes by testing for genetic linkages to known markers. Physical mapping involves determining the order of identifiable fragments of each chromosome. Sequencing the 3 billion nucleotide pairs of the haploid human genome may be the most time-intensive part of the project. The analysis of the genomes of other species will allow comparisons with the human genome. Information from the human genome project should contribute to the diagnosis, treatment, and prevention of genetic diseases, and provide insight into basic questions of molecular genetics and evolutionary biology.

Classical pedigree analysis of large families and many recently developed molecular techniques will contribute to the building of the human genetic map. The method called *chromosome walking* enables researchers to sequence the large number of restriction fragments produced from the very long eukaryotic chromosomes. Two samples of DNA are cut with two different restriction enzymes and libraries are produced from both sets of fragments. Starting with a gene or location that is already known, a probe is made that will then be used to search the other library for an overlapping fragment. A new probe is prepared from the end of that fragment, and the first library is searched for the next overlapping fragment.

Through polymerase chain reaction amplification of DNA from single sperm cells, the products of meiotic recombination can be directly analyzed and linkage maps can be developed based on the frequency of crossovers between genes.

Advances in DNA technology and computer software both will contribute to the success of the human genome project.

Medical Uses of DNA Technology PCR and labeled DNA probes are being used to identify difficult pathogens and to diagnose infectious diseases. DNA technology has led to the diagnosis of over 200 human genetic disorders. Genes have been cloned for a number of genetic diseases, making it possible to produce a probe that can locate the corresponding gene in DNA from individuals who are being tested. Restriction enzymes and gel electrophoresis are then used to compare the restriction fragment patterns formed from the sample DNA with patterns for the normal and mutant allele. Even if the gene has not yet been cloned, a disease gene may be diagnosed when closely linked with a RFLP marker. Comparisons of

blood samples within a family must be used to determine which variant of the RFLP marker is linked to the abnormal allele. Alleles for Huntington's disease have been detected in this manner.

Genetic engineering may provide the means for correcting genetic disorders in individuals by replacing or supplementing defective genes. New genes would be introduced into a few somatic cells of types that actively reproduce within the body. Thus the gene would be replicated when the cells are reinserted into the individual and the protein product of the introduced normal gene could then correct the biochemical defect. This type of gene therapy is currently being used successfully with several children with an immunodeficiency disease. The patient's own T lymphocytes are engineered to contain a normal ADA allele and then returned to the body.

Some of the technical problems involved with human gene therapy include how to get the proper control mechanisms to operate on the transferred gene, and when in development and into which tissues the gene should be introduced. The huge expense of gene therapy presents ethical and social questions concerning the availability of such treatments. The most difficult ethical question is whether germ cells should be treated to correct defects in future generations. Opponents fear that tampering with human genes will eventually lead to the practice of eugenics, the deliberate effort to control the genetic makeup of human populations.

Recombinant DNA techniques have been used to make large amounts of protein molecules from disease-causing viruses, bacteria, and other microbes, which can be used as vaccines if the protein subunit triggers an immune response against the pathogen. Genetic engineering methods can also be used to directly modify the genome of a pathogen so as to attenuate it (make it nonpathogenic). The vaccinia virus, which is the basis of the smallpox vaccine, has been modified to carry genes that induce immunity to other diseases.

Gene splicing has been used to produce hormones and proteins in bacteria. Insulin and human growth hormone were the first two polypeptide hormones made by recombinant DNA techniques to be approved for use in the United States. Genetically engineered human growth hormone is used to treat children born with hypopituitarism and may have other medical uses.

Two new approaches to fighting disease-causing viruses are under development. *Antisense nucleic acid* are single-stranded DNA or RNA molecules that would base-pair with and block the translation of viral RNA. Surface-receptor blockers are drugs that interfere with the binding of viral particles to receptors on cell membranes.

Forensic Uses of DNA Technology   Since the *DNA fingerprint* is unique for each individual (except identical twins), DNA technology using RFLP autoradiographs can be used in criminal cases. Forensic tests require only five or ten tiny regions of the genome that are known to be highly variable from one person to the next in order to identify victims and criminals.

Agricultural Uses of DNA Technology   Genetic engineering is providing means to improve the productivity of agricultural plants and animals. Vaccines, growth hormones, and antibodies are produced with recombinant DNA technology. *Transgenic organisms* containing genes from other species are being developed for potential agricultural use.

The ability to regenerate plants from single cells growing in tissue culture has made plant cells easier to manipulate than mammalian cells. A plasmid carried by the bacterium *Agrobacterium tumefaciens* is the best-developed DNA vector. In nature, this bacterium produces crown gall tumors in the plants it infects when its *Ti plasmid* (tumor inducing) integrates a segment of DNA into the plant chromosomes. Using DNA technology, foreign genes are inserted into Ti plasmids whose disease-causing properties have been eliminated. The recombinant plasmid can be either directly introduced into plant cells growing in culture or put back into *Agrobacterium* which is used to infect the cells. When these cells regenerate whole plants, the foreign gene is included in the plant genome.

Only dicotyledons are susceptible to infection by *Agrobacterium*. Techniques such as electroporation and the DNA particle gun are being used to transfer foreign genes into important monocots such as corn and wheat. Challenges still remain in identifying beneficial genes that will improve plant traits, particularly those traits that are polygenic.

Early results of genetic engineering in plants have been positive. Plant strains that carry a bacterial gene for resistance to herbicides have been developed, which will enable crops to be grown while weeds in their midst are killed. An antisense gene has been used to retard spoilage in tomatoes by producing mRNA that blocks the mRNA for a ripening enzyme. Crop plants are being engineered to be resistant to infectious pathogens and insect pests. Crop yields may be improved through the selective enlargement of plant parts or improvement to a plant's food value.

The production of bacteria with increased nitrogen-fixing potential and the engineering of plants that can fix nitrogen themselves are examples of a valuable potential use of recombinant DNA technology for improving plant productivity.

Safety and Ethical Issues

Scientists have worried that there might be dangerous consequences to DNA technology. One particular hazard concerns the production of new pathogens and their release into the environment. Early on, scientists working in this field developed a set of guidelines to deal with policy and safety issues. Several federal agencies set policy and regulate new developments in genetic engineering.

Ethical questions about human genetic information include who should have access to information about a person's genome and how that information should be used. Safety issues include concerns about potential harmful side effects of medical products. With genetically engineered agricultural products, there are potential dangers in the impact of such organisms on native species. Transgenic species may become "superweeds" or introduce some of their genes for resistance to herbicides, insect pests, and natural diseases into wild plants. Researchers are trying to develop mechanisms to control the escape of engineered plant genes.

Potential environmental and health hazards must be considered in the development of these powerful genetic techniques and remarkable products of biotechnology.

STRUCTURE YOUR KNOWLEDGE

1. Fill in the table on p. 147 on the basic tools of gene manipulation used in DNA technology.

2. There are many applications for DNA technology in basic research, in agriculture, and in medicine. Describe three or four examples in each of these categories.

3. The schematic diagram on p. 148 shows the steps required to clone a gene. Identify the structures labeled a through d.

   What is happening at step 1?

   What is happening at step 2?

   What is happening at step 3?

   How are the bacterial clones that have picked up the recombinant plasmid identified?

4. The gel on p. 147 was produced from four samples of single-stranded DNA fragments that were incubated with radioactively labeled primer, DNA polymerase, the four nucleotides, and a different one of the four dideoxy nucleotides. What is the sequence of nucleotides in the original single-stranded DNA fragment?

TECHNIQUE OR TOOL	BRIEF DESCRIPTION	USE IN DNA TECHNOLOGY
Restriction enzymes		
Gel electro-phoresis		
cDNA (comple-mentary DNA)		
DNA sequencing by Sanger method		
Polymerase chain reaction		
Labeled probe and auto-radiography		
RFLP analysis		
Chromosome walking		

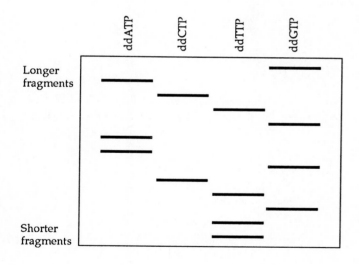

TEST YOUR KNOWLEDGE

MULTIPLE CHOICE: *Choose the one best answer.*

1. The role of restriction enzymes in DNA technology is to
   a. provide a vector for the transfer of the recombinant DNA.
   b. produce cDNA from mRNA.
   c. produce a staggered cut at specific recognition sequences on DNA.
   d. reseal "sticky ends" after base-pairing of complementary bases.

2. This segment of DNA has restriction sites I and II which create restriction fragments a, b, and c.

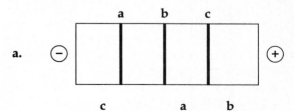

Which of the following gels produced by electrophoresis would represent the separation and identity of these fragments?

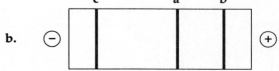

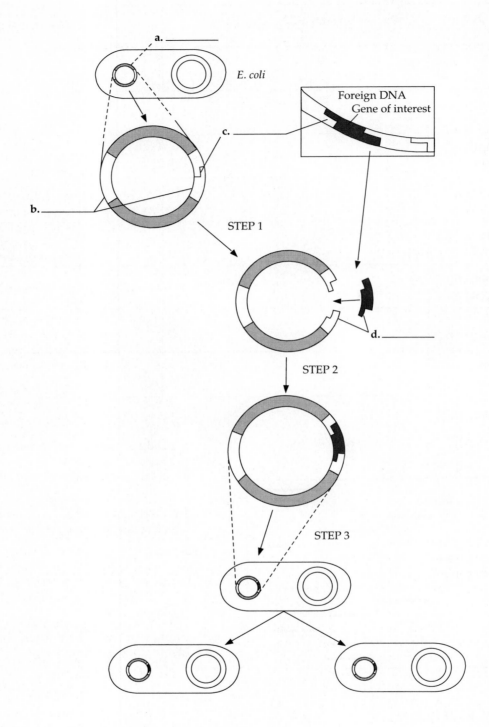

a. _____

E. coli

Foreign DNA
Gene of interest

c. _____

b. _____

STEP 1

d. _____

STEP 2

STEP 3

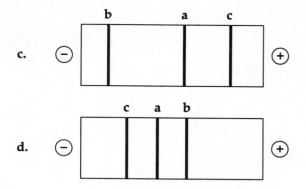

3. Plasmid DNA and bacteriophages
   a. are involved in cloning genes.
   b. are both vectors for transferring recombinant DNA into cells.
   c. can be treated so that they carry foreign genes in their DNA.
   d. all of the above

4. Which of the following DNA sequences would most likely be a restriction site?
   a. AACCGG
      TTGGCC
   b. GGTTGG
      CCAACC
   c. AAGG
      TTCC
   d. AATT
      TTAA

5. Genes for cloning may be chemically synthesized
   a. when the exact sequence of nucleotides is known.
   b. through the use of restriction enzymes and gel electrophoresis to separate restriction fragments.
   c. by the Sanger method.
   d. using cDNA.

6. Petroleum-lysing bacteria are being engineered for the removal of oil spills. What is the most realistic danger to the environment?
   a. mutations leading to the production of a strain pathogenic to humans
   b. extinction of natural microbes due to the competitive advantage of the "petro-bacterium"
   c. destruction of natural oil deposits
   d. none of the above

7. You are attempting to introduce a gene that imparts larval moth resistance to bean plants.

Which of the following vectors are you most likely to use?
   a. bacteriophage
   b. Ti plasmids
   c. gene gun
   d. either b or c

8. An attenuated virus
   a. is a virus that is nonpathogenic.
   b. in an elongated viral particle.
   c. can transfer recombinant DNA to its host cell.
   d. will not produce an immune response.

9. Difficulties in getting prokaryotic cells to express eukaryotic genes include the fact that
   a. the signals that control gene expression are different and prokaryotic promoter regions must be added to the recombinant DNA.
   b. the genetic code differs between the two because prokaryotes substitute the base uracil for thymine.
   c. prokaryotic cells cannot transcribe introns because their genes do not have them.
   d. the ribosomes of prokaryotes are not large enough to handle long eukaryotic genes.

10. Complementary DNA does not create as complete a gene library as the shotgun approach because
    a. it has eliminated introns from the genes.
    b. a cell produces mRNA for only a small portion of its genes.
    c. the shotgun approach produces more restriction fragments.
    d. cDNA is not as easily integrated into plasmids or phage genomes.

11. Which of the following is *not* true of recognition sequences?
    a. Modification by methylation of bases within them prevents restriction of bacterial DNA.
    b. They are usually symmetrical sequences of 4 to 8 nucleotides.
    c. They signal the attachment of RNA polymerase.
    d. There is a specific restriction enzyme for each recognition sequence.

12. The advantage of yeast for gene cloning is that it
    a. has RNA splicing machinery.
    b. has plasmids that can be genetically engineered.
    c. allows the study of eukaryotic gene regulation and expression.
    d. does all of the above.

CHAPTER 12
MEIOSIS AND SEXUAL LIFE CYCLES

Suggested Answers to Structure Your Knowledge

1.

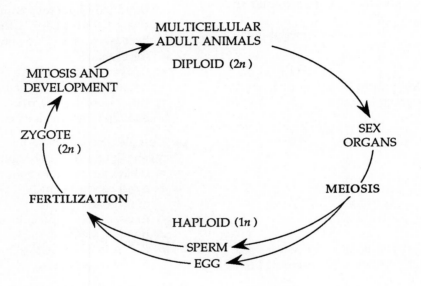

2.

a. Interphase I: chromosome replication, sister chromatids attached at centromere
b. Prophase I: synapsis of homologous pairs, crossing over at chiasmata, spindle forms
c. Metaphase I: homologous pairs line up at metaphase plate, with centromeres attached to spindle fibers from opposite poles
d. Anaphase I: homologous pairs of chromosomes separate and move toward opposite poles, centromeres do not split
e. Metaphase II: Chromosomes, consisting of two sister chromatids, align singly at metaphase plate
f. Anaphase II: Sister chromatids separate when centromeres split and single-stranded chromosomes move to opposite poles

3. Stages of meiosis diagrams in proper sequence:
   d. Interphase
   b. Prophase I
   e. Metaphase I
   c. Anaphase I
   a. Metaphase II
   f. Anaphase II

4.

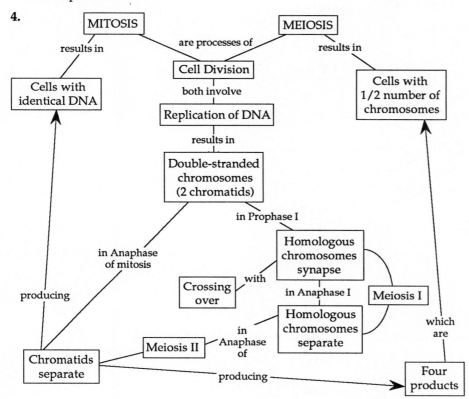

Answers to Test Your Knowledge

Multiple Choice:

1. b	**4.** a	**7.** c	**10.** b	**13.** c	**16.** b
2. d	**5.** d	**8.** c	**11.** a	**14.** d	
3. d	**6.** b	**9.** d	**12.** b	**15.** d	

CHAPTER 13
MENDEL AND THE GENE IDEA

Suggested Answers to Structure Your Knowledge

1. Mendel's law of segregation refers to the discrete nature of heritable traits. In anaphase I, gene pairs (alleles) segregate when homologous chromosomes move to opposite poles of the cell. The two cells formed from this division have one-half the number of chromosomes and one copy of each gene. Mendel's law of independent assortment relates to the lining up of synapsed chromosomes

at the equatorial plate in a random fashion during metaphase I. Chromosomes initially received from the mother and father may orient in either direction. The number of different chromosome arrangements in gametes is equal to 2^n where *n* is the haploid number of chromosomes.

2.

TtRr x TtRr

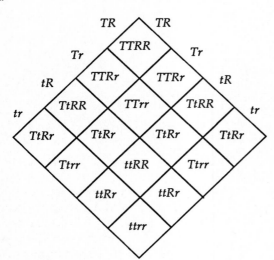

PHENOTYPE	PREDICTED RATIO
tall, red	3
tall, pink	6
tall, white	3
short, red	1
short, pink	2
short, white	1

Total = 16

3. It is important to remember that one allele does not "dominate" another. All alleles operate independently within a cell, coding for their gene products, usually enzymes, as specified by their particular sequence of DNA nucleotides. If both alleles code for functional enzymes, then the alleles may be codominant with both traits expressed in the heterozygote, as illustrated by MN or AB blood types. If the recessive allele codes for a nonfunctional enzyme, then the resulting recessive trait may only be obvious when homozygous (such as O blood type in which there is no carbohydrate molecule on the cell membrane). If one copy of an allele produces sufficient product that the dominant trait is expressed, then we have complete dominance where the heterozygote phenotype is indistinguishable from the homozygote dominant (smooth peas in which the enzyme converts sugar to starch). To understand how genotypes translate into phenotypes, we have to study biochemical pathways, which are controlled by enzymes coded for by genes, to see how molecular processes yield physical, physiological, and behavioral results.

Answers to Test Your Knowledge

Matching:

1. I	4. E	7. H	10. K
2. A	5. L	8. M	
3. F	6. J	9. C	

Multiple Choice:

1. b	4. c	7. c	10. d	13. b	16. d
2. d	5. b	8. c	11. b	14. d	17. a
3. b	6. c	9. b	12. d	15. c	18. c

Genetics Problems:

1.

Mother	Child	Man exonerated if
AB	A	no groups exonerated
O	B	A or O
A	AB	A or O
O	O	AB
B	A	B or O

2. Father's genotype must be *Pp* since polydactyly is dominant and he has had one normal child. Mother's genotype is *pp*. The chance of the next child having normal digits is 1/2 or 50% because the mother can only donate a *p* allele and there is a 50% chance that the father will donate a *p* allele.

3. *AB* gamete from *AABb* parent = 1/2 (1 x 1/2)
ab gamete from *AaBb* parent = 1/4 (1/2 x 1/2)
ABc gamete from *AABbcc* parent = 1/2 (1 x 1/2 x 1)
ABc gamete from *AaBbCc* parent = 1/8 (1/2 x 1/2 x 1/2)

4. probability of *AAbb* offspring = 1/2 (to get *AA*) x 1/2 (*bb*) = 1/4
probability of *aaBB* offspring = 1/4 (*aa*) x 1/2 (*BB*) = 1/8
probability of *AaBbCc* = 1 (*Aa*) x 1/2 (*Bb*) x 1 (*Cc*) = 1/2
probability of *aabbcc* = 1/4 (*aa*) x 1/4 (*bb*) x 1/2 (*cc*) = 1/32

5. The genotypes of the puppies were 3/8 *B_S_* (because *B* and *S* are dominant alleles, the "_" indicates that the second gene can be either *B* or *b*, *S* or *s*, and still produce a dominant phenotype), 3/8 *B_ss*, 1/8 *bbS_*, and 1/8 *bbss*. Because recessive traits show up in the offspring, both parents had to have had at least one recessive allele for both genes. Black occurs in a 6:2 or 3:1 ratio, indicating a heterozygous cross. Solid occurs in a 4:4 or 1:1 ratio, indicating a cross between a heterozygote and homozygous recessive. Parental genotypes were *BbSs* x *Bbss*.

6. The parental cross produced 25-cm tall F₁ plants, all *AaBbCc* plants with 3 units of 5 cm added to the base height of 10 cm. Of the 64 possible combination of gametes in the F₂, there will be 7 different phenotypic classes, varying from all dominant alleles, 5 dominant and 1 recessive, 4 dominant and 2 recessive, and so on, through all 6 recessive alleles.

7. It is good to decide first what phenotypes are produced by each genotype. *M_B_* individuals will be black, because both genes are dominant. *M_bb* individuals are brown. *mm__* individuals will all be white, because no melanin is produced to be laid down (the *B* genotype does not matter). All F₁ guinea pigs will be *MmBb* and black. The normal 9 *M_B_*, 3 *M_bb*, 3 *mmB_*, 1 *mmbb* ratio will be 9 black, 3 brown, and 4 white due to the epistatic effect of *mm* on the *B* gene.

8. The 1:1 ratio of the first cross indicates a cross between a heterozygote and a homozygote. The second cross indicates that the hairless hamsters cannot have had a homozygous genotype, because then only hairless hamsters should be produced. Let us say that hairless hamsters are *Hh* and normal-haired are *HH*. The second cross of heterozygotes should yield a 1:2:1 genotypic ratio. The 1:2 ratio indicates that the homozygous

recessive genotype may be lethal, with embryos that are *hh* never developing.

9. A ratio close to 1:2:1 ratio indicates a case of incomplete dominance. Medium-tailed pigs are probably *Tt*, and the heterozygous cross produced three *tt* stub-tailed pigs, six *Tt* medium-tailed pigs, and four *TT* long-tailed pigs.

10.

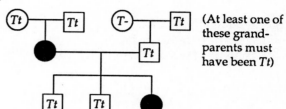

(At least one of these grandparents must have been *Tt*)

11. You can determine the genotype of the black parent to be *CC^{ch}*. The Himalayan parent could be either *C^hC^h* or *C^hc*. First list the alleles in order of dominance: *C, C^{ch}, C^h, c*. To approach this problem you should write down as much of the genotype as you can be sure about for each phenotype involved. The parents were black (*C_*) and Himalayan (*C^h_*). The offspring genotypes would be black (*C_*) and Chinchilla (*C^{ch}_*). Right away you can tell that the black parent must have been *CC^{ch}* because half of the offspring are Chinchilla; *C^{ch}* is dominant over *C^h* and *c* and could not have been present in the Himalayan parent. The genotype of this parent could be either *C^hC^h* or *C^hc*. A test cross would be necessary to determine this genotype.

You cannot definitely determine the genotype of the parents in the second cross. You can, however, eliminate some genotypes. The black parent could not be *CC* or *CC^{ch}*, because both the *C* and *C^{ch}* alleles are dominant to *C^h*, and Himilayan offspring were produced. The Chinchilla parent could not be *C^{ch}C^{ch}* for the same reason. One or both parents had to have *C^h* as their second allele.

12. This problem is challenging. First write out the phenotypes possible and the genotypes that could produce them. *BB_ _* and *_ _SS* genotypes are lethal. Bristled flies must be *Bbss*. Nonbristled flies could be *BbSs, bbSs,* or *bbss*. From the information given, the nonbristled parents must be *BbSb*. Now determine the probability of obtaining these genotypes.

bristled = *Bbss* = 1/2 x 1/4 = 1/8 or 2/16
nonbristled = *BbSs* (1/4) + *bbSs* (1/8) + *bbss* (1/16) = 7/16

The phenotypic ratio of the F_1 would be 2:7, bristled to nonbristled. The other genotypes are lethal.

A backcross of *Bbss* to *BbSs* would yield a 1:2 ratio of bristled to nonbristled.

bristled = *Bbss* = 1/2 x 1/2 = 1/4
nonbristled = *BbSs* (1/4) + *bbSs* (1/8) + *bbss* (1/8) = 4/8

13. Generation I
    Genotype of father = *Aa*
    Genotype of mother = *Aa*

Generation II
    Genotype of mate 1 = *AA* (probably)
    Genotype of mate 2 = *Aa*

Generation III
    Genotype of son 4 = *Aa*

The genotype of son 3 in generation II could be *AA* or *Aa*. If his wife is *AA*, then he could be *Aa* (both his parents are carriers) and the recessive allele never would be expressed in his offspring. Even if he and his wife were both carriers, there would be a 81/256 or 31.6% chance that all four children would be normally pigmented. (3/4 x 3/4 x 3/4 x 3/4)

CHAPTER 14
CHROMOSOMAL BASIS OF INHERITANCE

Suggested Answers to Structure Your Knowledge

1. When Morgan crossed the white-eyed male fly with red-eyed females, all the F_1 were red-eyed, indicating a simple case of dominance of red eyes. In the F_2 cross, the ratio of red to white eyes was the normal Mendelian 3:1. But Morgan noted that all female flies were red-eyed, whereas males were 1:1 red and white-eyed. Morgan hypothesized that this allele for eye color was carried on the X chromosome. F_1 males had only a wild-type red allele to pass on to their daughters and passed no allele for eye color on to male offspring. F_1 females were heterozygous, and thus half of the time they donated a wild-type allele and half of the time they donated a white allele to male offspring—yielding the 1:1 ratio of males. The association of the gene for eye color with the inheritance of the X chromosome gave evidence that Mendel's genes were located on chromosomes.

2. Genes that are not linked assort independently, and the ratio of offspring from a testcross with a dihybrid heterozygote should be 1:1:1:1 (*AaBb* x

aabb gives *AaBb, Aabb, aaBb, aabb* offspring). Genes that are linked and do not cross over should produce a 1:1 ratio (*AaBb* and *aabb*). If crossovers occur 50% of the time, the heterozygote will produce equal quantities of *AB, Ab, aB,* and *ab* gametes, and the genotypic ratio of offspring will be 1:1:1:1, the same as it is for unlinked genes. Because each 1% of crossovers is equal to 1 map unit, this measurement ceases to be meaningful at relative distances of 50 or more map units. However, crosses with intermediate genes on the chromosome could establish both that the genes *A* and *B* are on the same chromosome and that they are a certain distance apart.

3. If the gene for the mutant is sex-linked, you could assume it was on the X chromosome. A cross between mutant female flies and normal males should produce all normal female and mutant male flies.

   $X^m X^m$ x $X^+ Y$ produces $X^+ X^m$ and $X^m Y$ offspring

4.

NAME OF DISORDER OR SYNDROME	CHROMOSOMAL ALTERATION INVOLVED	SYMPTOMS OR ASSOCIATED TRAITS
Down Syndrome	Trisomy 21	Characteristic facial features, short stature, heart defects, respiratory infections, mental retardation
Klinefelter syndrome	XXY (or XXYY, XXXY, etc.)	Sterility, small testes, feminine body contours, normal intelligence or mental retardation
Turner Syndrome	Monosomy X (XO)	Short stature, sex organs do not mature, no secondary sex characteristics, normal intelligence
Cri du chat Syndrome	Deletion in Chromosome 5	Mental retardation, small head, unusual cry
CML	reciprocal translocation of chromosome 22 & 9	Chronic myelogenous leukemia—cancer affecting cells that give rise to white blood cells

The serious phenotypic effects that are associated with these chromosomal alterations indicate that normal development and functioning is dependent on genetic balance. Most genes appear to be vital to an organism's existence and extra copies of genes upset genetic balance. Inversions and translocations, which do not disrupt the balance of genes, can alter phenotype because of position effects on gene functioning. Aneuploidy involving sex chromosomes upset genetic balance less because so few genes are carried on the Y chromosome and extra X chromosomes are inactivated as Barr bodies.

Answers to Test Your Knowledge

Multiple Choice:

1. d	5. d	9. a	13. c	17. b
2. a	6. b	10. a	14. d	18. c
3. b	7. d	11. d	15. a	19. a
4. d	8. a	12. b	16. c	

Genetics Problems:

1. The gene is sex-linked, so a good notation is X^C, X^c, and Y so that you will remember that the Y does not carry the gene. Capital C indicates normal sight. Genotypes are:

   1. X^CY 3. X^CX^C or X^CX^c 5. X^cY
   2. X^CX^c 4. X^CX^c 6. X^cY
   7. X^CX^c

2. e a c b d

3. The genes appear to be linked, because the parental types appear most frequently in the offspring. Recombinant offspring represent 10 out of 40 total offspring for a recombination frequency of 25%, indicating that the genes are 25 map units apart.

4. Let X^L represent the dominant allele and X^l represent the recessive lethal allele. The mother is X^LX^l, the father must be X^LY. Male fetuses with the genotype X^lY would be expected one-half of the time and would spontaneously abort. Assuming an equal sex ratio at conception, the ratio of girl to boy children would be 2:1, or 6 girls and 3 boys.

5. Female flies produce v^+bw^+ gametes and males produce either vbw or Ybw gametes. All the F_1 are phenotypically wild-type red-eyed flies. Females are v^+vbw^+bw and males are v^+Ybw^+bw. In the reciprocal cross ($vvbwbw$ x $v^+Ybw^+bw^+$), all F_1 females are v^+vbw^+bw (red-eyed) and males are $vYbw^+bw$ (vermillion-eyed).

An F_2 is produced from the heterozygote red-eyed F_1 that resulted from the first cross. The following Punnett square shows the gametes that each parent can produce and the resulting phenotypes for the F_2.

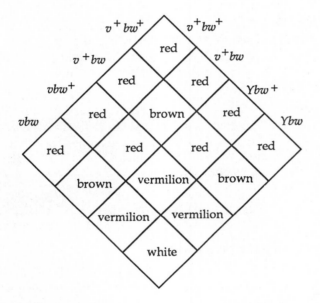

6. Females are the heterogametic sex in birds. For female chicks to be black, they must have received a recessive allele from the male parent. If all female chicks are black, the male parent must have been Z^bZ^b. If the male parent was homozygous recessive, then all male offspring will receive a recessive allele, and the female parent would have to be Z^BW to produce all barred males.

For female chicks to be both black and barred, the male parent must have been Z^BZ^b. If the female parent were Z^BW, only barred male chicks would be produced. To get an equal number of black and barred male chicks, the female parent must have been Z^bW.

PHENOTYPE	NUMBER
red-eyed female	6
brown-eyed female	2
red-eyed male	3
brown-eyed male	1
vermilion-eyed male	3
white-eyed male	1

CHAPTER 15
THE MOLECULAR BASIS OF INHERITANCE

Suggested Answers to Structure Your Knowledge

1.

INVESTIGATOR	ORGANISMS USED	TECHNIQUES, EXPERIMENTS	CONCLUSIONS
Griffith	S & R strains of bacteria; S caused pneumonia in mice	Injected mice with heat-killed S and live R cells	R cells transformed by genetic material from S cells. Heat-stable factor may not be protein.
Avery	S and R strains of bacteria	Purified chemicals from heat-killed S cells; tested which transformed R cells	Transforming agent was DNA.
Hershey & Chase	Viral phage and the bacterium *E. coli*	Radioactively-labeled DNA and protein of phage; centrifuged to find which molecule found in bacteria	DNA of phage injected into host; proteins stay outside; DNA is hereditary material.
Chargaff	DNA from several organisms	Compared quantity and ratio of nucleotides A, T, G, C	DNA composition is species-specific; Chargaff's rules: A=T and G=C.
Franklin	DNA	X-ray crystallography of DNA molecules	DNA is in shape of helix.
Watson & Crick	DNA	Information from X-ray photo; wire scale models	Double helix with rungs of A—T and G—C, sides of sugar-phosphate chain
Messelson & Stahl	*E. coli*	*E. coli* labeled with ^{15}N, grown on ^{14}N; centrifuged to find density	Confirmed proposed semiconservative replication of DNA

2.
a. helicase
b. single-strand binding protein
c. DNA polymerase
d. leading strand
e. lagging strand
f. DNA ligase
g. Okazaki fragment
h. RNA primer
i. primase
j. replication fork
k. 3' end of parental strand
l. 5' end

Answers to Test Your Knowledge

Multiple Choice:

1. c
2. a
3. b
4. a
5. d
6. b
7. a
8. b
9. a
10. d.
11. c
12. a
13. a
14. c
15. a
16. d
17. a
18. d
19. b
20. c

CHAPTER 16
FROM GENE TO PROTEIN

Suggested Answers to Structure Your Knowledge

1.

	TRANSCRIPTION	TRANSLATION
Template	DNA	RNA
Location	nucleus (cytoplasm in prokaryotes)	cytoplasm; ribosomes can be free or attached to ER
Molecules involved	RNA nucleotides, DNA, RNA polymerase, transcription factors	amino acids, tRNA, mRNA, ribosomes, enzymes, ATP, GTP, initiation and elongation factors
Enzymes needed	RNA polymerases, RNA processing enzymes	aminoacyl-tRNA synthetase, peptidyl transferase
Control— start & stop	transcription factors locate promoter region with TATA box, terminator sequence	initiation factors, AUG; stop codons, release factors
Product	mRNA	protein
Processing involved	RNA processing: 5′ cap and poly-A tail added; splicing of hnRNA— introns removed by snRNPs & spliceosomes, ribozymes	spontaneous folding, disulfide bridges; removal of some amino acids (signal sequence); cleaving; some form quaternary structure; modification with sugars, etc.
Energy source	ribonucleoside triphosphate	ATP and GTP

2. The genetic code includes the sequence of nucleotides on DNA that is transcribed into the codons found on mRNA and translated into their corresponding amino acids. There are 64 possible codons created from the four nucleotides used in the triplet code (4^3). Codons are needed only for 20 amino acids and stop codons. Redundancy of the code refers to the fact that several triplets may code for the same amino acid. Often these triplets only differ in the third nucleotide.

   The wobble hypothesis explains the fact that there are only about 45 different tRNA molecules that pair with the 64 possible codons. The third nucleotide of many tRNAs can pair with more than one base. Because of the redundancy of the genetic code, these wobble tRNAs still place the correct amino acid in position. The genetic code was thought to be universal, that each codon coded for exactly the same amino acid in all organisms. Recently, some exceptions to this universality have been found in a few ciliates and in the DNA of mitochondria and chloroplasts.

3.

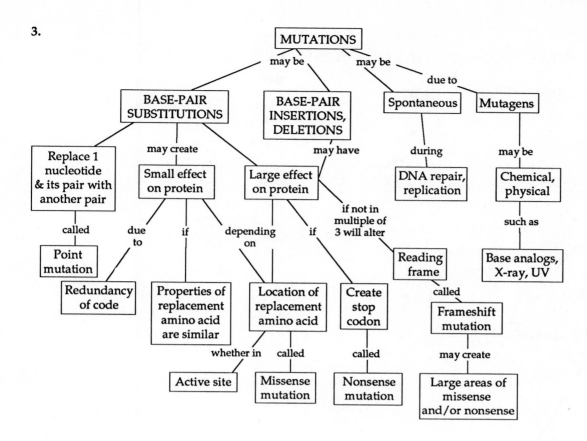

Answers to Test Your Knowledge

Multiple Choice:

1. d	**5.** b	**9.** a	**13.** d	**17.** c
2. a	**6.** d	**10.** b	**14.** d	**18.** b
3. b	**7.** a	**11.** b	**15.** c	
4. d	**8.** a	**12.** d	**16.** a	

Fill in the Blanks:

DNA codon	TAC	CAG	TTC	ATG
mRNA codon	AUG	GUC	AAG	UAG
Anticodon	UAC	CAG	UUC	none
Amino acid	methionine	valine	lysine	STOP

CHAPTER 17
MICROBIAL MODELS: THE GENETICS OF VIRUSES AND BACTERIA

Suggested Answers to Structure Your Knowledge

1.

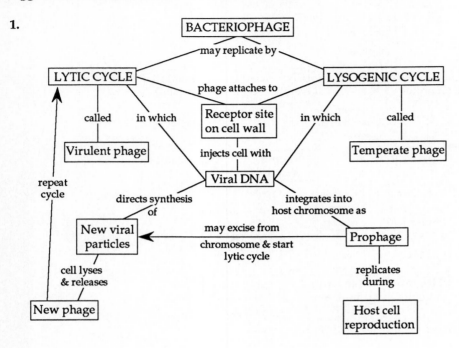

2. a. Regulatory gene
   b. Promoter
   c. Operator
   d. Structural genes
   e. Operon

   f. Corepressor
   g. Repressor
   h. RNA polymerase
   i. mRNA

3.

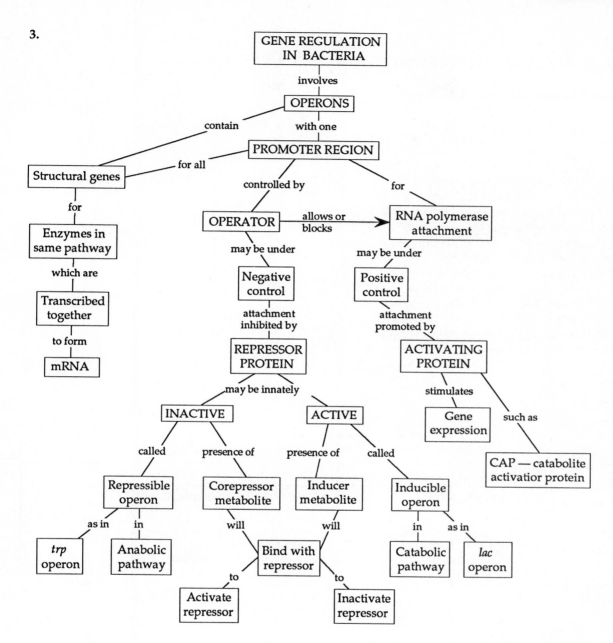

Answers to Test Your Knowledge

Matching:

1. C	**5.** A	**9.** K
2. G	**6.** M	**10.** I
3. H	**7.** L	
4. B	**8.** E	

Multiple Choice:

1. b	**7.** b	**13.** c
2. d	**8.** b	**14.** d
3. a	**9.** c	**15.** b
4. b	**10.** d	**16.** c
5. a	**11.** a	**17.** a
6. c	**12.** d	**18.** c

Matching:

1. C	**4.** E	**7.** A
2. D	**5.** F	**8.** H
3. G	**6.** B	

CHAPTER 18
GENOME ORGANIZATION AND EXPRESSION IN EUKARYOTES

Suggested Answers to Structure Your Knowledge

1. **a.** histone
   b. nucleosome
   c. 30 nm chromatin fiber
   d. looped domain
   e. condensed chromatin

2. Chromosomes have repetitive sequences called satellite DNA located within the centromere region. Nucleotide sequences associated with a gene may include recognition sequences for genes coding for enzymes in the same metabolic pathway and enhancers that may bind transcription factors and then interact with the promoter region to activate transcription. A gene includes a promoter region where RNA polymerase attaches with the aid of transcription factors. Exons are often separated by long introns, which may include numerous pseudogenes. Chromosomes may contain multigene families, which are copies of the same or similar genes which may be in tandem arrays or separated.

Answers to Test Your Knowledge

Multiple Choice:

1. b
2. a
3. d
4. b
5. a
6. c
7. c
8. d
9. c
10. d
11. c
12. a

3.

LEVEL OF CONTROL	TYPE OF CONTROL	EXAMPLES
Availability of genes	DNA packing & methylation	Heterochromatin in Barr body
	Gene amplification	rRNA genes in developing ovum
	Gene loss	genes of chromosomes lost in some insects
	Rearrangements in genome	transposons mating-type alleles in yeast immunoglobulin genes
Transcriptional control	Promoter region	transcription factors and RNA polymerase bind here
	Enhancer	binding of transcription factors activate gene expression, may involve steroids & other molecules
Post-transcriptional control	RNA processing & export	introns removed, mRNA gets 5' cap & poly-A tail, control over exit from nucleus
	mRNA degradation	lifespan of mRNA determines amount of protein made
	storage of inactive mRNA	need control signal for translation, mRNA stored in ovum cytoplasm
Translational control	initiation factor	necessary for mRNA & ribosomes to interact
Post-translational control	protein processing	cleavage of protein as in insulin, addition of chemical groups, addition of signal to target protein
	selective degradation of proteins	metabolic control of enzymes

CHAPTER 19
DNA TECHNOLOGY

Suggested Answers to Structure Your Knowledge

1.

TECHNIQUE OR TOOL	BRIEF DESCRIPTION	USE IN DNA TECHNOLOGY
Restriction enzymes	Bacterial enzymes that cut DNA at restriction sites, creating "sticky ends" that can base-pair with other fragments	Cut DNA sequences that can then join to make recombinant DNA (seal with ligase); form restriction fragments used for RFLP analysis and sequencing DNA
Gel electro-phoresis	Mixture of molecules applied to slab of gel in electric field; molecules separate, moving at different rates due to charge and size	Separate restriction fragments into pattern of distinct bands. Fragments can be sequenced, and put into gene libraries.
cDNA (complementary DNA)	mRNA isolated from cell is treated with reverse transcriptase to produce a complementary DNA strand, which then produces a double-stranded DNA gene	Creates artificial gene lacking introns which is easier to clone in bacteria, produces cDNA library of the genes that are active in a cell.
DNA sequencing by Sanger method	Single-stranded DNA fragments are incubated with 4 nucleotides, DNA polymerase, and 1 of 4 dideoxy nucleotides that interrupt synthesis. Samples are separated by gel electrophoresis.	Sequence of nucleotides is read from the four sets of bands on the gel. Electrophoresis will separate bands that differ by as little as one nucleotide.
Polymerase chain reaction	DNA is repeatedly melted and mixed with primers having complementary sequences for targeted DNA section, DNA polymerase, and nucleotides.	Rapidly produce multiple copies of a gene or section of DNA in vitro
Labeled probe and auto-radiography	Radioactively labeled single-stranded DNA or mRNA that will base-pair with complementary sequences of DNA	Locate gene in clone of bacteria, identify similar nucleic acid sequences, make cytological map of genes with *in situ* hybridization
RFLP analysis	Restriction fragments separated by gel electrophoresis, Southern blotting transfers single-stranded fragments to filter, radioactive probe added, and autoradiography reveals band pattern	Make DNA fingerprints for forensic use, map chromosomes using RFLP markers, diagnose genetic diseases
Chromosome walking	Prepare two libraries with different restriction enzymes, use probe from one library to locate overlapping fragment in the other and repeat process.	Determine sequence of fragments on a chromosome, showing rough order of genes or markers associated with those fragments.

2. Basic Research:
(1) Techniques such as the Sanger method are used to determine nucleotide sequences in DNA. (2) Radioactively labeled probes are used to map genes on chromosomes. (3) Molecules that control metabolism and development are often present in very small quantities. With cDNA, their genes can be cloned and the molecules produced in large enough quantities to study. (4) cDNA can be used as a probe to locate genes and study their noncoded regions such as introns and regulator sequences. (5) Using probes to locate genes with similar sequences, the evolutionary relationships between genes and between organisms can be studied.

Agricultural Applications:
(1) Vaccines, hormones, and antibodies, which will improve the health or productivity of livestock, can be produced using DNA technology. (2) The genome of agricultural plants and animals may eventually be directly altered and improved. (3) Improvements in the nitrogen-fixing capacity of bacteria or plants are being developed. (4) Plant varieties that have genes for resistance to diseases and herbicides are being developed.

Medical Applications:
(1) Vaccines are being developed, either through the production of virus subunits that produce immune responses or in the development of attenuated viruses. (2) Diagnosis of genetic diseases is possible by probing for a potentially defective gene with a normal gene and by RFLP analysis. (3) Treatment of genetic disorders may be possible through the replacement of a defective gene with a normal one. (4) Insulin, human growth factor, and several products of the immune system have been produced by biotechnology.

3. The following structures were indicated on the diagram: **a**. plasmid; **b**. identification genes (most likely for antibiotic resistance); **c**. restriction sites; **d**. sticky ends.

In the step labeled 1, the plasmid and foreign DNA are treated with the same restriction enzyme. In step 2, the sticky ends have joined by base pairing and are sealed by DNA ligase. In step 3, an *E. coli* cell has picked up the plasmid by transformation.

The clones that have incorporated the recombinant plasmid are identified as those that will grow on antibiotic plates for which the plasmid had the resistant gene, but will not grow on antibiotic plates for the gene that was disrupted by the restriction site and insertion of the new gene.

4. The sequence of nucleotides in the original fragment would be complementary to the sequence shown by the order of fragments. It would read, from the bottom up, A A C A G C T T C A G T C

Answers to Test Your Knowledge

Multiple Choice:

1. c	**4.** d	**7.** d	**10.** b
2. b	**5.** a	**8.** a	**11.** c
3. d	**6.** c	**9.** a	**12.** d

UNIT IV

MECHANISMS OF EVOLUTION

THE FAR SIDE By GARY LARSON

Great moments in evolution

CHAPTER 20

DESCENT WITH MODIFICATION: A DARWINIAN VIEW OF LIFE

FRAMEWORK

This chapter describes Darwin's formulation of evolution—descent from a common ancestor modified by the mechanisms of natural selection, resulting in the evolution of species adapted to their environments. The scientific and philosophical climate of Darwin's day was quite inhospitable to the implications of evolution, but most biologists accepted the theory of evolution quite rapidly. Only later was natural selection accepted as the mechanism of evolution, with the incorporation of genetic principles into the modern synthesis of evolution in the 1940s. Evidence for evolution is drawn from biogeography, the fossil record, taxonomy, comparative anatomy, comparative embryology, and molecular biology.

CHAPTER SUMMARY

Evolution refers to all the changes that have occurred in the history of life on Earth that have led to the present diversity of organisms. Darwin presented the first convincing case for evolution in his book, *On the Origin of Species by Means of Natural Selection*, published in 1859. Darwin made two major claims: species were not specially created in their present forms, but evolved from ancestral species, and natural selection provides the mechanism for evolution.

Pre-Darwinian Views

Darwin's theory was truly radical, for it challenged both the prevailing scientific views and the world view that had been held for centuries in Western culture.

The Scale of Life and Natural Theology The Greek philosopher Plato believed in two worlds, an ideal and eternal real world and the illusory world perceived by the senses. The variations in plant and animal populations were simply imperfect representatives of ideal forms. Plato's philosophy of ideal forms is known as idealism or essentialism.

His student Aristotle believed that all living forms could be arranged on a scale of nature of increasing complexity. Each group of organisms was fixed, permanent, and did not evolve.

The Judeo-Christian account of creation, the special design of each species during the week in which the Creator formed the universe, embedded the idea of the fixity of species in Western thought. Biology during Darwin's time was dominated by natural theology, the study of nature to reveal the Creator's plan.

One of the goals of natural theology was to classify species to reveal the rungs on the scale of life that God had created. Carl Linnaeus, working in the eighteenth century, developed both a binomial system for naming organisms according to their genus and species and a hierarchy of classifications for grouping species. *Taxonomy*, the branch of biology that names and classifies organisms, was begun by Linnaeus.

Cuvier, Fossils, and Catastrophism *Fossils* are remnants or impressions of organisms laid down in rock, usually *sedimentary rocks* such as sandstone and shale, that form through the compression of layers of sand and mud into superimposed layers called strata. Fossils reveal a succession of flora and fauna (plant and animal life).

Cuvier, the father of *paleontology*, the study of fossils, observed the differences between older fossils and modern life forms and the occurrences of extinctions. Adopting the view of history known as *catastrophism*, he speculated that the differences in fossil strata were the result of local catastrophic events such as floods or drought and not indicative of evolution.

Gradualism in Geology Gradualism, the idea that profound change is the cumulative result of slow but continuous processes, was used by Hutton in 1795 to explain the geological state of the earth. Lyell, the leading geologist of Darwin's time, extended gradualism to a theory of *uniformitarianism*, proposing that uniform rates and effects of geological processes balance out through time.

Darwin took two ideas from the observations of Hutton and Lyell: the Earth must be very old if geological change is slow and gradual, and very slow and subtle processes occurring over long periods of time can effect substantial change.

Lamarck's Theory of Evolution Lamarck, in 1809, was the first to publish a theory of evolution that explained how life evolves. Lamarck believed that evolution was driven by the tendency toward greater complexity and that as organisms evolved, they became better adapted to their environment. Lamarck explained the mechanism of evolution with two principles: the use or disuse of body parts leads to their development or deterioration, and the inheritance of acquired characteristics. In the creationist-essentialist climate of his time, Lamarck's views were dismissed and vilified; under present genetic knowledge, his idea of inheritance of acquired characteristics is sometimes ridiculed. Lamarck's theory, however, presented many key evolutionary ideas: that evolution is the best explanation for the fossil record and the current diversity of life, that Earth is very old, and that adaptation to the environment is the main result of evolution.

On the Origin of the Species

Natural theology, positing that each life form was specially created to fit perfectly in its environment, was the prevailing view when Darwin was born in 1809. Growing up with a strong interest in nature, Darwin was sent to medical school by his father, but ended up attending college to become a clergyman. Upon graduation, Henslow, his mentor and a botany professor, recommended him to the captain of the survey ship *HMS Beagle*, which was preparing for a voyage around the world.

The Voyage of the Beagle Darwin was 22 when he sailed from England as the naturalist on the *Beagle*. He spent the voyage collecting thousands of specimens of the fauna and flora of South America, observing the various adaptations of organisms living in very

diverse habitats, and making special note of the geographic distribution of species. He was particularly struck by the uniqueness of the fauna of the Galapagos Islands. Most of the animal species on the islands were unique to the Galapagos, although they resembled species from the nearby South American mainland.

Through his experiences on the *Beagle* and his reading of Lyell's *Principles of Geology*, Darwin came to doubt the church's orthodox view that the Earth was static and had been created only a few thousand years ago.

Darwin Frames His View of Life Upon learning that the 13 types of finches he had collected on the Galapagos were indeed separate species, Darwin began, in 1837, the first of several notebooks on the origin of species. He began to perceive that the origin of new species was closely related to the process of adaptation to different environments.

In 1844, Darwin wrote a long essay on the origin of species and natural selection but was reluctant to introduce his theory publicly. In 1858, Darwin received Wallace's manuscript describing a theory of natural selection identical to Darwin's. Wallace's paper and extracts of Darwin's unpublished essay of 1844 were jointly presented to the Linnean Society on July 1, 1858. Darwin quickly finished and published *The Origin of Species* the next year. Within a decade, Darwin's book and its defenders had convinced the majority of biologists that evolution was the best explanation for the forms of life on earth.

The Dual Meaning of Darwinism

Common Descent Darwin's concept of *descent with modification* included the notion that all organisms were related through descent from some unknown ancient prototype and had developed increasing modifications as they adapted to various habitats. Darwin's view of the history of life is analogous to a tree with a common ancestor at the fork of each new branch and modern species at the tips of the living twigs. Most branches are dead ends; about 99% of all species that have lived are extinct.

Natural Selection and Adaptation Darwin's book focused on how populations of individual species become adapted to their environments through natural selection. Ernst Mayr's description of Darwin's theory of *natural selection* includes the following facts and inferences:

Fact 1: Species have great potential fertility.
Fact 2: Most population sizes are stable.
Fact 3: Natural resources are limited.

Inference 1: Since only a fraction of offspring

survive, there is a struggle for limited resources.

Fact 4: There is variation between individuals in a population.

Fact 5: Much of this variation is inherited.

> **Inference 2:** Individuals whose inherited characteristics fit them best to the environment are likely to leave more offspring
>
> **Inference 3:** Unequal reproduction leads to gradual change in a population and the accumulation of favorable characteristics.

Natural selection is the differential reproductive success within a population that leads to greater adaptation to the environment.

Variation arises through chance events of mutation and genetic recombination, but natural selection is the result of definite environmental criteria for reproductive success. The excessive production of offspring sets up the struggle for life; only a small proportion can live to leave offspring of their own. Darwin's ideas were influenced by the writing of Malthus on the potential for the human population to grow much faster than the supply of food and resources.

Artificial selection used in the breeding of domesticated plants and animals provided Darwin with evidence that selection among the variations present in a population can lead to substantial changes. He reasoned that natural selection, working over hundreds or thousands of generations, could gradually create the modifications essential for the present diversity of life. Gradualism is basic to the Darwinian view of evolution.

Natural selection results in the evolution of populations, groups of interbreeding individuals of the same species in a common geographical area. Evolution is measured only as change in the relative proportions of variations in a population over time. Natural selection affects only those traits that are heritable—acquired, nongenetic characteristics can not evolve. And natural selection is a local and temporal phenomenon, depending on the specific environmental factors present in a region at a given time.

The best known example of natural selection in action is the English peppered moth, *Biston betularia*. Before the Industrial Revolution, the light colored variant of this moth was the one most commonly collected and dark moths were rare. As industrial pollution darkened tree trunks in the late 1800s, the light moths began to decrease in number and the dark variants to increase. This change in the color of moths in polluted areas is known as industrial melanism. Natural selection acts on the variation already present in a population. Predation by birds increases when the coloration of moths contrasts with their background. Predation was most likely one of the selection factors that contributed to the shift in the composition of peppered moth populations in industrial regions.

The evolution of protective coloration was documented by L. Brown's study of yellow mutant mice in a dark-colored granary. When not exposed to the selective pressure of predation by cats, the proportion of yellow mice was 46%. Following the introduction of cats to curtail the mouse population, only brown-colored mice were found.

P. and R. Grant documented the correlation between variations in the depth of finch beaks and the availability of small and large seeds during wet and dry years on a tiny islet of the Galapagos. This study and many others of natural populations, as well as laboratory experimental studies, have provided evidence for natural selection as the mechanism of evolution.

The Signs of Evolution

The validity of the evolutionary view of life is documented by signs from many sources: the geographic distribution of species and the fossil record, both cited by Darwin, as well as current discoveries, including those of molecular biology.

Biogeography The geographic distribution of species, or *biogeography*, first suggested common descent to Darwin. Islands have endemic species that are related to species on the nearest island or mainland. Widely separated areas having similar environments are more likely to have species taxonomically related to those of their region, regardless of environment, than to each other. Biogeographic distribution patterns make sense within evolution; modern species are found where they are because they evolved from ancestors who inhabited those regions.

The Fossil Record The succession of fossil forms supports the existence of the major branches of descent that were established with evidence from anatomy, biochemistry, molecular biology, and other sources. Paleontologists have discovered many transitional fossils linking modern species to their ancestral forms.

Taxonomy The taxonomy developed by Linnaeus provided a hierarchical organization of groups that suggested common descent and the branching genealogy of the tree of life. Genetic analysis now shows that taxonomy reflects evolutionary relatedness.

Comparative Anatomy The anatomical similarities among species grouped in the same taxonomic category provide evidence of common descent. The same skeletal elements make up the forelimbs of all mammals regardless of function or external shape. These forelimbs are *homologous structures*, similar because of their common ancestry. Comparative anatomy illustrates that evolution is a remodeling process in which ancestral structures are modified for new functions.

Vestigial organs are rudimentary structures, of little or no value to the organism, that are historical remnants of ancestral structures. The skeletons of some snakes retain vestigial pelvic and leg bones.

Comparative Embryology Comparative embryology shows that closely related organisms have similar stages in their embryonic development. Early embryos of all vertebrates pass through a stage in which they have gill slits; these slits develop into gills in fish, but into different structures in other vertebrates.

In the late 19th century, many embryologists adopted the view that "ontogeny recapitulates phylogeny," stating that the embryonic development (*ontogeny*) of an organism is a replay of its evolutionary history (*phylogeny*). Although the recapitulation theory is an overstatement, ontogeny can provide evidence of homology between structures that are very different in their adult forms.

Molecular Biology An organism's DNA reflects its ancestry; closely related species should have a larger proportion of DNA and proteins in common than do more distantly related species.

Darwin's hypothesis—that all forms of life descended from the earliest organisms and are thus related—is supported through molecular biology. For example, all aerobic species have the respiratory protein, cytochrome *c* (with some variations in the amino acid sequences, but with the same essential structure and function). A common genetic code, passed along through all branches of evolution, is also important evidence that all life is related. Evolution is the key to both the unity and diversity of life.

Just a Theory?

The evolution of modern species from ancestral forms is supported by historical facts such as fossils, biogeography, and molecular biology. The second of Darwin's claims, that natural selection is the main mechanism of evolution, is a theory that explains the historical facts of evolution. A scientific "theory" is a unifying concept with broad explanatory power and predictions that have been tested by experiments and observations.

STRUCTURE YOUR KNOWLEDGE

1. Briefly state the main components of Darwin's theory of evolution.

2. Darwin and present-day biologists have drawn on six sources of evidence for evolution. Briefly describe the contributions from each of these areas.

TEST YOUR KNOWLEDGE

MATCHING: *Match the theory or philosophy and its proponent(s) with the following descriptions.*

A. catastrophism
B. essentialism
C. inheritance of acquired characteristics
D. gradualism
E. natural selection
F. natural theology
G. ontogeny recapitulates phylogeny
H. scale of nature
I. uniformitarianism

a. Aristotle
b. Cuvier
c. Hutton
d. Lamarck
e. Linnaeus
f. Lyell
g. some embryologists
h. Plato
i. Wallace

THEORY   PROPONENT

1. _____   _____   discovery of the Creator's plan through the study of His works

2. _____   _____   history of earth marked by floods or droughts that resulted in extinctions

3. _____   _____   replication of ancestry of organism during its development

4. _____   _____   early explanation of mechanism of evolution

5. _____   _____   profound change is the cumulative product of slow but continuous processes

6. _____   _____   fixed species on a continuum from simple to complex

7. _____   _____   different reproductive success leads to adaptation to environment and evolution

8. _____   _____   ideal world with perfect forms of which world of senses is imperfect representation

MULTIPLE CHOICE: *Choose the one best answer.*

1. The classification of organisms into hierarchical groups is called
   a. the scale of nature.
   b. taxonomy.
   c. natural theology.
   d. ontogeny.

2. The study of fossils is called
   a. phylogeny.
   b. gradualism.
   c. paleontology.
   d. fossilogy.

3. To Cuvier, the differences in fossils from different strata were evidence for
   a. changes occurring as a result of cumulative but gradual processes.
   b. divine creation.
   c. evolution by punctuated equilibrium.
   d. local catastrophic events such as droughts or floods.

4. Darwin proposed that new species evolve from ancestral forms by
   a. the gradual accumulation of adaptations to changing environments.
   b. the inheritance of acquired adaptations to the environment.
   c. the struggle for limited resources.
   d. the accumulation of mutations.

5. Artificial selection
   a. was used by Darwin as evidence for changes possible with natural selection.
   b. involves the artificial insemination of females of a species.
   c. resulted in the change in the English peppered moth population.

d. all of the above.

6. The best description of natural selection is
   a. the survival of the fittest.
   b. the struggle for existence.
   c. the reproductive success of the members of a population best adapted to the environment.
   d. the overproduction of offspring in environments with limited natural resources.

7. The biogeographic distribution of species
   a. provides evidence for evolution.
   b. provides evidence for natural selection.
   c. shows that many endemic island species are related to species on the nearest mainland.
   d. all of the above.

8. The remnants of pelvic and leg bones in a snake
   a. are vestigial structures.
   b. provide support for the fact that ontogeny recapitulates phylogeny.
   c. are homologous structures.
   d. provide evidence for orthogenesis.

9. Which of the following would provide the best information for distinguishing phylogenetic relationships between several very similarly appearing organisms?
   a. the fossil record
   b. homologous structures
   c. comparative embryology
   d. genetic analyses and protein comparisons

10. The best evidence for Darwin's claim that all of life has descended from common ancestors is
    a. the fossil record.
    b. comparative embryology.
    c. taxonomy.
    d. molecular biology.

CHAPTER 21

HOW POPULATIONS EVOLVE

FRAMEWORK

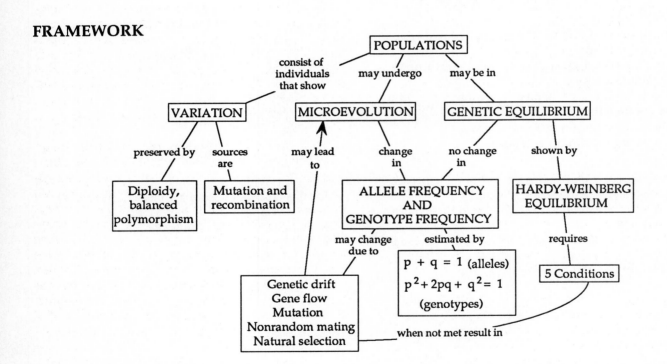

CHAPTER SUMMARY

Although it is individuals that are selected for or against by natural selection, it is populations that actually evolve. The accumulated effect of differential reproductive success over generations leads to changes in the proportions of phenotypes in a population.

The Modern Evolutionary Synthesis

Most biologists rapidly accepted evolution, but not Darwin's suggestion of natural selection as the mechanism of evolution. Without a theory of genetics, the perpetuation of parental traits as well as chance variations in offspring could not be explained.

Although Mendel was a contemporary of Darwin, his contribution to the theory of inheritance and natural selection was not recognized. When Mendel's work was rediscovered, many geneticists believed that Darwin's focus on the inheritance of quantitative traits that vary on a continuum could not be explained by the inheritance of discrete Mendelian traits.

During the 1920s, research focused on rapid phenotypic change caused by mutations as an alternative mechanism to explain evolution. Also many scientists supported the idea of goal-oriented evolution, or orthogenesis, as opposed to Darwin's mechanistic explanation based on natural selection.

The emergence of *population genetics* in the 1930s, with its emphasis on quantitative inheritance and genetic variation within populations, reconciled Mendelism with Darwinism.

In the early 1940s, a comprehensive theory of evolution, known as the *modern synthesis*, or neo-Darwinism, was developed. This theory, advanced by Dobzhansky, Mayr, Simpson, and Stebbins, emphasizes the importance of populations as the units of evolution, the essential role of natural selection, and the gradualness of evolution. Many evolutionists are now challenging some aspects of the modern synthesis.

The Genetics of Populations

A population is a localized group of individuals of the same species. A *species* is a group of populations that have the ability to interbreed in nature. Within the geographic range of a species, populations may be isolated or overlapping. Individuals within a population are likely to be more closely related than individuals from different populations.

The Gene Pool and Microevolution The *gene pool* is the term for all the genes present in a population at any given time. For individuals of a diploid species, the pool includes two alleles for each gene locus. If all individuals are homozygous for the same allele, the allele is said to be *fixed*. More often, two or more alleles are present in the gene pool in some relative proportion or frequency. *Microevolution* is a change in the relative frequencies of alleles in a population over a succession of generations.

The Hardy-Weinberg Theorem In the absence of selection pressure and other agents of change, the allele frequencies within a population will remain constant from one generation to the next, in spite of the shuffling of alleles by meiosis and random fertilization. This stasis is formulated as the *Hardy-Weinberg theorem*, named for its originators.

The allele frequency within a population determines the proportion of gametes that will contain an allele. The random combination of gametes will yield offspring with genotypes that reflect and reconstitute the allele frequencies. The frequencies of both alleles and genotypes will remain stable in a population that is in *Hardy-Weinberg equilibrium*.

With the Hardy-Weinberg equation, the frequencies of alleles within a population can be estimated from the genotype frequencies and vice versa. The letters p and q represent the proportions of the two alleles within a population. The combined frequencies of the alleles must equal 100% of the genes for that locus within a population: $p + q = 1$.

The frequencies of the genotypes in the offspring reflect the frequencies of the alleles and the probability of each combination. According to the rule of multiplication, the probability that two gametes containing the same allele will come together in a zygote is equal to ($p \times p$) or p^2, or ($q \times q$) or q^2. For a p and q allele to combine, the p allele could come from either parent; therefore the frequency of a heterozygous offspring is equal to $2pq$. The sum of the frequencies of all possible genotypes in the population adds up to one: $p^2 + 2pq + q^2 = 1$.

Allele frequencies can be determined from genotype frequencies. If the frequencies of the homozygous genotype (p^2) and the heterozygous genotype ($2pq$) are known, then the frequency of the p allele can be determined. All the gametes from the homozygotes and one-half the gametes from the heterozygotes will contain p. The frequency of p in the gene pool will equal the frequency of the homozygous dominant genotype and one-half the frequency of heterozygous genotypes. If the frequency of homozygous recessive individuals is known (q^2), then the frequency of q may be determined as the square root of q^2 (assuming the population is in Hardy-Weinberg equilibrium for that gene).

As an example, if p = frequency of allele A = 0.8, and q = frequency of allele a = 0.2, then, according to $p^2 + 2pq + q^2 = 1$, the frequencies of genotypes in the next generation would be AA = 0.64, Aa = 0.32, and aa = 0.04. To determine allele frequencies from that population, p = frequency AA + 1/2 frequency Aa, or 0.8 (0.64 + 0.16), and q = 1/2 frequency Aa + frequency aa, or 0.2 (0.16 + 0.04). Alternately, $q = \sqrt{aa} = 0.2$. The following figure shows the conversion of genotype frequencies to alleles and back to genotypes.

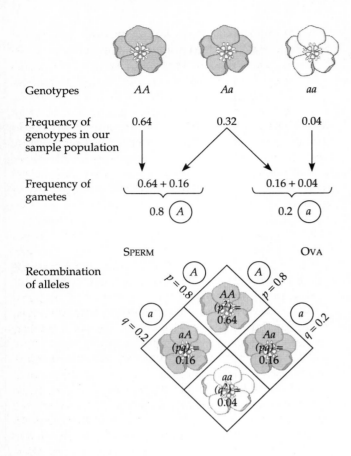

The Hardy-Weinberg equilibrium is maintained only if all of the following five conditions are met: (1) a large population, (2) no migration into or out of the population, (3) no net changes in the gene pool due to mutation, (4) random mating, and (5) equal reproductive success of all genotypes. The Hardy-Weinberg predictions can serve as a baseline for comparison with actual populations, in which these five conditions are almost never met and gene pools are changing.

Causes of Microevolution

Five sources of microevolution arise from deviations from one or more of the five conditions of Hardy-Weinberg equilibrium. Natural selection, leading to the unequal reproductive success of different genotypes, tends to increase the fitness of a population to its environment; the other four agents are chance events and usually nonadaptive.

Genetic Drift   Chance deviations from expected results are more likely to occur in a small sample due to *sampling error*. If a population is small, the random drawing of alleles to form the next generation may not represent the allele frequencies in the gene pool. *Genetic*

drift is a chance change in the gene pool of a small population and is likely to play a role in the microevolution of populations of less than 100 individuals.

The *bottleneck effect* occurs when some disaster or other factor reduces the population size dramatically and the few surviving individuals are unlikely to represent the genetic makeup of the original population. Genetic drift will remain a factor in the population until it is large enough for chance events to be less significant. A bottleneck usually reduces variability because some alleles are lost from the gene pool. The northern elephant seal and South African cheetah populations show reduced genetic variation due to bottleneck events.

Genetic drift that occurs when only a few individuals colonize a new area is known as the *founder effect*. Allele frequency in the small sample is unlikely to be representative of the parent population, and genetic drift will affect the gene pool of the new population until it is larger.

Gene Flow   The migration of individuals or the transfer of gametes between populations may result in the gain or loss of alleles, a phenomenon known as *gene flow*. Differences in allele frequencies between populations, which may have developed by natural selection or genetic drift, tend to be reduced by gene flow.

Mutation   The altering of allele frequency due to mutation is probably of little importance in microevolution due to very low mutation rates for most gene loci (one mutation in every 10^5 to 10^6 gametes). Mutation is central to evolution, however, because it is the original source of genetic variation.

Nonrandom Mating   Individuals tend to mate more often with close neighbors than with more distant population members. The effect of this *inbreeding* is that genotype frequencies may vary from those predicted from Hardy–Weinberg equilibrium. The proportion of homozygotes will increase, and thus more individuals will express recessive phenotypes. Although inbreeding may change the ratios of genotypes and phenotypes, allele frequencies within the population will remain the same.

In *assortative mating*, the choice of mates reflects a preference for like individuals. This nonrandom mating also results, as does inbreeding, in fewer heterozygous individuals, but does not change allele frequencies in the gene pool.

Natural Selection   For the Hardy-Weinberg equilibrium to be maintained, there must be no differential success in reproduction, a condition that is probably never met. Some individuals are more successful in producing viable, fertile offspring and thus pass their alleles to the next generation in disproportionate number. Natural selection is likely to be adaptive; favorable genotypes are maintained in a population. As the environment changes, selection favors genotypes that are adapted to the new conditions.

The Genetic Basis of Variation

The Nature and Extent of Genetic Variation   Individual variation, the slight differences between individuals as a result of their unique genomes, is the raw material for natural selection. Polygenic traits, those that are additively affected by two or more gene loci, provide much of the heritable variation within a population. Some traits vary categorically as distinct phenotypes and may be determined by a single gene locus. *Polymorphism* occurs when two or more discrete forms, or *morphs*, are evident in a population. Human blood groups are an example of a polymorphism.

The extent of genetic variation is evident in the molecular differences found by using biochemical methods such as electrophoresis to compare the protein products of specific gene loci among individuals in a population. In *Drosophila*, any two flies in a population may differ genetically at about 25% of their loci.

Geographic variations are regional differences in allele frequencies among populations of a species. These variations may be due to differing environmental selection factors or simply to genetic drift. If an environmental parameter changes gradually across a distance, there may be graded variations within a species along a geographic axis, called a *cline*, that parallel the environmental gradient. A cline may also exist along the graded overlap between neighboring populations.

Sources of Genetic Variation   New alleles originate by mutations, most of which occur in somatic cells and cannot be passed on to the next generation. Point mutations that alter a protein enough to affect its function are more often harmful than beneficial. Rarely, however, a mutation may result in an individual better adapted to the environment and more reproductively fit or a mutation already present in the population may be selected for when the environment changes.

Chromosomal mutations are most often deleterious. Occasionally, a translocation of a chromosomal piece may bring alleles together that are beneficial in combination. Duplication of chromosomal segments is usually harmful, but those that do not upset the genetic balance within cells may provide an expanded genome with extra loci that could eventually take on new functions by mutation. The shuffling of exons may also result in new genes.

Mutation is a source of genetic variation for bacteria and microorganisms with very short generation times. A new beneficial mutation can increase in frequency rapidly in a bacterial population that is growing by the asexual expansion of clones.

Most of the genetic variation present in a population of plants or animals, however, is due to unique recombinations of existing alleles that each individual inherits from the gene pool. Crossing over and independent assortment produce gametes with a great deal of genetic variation, and each zygote has a unique assortment of genes from two parents.

How Genetic Variation Is Preserved   Natural selection selects for favorable genotypes and tends to eliminate others, setting a trend toward genetic uniformity. Several mechanisms help to preserve variation.

Diploidy in most eukaryotes maintains genetic variation by hiding recessive alleles in heterozygotes, enabling them to perpetuate and be selected for, should the environment change.

Balanced polymorphism occurs when natural selection maintains variation at some gene loci. When individuals heterozygous at a certain gene locus are reproductively more fit, this *heterozygote advantage* tends to maintain two or more alleles at this locus. In the case of sickle cell anemia, a lethal recessive allele is maintained at a higher proportion in the population in countries with malaria due to the malaria-resistant advantage of the heterozygote.

Hybrid vigor may be seen in plants when two highly inbred varieties are crossbred. The crossbreeding may mask harmful recessive alleles that were homozygous in the inbred varieties and produce a heterozygote advantage at other loci.

Balanced polymorphism may be maintained in a population that ranges across a patchy environment. In the land snail, *Cepaea nemoralis*, each of the distinctively colored morphs is camouflaged in a particular patch of the habitat. In *frequency-dependent selection*, a morph's reproductive success declines if it becomes too common. The polymorphism of female African swallowtail butterflies that resemble several different species of noxious butterflies increases the effectiveness of this mimicry. Should any one morph become too common, birds would stop associating that coloration with bad taste.

Is All Genetic Variation Adaptive?   Some of the diversity seen in populations may be *neutral variations* that do not confer a selective advantage or disadvantage. By this *neutral theory of molecular evolution*, the relative frequencies of the alleles for the beta chain of human hemoglobin, for example, will not be affected by natural selection, but will change randomly by genetic drift.

There is no consensus on how much variation, if any, is truly neutral, since it cannot be shown that an allele confers no benefit at all to an organism.

Adaptive Evolution

Adaptive evolution is a combination of chance and sorting: the chance occurrence of new genetic variation by mutation and sexual recombination, and the sorting or selecting of those variations most fit for the environment.

Fitness The phrases "survival of the fittest" and "struggle for existence" are misleading. Darwinian fitness is measured only by the relative contribution an individual makes to the gene pool of the next generation. Success is determined not simply by survival of an individual, but by the number of offspring produced.

Population geneticists speak in terms of the *relative fitness* of a genotype as its contribution to the next generation as compared to the contribution of other genotypes. The most fecund variants are said to have a relative fitness of 1, whereas the fitness of another genotype is the percentage of offspring it produces in comparison. The *selection coefficient* is the difference between the two fitness values, a measure of the selection against the inferior variant. The larger the selection coefficient, the more disadvantageous the genotype; a lethal genotype would have a selection coefficient of 1.0.

The rate at which a deleterious allele declines in a population depends both on the selection coefficient and on whether the allele is dominant or recessive. Harmful recessives are rarely eliminated due to heterozygote protection; the more rare the allele becomes, the higher the percentage of the remaining alleles that are hidden in heterozygotes. Likewise, beneficial recessives increase slowly because they must occur as a homozygote to be selected for. Dominant alleles can be selected for and against much more rapidly.

What Does Selection Act On? Selection acts on phenotype, the physical traits of an organism, and indirectly adapts a population to its environment by selecting for and maintaining favorable genotypes in the gene pool.

Genotypes may be pleiotropic—having more than one effect. Some of these effects may be positive whereas others may be negative. Many traits are polygenic, influenced by several genes. Indeed most variation that natural selection acts upon is due to quantitative traits. Selection acts on an organism that is an integrated composite of many phenotypic features. The fitness of any genotype depends on the entire genetic context of the individual.

Modes of Natural Selection The frequency of a trait, especially a quantitative trait determined by many gene loci, may be affected by three modes of natural selection. *Stabilizing selection* acts against extreme phenotypes and favors more intermediate forms, tending to reduce phenotypic variation in stable environments. *Directional selection* occurs most frequently during periods of environmental change when individuals deviating in one direction from the average for some phenotypic trait may be favored. *Diversifying selection*, or disruptive selection, occurs when the environment favors individuals on both extremes of a phenotypic range. Balanced polymorphism may result from diversifying selection.

Sexual Selection *Sexual dimorphism* is the distinction between males and females on the basis of secondary sexual characteristics such as size, plumage, or antlers. In vertebrates, the male is usually the showier sex, and these characteristics may serve to attract females or compete with other males for females. *Sexual selection* is the selection for traits that may not be adaptive to the environment, but do enhance reproductive success by increasing an individual's success in attracting a mate. Females, through their choice of mates, play an important role in the evolution of such traits.

Does Evolution Fashion Perfect Organisms? There are at least four reasons why evolution does not produce perfect organisms. First, each species has evolved from a long line of ancestral forms, many of whose structures have been co-opted for new situations. Second, adaptations are often compromises between the need to do several different things, such as swim and walk, be agile and strong. Third, the evolution that occurs as the result of chance events, such as genetic drift, is not adaptive. Fourth, natural selection can act on only those variations that are available; new alleles do not arise when needed.

STRUCTURE YOUR KNOWLEDGE

1. **a.** What is the Hardy-Weinberg theorum?
   b. What is meant by Hardy-Weinberg equilibrium? Under what environmental conditions do we expect to find Hardy-Weinberg equilibrium?

c. Write the Hardy-Weinberg equation and define each of the variables. Make sure you understand how to use the Hardy-Weinberg equation to determine both allele and genotype frequencies.

2. Create a concept map that organizes your understanding of the possible causes of microevolution.

3. Natural selection tends to work toward genetic unity; the genotypes that are most fit produce the most offspring and increase the frequency of adaptive genotypes in the population. Yet there remains a great deal of variability within the populations of a species. Describe some of the factors that contribute to this genetic variability.

4. You collect 100 samples from a large butterfly population. Fifty specimens are dark brown, 20 are speckled, and 30 are white. Coloration in this species of butterfly is controlled by one gene locus: *BB* individuals are brown, *Bb* are speckled, and *bb* are white. What are the allele frequencies for the coloration gene in this population? Is this population in Hardy-Weinberg equilibrium? Explain your answer.

TEST YOUR KNOWLEDGE

MULTIPLE CHOICE: *Choose the one best answer.*

1. The modern evolutionary synthesis emphasizes
   a. that populations are the units in which evolution occurs.
   b. the essential role of natural selection in evolution.
   c. the gradualness of the evolutionary process.
   d. all of the above.

2. According to the Hardy-Weinberg theorem,
   a. the gene pool should remain constant from one generation to the next if five conditions are met.
   b. only natural selection, resulting in unequal reproductive success, will cause evolution.
   c. the square root of the frequency of individuals showing the recessive trait will always equal the frequency of q.
   d. all of the above are correct.

3. If a population has the following genotype frequencies, $AA = 0.42$, $Aa = 0.46$, $aa = 0.12$, what are the allele frequencies?
   a. $A = 0.42$, $a = 0.12$
   b. $A = 0.88$, $a = 0.12$
   c. $A = 0.65$, $a = 0.35$
   d. $A = 0.6$, $a = 0.4$

4. In a population with two alleles, *B* and *b*, the allele frequency of *B* is 0.8. What would be the frequency of heterozygotes if the population is in Hardy-Weinberg equilibrium?
   a. 0.16
   b. 0.32
   c. 0.64
   d. 0.8

5. In a population that is in Hardy-Weinberg equilibrium, 16% of the population show a recessive trait. What percent show the dominant trait?
   a. 48%
   b. 60%
   c. 84%
   d. 96%

6. Genetic drift is likely to be seen in a population
   a. that has a high migration rate.
   b. that has a high mutation rate.
   c. in which there is assortative mating.
   d. that is very small.

7. Cheetahs are believed to have experienced a bottleneck effect which resulted in
   a. reduced genetic variability.
   b. many alleles that are fixed in their population.
   c. a small population that is then subject to genetic drift.
   d. all of the above.

8. Gene flow often results in
   a. populations that are better adapted to the environment.
   b. an increase in sampling error in the formation of the next generation.
   c. a reduction of the allele frequency differences between populations.
   d. nonassortative matings.

9. The existence of two distinct phenotypic forms in a species is known as
   a. geographic variation.
   b. sexual selection.
   c. heterozygote advantage.
   d. polymorphism.

10. Assortative mating will most likely result in
    a. a change in allele frequency.
    b. an increase in the gene loci that are homozygous.
    c. sexual selection.
    d. neutral variations.

11. Variations in a species existing over a large temperature range may
    a. result in stabilizing selection.
    b. be a cline if the variations parallel the temperature change.

c. result in directional selection, favoring the milder climate.

d. all of the above.

12. Mutations are rarely the cause of microevolution because
   a. they are most often harmful and do not get passed on.
   b. they may be masked in diploid individuals and not able to be selected for.
   c. they occur very rarely.
   d. of all of the above.

13. In a study of a population of field mice, you find that 48% of the mice have a coat color that indicates that they are heterozygous for a particular gene. What would be the frequency of the dominant allele in this population?
   a. 0.4
   b. 0.5
   c. 0.7
   d. you cannot estimate allele frequency from this information.

14. In a random sample of a population of Shorthorn cattle, 73 animals were red ($C^R C^R$), 63 were roan ($C^R C^r$—a mixture of red and white), and 13 were white ($C^r C^r$). Estimate the allele frequencies of C^R and C^r and determine whether the population is in Hardy-Weinberg equilibrium.
   a. $C^R = 0.7$, $C^r = 0.3$; the genotypic ratio is what would be predicted from these frequencies and the population is in equilibrium.
   b. $C^R = 1.04$, $C^r = 0.44$; the allele frequencies add up to greater than 1 and the population is not in equilibrium.
   c. $C^R = 0.64$, $C^r = 0.36$; because the population is large and a random sample was chosen, the population is in equilibrium.
   d. you cannot estimate allele frequency from this information.

15. The rate at which a harmful allele declines in a population depends upon
   a. whether the allele is dominant or recessive.
   b. the selection coefficient for the genotypes in which it occurs.
   c. the frequency of the allele in the population.
   d. all of the above.

16. A scientist observes that the height of a certain species of asters decreases as the altitude on a mountainside increases. She gathers seeds from samples at various altitudes, plants them in a uniform environment, and measures the height of the new plants. All of her experimental asters grow to approximately the same height. From this she concludes that
   a. height is not a quantitative trait.
   b. the cline she observed was due to genetic variations.
   c. the height variation she initially observed was due to environmental factors.
   d. the differences in the parent plant's heights were due to directional selection.

17. Disruptive selection may result in
   a. balanced polymorphism.
   b. one phenotype gradually disappearing from the population.
   c. maladaptive evolution.
   d. a high selection coefficient.

18. Sexual selection will
   a. result in individuals better adapted to the environment.
   b. increase assortative mating.
   c. select for traits that enhance an individual's chance of mating.
   d. result in stabilizing selection.

19. The greatest source of genetic variation in most populations is from
   a. mutations.
   b. recombination.
   c. selection.
   d. polymorphism.

20. As a lethal recessive allele decreases in frequency in a population,
   a. the proportion of alleles hidden in the heterozygote increases.
   b. the selection coefficient approaches 1.
   c. its relative fitness increases.
   d. it becomes a neutral variation.

21. A plant population is found in an area that is becoming more arid. The average surface area of leaves has been decreasing over the generations. This is an example of
   a. stabilizing selection.
   b. directional selection.
   c. disruptive selection.
   d. heterozygote superiority.

22. If individuals with a particular genotype have a relative fitness of 0.7, they
   a. have a selection coefficient of 0.7.
   b. are carrying a lethal gene.
   c. are reproductively less fit in comparison with a genotype with a selection coefficient of 0.2.
   d. leave 30% as many offspring as do individuals with the most fecund genotype.

CHAPTER **22**

THE ORIGIN OF SPECIES

FRAMEWORK

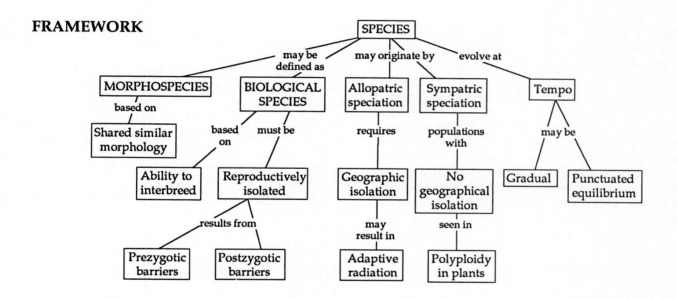

CHAPTER SUMMARY

The origin of species, or *speciation*, is the basis of the evolution of biological diversity. *Anagenesis*, or *phyletic evolution*, involves the transformation of a lineage of organisms into forms different enough to be considered new species. In *cladogenesis*, or *branching evolution*, new species arise from parent species that continue to exist. Cladogenesis is both the more common pattern of evolution and the process that increases biological diversity. Evolutionary theory attempts to determine the mechanisms by which new species originate.

The Species Problem

Taxonomists often find that their classification of local species corresponds to the folk taxonomy of a region.

Two Concepts of Species *Species* are most often characterized by their physical form or morphology. Species determined on the basis of anatomical features are known as *morphospecies*. The concept of *biological species*, developed by Mayr in 1942, goes beyond the physical differences between species and considers reproductive, and thus genetic, isolation to be the basis for separating species.

A biological species is a population or group of populations of individuals that have the potential to interbreed with each other in nature and produce fertile offspring, but which do not interbreed with other species in nature. A biological species is the largest unit in which gene flow is possible. Members of a species are said to be *conspecific.*

Limitations of the Biological Species Concept The biological species concept does not work for species that are completely asexual, such as prokaryotes and some protists and fungi. Extinct species also cannot be grouped based on the criterion of interbreeding.

We cannot tell whether populations that are geographically isolated are able to interbreed in nature. Even in some sexually reproducing, geographically neighboring species, it may be difficult to apply the biological species concept. Phenotypically distinct populations that are separated geographically, but presumably capable of interbreeding, are sometimes called subspecies. Gene flow among subspecies may be so circuitous or slight that it is difficult to decide whether some subspecies should be designated as separate species. These groups may be in the midst of the gradual evolution of populations into new species. Even if not universally applicable, however, the concepts of morphospecies and biological species usually result in the same classification of groups as species.

Reproductive Barriers

Any intrinsic mechanism that prevents two species from producing fertile hybrids is a reproductive barrier serving to preserve the genetic integrity of a species. *Prezygotic barriers* function before the formation of a zygote by preventing mating between species or the successful formation of a zygote. Should a hybrid zygote form, *postzygotic barriers* prevent it from developing into a fertile adult.

Prezygotic Barriers Prezygotic barriers include *habitat isolation*, in which two species may live in the same area but occupy different habitats; *temporal isolation*, in which two species breed at different times; *behavioral isolation*, in which courtship rituals and behavioral signals attract only conspecific mates; *mechanical isolation*, in which anatomical incompatibility or mechanical barriers prevent mating with nonconspecifics; and *gametic isolation*, in which the gametes of different species fail to fuse due to inhospitable female reproductive tract environment or lack of specific recognition molecules on the surfaces of gametes.

Postzygotic Barriers Postzygotic barriers include *hybrid inviability*, in which a hybrid zygote fails to survive embryonic development due to genetic incompatibility; *hybrid sterility*, in which a viable hybrid individual is sterile, often due to the inability to produce normal gametes in meiosis; or *hybrid breakdown*, in which the hybrids are viable and fertile, but their offspring are feeble or sterile.

Introgression When fertile hybrids do successfully mate with one of their parent species, genes may pass between species in a process called *introgression*. This small amount of gene transplantation increases the reservoir of genetic variation present in a species without threatening its integrity as a distinct species.

The Biogeography of Speciation

The evolution of reproductive barriers is the critical event in the origin of new species. When the gene pool of a population becomes separated from other populations, the isolated population can follow its own evolutionary course as a result of selection, genetic drift and mutation. Speciation mechanisms can be grouped by biogeographical factors: *Allopatric* speciation occurs when the gene pool of a population becomes segregated geographically from other populations. *Sympatric* speciation occurs when a subpopulation becomes reproductively isolated in the midst of its parent population.

Allopatric Speciation Geological change can separate and isolate populations. The extent of the geographical barrier necessary to maintain genetic separation depends on the ability of the organisms to disperse. The numerous species of pupfishes living in springs in Death Valley probably descended from an ancestral species whose range was fragmented when the region became arid.

Evidence shows that allopatric speciation occurs much faster in small populations, perhaps within only hundreds to thousands of generations.

Geographic isolation of a small population at the fringe of the parent population's range is most likely to result in speciation because of three factors: First, the gene pool of the peripheral isolate probably represents the extremes of any clines in the original population. If the isolated population is small, the founder effect may produce a population whose gene pool differs from the parent pool. Second, genetic drift in the small gene pool may cause phenotypic divergence from the parent population by chance. Third, selection pressures are likely to be different, and possibly more severe, on the fringe of a population's range, pushing the gene pool of the peripheral isolate in a different direction than that of the parent population. These factors do not guarantee successful speciation, and most pioneer populations probably become extinct.

Allopatric speciation may occur on island chains when small founding populations evolve in isolation and under somewhat differing environmental conditions. A single ancestral species of finch probably gave rise to the 13 species of finches now found on the Galapagos through the process of isolation and recolonization of islands. *Adaptive radiation* is the formation of numerous species from an ancestral species that has been confronted with new and diverse habitats. The Hawaiian Archipelago, a series of relatively young, isolated, and diverse islands, is home to thousands of endemic species, examples of adaptive radiation resulting from multiple invasions and allopatric speciations.

Sympatric Speciation   In sympatric speciation, a radical change in the genome of a subpopulation may result in its reproductive isolation in the midst of the parent population. Such events are more common in plants than in animals. In *autopolyploidy*, a single species doubles its chromosome number to the tetraploid (*4N*) state through the fusion of diploid gametes. Tetraploids can fertilize themselves or mate with other tetraploids, but cannot mate with diploids from the parent population, resulting in reproductive isolation in just one generation. De Vries documented this form of speciation in the evening primrose.

Polyploid species arise more commonly through *allopolyploidy*, which is interspecific hybridization. Such hybrids are usually sterile due to difficulties in the meiotic production of gametes, but one of two mechanisms may result in the production of a fertile polyploid: (1) mitotic nondisjunction in the reproductive tissue allows the hybrid to produce gametes with complete sets of chromosomes or (2) an unreduced diploid gamete from one plant may fuse with a normal haploid gamete of another, creating a sterile triploid hybrid. If a later nondisjunction in an asexual clone produces a triploid gamete that unites with a normal haploid gamete of the second plant, a fertile hybrid results. The chromosome number of an allopolyploid is equal to the sum of the diploid chromosome numbers of the parent species.

Speciation of polyploids, especially allopolyploids, has been frequent and important in plant evolution. Many of our agricultural plants are polyploids. Plant geneticists are now hybridizing plants and using chemicals to induce nondisjunction in order to create new, specially adapted polyploid species.

Sympatric speciation in animals may involve isolation within the geographic range of the parent population based on different resource usage. Assortative mating within a population that had a balanced polymorphism could also lead to sympatric speciation.

Genetic Mechanisms of Speciation

Templeton and other population geneticists have recommended grouping speciation mechanisms according to genetic processes, rather than biogeographical factors.

Speciation by Divergence   Reproductive barriers may arise coincidentally as two populations accumulate genotypic disparities as they adapt to different environments. In evolution by divergence, natural selection does not necessarily favor reproductive barriers; genetic isolation may be a secondary consequence of adaptations to different environments. Postzygotic reproductive barriers that affect the viability of hybrids may result from pleiotropic effects of important genes controlling development. As populations diverge in their adaptation to specific environments, prezygotic habitat barriers can also evolve.

Sexual selection, which enhances an organism's reproductive success with its own species, may lead indirectly to reproductive barriers. The *recognition concept of species* emphasizes that natural selection acts on characteristics that maximize successful mating within a species, as opposed to driving the development of reproductive barriers in an isolated population that may be speciating.

Speciation by Peak Shifts   According to Wright's "adaptive landscape" metaphor, *adaptive peaks* occur where a population's gene pool is at an equilibrium that maximizes fitness. Valleys separating the peaks represent low average fitness of individuals. Natural selection will tend to push the population back to the former peak when there has been some slight nonadaptive change in allele frequency. When environmental conditions change, the adaptive landscape itself changes, creating new peaks and valleys.

Speciation by peak shifts, associated with nonadaptive changes in the gene pool, can be initiated by a bottleneck in a population. Genetic drift can knock a small population off its adaptive peak, and natural selection can then push it to a new adaptive peak. The adaptive landscape is redefined by a change in location. Following a founder effect and resulting genetic drift, adaptive evolution can push a destabilized gene pool to a new adaptive peak in a new environment.

How Much Genetic Change Is Required for Speciation?
Substantial genetic change is not required for speciation. Reproductive isolation may occur by a change

at just a few gene loci or by cumulative divergence at many gene loci.

Gradual and Punctuated Interpretations of Speciation

The traditional evolutionary concept of the origin of species involves the gradual divergence of populations by microevolution, with each newly formed species continuing to evolve over long periods of time. The fossil record, however, provides few cases of transitional forms. Rather, new forms appear rather suddenly, persist unchanged for a long time, and then disappear.

In 1972 Eldredge and Gould proposed the theory of *punctuated equilibrium*, in which long periods of stasis are punctuated by episodes of relatively rapid change and speciation. Speciation by polyploidy in plants and by small allopatric populations may occur fairly rapidly. In geological time, a few thousand years for a species to evolve is small compared to the few millions of years a successful species may remain in existence.

The theory of punctuated equilibrium posits that a species may change very little after its origin if it remains in a stable environment. Long periods of stasis may result from the tendency of stabilizing selection to maintain a population at an adaptive peak.

Some gradualists maintain, however, that fossils show stasis only in external anatomy and that changes in internal anatomy, physiology, and behavior go unrecorded. Even reports of long periods of morphological stasis are debatable: After analyzing a particularly complete set of trilobite fossils, Sheldon found a gradual change in aspects of their morphology, which challenged earlier categorizations of the youngest and oldest fossils in each evolutionary lineage as different species. Many additional studies of fossil lineages will be needed to determine the roles of gradual and punctuated tempos in the origin of species.

STRUCTURE YOUR KNOWLEDGE

1. How are speciation and microevolution different?

2. What is probably the key event in the origin of a species? How might this event occur?

3. Compare the gradual and punctuated equilibrium theories of evolution and give evidence to support each theory.

TEST YOUR KNOWLEDGE

MULTIPLE CHOICE: *Choose the one best answer.*

1. Most new species probably have arisen by
   a. anagenesis.
   b. cladogenesis.
   c. phyletic evolution.
   d. both a and c.

2. Which of the following is *not* a type of intrinsic reproductive isolation?
   a. mechanical isolation
   b. behavioral isolation
   c. geographical isolation
   d. gametic isolation

3. A morphospecies and a biological species may differ because
   a. organisms that look different may be able to interbreed.
   b. organisms that do not breed in nature may do so in captivity.
   c. hybrids formed from the mating of two morphospecies are sterile.
   d. anatomical features do not provide a reliable basis for grouping organisms.

4. Gene flow can occur within all of the following groups of organisms except
   a. a population.
   b. a species.
   c. a genus.
   d. a subspecies.

5. For which of the following is the biological species concept least appropriate?
   a. plants
   b. bacteria
   c. mules
   d. field biologists

6. Subspecies are usually
   a. separated geographically.
   b. phenotypically distinct.
   c. assumed to be capable of interbreeding.
   d. all of the above.

7. If two species are able to breed but produce sterile hybrids, their genetic integrity is maintained by
   a. gametic isolation.
   b. a prezygotic barrier.
   c. hybrid inviability.
   d. a postzygotic barrier.

8. Introgression occurs when
   a. a hybrid successfully breeds with an individual of a parent species.

b. hybrids successfully breed with each other.

c. an allopolyploid is produced.

d. all of the above take place.

9. Speciation is most likely to occur in allopatric populations when

a. the splinter population is small.

b. a peripheral isolate is exposed to more severe selection pressures at the boundary of the population's range.

c. a small subset of the population becomes adapted to a new food source.

d. both a and b.

10. A tetraploid plant species is probably the result of

a. allopolyploidy.

b. autopolyploidy.

c. hybridization and nondisjunction.

d. allopatric speciation.

11. Sexual selection may lead indirectly to reproductive barriers because

a. isolated populations are exposed to different selection pressures.

b. natural selection could act on characteristics that maximize successful mating within a species.

c. hybrids may be reproductively more fit.

d. prezygotic barriers are more likely to evolve before postzygotic barriers.

12. Adaptive radiation may occur when

a. a small population is reunited with its parent population when a geographical barrier is removed.

b. numerous invasions and allopatric speciations take place on islands that provide a diversity of environmental opportunities and problems.

c. a segment of the population develops a reproductive isolating mechanism.

d. all of the above.

13. According to Wright's evolutionary metaphor,

a. a population has only one adaptive peak and is maintained there by natural selection.

b. a small population that drifts into a valley surrounding an adaptive peak will either become extinct or evolve into a new species.

c. an environmental change may create a new adaptive landscape and natural selection can push a destabilized gene pool to a new adaptive peak.

d. the valleys between peaks represent changes in environmental conditions and populations must move through these valleys to become a new species.

14. The punctuated equilibrium theory maintains that

a. long periods of stasis are punctuated by episodes of relatively rapid speciation and change.

b. microevolution is the driving force of speciation.

c. most rapid speciation events involve polyploidy in plants.

d. evolution occurs rapidly when the adaptive landscape changes.

15. A new plant species formed from hybridization of a plant with a diploid number of 18 with a plant with a diploid number of 12 would probably have a gamete chromosome number of

a. 12.

b. 15.

c. 18.

d. 30.

16. Sheldon's analysis of huge numbers of trilobite fossils provides support for

a. the punctuated equilibrium model of evolution.

b. the gradual tempo of the evolution of new species.

c. the need for additional exhaustive studies of fossil lineages.

d. both b and c.

CHAPTER 23

TRACING PHYLOGENY: MACROEVOLUTION, THE FOSSIL RECORD, AND SYSTEMATICS

FRAMEWORK

This chapter considers the major events and evolutionary trends that led to the biological diversity of today. Mechanisms for macroevolution include the gradual modification of preadapted structures for new functions, evolutionary trends perhaps as a result of species selection, and adaptive radiations as new adaptive zones appear when evolutionary novelties develop or mass extinctions occur. Continental drift and other major geological events affect the course of macroevolution.

A goal of systematics is to determine the phylogenetic history of species. The branch of systematics called taxonomy names and classifies species according to their presumed evolutionary relationships. Phylogenetic trees are constructed from evidence gathered from the fossil record and from anatomical and molecular homologies.

The relationship between microevolution and macroevolution and the role of natural selection and chance in macroevolution are controversial topics in contemporary evolutionary theory.

CHAPTER SUMMARY

Macroevolution is the origin of taxonomic groups higher than species. When biologists study macroevolution, they consider the major events revealed by the fossil record: large diversifications of taxonomic groups, the origin of novel biological designs, major extinctions, and the mechanisms that may have produced these evolutionary developments.

The Record of the Rocks

Paleontologists reconstruct evolutionary history by studying the succession of organisms found in the fossil record.

How Fossils Form   Fossils are preserved impressions or remnants of past organisms. Sedimentary rocks formed from the compression of sand and mud deposits are the richest sources of fossils. Hard parts of animals, such as bones, teeth, or shells, may remain as fossils. Petrification, the replacement of organic tissue with dissolved minerals, may turn the fossil to stone. Sometimes enough organic material is retained in the fossil that biochemical analysis and electron microscopy on cells can be done.

Molds of organisms, left when their bodies decay and minerals dissolved in water crystallize in the cast, are a common type of fossil. Trace fossils formed in footprints, burrows, or other impressions of animal activities, are like fossilized behavior, giving clues to how animals moved or lived. Some fossils are the entire body of an organism preserved in a substance such as ice, amber, or bogs that retard decomposition.

Limitations of the Fossil Record   The formation of a fossil is an unlikely occurrence. The incompleteness of the fossil record is understandable considering that a large number of species that lived probably left no fossils, most fossils that form are destroyed, and only a fraction of existing fossils have been found. Organisms that were abundant, widespread, existed over a long period of time, and had hard shells or skeletons are overrepresented in the fossil record. Even with these limitations, fossils leave a detailed register of macroevolution.

The Fossil Record and the Geological Timetable
Layers, or strata, of rock form during periods of sedimentation within bodies of water. The order in which fossils appear in the strata of sedimentary rocks indicates their relative age. Gaps may appear in the sequence, indicating that an area was above sea level or underwent erosion during some period of geological time.

Index fossils, such as shells of widespread animals, are used to correlate strata from different locations and allow geologists to develop a composite picture of a consistent sequence of geological periods. There are four geological eras—the Precambrian, Paleozoic, Mesozoic, and Cenozoic—that are delineated by major transitions, such as mass extinction and extensive radiation. The eras are subdivided into periods and epochs, which are associated with particular evolutionary developments shown in the fossils found in the strata.

Radioactive dating is used to determine the age of rocks and fossils. During an organism's lifetime, it accumulates radioactive isotopes in proportions equal to the relative abundance of the isotopes in the environment. After the organism dies, the isotopes decay at a fixed rate, known as the *half-life* or the number of years it takes for one-half of the radioactive isotopes in the specimen to decay. Carbon-14 dating is used for determining the age of relatively young fossils; potassium-40, with a half-life of 1.3 billion years, can be used to date rocks hundreds of millions of years old.

Racemization, or the chemical conversion of *L*-amino acids to *D*-amino acids, is another means of dating organic remains. During an organism's life, only *L*-amino acids are synthesized, but following death *L*-amino acids are slowly converted to *D*-amino acids. The proportion of *L*- and *D*-amino acids can be used to date some fossils.

Mechanisms of Macroevolution

Origins of Evolutionary Novelties Evolutionary novelties that define higher taxa may evolve by the gradual modification of existing structures for new functions. *Preadaptation* is the term for structures that evolved and functioned in one setting and were then co-opted for a new function. Feathers and wings may have first developed as structures for capturing prey and were then shaped into structures adapted for flight. Preadaptation, together with the Darwinian concept of gradualism, provide a mechanism by which novel designs can arise slowly through many small changes in current structures.

Slight changes in the function of genes that control development may create major changes in adult morphology. *Allometric growth*, the different rates of growth in various parts of the body, results in the final shape of the organism. A minor genetic alteration that affects allometric growth can have a major morphological effect.

Novel organisms can also be produced by genetic changes that affect the timing of developmental events. *Paedomorphosis* is the retention in the adult of juvenile traits of ancestral organisms. The continuation of growth of the human brain for several years longer than the chimpanzee brain is a prolonging of a juvenile process that has greatly influenced human intelligence.

Both *heterochrony*, evolutionary changes in the rate or timing of development, and *homeotic* changes that alter the spatial arrangement of body parts, have been important in the macroevolution of new forms of organisms.

The Difficulty of Interpreting Evolutionary Trends
Evolutionary trends rarely occur in the fossil record as a sequence of gradually changing intermediate forms. The modern horse, *Equus*, for example, descended from its much smaller ancertor, *Hyracotherium*, through a series of speciation episodes including several adaptive radiations producing many different species that appeared in the fossil record, remained fairly unchanged, and then became extinct. *Equus* did not phyletically evolve as the direct result of trends of increasing size, reduction in the number of toes, and changes in dentition.

Evolutionary trends that are evident in the fossil record are most often the result of punctuated equilibrium, in which new species change by increments as they branch from ancestral ones.

According to Stanley's model of *species selection*, an evolutionary trend is analogous to a trend in a population produced by natural selection, where the best-adapted individuals are most successful reproductively. The species that live the longest before extinction and speciate most (i.e. produce the most offspring) determine the direction of the trend. Evolutionary trends are ultimately dictated by environmental conditions; if conditions change, an evolutionary trend may end or change.

Continental Drift and the Biogeography of Macroevolution Biogeography, the geographic distribution of species, correlates with the geological history of the Earth. The continents rest on great plates of crust that float on the molten mantle. These plates shift and move, creating earthquakes, volcanoes, and mountains in regions where plates abut.

Large-scale continental drift brought all the land masses together into a supercontinent named *Pangaea* about 250 million years ago, near the end of the Paleozoic era. This tremendous change undoubtedly had a great environmental impact as shorelines were eliminated, oceans deepened, shallow coastal areas were drained, and the climate of the land mass changed. Species that had evolved in isolation came together and competed, many species became extinct, and new opportunities for remaining species became available.

About 180 million years ago, during the early Mesozoic era, Pangaea broke up and the continents drifted apart, creating a huge geographic isolation event. Current biogeography, including the marsupial diversification in Australia, reflects this separation.

Punctuations in the History of Biological Diversity
Major adaptive radiations occurred early in the history of some taxa, when the evolution of a novel characteristic opened a new *adaptive zone*, or way of life with unexploited opportunities. Nearly all the animal phyla that exist today evolved during the first 10 to 20 million years of the Cambrian, the first period of the Paleozoic era. The origin of shells and skeletons in a few taxa opened an adaptive zone by making more complex designs possible and changing predator-prey relationships.

An empty adaptive zone can be exploited only if appropriate evolutionary novelties arise, and novelties that do arise cannot enable organisms to move into adaptive zones that do not exist or are filled. Mammals existed at least 75 million years before their first major adaptive radiation, which has been linked to the ecological void left by the extinction of the dinosaurs. Mass extinctions are often followed by new adaptive radiations.

A species may become extinct due to a change in its physical or biological environment. Extinction, inevitable in a changing world, usually occurs at a rate of between 2.0 and 4.6 families per million years. During periods of major environmental change, mass extinctions may occur.

About a dozen mass extinctions have been found in the fossil record. The Permian extinctions, occurring at the boundary between the Paleozoic and Mesozoic eras about 250 million years ago, claimed over 90% of marine animal species. This extinction coincided with the formation of Pangaea and may be related to environmental changes associated with that event.

The Cretaceous extinction, which marks the boundary between the Mesozoic and Cenozoic eras about 65 million years ago, claimed over one-half the marine species, many families of terrestrial plants and animals, and the dinosaurs. The climate was cooling during that time and shallow seas were receding from continental lowlands.

The climate change may have been caused by increased volcanic activity which blocked sunlight from reaching the Earth. There is also evidence that an asteroid or comet collided with Earth during the Cretaceous extinctions. The thin layer of clay enriched in iridium that separates Mesozoic from Cenozoic sediments may have been the fallout from a huge cloud of dust created when an asteroid hit Earth. This cloud, similar to what is projected in a nuclear winter, would have blocked light and severely affected weather for several months.

Systematics: Tracing Phylogeny

Phylogeny is the evolutionary history of a species. A phylogenetic tree is a diagram of proposed evolutionary relationships of various groups. *Systematics* is the branch of biology concerned with the diversity of life and its phylogenetic history.

Taxonomy   Taxonomy involves the identification and classification of species. Linnaeus developed a system of taxonomy that assigned each species a two-part Latin name—a *binomial*. The first word is the genus name and the second is the *specific epithet* or species name. Linnaeus also developed a hierarchy of classifications that organizes similar groups into more general categories, proceeding from species to *genus, family, order, class, phylum*, and finally to *kingdom*.

Taxonomy names and describes diagnostic characteristics for new species. Species are ordered into *taxa*, or units of increasingly broader taxonomic categories. Both words of the binomial are italicized, and the names for all taxa at the genus level and higher are capitalized.

Only species exist in nature as biologically identifiable groups, connected by interbreeding and reproductively isolated from other groups. The assignment of taxa to each broader category is often subjective, depending upon the distinctions a taxonomist deems important. A goal of systematics is that classification reflects evolutionary relationships of species. Each taxon should be *monophyletic*, meaning that it should contain only species that are derived from a single ancestor. A *polyphyletic* taxon includes groups having different ancestry, and a *paraphyletic* taxon excludes species that share a common ancestor with species in the taxon.

Sorting Homology from Analogy   Species are generally classified into higher taxa based on similarities in morphology and other characteristics. *Homology* is likeness based on a shared ancestry; *analogy* is similarity due to evolutionary convergence. In *convergent evolution*, unrelated species develop similar features because they have similar ecological roles and natural selection has resulted in analogous adaptations.

Phylogenetic trees are built on homologous similarities. In general, the more homology between two species, the more closely they are related. The issue can be confused when adaptation leads to large differences and convergence creates misleadingly similar structures. When two similar structures are very complex, they are more likely to be homologous and have a shared origin.

Molecular Systematics A powerful taxonomic tool is the comparison of sequences of amino acids in proteins and nucleotides in DNA as evidence of shared ancestry. The protein cytochrome *c* is found in all aerobic organisms in slightly different forms. Comparisons of the amino acid sequences of this protein from many species show that differences increase as the groups are more taxonomically distant. Phylogeny based on cytochrome *c* is consistent with comparative anatomy and fossil evidence.

DNA-DNA hybridization is a technique that measures the extent of base pairing between single strands of DNA from two species by the temperature at which hybridized strands melt. The more extensive the pairing, the tighter the strands are bound and the higher the temperature needed to melt them, indicating a greater homology between the DNA sequences and a closer phylogenetic relationship between the species. Evidence from this technique generally supports phylogenetic relationships established by other methods and has been used to settle some old taxonomic disputes.

Restriction mapping of DNA compares segments of DNA from different species produced by restriction enzymes used in recombinant DNA technology. If two species have diverged greatly, their collections of restriction fragments will not match closely. Restriction mapping of mitochondrial DNA, which mutates about ten times faster than nuclear DNA, provides phylogenetic information for closely related species.

DNA sequencing, determining the actual order of nucleotides in DNA, provides the most accurate information on genetic divergence between species. The sequencing of ribosomal RNA, whose genes change slowly compared to other DNA, can be used to determine early phylogenetic relationships.

Molecular systematics have now been applied to DNA extracted from fossils. Small quantities of DNA are amplified using polymerase chain reaction and tested for homology with modern species. Analysis of DNA from human fossils will help anthropologists to determine human genealogy.

Some proteins seem to change or evolve at consistent rates. The number of amino acid substitutions in homologous proteins is proportional to the time elapsed since two species diverged. DNA comparisons also have been used as molecular clocks to date phylogenetic branchings. Dates obtained by these methods are generally consistent with the fossil record and may be more closely correlated with the time species have been separated than are morphological differences.

The consistent rate of protein change and the rate of DNA divergence indicate that the accumulation of neutral mutations may change the genome as a whole more than do changes brought about by natural selection. Disagreement about the extent of neutral mutations has caused some evolutionists to rely on molecular systematics to determine the sequence of branches in phylogeny, but not the actual dates for the origin of taxa.

Schools of Taxonomy Phylogenetic trees indicate the relative time of origin of different taxa and the degree of divergence that develops between branches. *Phenetics* is a school of taxonomy that determines taxa strictly on the basis of phenotypic similarities and differences for as many characteristics as possible without making *a priori* assumptions about phylogenetic relationships. Computers are used to make quantitative comparisons between species.

Cladistics is a taxonomic school that classifies organisms according to the order that clades, or evolutionary branches, arose. Phylogeny is diagrammed on a cladogram, a series of dichotomous forks, each of which is defined by novel homologies for the species on that branch. A primitive character, or *plesiomorphic character*, is common for all the species to be grouped, and derived characters, or *apomorphic characters*, are identified as homologies that evolved after a branch point and define that branch.

In the cladistic approach, birds are closer relatives of crocodiles than crocodiles are of lizards and snakes. Birds and crocodiles share apomorphic characters not found in snakes and lizards. According to cladistics, the class Aves does not exist because birds are found within the clade of reptiles. The degree of morphological difference between branches is not a consideration in cladistics, only the time and sequence of evolutionary origin.

Classical evolutionary taxonomy, which predates both phenetics and cladistics, considers both the homology of structures and the sequence of branching in developing phylogenetic trees. In cases of conflict between these two characteristics, a subjective decision is made. According to classical taxonomy, even though birds may share a close phylogenetic branch with crocodiles, they are assigned to their own class on the basis of the major adaptive divergence that has resulted from the ability to fly.

Is a New Evolutionary Synthesis Necessary?

The modern synthesis, which has dominated evolutionary theory for 50 years, combines contributions from paleontology, biogeography, systematics, and population genetics, along with recent discoveries in molecular biology, to reaffirm the Darwinian view of life. This paradigm maintains that the gradual accumulation of many small changes occurring over vast periods of time can result in large-scale evolutionary changes. Microevolution, changes in gene frequencies in populations, is sufficient to explain most macroevolution, and natural selection is seen as the major cause of evolution at all levels.

A current evolutionary debate concerns both the rate of evolution (gradualism versus punctuated equilibrium) and the relative contribution of microevolution to macroevolution. Some scientists favor a hierarchical theory, which maintains that events that lead to speciation and episodes of macroevolution may have little to do with adaptation and a lot to do with *contingency*, the occurrence of chance events. According to this theory, the beginnings of most new species result from chance geographic isolation, genetic accidents, and genetic drift in small, isolated populations. Macroevolution related to continental drift and mass extinctions is believed to have affected biological diversity at least as much as has the gradual adaptation that results from selection.

The modern synthesis does not claim that evolution is always gradual or that chance events do not change gene pools. The major difference between the modern synthesis and the hierarchical theory is the relative importance of the different evolutionary mechanisms of natural selection and chance events, and the connection between microevolution and macroevolution. Evolutionists from both schools, however, agree that natural selection is the mechanism of adaptation that fine-tunes a population to its environment.

STRUCTURE YOUR KNOWLEDGE

1. Organize a concept map that describes systematics, the major objectives of taxonomy, and the three schools of taxonomic thought.

2. Compare microevolution and macroevolution. To what extent are these two related?

3. Continental Drift and Biogeography: Match the maps of the world with their appropriate points on the geological time scale. Describe the significance of each of these stages of continental drift to the biogeography of the earth.

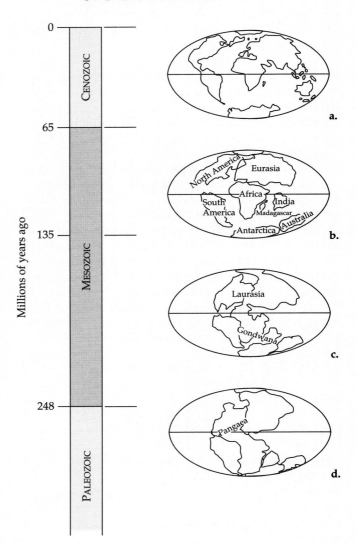

TEST YOUR KNOWLEDGE

MULTIPLE CHOICE: *Choose the one best answer.*

1. The richest source of fossils is found
   a. in coal and peat moss.
   b. along gorges.

    c. within sedimentary rock strata.
    d. encased in volcanic rocks.

2. Which of the following is least likely to leave a fossil?
    a. a soft-bodied land organism such as a slug
    b. a marine organism with a shell such as a mussel
    c. a vascular plant embedded in layers of mud
    d. a fresh-water snake

3. Index fossils are fossils of
    a. unique organisms that are used to determine the relative rates of evolution in different areas.
    b. widespread organisms that allow geologists to correlate strata of rocks from different locations.
    c. transitional forms that link major evolutionary groups.
    d. extinct organisms that mark the separation of different eras.

4. The half-life of carbon-14 is 5600 years. A fossil that has one-eighth the normal proportion of C-14 to C-12 is probably
    a. 2800 years old.
    b. 11,200 years old.
    c. 16,800 years old.
    d. 22,400 years old.

5. The Permian extinctions between the Paleozoic and Mesozoic eras
    a. resulted in mass extinction of many marine invertebrates.
    b. was not well chronicled because most of the species that became extinct were soft bodied.
    c. coincided with the breaking apart of Pangea.
    d. made way for the adaptive radiation of mammals, birds, and pollinating insects.

6. The relatively youngest fossils
    a. will have the greatest amount of *D*-amino acids.
    b. will be found in the highest strata.
    c. will have the greatest amount of DNA-DNA hybridization.
    d. will have less radioactivity.

7. In the binomial, *Homo Sapiens*,
    a. *Homo* is the name of the genus.
    b. *Sapiens* should not be capitalized.
    c. the words are italicized because they are Latin.
    d. all of the above.

8. Related families are grouped into the next highest taxon called a
    a. class.

    b. order.
    c. phylum.
    d. genus.

9. The only biologically identifiable group in nature is the
    a. kingdom.
    b. phylum.
    c. genus.
    d. species.

10. When two structures are homologous, they
    a. indicate that the species have evolved from a common ancestor.
    b. share common embryological development patterns.
    c. still may be adapted for different functions in the two species.
    d. all of the above.

11. Species selection
    a. can cause evolutionary trends when the most successful species exist for long periods and speciate often.
    b. is not a gradual process because most species form, change little, and then become extinct.
    c. is the result of chance events, not natural selection.
    d. can explain the rapid speciation events following mass extinctions.

12. Convergent evolution may result
    a. when species have similar ecological roles.
    b. when homologous structures are adapted for different functions.
    c. as a result of adaptive radiation.
    d. when species are widely separated geographically.

13. *Lystrasaurus*, a fossil dinosaur, is found on most continents. This distribution is best explained by
    a. the relative ease of finding dinosaur bones.
    b. divergent evolution.
    c. continental drift.
    d. dispersal capabilities of ancient reptiles.

14. The development of feathers in the ancestors of birds is an example of
    a. convergent evolution.
    b. divergent evolution.
    c. preadaptation.
    d. adaptive radiation.

15. Shared derived characters are
    a. used to characterize all the species on a branch of a cladogram.
    b. determined through a computer comparison of

all anatomical characteristics of a group of species.

c. homologous structures that develop during adaptive radiation.

d. used to indicate the degree of morphological differences between clades.

16. Many hypotheses have been suggested for the extinction of the dinosaurs. The explanation that probably accounts for most of the decline in species diversity at the end of the Mesozoic is
   a. the collision of an asteroid with Earth that wiped out populations of many species.
   b. mammalian competition with poorly adapted reptiles.
   c. ecological shifts due to climatic change.
   d. overpopulation of the huge reptiles.

17. Allometric growth
   a. is the uneven growth of different body parts.
   b. involves paedomorphosis.
   c. results in an evolutionary trend of increasing body size.
   d. is an example of a homeotic alteration.

18. DNA—DNA hybridization compares the genomes of two species by
   a. determining the temperature at which hybrid DNA melts and the strands separate.
   b. determining the extent of base pairing between single-stranded DNA of two species.
   c. establishing the degree of DNA homology and thus the degree of relatedness between two species.
   d. all of the above.

TRUE OR FALSE: *Indicate T or F, and then correct the false statements.*

1. _____ A monophyletic taxon includes only species that share a common ancestor.

2. _____ The more the sequences of amino acids in homologous proteins vary, the more recently the two species have diverged.

3. _____ Phylogenetic trees determined on the basis of similar structures may be inaccurate when adaptive radiations have created large differences or when convergent evolution has created misleading similarities.

4. _____ Phenetics is the school of taxonomy that is concerned with the order in which new groups branch over time on the phylogenetic tree.

5. _____ The retention in the adult of juvenile traits of ancestors is called paedomorphosis.

6. _____ Carbon-14 dating is not a good technique for sequencing fossils from the Carboniferous Period because carbon is too abundant in that period to allow accurate dating.

7. _____ The layer of clay enriched in iridium between the sediments of the Mesozoic and Cenozoic eras is indicative of a nuclear winter.

8. _____ According to the modern synthesis, natural selection is seen as the major cause of evolution at all levels.

ANSWER SECTION

CHAPTER 20
DESCENT WITH MODIFICATION: A DARWINIAN VIEW OF LIFE

Suggested Answers to Structure Your Knowledge

1. The two major components of Darwin's evolutionary theory are that all of life has descended from a common ancestral form and that the modification of that form has been the result of natural selection by the reproductive success of the individuals best adapted to the environment. The theory of natural selection is based on several facts and assumptions. The overproduction of offspring in conjunction with limited resources leads to a struggle for existence and differential reproductive success of those organisms best suited to the local environment. The unequal survival and reproduction of the most fit individuals in a population leads to the gradual accumulation of adaptive characteristics in a population within a particular environment.

2. Sources of evidence for evolution:
   (a) The distribution of species geographically (*biogeography*) supports evolutionary theory because of the similarities among species in close proximity and the differences in species that are separated by great distances, even if their habitats are similar.
   (b) The *fossil record* documents the changes that have occurred in species throughout time and provides transitional forms that link different phylogenetic groups.
   (c) The *taxonomic groupings* of species into a hierarchy of related groups points to the evolution of species from common ancestors.

   (d) *Comparative anatomy* illustrates the relatedness of groups based on their similar structures. Homologous structures and vestigial structures provide evidence of evolution.
   (e) *Comparative embryology* illustrates the commonalties of developmental patterns in related groups. It also helps to identify the homology of structures that may appear different in adult form.
   (f) *Molecular biology*—the similarities in the nucleotide sequences of DNA and protein products among related species and the common genetic code give evidence for descent from a common ancestor.

Answers to Test Your Knowledge

Matching:

1. F	e		5. D	c
2. A	b		6. H	a
3. G	g		7. E	i
4. C	d		8. B	h

Multiple Choice:

1. b		5. a		9. d	
2. c		6. c		10. d	
3. d		7. d			
4. a		8. a			

CHAPTER 21
HOW POPULATIONS EVOLVE

Suggested Answers to Structure Your Knowledge

1. **a.** The Hardy-Weinberg theorum states that allele and genotype frequency within a population will remain constant from one generation to the next as long as the population is large, mutation and migration are neglible, and no selective pressure operates.

b. A population in Hardy-Weinberg equilibrium does not show a change in allele or genotype frequency from one generation to the next. Such an equilibrium would most likely occur in populations in stable, unchanging environments.

c. $p^2 + 2pq + q^2 = 1$ In the Hardy-Weinberg equation, p and q refer to the frequencies of two alleles (A and a) in the gene pool. The frequency or proportion of gametes containing the A allele is equal to p, and the frequency of gametes containing the a allele is equal to q. The proportion of offspring resulting from gametes containing two A alleles is ($p \times p$) or p^2. Likewise, aa offspring will equal q^2. Aa individuals can be formed in two ways, depending on whether the ovum or sperm carries the A or a allele. Thus the frequency of heterozygous Aa offspring is equal to $2pq$.

2.

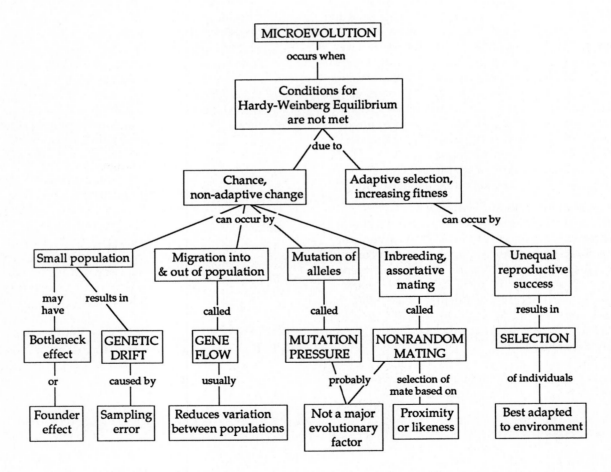

3. Genetic variation is retained within a population by diploidy and balanced polymorphism. Diploidy masks recessive alleles from selection when they occur in the heterozygote. Thus, less adaptive or even harmful alleles are maintained in the gene pool and are available should selec-tion pressures change and they become relatively better adapted to the new environment. Balanced polymorphism results in several alleles for a gene locus being maintained in a population. In situations in which there is heterozygote advantage, two alleles may be simultaneously selected for and retained in roughly equal frequencies within the gene pool. A patchy environment can favor different morphs within the geographic range of a species. Frequency-dependent selection, in which morphs present in higher numbers are selected against by predators, is another cause of balanced polymorphism.

4. The allele frequencies estimated from this sample are $B = 0.6$ and $b = 0.4$. The frequency of B gametes contributed to the allele pool by BB individuals would be 0.5, and the Bb individuals would contribute 0.1 B gametes and 0.1 b gametes. The bb individuals would contribute b gametes at a 0.3 frequency. Assuming this sample truly represents the individuals in the larger population, then we must say that the population is not in Hardy-Weinberg equilibrium. If it were, we would expect to find

36% BB, 48% Bb, and 16% bb individuals. The larger numbers of both homozygous genotypes indicates that assortative mating may be affecting the genotypic frequencies in the population. Other possible explanations for these data are incorrect sampling techniques or the action of some selection agent on speckled butterflies that reduces the numbers of heterozygotes in the population.

Answers to Test Your Knowledge

1. d	6. d	11. b	16. b	21. b
2. a	7. d	12. d	17. a	22. c
3. c	8. c	13. d	18. c	
4. b	9. d	14. a	19. b	
5. c	10. b	15. d	20. a	

CHAPTER 22
THE ORIGIN OF SPECIES

Suggested Answers to Structure Your Knowledge

1. Speciation is the process by which a new, reproductively isolated species evolves from its predecessor. It is part of the process of evolution and the increase in biological diversity. Microevolution is the process by which changes occur within the gene pool of a population, as a result of either chance events or natural selection. If the makeup of the gene pool changes enough, microevolution may lead to speciation.

2. Reproductive isolation is probably the key event in the origin of a species. Isolating mechanisms may develop as the result of chance events. Natural selection then may work on the newly-isolated gene pool to develop a population adapted for its environment. These chance events may occur because of geographic separation of a small population and genetic drift, or mutations, as in polyploidy in plants. Sexual selection, by which natural selection reinforces mechanisms that maximize successful mating within a species, may lead indirectly to reproductive barriers between speciating populations.

3. In the gradual tempo theory of evolution, small changes accumulate within populations as a result of chance events and natural selection. These changes may lead, particularly as environmental conditions change, to the gradual evolution of new life forms. The punctuated equilibrium theory holds that evolution occurs in spurts of relatively rapid change inserted within large periods of stasis. The fossil record seems to indicate that new species appear fairly rapidly, with greatest change early in their existence, and then remain relatively unchanged during their duration on Earth. The lack of transitional forms, intermediate between species or between higher taxonomic groups, is further evidence for the punctuated pace of evolution. Additional studies of fossil lineages should help to distinguish between the gradual or punctuated tempos in the evolutionary origin of species.

Answers to Test Your Knowledge

1. b	5. b	9. d	13. c
2. c	6. d	10. b	14. a
3. a	7. d	11. b	15. b
4. c	8. a	12. b	16. d

CHAPTER 23
TRACING PHYLOGENY: MACROEVOLUTION, THE FOSSIL RECORD, AND SYSTEMATICS

Suggested Answers to Structure Your Knowledge

1.

2. Microevolution consists of changes in gene frequencies that occur within a population over generations due to chance events such as genetic drift, gene flow, and mutation pressure, and to differential reproductive success as a result of natural selection. New species may form as the result of microevolution or as the result of geographical separation or mutation. Macroevolution includes the major events and trends that have occurred in the history of life as taxa higher than species evolve.

Macroevolution can result from the gradual modification of preadapted structures for new functions through natural selection (as in microevolution), or more likely, from chance-related mechanisms such as major geological changes resulting in geographic rearrangements of species, genetic changes that affect temporal or spatial aspects of development, and adaptive radiations as new adaptive zones result from the evolution of novel characteristics or from mass extinctions. Microevolution and macroevolution both involve natural selection and the element of chance. Natural selection appears to play a larger role in microevolution, while chance may play the largest role in macroevolution.

3. Continental drift and biogeography

    Globe C, placed at the border between the Paleozoic and Mesozoic eras about 248 million years ago, represents the formation of Pangaea. This event coincided with the Permian extinctions, which may have been related to major environmental changes associated with the reduction in shoreline and drainage of coastal seas.

    Globe D should be placed in the middle of the Mesozoic era about 135 million years ago, representing the breakup of Pangaea and the start of a major biogeographical separation that has led to the geographical distribution of organisms we see today.

    Globe A should be placed at 65 million years ago at the boundary between the Mesozoic and Cenozoic eras, the time of the Cretaceous extinctions. The continued separation of the continents may have affected climatic conditions that contributed to those extinctions.

    Globe B should be placed at the present.

Answers to Test Your Knowledge

Multiple Choice:

1. c	5. a	9. d	13. c	17. a
2. a	6. b	10. d	14. c	18. d
3. b	7. d	11. a	15. a	
4. c	8. b	12. a	16. c	

True or False:

1. True
2. False, change the *more* to the *less* they vary
3. True
4. False, change *phenetics* to *cladistics*
5. True
6. False, change to: because *the half-life of carbon-14 is too short to date fossils that are that old.*
7. False, change to: indicative of *an asteroid having crashed into the earth.*
8. True

UNIT V

EVOLUTIONARY HISTORY OF BIOLOGICAL DIVERSITY

Ed and Barbara are visited by the insects of the Amazon Basin.

CHAPTER 24

EARLY EARTH AND THE ORIGIN OF LIFE

FRAMEWORK

This chapter describes the formation of Earth and presents a possible scenario for the chemical evolution of life between 4.1 and 3.5 billion years ago. Conditions on the primitive Earth are thought to have favored the spontaneous formation of organic monomers, the linking of these monomers into polymers, the grouping of aggregates of organic molecules into droplets called protobionts which became capable of metabolism and reproduction, and the development of self-replicating genetic information capable of directing metabolism and reproduction.

From this proposed beginning, an incredible biological diversity has evolved. This heterogeneous group of organisms is now classified into five kingdoms: Monera, Protista, Plantae, Fungi, and Animalia.

CHAPTER SUMMARY

Life on Earth has been evolving for 3.5 billion years. Geological events have altered the course of biological evolution and life has changed the Earth. The fossil record documents, however incompletely, the development of much of the diversity of life, a development that has often been episodic. However, the origin of life on the young planet Earth remains unrecorded and a matter of speculation.

The Antiquity of Life

Precambrian fossils provide evidence of animals dating back 700 million years and a succession of primarily prokaryotic microorganisms spanning nearly 3 billion years. Fossils that appear to be prokaryotes have been discovered in a rock formation called the Fig Tree Chert, which is 3.4 billion years old. Fossils resembling spherical and filamentous prokaryotes have been found in rocks called *stromatolites* that are 3.5 billion years old. Stromatolites are banded domes of sediment that form around the jellylike coats of microbes. It is possible that the earliest life forms may have emerged as long as 4 billion years ago.

The Origin of Life

Between 4.1 billion years ago, when Earth's crust began to solidify, and 3.5 billion years ago, when records of stromatolites are found, life began. Most biologists believe that life developed on Earth from nonliving materials that became ordered into molecular aggregates capable of self-replication and metabolism.

According to one hypothesis, a chemical evolution in four stages produced the first organisms: (1) the abiotic (non-living) synthesis and accumulation of small organic molecules; (2) the joining of organic monomers into polymers; (3) the aggregation of molecules into droplets, called *protobionts*; and (4) the origin of heredity.

Abiotic Synthesis of Organic Monomers Oparin and Haldane independently postulated that conditions on the primitive Earth, in particular the reducing atmosphere and lightning and intense ultraviolet radiation, favored the synthesis of organic compounds from inorganic precursors available in the atmosphere and seas.

In 1953, Miller and Urey tested the Oparin–Haldane hypothesis with an apparatus that simulated early Earth conditions. After a week of applying sparks (lightning) to a warmed flask of water (the primeval sea) in an atmosphere of H_2O, H_2, CH_4, and NH_3,

Miller and Urey found a variety of amino acids and other organic compounds in the flask.

Laboratory replications of the early Earth have been able to produce all 20 common amino acids, several sugars, lipids, the purine and pyrimidine bases of DNA and RNA, and even ATP. The abiotic synthesis and accumulation of organic monomers could have been a natural process on primeval Earth.

Abiotic Synthesis of Polymers   The abiotic synthesis of polymers may have occurred when dissolved organic monomers splashed onto hot rocks. Fox has created polypeptides he calls proteinoids by dripping dilute solutions of organic monomers onto hot sand or rock in the laboratory.

Clay may have been an important substratum for polymerization because of its ability to concentrate organic monomers when they bind to charged clay particles. Metal atoms, such as iron and zinc, at some of the binding sites may have functioned as catalysts facilitating the dehydration reactions that link monomers. Iron pyrite has also been proposed as a substratum for organic synthesis.

Formation of Protobionts   Protobionts, aggregates of abiotically produced organic molecules that exhibit some of the properties associated with life, probably preceded living cells. Protobionts may have developed such capabilities as metabolism, excitability, and the ability to maintain an internal chemical environment different from the surroundings.

Laboratory experiments indicate that protobionts could have formed spontaneously: Proteinoids, when mixed with cool water, self-assemble into tiny droplets called microspheres, which are coated by a selectively permeable membrane across which a membrane potential may develop. Protobionts can discharge this voltage in nervelike fashion.

When the organic ingredients include certain lipids, droplets called liposomes, surrounded by a lipid bilayer, may spontaneously form. Oparin has made coacervates, which are colloidal droplets that form when a solution of polypeptides, nucleic acids, and polysaccharides is shaken. If enzymes are included in the mix, the coacervates are capable of absorbing substrates, catalyzing reactions, and releasing the products.

Ancient protobionts would not have had enzymes, but some abiotically produced molecules have catalytic capabilities that could have allowed protobionts to develop a simple metabolism.

The Origin of Genetic Information   The last step necessary for the evolution of life was the origin of genetic information. Successful protobionts would need not only to grow and divide, but also to develop a mecha-

nism for replicating their successful characteristics, for creating instructions for making their key molecules. One hypothesis maintains that the first genes were short strands of RNA, which have been shown in the laboratory to be capable of self-replication. Cech's recent discovery of RNA catalysts, called *ribozymes*, which remove introns from RNA and catalyze RNA synthesis in modern cells, indicates that RNA molecules may have been autocatalytic and self-replicating in the prebiotic world.

Natural selection has been observed operating on populations of RNA molecules in test tubes. Early populations of RNA molecules may have evolved as those molecules with the most stable three-dimensional conformation and greatest autocatalytic activity within a particular environment successfully competed for monomers and generated families of similar sequences.

RNA-directed protein synthesis may have begun with the weak binding of specific amino acids to bases along RNA molecules and the linkage of these amino acids, aided by a catalyst such as zinc, to form a short polypeptide. This polypeptide may have then behaved as an enzyme to help the RNA molecule replicate.

When these early RNA and polypeptide molecules became packaged into protobionts, molecular cooperation could become more efficient due to the concentration of components and the potential for the protobiont to evolve as a unit. If primitive enzymes contained within a membrane-bound protobiont developed the ability to extract energy from an organic fuel, that energy would be available for the other reactions within the protobiont and a simple metabolism could evolve.

This four-step scenario results in a hypothetical antecedent of a cell, in which an aggregate of molecules selectively incorporates monomers from its surroundings and uses enzymes produced by genes to make polymers and to carry out chemical reactions. The protobiont could grow and split and distribute copies of its genes to offspring. Errors in the copying of RNA would lead to variations, some of which could have metabolic or genetic advantages. DNA, a more stable genetic molecule, eventually replaced RNA as the carrier of genetic information.

Laboratory simulations cannot establish that this sequence actually occurred in the evolution of life, but they have shown that some of the key steps are possible. Alternate explanations include the speculation, called *panspermia*, that some organic compounds may have reached the early Earth in meteorites and comets. Modern meteorites contain amino acids and other organic compounds which have been shown to produce small vesicles when mixed with water.

Because the surface of early Earth was such an inhospitable environment, bombarded by asteroids and other cosmic debris, some scientists speculate that life began on the sea floor, perhaps at deep-sea vents. Other biologists suggest that RNA is too complicated to have formed the first genetic system and that simpler self-replicating organic molecules were involved.

Via whatever route, at some point membrane-enclosed compartments capable of metabolism and genetic replication evolved past the grey border separating them from true cells so that by 3.5 billion years ago prokaryotes were already flourishing.

The Kingdoms of Life

The kingdom is the most inclusive taxonomic category. Intuitively and historically, humans have divided the diversity of life into two kingdoms—plants and animals. The plant kingdom was stretched to make room for bacteria, unicellular eukaryotes with chloroplasts, and fungi, whereas unicellular creatures that move and ingest food were placed in the animal kingdom.

Whittaker's proposal for a five-kingdom system, made in 1969, has slowly gained favor among biologists. The five kingdoms are Monera, Protista, Plantae, Fungi, and Animalia. The prokaryotes are set apart from the eukaryotes and placed in the kingdom Monera. The kingdoms Plantae, Fungi, and Animalia are multicellular eukaryotes, defined by characteristics of nutrition, structure, and life cycle. Plants are autotrophic organisms; animals are heterotrophic organisms that ingest and digest their food; and fungi are absorptive heterotrophic organisms. Protists are unicellular eukaryotes or simple multicellular organisms that descended from them.

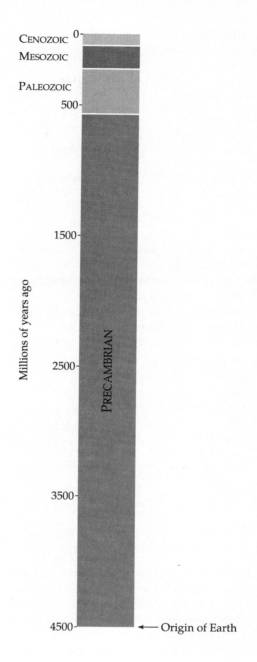

STRUCTURE YOUR KNOWLEDGE

1. What do scientists think the primitive Earth was like? How could life possibly evolve in such an inhospitable environment?

2. Describe the significance of the four steps that are proposed as a possible route for the evolution of protobionts.

3. Develop a simple concept map of the five kingdoms of life, indicating the key characteristics that differentiate the kingdoms.

4. On the following time line, indicate the age of the oldest fossils of prokaryotes. Show the approximate time when life on Earth originated. Continue to fill in key events on this time line as you work through the chapters on the five kingdoms of life.

TEST YOUR KNOWLEDGE

MULTIPLE CHOICE: *Choose the one best answer.*

1. The primitive atmosphere of Earth may have favored the synthesis of organic molecules because
   a. it was highly oxidative.
   b. it was reducing and had energy sources in the form of lightning and UV radiation.
   c. it had a great deal of methane and organic fuels.
   d. it had plenty of water vapor, carbon, and nitrogen.

2. Life on Earth is thought to have begun
   a. 700 million years ago.
   b. 3 billion years ago.
   c. between 3.5 and 4.1 billion years ago.
   d. 5 billion years ago.

3. Stromatolites are
   a. the earliest fossils.
   b. similar to the layered mats formed by cyanobacteria.
   c. banded domes of sediment formed around motile microbes.
   d. all of the above.

4. Coacervates are
   a. droplets of molecules covered with a selectively permeable membrane.
   b. colloidal droplets formed from polypeptides, nucleic acids, and polysaccharides.
   c. polypeptides formed in the laboratory by dripping organic monomers onto hot rocks.
   d. organic molecules associated with clay crystals.

5. A proposed hypothesis for the origin of genetic information is that
   a. early DNA molecules coded for RNA, which then catalyzed the production of proteins.
   b. early polypeptides became associated with RNA bases and a catalyst such as zinc could have linked the bases into RNA molecules.
   c. short RNA strands were capable of self-replication and evolved by the natural selection of molecules that were most stable and autocatalytic.
   d. as protobionts grew and split, they distributed copies of their molecules to their offspring.

6. Evidence that protobionts may have formed spontaneously comes from
   a. the discovery of ribozymes, showing that prebiotic RNA molecules may have been autocatalytic.
   b. the laboratory synthesis of microspheres, liposomes, and coacervates.
   c. the fossil record found in the Fig Tree Chert.
   d. the abiotic synthesis of polymers.

7. If you examined one of the early Precambrian fossils in detail you would be likely to observe impressions of structures resembling
   a. cell walls.
   b. nuclear membranes.
   c. mitochondria.
   d. xylem cells.

8. Polymer formation from organic monomers may have been facilitated by

a. the electrification of monomers by lightning.
b. proteinoid formation by pH changes.
c. amino acid concentration on clay particles.
d. sedimentary accumulation on stromatolites.

9. The earliest life forms may have exhibited all of the following characteristics except
   a. nerve-like action potentials.
   b. sexual reproduction.
   c. membranes formed from lipid bilayers.
   d. metabolism of chemical compounds.

10. The hypothesis that life originated in the sea is based on
    a. that fact that most species are marine.
    b. the discovery of deep-sea vents.
    c. the inhospitable asteroid-crashing environment of early Earth.
    d. both b and c.

11. In the five-kingdom system of classification,
    a. prokaryotes are placed in the kingdom Monera.
    b. the fungi include absorptive heterotrophs.
    c. single-celled or simple multicellular eukaryotes are placed in the Protista.
    d. all of the above are correct.

12. In order to place a newly identified species into one of the five kingdoms, the first piece of information you would need to acquire is
    a. the way in which it acquires food.
    b. the absence or presence of a nuclear membrane.
    c. whether the organism is multicellular.
    d. its life cycle.

MATCHING: *Match the scientist with the theory or experiment.*

1. _____   produced organic compounds in a lab apparatus   A. Fox

2. _____   discovered RNA catalysts called ribozymes   B. Oparin and Haldane

3. _____   developed the five-kingdom system   C. Crick

4. _____   claimed that conditions of primitive Earth favored synthesis of organic compounds   D. Cech

                                                                                                        E. Whittaker

5. _____   produced proteinoids by dripping monomers onto hot sand in the laboratory   F. Miller and Urey

CHAPTER 25

PROKARYOTES AND THE ORIGINS OF METABOLIC DIVERSITY

FRAMEWORK

This chapter describes the bacteria and cyanobacteria of Kingdom Monera. It reviews the morphology of the prokaryotic cell and presents an overview of the groups archaebacteria and eubacteria. Every mode of nutrition and most metabolic pathways that are found today evolved within the prokaryotes. Prokaryotes exist in all conceivable habitats and in various symbiotic relationships with other organisms. These most numerous and diverse of all organisms are essential to the chemical cycles necessary to maintain life on Earth.

CHAPTER SUMMARY

The history of prokaryotes or *bacteria* spans at least 3.5 billion years. The kingdom Monera is characterized by the prokaryotic cell. Prokaryotes outnumber all eukaryotes combined and flourish in all habitats, including ones that are too harsh for any other forms of life. The collective impact of these microscopic organisms on Earth is huge.

Many prokaryotes live in symbiotic relationships. These arrangements may have led to the origin of mitochondria and chloroplasts and, indeed, the evolution of the other four kingdoms.

Prokaryotic Form and Function

Morphology of Prokaryotes Prokaryotes exist primarily as single cells, although some species may form permanent aggregates of identical cells called *colonies* and some exhibit a simple multicellular form. Prokaryotic cells are usually 1-5μm in diameter, one-tenth the size of eukaryotic cells. The three most common shapes of prokaryotes are spheres (cocci), rods (bacilli) and spirals (spirilla).

The Cell Surface Nearly all prokaryotes have cell walls of *peptidoglycan*, a matrix composed of polymers of sugars cross-linked by short polypeptides. The *gram stain* is an important tool for identifying bacteria as *gram-positive* (bacteria with walls containing a thick layer of peptidoglycan) or *gram-negative* (bacteria with more complex walls including an outer lipopolysaccharide membrane). Pathogenic gram-negative bacteria are often more harmful because their lipopolysaccharides may be toxic. This outer membrane protects them from the defenses of their hosts and from antibiotics. Many antibiotics, such as penicillin, inhibit the synthesis of cross-links in peptidoglycan and prevent wall formation.

Many prokaryotes secrete a sticky *capsule* outside the cell wall that serves for protection and as a glue for adhering to a substratum or each other. Bacteria may also attach by means of surface appendages called *pili*. Pili may be specialized for transfer of DNA during conjugation.

Motility of Prokaryotes Many bacteria are equipped with flagella, either scattered over the cell surface or concentrated at one or both ends of the cell. These flagella differ from eukaryotic flagella in size, lack of a plasma membrane cover, and structure. Spirochetes have filaments attached at either end of the cell that spiral around the cell under the cell wall and move past each other, causing the cell to move in a corkscrew fashion. Other motile bacteria glide on a slime that they secrete. Many motile bacteria exhibit *taxis*, an oriented movement in response to chemical, light, magnetic, or other stimuli. Responding to concentrations sensed by surface receptor molecules, a bacterium may use alternating runs and tumbles to move along a chemical gradient.

Internal Membranous Organization Extensive internal membrane systems are not found in prokaryotic cells, although there may be some infoldings of the plasma membrane that function in respiration. Thylakoid membranes in prokaryotes called cyanobacteria function in photosynthesis.

The Prokaryotic Genome The haploid genome of prokaryotes contains only 1/1000 as much DNA as a eukaryote. The circular double-stranded DNA chromosome, often called a genophore to distinguish it from eukaryotic chromosomes, is concentrated in a *nucleoid region* and has very little protein associated with it. Smaller rings of DNA, called plasmids, may carry genes for antibiotic resistance, metabolism of unusual nutrients, or other functions. They replicate independently of the genophore and may be transferred between bacteria during conjugation.

Prokaryotic ribosomes are smaller than eukaryotic ribosomes and differ in their protein and RNA content. Some antibiotics work by binding to prokaryotic ribosomes and blocking protein synthesis, but they do not inhibit eukaryotic ribosomes.

Growth, Reproduction and Gene Exchange Prokaryotes reproduce asexually by binary fission, producing a *colony* of progeny. Each species has an optimal growing environment. Some bacteria form *endospores*, which are tough walled cells that can resist even boiling water. Microbiologists must therefore use autoclaves to sterilize laboratory equipment and media. Endospores can remain dormant for centuries and still hydrate and reproduce once the environment is favorable. Some prokaryotes, such as thermoacidophiles, actually thrive in harsh environments.

Bacterial growth in an optimal environment is geometric, with generation times from 20 minutes to three hours. The growth of bacterial colonies usually stops at some point due to the exhaustion of nutrients or the toxic accumulation of wastes. Bacteria, as well as protists and fungi, release *antibiotics*, to inhibit the growth of competitors. Humans have used some of these compounds to control pathogenic bacteria.

Despite the lack of meiosis and sexual cycles, there are three mechanisms of genetic recombination in bacteria: *transformation*, the uptake of genes from the environment; *conjugation*, the direct transfer of genetic material from one bacterium to another; and *transduction*, the injection of genes by a viral bacteriophage. Mutations are the major source of genetic variation in prokaryotes. Due to short generation times, favorable mutations are rapidly spread.

Metabolic Diversity Nutrition refers to how an organism obtains energy and the carbon it uses for synthesizing organic compounds. Species may be phototrophs, using light for energy, or chemotrophs, obtaining energy from chemicals. If CO_2 is the only carbon source needed, organisms are called autotrophs; when organic nutrients are needed, organisms are called heterotrophs.

Thus, there are four major nutritional categories: (1) *photoautotrophs*, cyanobacteria and other photosynthetic prokaryotes that use light energy and CO_2 to synthesize organic compounds; (2) *photoheterotrophs*, which use light energy to generate ATP, but obtain carbon in organic form; (3) *chemoautotrophs*, which obtain energy by oxidizing inorganic substances (such as H_2S, NH_3, or Fe^{2+}) and need only CO_2 as a carbon source; and (4) *chemoheterotrophs*, which use organic molecules as both an energy and carbon source. Plants and some protists are also photoautotrophs, and fungi, animals, and other protists are chemoheterotrophs.

There are three subcategories of chemotrophs. The majority of bacteria are either *saprobes*, which decompose and absorb nutrients from dead organic matter, or *parasites*, which absorb nutrients from living hosts. The third subcategory is *phagotrophs*, which ingest their food, but there are no known prokaryotic phagotrophs. Some fastidious bacteria require all 20 amino acids, vitamins, and other organic compounds in order to grow, whereas others may flourish with glucose as their only organic nutrient. There is such diversity of chemoheterotrophs that almost any organic molecule can serve as food for some species. The few synthetic organic compounds that cannot be broken down by any bacteria are called nonbiodegradable.

Obligate aerobes need oxygen for cellular respiration; *facultative anaerobes* can use oxygen but also can grow in anaerobic conditions using fermentation; *obligate anaerobes* cannot use—and are poisoned by—oxygen.

Nitrogen, an essential component of proteins and nucleic acids, is cycled through the ecosystem with the aid of bacteria. Most nitrogenous compounds are metabolized by some prokaryotes. Through *nitrogen fixation*, diverse species of bacteria convert atmospheric N_2 to ammonia.

The Diversity of Prokaryotes

Molecular systematics, using comparisons of amino acid sequences and base sequences in DNA and RNA, particularly ribosomal RNAs, provide the best approach to the phylogenetic classification of this immense and heterogeneous kingdom that diversified so long ago. Using these methods, the kingdom

Monera can be divided into about 20 major groups. These comparisons indicate that prokaryotes split into at least two divergent lineages very early in the history of life: the *archaebacteria* and the *eubacteria*.

Archaebacteria The archaebacteria, of ancient origin, have several distinctive traits: their cell walls lack peptidoglycan; their plasma membranes contain a unique lipid composition; and their RNA polymerase and a ribosome protein are unlike the eubacteria, resembling those of eukaryotes. Archaebacteria are found living in extreme environments, perhaps resembling habitats on early Earth. The archaebacteria presently consist of three subgroups: the methanogens, the extreme halophiles, and the thermoacidophiles.

Methanogens have a unique energy metabolism in which H_2 is used to reduce CO_2 to methane (CH_4). These strict anaerobes live in swamps and marshes, bubbling up marsh gas or methane. They are important decomposers in sewage treatment and have been used to convert manure to methane on some farms. Methanogens help decompose cellulose in the guts of cattle and other herbivores.

The *extreme halophiles* (salt-loving) live in saline places such as the Great Salt Lake and the Dead Sea. They have a photosynthetic pigment called bacteriorhodopsin that gives them their pink color.

Thermoacidophiles need an environment that is both hot and acidic. They may be found oxidizing sulfur for energy in the hot sulfur springs in Yellowstone National Park.

Woese has proposed that a new taxonomic category called the *domain* be used to recognize the early evolutionary split within prokaryotes. The three domains would be Bacteria (eubacteria), Archaea (archaebacteria), and Eucarya (eukaryotes). Molecular evidence indicates that archaebacteria may be more closely related to eukaryotes than to eubacteria.

Eubacteria Eubacteria are found in a variety of habitats and show great diversity in their modes of nutrition. The cyanobacteria, or blue-green algae, have a plantlike mechanism of photosynthesis, with the components of the light reaction built into thylakoid membranes. Some filamentous cyanobacteria have heterocysts, specialized cells with nitrogen-fixing enzymes. Other photoautotrophs include purple and green sulfur bacteria that use electrons from H_2S rather than H_2O and are strict anaerobes.

Enteric bacteria inhabit the intestinal tracts of animals. Many are harmless permanent residents, but a few may cause diseases. Mycoplasmas, perhaps the smallest of all cells, are the only prokaryotes that lack cell walls. Myxobacteria cells glide through the soil on a secreted slime. Under adverse conditions, the cells congregate to form an erect fruiting body in which durable spores are formed.

The pseudomonads are found in almost all aquatic and soil habitats and include members capable of utilizing a wide range of organic nutrients. Rickettsias are small intracellular parasites of animals. Spirochaetes are spiral-shaped saprobes or parasitic bacteria that move in a corkscrew fashion. A spirochaete is the cause of syphilus and Lyme disease.

The Origins of Metabolic Diversity

All major metabolic pathways evolved in prokaryotes before there were eukaryotes, probably within the first billion years of life on Earth. Hypotheses about metabolic evolution, including the following proposed sequence of events, are based upon the geological evidence of conditions of early Earth, molecular systematics, and the energy metabolism found in existing groups.

The Origin of Glycolysis The first prokaryotes probably were chemoheterotrophs that absorbed abiotically synthesized organic compounds. ATP must have become an important energy molecule early in the history of life, based on its universal role in energy exchange in all modern organisms. Selection of cells with enzymes that could regenerate ATP from ADP may have resulted in the gradual evolution of glycolysis, the metabolic pathway that breaks down organic molecules and generates ATP by substrate phosphorylation. As the only metabolic pathway common to all modern organisms, glycolysis also must have developed relatively early in metabolic evolution.

Glycolysis and fermentation do not require molecular oxygen. The archaebacteria and other obligate anaerobes that use fermentation are believed to have forms of nutrition most like that of the original prokaryotes that developed on anaerobic Earth.

The Origin of Electron Transport Chains and Chemiosmosis The chemiosmotic synthesis of ATP is common to all five kingdoms, again implying an early origin. Electrons are transferred through a chain of membrane proteins, pumping hydrogen ions across a membrane and synthesizing ATP using a proton gradient. Transmembrane proton pumps, driven by ATP, may have originally functioned to help regulate internal pH of early prokaryotes. An electron transport chain that coupled the oxidation of organic acids with the transport of H^+ out of the cell would have conserved ATP. Excess H^+ could then diffuse back, reverse the proton pump, and thus generate ATP. This type of anaerobic respiration persists in some modern bacteria.

The Origin of Photosynthesis Light-absorbing pigments may have protected cells from excess light and become coupled with electron transport systems to drive ATP synthesis. A simple mechanism of photophosphorylation has been found in some extreme halophiles in which bacteriorhodopsin, a membrane-bound pigment molecule, uses light energy to pump protons out of the cell. The resulting gradient is used for ATP synthesis.

The development of pigments and photosystems coopting components of electron-transport chains may have allowed some prokaryotes to use light energy to drive electrons from H_2S to $NADP^+$ and thus generate reducing power to fix CO_2 and make their own organic molecules. The nutrition of the green and purple sulfur bacteria (anaerobic phototrophs) probably is most like that of the early photosynthetic prokaryotes.

Cyanobacteria, the Oxygen Revolution, and the Origins of Cellular Respiration The first cyanobacteria evolved a mechanism that reduced CO_2 using water as a source of electrons and hydrogen. They began to release O_2 as a byproduct of their photosynthesis.

Cyanobacteria evolved at least 2.5 billion years ago, contributing to the fossil stromatolites that have been found all over the world. Banded iron formations are found in marine sediments from that same time. The oxygen evolved by the cyanobacteria probably precipitated dissolved iron ions as iron oxide. When the dissolved iron was finally exhausted, oxygen began to accumulate in the seas and bubble out into the atmosphere. Oxidized iron is found in terrestrial rocks from about 2 billion years ago.

The change to a more oxidizing atmosphere probably caused the extinction of many bacteria, while others survived in anaerobic habitats where they are still found today. Some bacteria evolved antioxidant mechanisms that protected them from rising oxygen levels. Some photosynthetic prokaryotes, however, began using the oxidizing power of O_2 to pull electrons from organic molecules down existing transport chains, leading to the development of aerobic respiration by the coopting of transport chains from photosynthesis. Purple nonsulfur bacteria are photoheterotrophs that still use a hybrid photosynthetic and respiratory electron transport system. Several bacterial lines gave up photosynthesis and their electron transport chains became adapted to function exclusively in aerobic respiration.

Every type of nutrition and energy metabolism evolved in the prokaryotes on ancient Earth.

The Importance of Prokaryotes

Prokaryotes and Chemical Cycles Prokaryotes are indispensable in recycling chemical elements between the biological and physical worlds. Bacteria, along with fungi, are *decomposers* of dead organisms and of the waste products of living ones. They return carbon, nitrogen and other elements to the environment for assimilation into new living forms. Autotrophic bacteria bring carbon from CO_2 into the food chain and release O_2 to the atmosphere, and cyanobacteria fix atmospheric nitrogen and supply plants with the nitrogen they need to make proteins.

Symbiotic Bacteria *Symbiosis* is an ecological relationship involving direct contact between organisms of two different species. The organisms are called *symbionts* and, if one is much larger than the other, it is called the *host*. There are three kinds of sybiosis. In *mutualism*, both symbionts benefit. *Nodules* on the roots of legumes house symbiotic bacteria that fix nitrogen for the host and, in return, receive nutrients. In *commensalism*, one symbiont benefits while the other is neither harmed nor helped. Many of the bacteria found in the human body are commensal, although some may be mutualistic. In *parasitism*, the symbiont, now called a parasite, benefits at the expense of the host. Parasitic bacteria that cause disease are called pathogens.

Symbiosis among prokaryotes probably played an important role in evolution and is most likely still common, although little studied or understood.

Bacteria and Disease About one-half of all human diseases are caused by pathogenic bacteria that manage to invade the body, resist internal defenses, and grow enough to harm the host. *Opportunistic* bacteria are normal inhabitants of the human body that cause illness when the body's defenses are weakened.

Koch, the first to connect particular diseases to specific bacteria, suggested four criteria for establishing this connection. Called *Koch's postulates*, they include the following: (1) find the same pathogen in each diseased individual; (2) isolate and grow the pathogen in a pure culture; (3) induce the disease in experimental animals using the cultured pathogen; (4) isolate the same pathogen from the experimental animal after it develops the disease.

Some bacteria cause disease when they invade host tissues. Examples are the rickettsias that cause Rocky Mountain spotted fever and typhus, the chlamydias that causes nongonococcal urethritis, and the actinomycetes that cause tuberculosis and leprosy.

Pathogenic bacteria more commonly cause disease by producing toxins. *Exotoxins*, among the most potent poisons known, are proteins secreted by the bacteria that can induce various symptoms and cause such diseases as botulism and cholera. *Endotoxins*, which are components of the outer membrane of certain gram-negative bacteria, all produce fever and aches in the host. Examples are typhoid fever and *Salmonella* food poisoning.

In the past century, improved hygiene, sanitation, and the development of antibiotics have decreased the incidence and severity of bacterial disease, reduced infant mortality and extended life expectancy in developed countries. More than half of our antibiotics come from the soil bacteria genus *Streptomyces*, an actinomycete.

Putting Bacteria to Work   The diverse metabolic capabilities of prokaryotes have been used to digest organic wastes, produce chemical products, make vitamins and antibiotics, and produce food products such as yogurt and cheese. Research using *Escherichia coli* and other bacteria has expanded our understanding of molecular biology, and recombinant DNA techniques using bacteria develop both basic knowledge and practical applications.

2. Identify the large structure inside this bacterial cell. What is its significance?

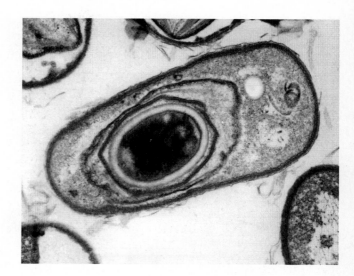

STRUCTURE YOUR KNOWLEDGE

1. Fill in the following chart with a brief description of the characteristics of prokaryotic cells.

CHARACTERISTIC	DESCRIPTION
Cell shape	
Cell size	
Cell surface	
Motility	
Internal Membranes	
Genome	
Growth and Reproduction	

3. Identify the spherical cell in this filamentous cyanobacteria. What is the function of this cell?

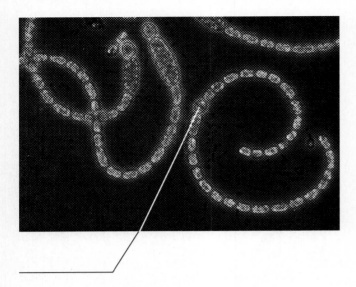

4. Create a concept map that organizes the various modes of nutrition (along with their associated terminology) found in living organisms.

5. List the five or six stages that may have occurred in the early evolution of nutrition and metabolism in the prokaryotes.

6. Describe four positive ways in which prokaryotes impact on our lives and on the world around us.

7. Add to the time line in Chapter 24 the approximate time when oxygen from photosynthetic cyanobacteria began to accumulate in the atmosphere.

TEST YOUR KNOWLEDGE

MULTIPLE CHOICE: *Choose the one best answer.*

1. Which of the following is *not* true of plasmids?
   a. They replicate independently of the main chromosome.
   b. They may carry genes for antibiotic resistance or special enzymes.
   c. They are essential for the existence of bacterial cells.
   d. They may be transferred between bacteria during conjugation.

2. Gram-positive bacteria
   a. have peptidoglycan in their cell walls, whereas gram-negative bacteria do not.
   b. have an outer membrane around their cell walls.
   c. have lipopolysaccharides in their cell walls and thus may be less pathogenic than gram-negative bacteria.
   d. have simpler, thick peptidoglycan cell walls.

3. Because prokaryotes have ribosomes different from those of eukaryotes,
   a. some selective antibiotics can block protein synthesis of bacteria without harming the eukaryotic host.
   b. it is believed that eukaryotes did not evolve from prokaryotes.
   c. protein synthesis can occur at the same time as transcription.
   d. their rate of protein synthesis is much slower.

4. Many prokaryotes secrete a sticky capsule outside the cell wall that
   a. allows them to glide through a slime layer.
   b. serves as protection and glue for adherence.
   c. reacts with the gram stain.
   d. is used for attaching cells during conjugation.

5. Chemoautotrophs
   a. are photosynthetic.
   b. use organic molecules for an energy and carbon source.
   c. oxidize inorganic substances for energy and use CO_2 as a carbon source.
   d. use light to generate ATP but need organic molecules for a carbon source.

6. The major source of genetic variation in prokaryotes is
   a. binary fission.
   b. mutation.
   c. conjugation.
   d. plasmid exchange.

7. A symbiotic relationship in which the symbiont benefits and the host is neither harmed nor helped is called
   a. saprocism.
   b. opportunism.
   c. mutualism.
   d. commensalism.

8. According to Koch's postulates, a specific bacteria can be linked to a disease if
   a. the same pathogen is isolated from each diseased individual.
   b. the cultured pathogen causes the disease when introduced to experimental animals.
   c. the same pathogen can be isolated from the experimental animal after it develops the disease.
   d. all of the above are done.

9. The first form of nutrition to evolve was probably that of

a. photoautotrophs that used light energy to reduce CO_2 with electrons from H_2S.
b. chemoheterotrophs that used abiotically made organic compounds.
c. anaerobic chemoautotrophs.
d. photoheterotrophs that used light for energy and abiotically made organic compounds for a carbon source.

10. Evidence of the early evolution of cyanobacteria comes from
    a. stromatolites.
    b. banded iron formations in marine sediments.
    c. oxidized iron layers in terrestrial rocks.
    d. all of the above.

11. Glycolysis
    a. is the only metabolic pathway common to all modern organisms and may have been the first metabolic pathway to evolve.
    b. breaks down organic molecules and generates ATP by creating a proton gradient that reverses transmembrane proton pumps.
    c. uses the electron transport chain to produce ATP.
    d. All of the above are correct.

12. Aerobic respiration may have originated when
    a. some bacteria evolved antioxidant mechanisms to protect them from rising oxygen levels.
    b. photosynthetic bacteria produced an excess of organic molecules.
    c. some bacteria used modified electron transport chains from photosynthesis to pass electrons from organic molecules to O_2.
    d. some bacteria gave up photosynthesis and reverted to chemoheterotrophic nutrition.

MATCHING: *Match the bacterial group with its characteristics.*

1. _____ grow in hot and acidic habitats   **A.** cyanobacteria

2. _____ small parasitic bacteria, one causes Rocky Mountain spotted fever   **B.** extreme halophiles

3. _____ bacteria found in animal intestinal tracts   **C.** streptomycetes

4. _____ bacteria from which antibiotics are obtained   **D.** thermacidophiles

5. _____ spiral-shaped bacteria; one causes syphilis   **E.** mycoplasmas

6. _____ photosynthetic bacteria, have thylakoid membranes, evolve O_2 **F.** myxobacteria

7. _____ prefer saline environment, simple photophosphorylation   **G.** methanogens

8. _____ anaerobes that produce marsh gas   **H.** spirochaetes

9. _____ bacteria that can survive heat, cold and desiccation in resistant stage; one causes botulism   **I.** rickettsias

10. _____ photoautotrophic bacteria that use H_2S rather than H_2O to reduce $NADP^+$ **J.** green and purple sulfur bacteria

K. endospore forming bacteria

L. enteric bacteria

FILL IN THE BLANKS

1. _____ the name for spherical bacteria

2. _____ region in which the prokaryotic chromosome is found

3. _____ common laboratory technique for identifying bacteria.

4. _____ surface appendages of bacteria used for adherence to substrate

5. _____ an oriented movement in response to light or chemical stimuli

6. _____ nutrition in which organic molecules are used for both energy and carbon source

7. _____ type of nutrition of purple and green sulfur bacteria

8. _____ organism that can use oxygen or grow anaerobically

9. _____ organisms that decompose and absorb nutrients from dead organic matter

10. _____ proteins that are secreted by bacteria and are potent poisons

11. _____ thickenings on roots that house nitrogen-fixing bacteria

12. _____ bacterial group that lacks peptidoglycan in the cell walls and is found in harsh habitats

CHAPTER 26

PROTISTS AND THE ORIGIN OF EUKARYOTES

FRAMEWORK

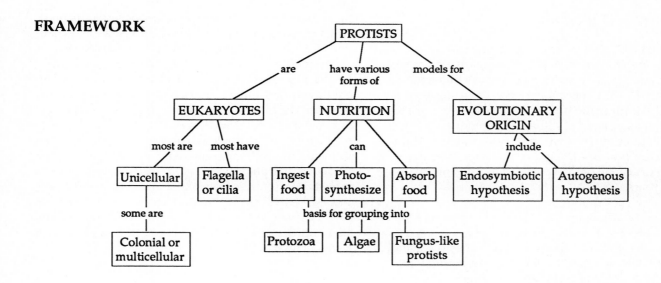

CHAPTER SUMMARY

The kingdom Protista consists primarily of eukaryotic unicellular organisms, with a few colonial and multicellular organisms. Ancient protists, probably the first eukaryotes to evolve from prokaryotes, were the ancestors of modern protists, plants, fungi, and animals.

Characteristics of Protists

Protists abound almost anywhere there is water—they occupy freshwater, marine, and moist terrestrial habitats or live symbiotically within the bodies of hosts. They form an important part of the *plankton*, the drifting community of organisms found in bodies of water.

Nearly all protists are aerobic and use mitochon-dria for their respiration. Nutritionally, protists can be photoautotrophs, heterotrophs, or *mixotrophs*, which both photosynthesize and eat organic molecules. Mutualism between photoautotrophic and heterotrophic protists is common. Based on their mode of nutrition and not phylogeny, protists are often grouped for convenience into plant-like protists (*algae*, singular *alga*), animal-like protists (*protozoa*, singular *protozoan*), and fungus-like protists.

Most protists have flagella or cilia at some point during their lifetime. These structures, which differ from bacterial flagella that attach to the cell surface, are extensions of the plasma membrane that encase cytoplasm and bundles of microtubules. Cilia and flagella both have a "9 + 2" ultrastructure of microtubules, but cilia are more numerous and shorter.

Mitosis occurs in most phyla of protists, but details

of the process vary. All protists can reproduce asexually; some also have sexual reproduction; and others may use meiosis and *syngamy* (the union of gametes) to exchange genes between two otherwise asexually reproducing individuals. The life histories of protists include the three basic types of sexual life cycle that differ in the timing of meiosis and syngamy, as well as variations on those types. Many protists can form resistant *cysts* to withstand harsh conditions.

These unicellular organisms are exceedingly complex at the cellular level. A single cell must be capable of performing all the basic functions that are divided among the specialized cells of plants and animals.

The Origin of Eukaryotes

The increase in complexity and organization of prokaryotes followed three separate trends: the move toward multicellular prokaryotes, such as filamentous cyanobacteria with specialized cells; the evolution of bacterial communities in which species benefited from the metabolic specialities of other species; and the trend to compartmentalize different functions within a cell that led to eukaryotic cells. Protists, as the first eukaryotic organisms, have been evolving for 2 billion years.

The Antiquity of Eukaryotes   The oldest putative fossils of eukaryotes, found in rocks 1.5 billion years old, are *acritarchs*, Precambrian structures that resemble ruptured coats of cysts similar to those of algal protists. Almost all eukaryotes require oxygen. The oxygen revolution created by cyanobacteria probably occurred about 2.5 billion years ago, and eukaryotic life may date back almost that far.

Models of Eukaryotic Origins   There are two hypothetical models for the origin of the eukaryotic cell. According to the *autogenous model*, eukaryotic cells are believed to have evolved by the specialization of internal membranes derived from invaginations of the plasma membrane.

In the *endosymbiotic model*, developed by L. Margulis, eukaryotic cells evolved from symbiotic combinations of prokaryotic cells. Chloroplasts are thought to be descendants of photosynthetic prokaryotes that became endosymbionts within larger cells. Mitochondria may have developed from aerobic heterotrophic bacteria that became incorporated into larger cells.

Various similarities exist between modern eubacteria and the chloroplasts and mitochondria of eukaryotes, including their size, membrane enzymes and transport systems, circular DNA molecules not associated with proteins, and the process of division. Chloroplasts and mitochondria contain ribosomes and other equipment needed to transcribe and translate their DNA into proteins. The ribosomes of chloroplasts and mitochondria more closely resemble prokaryotic ribosomes than they do cytoplasmic ribosomes. These similarities, as well as the present-day existence of endosymbiotic relationships, are cited as evidence of the endosymbiotic hypothesis. Molecular systematics, comparing base sequences of ribosomal RNA, also suggest eubacterial origins for chloroplasts and mitochondria.

Even though chloroplasts and mitochondria are not genetically autonomous as are bacteria, the development of nuclear control over the endosymbionts probably increased during the billions of years of coevolution as a result of mutations and DNA transposons.

The autogenous and endosymbiotic models need not be mutually exclusive. Chloroplasts and mitochondria may have originated as endosymbionts, whereas the components of the endomembrane system may have developed as modifications of the plasma membrane. Multiple separate events may have led to increasing cellular complexity, as indicated by the evidence that photosynthetic protists evolved at least three times from separate ancestors. A comprehensive theory for the development of the eukaryotic cell has yet to explain the evolution of the 9 + 2 flagella and cilia and the processes of mitosis and meiosis, which also rely on microtubules.

The Boundaries of Kingdom Protista

There is disagreement over which groups to include in this kingdom. This book includes the unicellular eukaryotes Whittaker originally assigned to the kingdom Protista as well as some organisms that lack the distinctive traits of plants, fungi, or animals. Some biologists have advocated the name Protoctista to indicate that they are including some phyla of multicellular organisms that seem to be more closely related to unicellular eukaryotes. This book uses the more accepted name, kingdom Protista, incorporating the multicellular algae and some funguslike phyla into the kingdom.

The morphology and life styles of the protists differ widely. Protistan taxonomy is in a state of flux because the evolutionary relationships among these diverse organisms are not clear. The informal classification scheme followed in this book for surveying the diversity of protists uses the three general categories of protozoa, algae, and fungus-like protists.

Protozoa

Protists that ingest food are informally grouped as protozoans, which are subdivided into phyla on the basis of how they feed or move.

Rhizopoda   The *amoebas* are naked or shelled simple protists without flagella that move and feed with *pseudopodia*. Microtubules and microfilaments of the cytoskeleton function in the formation of these cellular extensions. Nontypical forms of mitosis occur in the asexual reproduction of this group and there is not sexual reproduction. Most amoebas are found freeliving in freshwater, marine, or soil habitats. Some are parasites, such as *Entamoeba histolytica* which causes amoebic dysentery in humans.

Actinopoda   The slender projections called axopodia, which radiate from these protists, help these planktonic organisms to remain buoyant and to phagocytize microscopic food organisms. *Heliozoa* are freshwater actinopods. *Radiolarians* are primarily marine and have delicate silica shells.

Foraminifera   *Forams* are marine organisms known for their porous, calcareous shells. Cytoplasmic strands, extending through the pores, function in swimming, shell formation, and feeding. Many forams derive nourishment from symbiotic algae. Foram fossils are useful as index fossils for dating sedimentary rocks.

Apicomplexa   These animal parasites have complex life cycles that often include several host species. Infections are spread by tiny infectious cells called *sporozoites* (the group was formerly called Sporozoa). Malaria, which affects over two hundred million people a year, is caused by the apicomplexan *Plasmodium*. Control of malaria is complicated by the development of insecticide resistance in *Anopheles* mosquitos, which spread the disease, and drug resistance in *Plasmodium*, as well as the sequestering of the parasite within human liver and blood cells, and the ability of the parasite to change the surface proteins presented to the infected person's immune system.

Zoomastigophora   These protozoa are characterized by many whiplike flagella and are also called *zooflagellates*. They are heterotrophic organisms that may be freeliving or symbiotic. Mutualistic flagellates that live in the gut of termites digest the cellulose eaten by the host. *Trypanosoma*, a parasitic zoomastigote, causes African sleeping sickness.

Ciliophora   These protists are characterized by the cilia they use to move and feed. Their numerous cilia, coordinated by a submembrane system of microtubules, may be widespread or organized into leglike cirri or tightly packed rows that function as locomotory membranelles. Most *ciliates* are complex unicellular organisms found in fresh water. Ciliates have two types of nuclei—a large *macronucleus* that controls everyday functions of the cell and splits during binary fission (asexual reproduction) and many small *micronuclei*, which are exchanged in a sexual process called *conjugation*. In *Paramecium*, a micronucleus in each of two attached protists undergoes meiosis, and a haploid micronucleus from each cell is exchanged. Syngamy, or nuclear fusion, generates genetic variation and is followed by several mitotic divisions and the production of new individuals with genomes differing from the original conjugating cells.

Algae

Most protists of the algal phyla are photosynthetic. Except for cyanobacteria (formerly called blue-green algae), all organisms generically called algae are included in the kingdom Protista in Campbell's classification system. Some biologists place the green, red, and brown algae in the plant kingdom. An important ecological group, algal protists generate about half of all photosynthetically-produced organic material and form the base of aquatic food webs.

All algae have chlorophyll *a*, as do the cyanobacteria and plants. Accessory pigments vary among algal protists and are used—along with cell-wall chemistry, stored food products, number and position of flagella, and chloroplast structure—as classification characteristics.

Dinoflagellata   *Dinoflagellates* make up a large proportion of the phytoplankton at the base of most marine food chains. Blooms of dinoflagellates are responsible for the often harmful red tides in coastal waters. The photosynthesis of symbiotic dinoflagellates, living in small coral cnidarians, is the main food source for coral reef communities. Some dinoflagellates are parasitic or carnivorous heterotrophs.

Dinoflagellates have brownish red plastids containing xanthophyll accessory pigments. The beating of two flagella in perpendicular grooves between the submembrane cellulose plates of the cell produces a characteristic spinning movement. The division of the dinoflagellate nucleus during asexual reproduction is unusual.

Chrysophyta   The color of *golden algae* results from yellow and brown carotenoid and xanthophyll accessory pigments. There are many colonial forms of these biflagellated, freshwater plankton. Golden algae may form resistant cysts during environmental stress.

Bacillariophyta   *Diatoms* are common unicellular marine and freshwater plankton with unique glasslike shells. Although diatoms have the same pigments as golden algae, most phycologists (algae specialists) classify them in a separate phylum on the basis of their unique cell structure and life cycle. Diatoms usually reproduce asexually, although eggs and flagellated sperm may be produced. Food reserves in the form of an oil contribute to buoyancy. Massive quantities of fossilized shells make up diatomaceous earth.

Euglenophyta   The best known member of this phylum is *Euglena*, a common green flagellate found in pond water. Euglenophytes have the same pigments as green algae, but they lack cellulose cell walls and use flexible, internal protein plates for support. *Euglena* photosynthesizes in the light, but can ingest food particles in the dark. Some euglenophytes are entirely heterotrophic.

Chlorophyta   The chloroplasts of *green algae* resemble those of plants, and most botanists believe that plants evolved from chlorophytes. Most species of green algae live in freshwater, although many are marine. Various unicellular forms may be planktonic, inhabitants of damp soil, symbionts in invertebrates and protozoa, or mutualistic partners with fungi—forming associations known as *lichens*. Colonial species may be filamentous. Some large multicellular forms resemble plants, but are more closely related to the green algae.

Three separate evolutionary trends within the green algae have led to increasingly complex flagellated colonies, as in *Volvox*; multinucleated filaments, as in *Bryopsis*; and definite multicellular forms, as seen in *Ulva*.

The life cycles of most green algae include sexual (with biflagellated gametes) and asexual stages. *Conjugating algae*, such as *Spirogyra*, produce amoeboid gametes.

In the life cycle of the unicellular flagellated *Chlamydomonas*, the haploid mature organism reproduces asexually by dividing mitotically to form four zoospores, which grow into mature haploid cells. Under harsh conditions, the cell may produce many haploid gametes, which pair off and fuse with similar-looking gametes (*isogamy*) of opposite mating strains. The resulting diploid zygote secretes a protective coat. Following dormancy, the zygote undergoes meiosis, forming four haploid cells which grow into mature individuals. Some species of green algae produce

gametes that differ in size and morphology (*anisogamy*) and may exhibit *oogamy*, in which a flagellated sperm fertilizes a larger nonmotile egg.

Some multicellular green algae have an *alternation of generations*, in which a haploid multicellular individual (called the *gametophyte*) mitotically produces haploid gametes that fuse to form a diploid zygote. The zygote develops into the multicellular *sporophyte*, which produces haploid reproductive cells called spores by meiosis. Zoospores grow into the haploid gametophyte generation. Alternation of generations is common in green, brown, and red algae. Such a cycle is shown by the life cycle of *Ulva* on p. 215.

Seaweeds   Large marine green, brown, and red algae are called seaweeds. Unique anatomical and biochemical adaptations allow seaweeds to survive in the challenging intertidal zone. Some seaweeds have tissues and organs that are analogous to those found in plants, but they evolved independently. Seaweeds lack true roots, stems, and leaves, but may have a plant-like *thallus*, or body, consisting of a rootlike *holdfast* and a stemlike *stipe* that supports leaflike *blades*. Some brown algae have floats, which keep the blades near the surface.

Biochemical adaptations to intertidal and subtidal conditions include cell walls containing gel-forming polysaccharides that protect from abrasion and drying, and calcium carbonate-encrusted cells walls that protect from invertebrate grazers. Humans use some seaweeds for food. Commercial uses of seaweeds include thickeners for processed foods, lubricants, and the culture media agar.

Phaeophyta   The mostly marine *brown algae* are the largest and most complex protists. Brown algae are common along temperate coasts. Giant, fast-growing seaweeds known as kelps are brown algae that fasten to the sea floor by holdfasts and transport photosynthetic products from the blades down the stipe through tubular cells.

In most brown algae, there is an alternation of generations. The gametophyte and sporophyte may be similar in appearance (*isomorphic*) or distinct (*heteromorphic*).

Rhodophyta   The color of *red algae* is due to the accessory pigment phycoerythrin. These mostly marine algae also may have diverse accessory pigments that best absorb the wavelengths of light that penetrate into deep water. Individuals of the same species can change pigmentation to optimize photosynthesis at different depths. Most red algae are multicellular, often with filamentous, delicately branched thalli.

Alternation of generations is common, although life cycles are diverse. Unlike other algal protists, they have no flagellated cells in their life cycles.

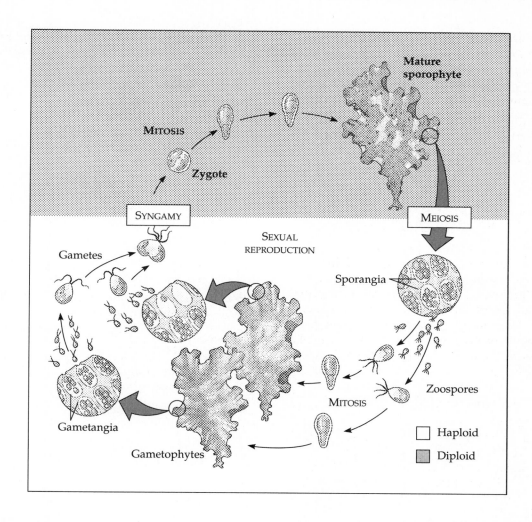

Fungus-like Protists

General consensus places two phyla of slime molds as well as water molds (Oomycota) and Chytrids in the kingdom Protista. Although the slime molds resemble fungi in appearance and life style and they are still studied by *mycologists* (biologists who study fungi), their cellular organization, reproduction, and life cycles are different enough to suggest that they are more closely related to amoeboid protists.

Myxomycota *Plasmodial slime molds* are heterotrophic, engulfing food particles by phagocytosis as they grow through leaf litter or rotting logs. A coenocytic (multinucleate) amoeboid mass called a *plasmodium* is the feeding stage. During the sexual stage, erect stalks produce resistant spores by meiosis. Spores germinate, the amoeboid or flagellated haploid cells fuse, and the diploid nucleus repeatedly divides without cytoplasmic divisions.

Acrasiomycota During the feeding stage of the life cycle, *cellular slime molds* consist of solitary amoeboid cells. In the absence of food, the cells congregate into a mass resembling a plasmodial slime mold, except that it is composed of cells separated by a membrane and therefore is not coenocytic. Cellular slime molds are haploid and the fruiting bodies produce resistant spores asexually by mitosis. In sexual reproduction, two amoebas fuse to form a zygote which develops into a giant cell enclosed in a protective wall. Following meiosis and mitosis, haploid amoebas are released.

Oomycota The water molds, white rusts, and downy mildews of this phylum resemble fungi in appearance and nutrition, but they have cell walls made of cellulose, a predominant diploid stage in their life cycles, and flagellated cells—all characteristics unlike those of true fungi.

The large egg cell of a water mold is fertilized by a smaller sperm nuclei. Zygotes, following dormancy within resistant walls, germinate to form coenocytic hyphae, which are tipped by zoosporangia that asexually produce flagellated zoospores. The saprophytic water molds are important decomposers in aquatic ecosystems. White rusts and downy mildews can be destructive parasites of land plants.

Chytridiomycota Members of this phylum include saprophytic and parasitic organisms that range from simple one-celled, coenocytic chytrids to more complex members with branched, coenocytic hyphae that resemble fungi. True fungi may have evolved from Chytridomycota. Chytrids differ from fungi in producing motile spores and flagellated gametes.

Evolutionary Relationships Among Algae Chloroplasts appear to have had at least three separate origins, as evidenced by three lineages seen in algal protists. In the red line, the chloroplasts of red algae show similarities to the thylakoid arrangement and pigment composition of cyanobacteria. The chloroplasts of green algae, the green line, are more similar to a eubacterium named *Prochlorothrix*, which is the only known prokaryote with chlorophyll *b* as well as *a*. The thylakoids of the brown line generally are in stacks of three, and chlorophyll *c* and xanthophylls are accessory pigments. A modern prokaryote that resembles a possible ancestor of the prokaryotic symbiont of the brown plastids has not been found.

The Origins of Multicellularity

Eukaryotic organization allowed for more complex structure and diverse functions. The step from unicellular protists to multicellular forms opened new opportunities for specialization and adaptation. Multicellularity evolved in the kingdom Protista several times, creating the ancestors of algae, plants, fungi, and animals. Colonial aggregations of cells probably became more interdependent, leading to the specialization of cells and the division of labor. Some cells in the colony may have lost their flagella and eventually become specialized for other functions, such as reproduction.

STRUCTURE YOUR KNOWLEDGE

1. (a) What are the criteria or characteristics of the organisms placed in the kingdom Protista? (b) Why is there a controversy over which groups belong in this kingdom?

2. Describe the two hypotheses for the origin of eukaryotic cells. What evidence is given to support the endosymbiotic model?

3. Draw a concept map to show the various types of life cycles found in the protists.

4. Label the indicated structures in the diagram of a *Paramecium* on p. 217. To what phylum does this organism belong? _____

5. Indicate on the time line in Chapter 24 the age of the oldest known eukaryotic fossils.

TEST YOUR KNOWLEDGE

MATCHING: *Match the protistan phyla with their descriptions.*

1. _____ Naked and shelled amoebas   **A.** Acrasiomycota

2. _____ Photosynthetic or heterotrophic flagellates   **B.** Actinopoda

3. _____ Heliozoa, radiolarians, siliceous skeletons   **C.** Apicomplexa

4. _____ Plasmodial slime molds, coenocytic   **D.** Bacillariophyta

5. _____ Marine, with calcareous porous shells   **E.** Chlorophyta

6. _____ Green algae, probable ancestors of plants   **F.** Chrysophyta

7. _____ Brown algae, large, complex seaweeds   **G.** Chytridiomycota

8. _____ Golden algae, biflagellated, freshwater plankton   **H.** Ciliophora

9. _____ Diatoms, two-piece shells of silica   **I.** Dinoflagellata

10. _____ Red algae, pigment phycoerythrin   **J.** Euglenophyta

11. _____ Complex, unicellular, ciliated, macro- and micronuclei   **K.** Foraminifera

12. _____ Water molds, white rusts, downy mildews   **L.** Myxomycota

13. _____ Many whiplike flagella; free-living, mutualistic, parasitic   **M.** Oomycota

14. _____ Parasites, complex life cycles, sporozoites; malaria   **N.** Phaeophyta

15. _____ Whirling movement, flagella in grooves in shell, cause red tides   **O.** Rhizopoda

16. _____ Cellular slime molds, unicellular feeding stage, form asexual fruiting body   **P.** Rhodophyta

  Q. Zoomastophora

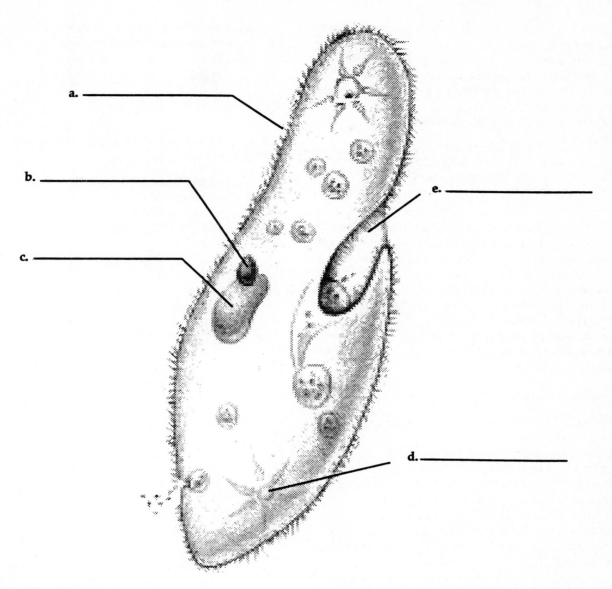

a.

b.

c.

e.

d.

MULTIPLE CHOICE: *Choose the one best answer.*

1. A coenocytic organism
   a. uses an amoeboid type of movement.
   b. consists of a thallus with no true roots, stems, or leaves.
   c. is multinucleate.
   d. produces asexual fruiting bodies.

2. The slime molds and multicellular algae are included in the kingdom Protista because
   a. they appear to be more closely related to unicellular eukaryotes.
   b. they lack important characteristics of the fungi and plants.
   c. kingdom Protista includes eukaryotic organisms that do not clearly belong in the other three kingdoms.
   d. of all of the above.

3. Genetic variation is generated in the ciliates when
   a. a micronucleus replicates its genome many times and becomes a macronucleus.
   b. isogametes, in the form of zoospores, fuse.
   c. micronuclei are exchanged in conjugation.
   d. oogamous sexual reproduction occurs.

4. Phytoplankton
   a. form the basis of most marine food chains.
   b. include the plantlike green, red, and brown algae.
   c. are mutualistic symbionts that provide food for coral reef communities.
   d. are unicellular protozoans.

5. Mycologists are specialists in the study of
   a. algae.
   b. fungi.
   c. protists.
   d. phylogeny.

6. The chlorophyta are believed to be the ancestors of plants because
   a. they are the only multicellular algal protists.
   b. they do not have flagellated gametes.
   c. of similarities in chloroplasts and pigment composition.
   d. they exhibit an alternation of generations.

7. The oldest reputed eukaryotic fossils
   a. date from 1.5 billion years ago.
   b. resemble the ruptured coats of algal cysts.
   c. are acritarchs.
   d. All of the above are correct.

8. According to the endosymbiont theory,
   a. multicellularity evolved when primitive cells incorporated prokaryotic cells that then took on specialized functions.
   b. the symbiotic associations found in lichens resulted from the incorporation of algal protists into the ancestors of fungi.
   c. the chloroplasts and mitochondria of eukaryotic cells began as prokaryotic endosymbionts.
   d. the infoldings and specializations of the plasma membrane led to the evolution of eukaryotic cells.

9. Examples of mutualistic symbiotic relationships include
   a. termites and zoomastigotes.
   b. dinoflagellates living in small coral cnidarians.
   c. chlorophytes and fungi in lichens.
   d. all of the above.

10. The Precambrian fossil evidence of protists, in the form of cysts called acritarchs, indicates that the atmosphere at that time
    a. was fairly high in oxygen because most protists are aerobic.
    b. was still low in oxygen because most protists are anaerobic.
    c. had a high sulfur content due to chemoautotrophic protists.
    d. cannot be determined because protists are strictly aquatic.

CHAPTER 27

PLANTS AND THE COLONIZATION OF LAND

FRAMEWORK

This chapter details the evolution of plants and their adaptations to terrestrial habitats. The kingdom Plantae includes multicellular, photoautotrophic eukaryotes that develop from embryos retained in a protective jacket of cells. Plants exhibit an alternation of generations in which the diploid sporophyte is the more conspicuous stage in all divisions except the bryophytes. The development of a cuticle and jacketed reproductive organs, vascular tissue, seeds and pollen, and flowers are linked to the major periods of plant evolution in which mosses, ferns, conifers, and flowering plants appeared and radiated.

CHAPTER SUMMARY

For the first 3 billion years, life was confined to marine and aquatic environments. Plants began the movement of life onto land about 425 million years ago. The evolutionary history of the plant kingdom involves the increasing adaptation to changing terrestrial conditions.

Introduction to the Plant Kingdom

General Characteristics of Plants Plants are multicellular, photoautotrophic eukaryotes. Nearly all plants are terrestrial, although some have returned to water during their evolution. Adaptations to life on land include a waxy *cuticle* on the stems and leaves that prevents desiccation and *stomata* or pores to allow for the exchange of gases through the cuticle. Plant chloroplasts contain chlorophylls *a* and *b* and a variety of carotenoids. The cell wall of plants consists of cellulose; starch is the food storage compound.

A Generalized View of Plant Reproduction and Life Cycles Nearly all plants reproduce sexually; most can also propagate asexually. Gametes are produced and the egg is fertilized and develops into an embryo within *gametangia*—organs with a protective jacket of cells.

The life cycle of plants involves the alternation of generations between the haploid gametophyte and the diploid sporophyte. The gametophyte produces gametes by mitosis. The diploid zygote develops into the sporophyte stage, which produces spores by meiosis. Germinating spores grow into the gametophyte. In all extant plants, these generations are *heteromorphic*. In all but the bryophytes, the sporophyte is the more conspicuous stage. Evolutionary trends include the reduction of the haploid generation with increasing dominance of the diploid and adaptations to terrestrial life, such as the replacement of flagellated sperm by pollen.

Some Highlights of Plant Evolution Four major periods of plant evolution are linked to the evolution of structures that opened new adaptive zones on land. The first period involved the origin of plants, probably from green algae, during the mid-Silurian period about 425 million years ago. Terrestrial adaptations included a cuticle, jacketed reproductive organs, and *vascular tissue*—specialized cells joined into tubes for transport throughout the plant. Most mosses lack vascular tissue; evidence from molecular systematics suggests an early divergence between the non-vascular mosses and the lineages leading to vascular plants.

The second period of plant evolution took place 400 million years ago during the early Devonian period with the adaptive radiation of vascular plants.

The third major period of plant evolution, about 360 million years ago, was marked by the origin of vascular plants with *seeds*, which are embryos enclosed with a store of food inside a protective coat.

Early seed plants gave rise to *gymnosperms* (naked seed plants), such as the conifers. Ferns, which are seedless plants, dominated the landscape along with gymnosperms for over 200 million years.

The emergence of flowering plants about 130 million years ago, during the early Cretaceous period, started the fourth major evolutionary episode. The majority of contemporary plants are *angiosperms*, or flowering plants, which bear seeds in protective chambers called ovaries.

Classification of Plants   Plant biologists use the term *division* in place of *phylum* for the major taxonomic groups within the plant kingdom. This textbook recognizes 12 divisions within the kingdom Plantae.

The Move onto Land

The Case for Green Algae as the Ancestors of Plants
Plants probably evolved from green algae, most of which share the following characteristics with plants: (1) chlorophyll *a* and the accessory pigments chlorophyll *b* and beta-carotene; (2) thylakoid membranes stacked into grana; (3) cell walls of cellulose; (4) starch as the carbohydrate reserve; (5) cell plate formation (in some green algae) in cytokinesis.

Filamentous chlorophytes along the edge of bodies of water may have been the first green algae to colonize the land. Fluctuations in water levels may have resulted in selection for species that could survive exposed periods. The development of waxy cuticles and jacketed reproductive organs opened the adaptive zone of land.

Bryophytes: Divisions Bryophyta, Hepatophyta, and Anthocerophyta   The mosses, liverworts and hornworts, formerly grouped into a single division, have been separated to reflect their lack of close relationship. They share many characteristics, however, and are still commonly referred to as bryophytes. Bryophytes have cuticles and protective gametangia, but they are restricted to damp, shady places due to both their flagellated sperm and, with a few exceptions, their lack of vascular tissue. Sperm are produced in *antheridia*. One egg is produced and fertilized within the *archegonium*, where the zygote develops into an embryo.

Mosses of the division Bryophyta grow as water-absorbing mats with many plants in a tight pack. Each gametophyte plant attaches to the soil with elongated cells called rhizoids. Most photosynthesis occurs in small leaflike appendages. The embryo grows into the sporophyte generation while still attached within the archegonium, and the sporophyte depends on the gametophyte for water and nutrients. At the tip of the sporophyte stalk, a *sporangium* produces haploid spores by meiosis. Germinating spores grow into a small, green, threadlike protonema from which the gametophyte plant develops.

Some *liverworts*, division Hepatophyta, have bodies that are divided into lobes. Their life cycle is similar to that of mosses, although they also reproduce asexually by the formation of gemmae that grow in cups on the surface of the gametophyte. *Hornworts*, division Anthocerophyta, resemble liverworts, but their sporophytes are elongated capsules that grow like horns from the matlike gametophyte.

The three divisions of bryophytes date back at least 400 million years and are successfully adapted to moist terrestrial habitats.

Terrestrial Adaptations of Vascular Plants   During the evolution of vascular plants, there was increasing differentiation of an underground root system which absorbs water and minerals and an aerial shoot system of stems and leaves which receives light for photosynthesis. *Lignin*, a hard material embedded in the cellulose of plant cell walls, was an important adaptation which provides support for the aerial parts of the plant. The vascular system of *xylem* and *phloem* provides the conducting vessels to connect roots and leaves. Hollow xylem cells carry water and minerals up from the roots; their lignified walls also provide support. The living cells of phloem distribute organic nutrients throughout the plant. In some vascular plant groups, seeds, pollen, and the increasing dominance of the diploid sporophyte were additional terrestrial adaptations.

The Earliest Vascular Plants   Fossils of vascular plants are present in the sedimentary rocks of the late Silurian and early Devonian periods. *Cooksonia*, a simple, dichotomous branching plant, is the oldest petrified plant that has been discovered. *Zosterophyllum* and *Rhynia* are characteristic genera of early Devonian species that are the probable ancestors, respectively, of the division Lycophyta and the other divisions of vascular plants.

Seedless Vascular Plants

Four extant divisions have retained the primitive seedless condition.

Division Psilophyta   *Psilotum*, known by the common name *whiskfern*, is one of two genera of relatively simple, primitive vascular plants in this division.

Rhizomes bear tiny rhizoids. The scales emerging from the dichotomously branching stems lack vascular tissue and thus are not true leaves. Sporangia along the stems release haploid spores. The nonphotosynthetic gametophyte depends on symbiotic fungi.

Division Lycophyta Lycopods were a major part of the landscape during the Carboniferous period (340 to 280 million years ago). The division Lycophyta split into two evolutionary lines: one that led to huge woody trees and a second that remained small and herbaceous. The giant lycopods became extinct when the Carboniferous swamps dried up. The small lycopods are represented today by the genera *Lycopodium* and *Selaginella*, commonly called club mosses or ground pines. Many tropical species grow on trees as *epiphytes*—plants that anchor to other organisms but are not parasites.

In the club moss, the sporangia are borne on *sporophylls*—leaves specialized for reproduction. Spores germinate and grow into inconspicuous, subterranean gametophytes which depend on symbiotic fungi for nutrition. These gametophytes produce both archegonia and antheridia. *Lycopodium* is said to be *homosporous* because it produces only one kind of spore, which develops into bisexual gametophytes. *Selaginella* is *heterosporous*; it produces *megaspores* that develop into female gametophytes bearing archegonia and *microspores* that develop into male gametophytes with antheridia that make flagellated sperm.

Division Sphenophyta This division originated in the Devonian radiation and produced tall plants during the Carboniferous period. Today only the genus *Equisetum* is found. Commonly called horsetails, these homosporous plants grow in damp locations. The gametophyte is minute, but freeliving and photosynthetic.

Division Pterophyta Ferns were also found in the great forests of the Carboniferous period and are the most numerous of seedless plants in the modern flora. Fern leaves, commonly called fronds, are much larger than are those of lycopods. The leaves of lycopods probably evolved as emergences from the stem containing a single strand of vascular tissue and are called *microphylls*. Fern leaves, called *megaphylls*, have branching systems of veins. One theory holds that megaphylls evolved by the formation of webbing between closely-growing separate branches.

Some fern leaves are specialized sporophylls with sporangia, arranged into clusters called sori, on their undersides. Most ferns are homosporous, although the archegonia and antheridia on the small, photosynthetic gametophyte mature at different times. Sperm swim to fertilize the egg in neighboring archegonia, and the young sporophyte grows out from the archegonium.

The Coal Forests The seedless plants of the Carboniferous forests left behind extensive beds of coal. Dead plants did not completely decay in the stagnant swamp waters and great accumulations of peat developed. When the sea later covered the swamps and marine sediments piled on top, heat and pressure converted the peat to coal.

Terrestrial Adaptations of Seed Plants

Three life-cycle modifications contributed to the success of seed plants on land: (1) the further-reduced gametophyte generation was retained and protected within the sporophyte plant; (2) pollination replaced the swimming of sperm to egg; (3) the seed, with embryo surrounded by a food supply within a seed coat, maintained dormancy through harsh conditions and functioned in dispersal. The dominance of the diploid sporophyte may also have provided spare alleles should an allele be damaged by the more intense, mutagenic radiation on land.

Gymnosperms

Of the two groups of seed plants, gymnosperms appear earlier in the fossil record. Three of the four divisions are relatively small: Cycadophyta, which includes the palmlike cycads; Ginkgophyta, with the deciduous, ornamental ginkgo tree; and Gnetophyta, which includes the bizarre *Welwitschia* as one of its three genera.

Division Coniferophyta The name *conifer* refers to the reproductive structure, the cone. Most conifers are evergreens, with needle-shaped leaves covered with a thick cuticle. Coniferous trees are among the tallest, largest, and oldest living organisms.

The pine tree is a heterosporous sporophyte. Pollen cones consist of many tiny sporophylls that bear sporangia. Meiosis gives rise to microspores that develop into pollen grains—the immature male gametophyte. Scales of the ovulate cone hold ovules, each of which contains a sporangium, called a nucellus, within a protective integument. A megaspore mother cell undergoes meiosis, and one of the resulting megaspores undergoes repeated divisions to produce a female gametophyte in which a few archegonia develop.

Pollination occurs when a pollen grain is drawn through the micropyle, an opening in the integument around the nucellus. A pollen tube grows and digests its way through the nucellus. Fertilization occurs when a sperm nucleus joins with the egg nucleus. The zygote develops into a sporophyte embryo, which is nourished by the remaining female gametophyte tissue and enclosed in a seed coat. Seed production takes three years.

The History of Gymnosperms Gymnosperms probably developed from a group of plants that had evolved seeds by the end of the Devonian Period. The divisions of gymnosperms arose during the Carboniferous and early Permian period. As the Permian progressed, continental interiors became warmer and drier, giving selective advantage to the conifers and cycads which replaced the lycopods, horsetails, and ferns that dominated the Carboniferous swamps. The end of the Permian marked the boundary between the Paleozoic and Mesozoic eras. The Mesozoic is sometimes called the age of dinosaurs. Conifers and the great palmlike cycads supported these giant reptiles. When the climate became cooler at the end of the Mesozoic, the dinosaurs and many plants became extinct, but some gymnosperms, particularly conifers, persisted.

Angiosperms

Angiosperms, or flowering plants, are the most diverse and widespread of modern flora. The division Anthophyta is divided into two classes: Monocotyledones and Dicotyledones.

Most angiosperms rely on insects or other animals for transferring pollen to female sex organs, an advance over the more random wind pollination of gymnosperms. Gymnosperms transport water through *tracheids*, relatively primitive, tapered cells that also function in mechanical support. In angiosperms, shorter, wider *vessel elements* are arranged end-to-end to form more specialized tubes for water transport. *Fibers*, which have thick lignified walls that add support, also evolved from tracheids and are found in both conifers and angiosperms.

The Flower The *flower*, the reproductive structure of an angiosperm, is a compressed shoot with four whorls of modified leaves: *sepals*, *petals*, *stamens*, and *carpels*. Sepals are usually green, whereas petals are brightly colored in most flowers that are pollinated by insects and birds. A stamen consists of a stalk, called a *filament*, and an *anther*, where pollen is produced. The carpel has a sticky *stigma*, which receives pollen, and a *style*, which leads to the *ovary*. The ovary contains ovules that develop into seeds.

Four evolutionary trends in angiosperms include: (1) reduction of the number of floral parts; (2) fusion of floral parts (compound carpels are common—the term *pistil* refers to single or fused carpels); (3) change from radial to bilateral symmetry; and (4) lowering the position of the ovary to below the petals and sepals for better protection.

The Fruit A *fruit* is a mature ovary, which functions in the protection and dispersal of seeds. Simple fruits develop from a single ovary; aggregate fruits come from several ovaries of the same flower; and multiple fruits, such as a pineapple, develop from several separate flowers. Fruits may be modified in various ways to disperse seeds.

Life Cycle of an Angiosperm Immature male gametophytes, consisting of two haploid cells, are *pollen grains*, which develop within the anthers. *Ovules* include the female gametophyte, which consists of an *embryo sac* with seven haploid cells (one of which contains two nuclei).

Most flowers have some mechanism to ensure *cross-pollination*, the transfer of pollen from one plant to another. A pollen grain germinates on the stigma and extends a pollen tube down the style to the ovule, where it releases two sperm cells into the embryo sac. In a process called *double fertilization*, one sperm unites with the egg to form the zygote and the other sperm nucleus fuses with the large cell containing two nuclei in the center of the embryo sac. The *endosperm*, which develops from this triploid (3N) nucleus, serves as a food reserve for the embryo. The embryo consists of a rudimentary root and one (in monocots) or two (in dicots) seed leaves, called *cotyledons*. The seed is a mature ovule containing an embryo, endosperm, and a seed coat derived from the integuments.

The Rise of Angiosperms Angiosperms appear rather suddenly in the fossil record about 120 million years ago in the early Cretaceous period, with no transitional links to ancestors. By the end of the Cretaceous, 65 million years ago, angiosperms had radiated and become the dominant plants, as they are today. Nearly all paleobotanists agree that the angiosperms must have evolved from some group of gymnosperms. The abruptness of their appearance may be due to an imperfect fossil record or perhaps to the punctuated nature of evolution.

Angiosperms rose to prominence during the Cretaceous, a period of climatic change and extinctions. Dinosaurs and many cycads and conifers disappeared and were replaced by mammals and flowering plants. The end of the Cretaceous marks the boundary between the Mesozoic and Cenozoic eras.

Relationships between Angiosperms and Animals
Animals influenced the evolution of plants and vice versa. Animal predation on plants may have provided selective pressure for plants to keep spores and vulnerable gametophytes on the plant. As flowers and fruits evolved, some predators became beneficial as pollinators and seed dispersers. The *coevolution*, or mutual evolutionary influence, of angiosperms and their pollinators is seen in the diversity of flowers, whose color, fragrance, and shape are usually matched to the pollinator's sense of sight and smell or its particular morphology. As seeds mature, fruits soften and increase in sugar content, attracting bird and mammal seed dispersers.

All of our fruit and vegetable crops are angiosperms. Grains are grass fruits; their endosperm is the main food source for most people and their domesticated animals. Humans have intervened in plant evolution by selective breeding. Agriculture is a unique relationship between plants and the humans who cultivate them.

The Value of Plant Diversity

The growing human population and its demand for space, food, and natural resources are leading to the extinction of hundreds of species each year. Tropical rain forests, where plant diversity is greatest, could be eliminated within 25 years if the pace of destruction continues. Humans are destroying their potential supply of new food crops and medicines. Economic and political solutions are required to preserve plant diversity.

STRUCTURE YOUR KNOWLEDGE

1. The evolution of plants shows a trend of increasing adaptations to a terrestrial habitat. In the table below, list the plant characteristics that contribute to survival on land and check which of the major plant groups have those characteristics.

PLANT GROUPS	Bryophyta	Lycophyta Sphenophyta	Pterophyta	Coniferophyta	Anthophyta
COMMON NAMES					
ADAPTIVE CHARACTERISTICS					
Cuticle, stomata					

2. Draw a phylogenetic tree showing the major lines of plant evolution on this geological time frame. Note the time of origin of plants, bryophytes, vascular plants, seedless plants, gymnosperms, and angiosperms. Indicate the periods of dominance for the major groups. Discuss the impact of the major climatic changes that occurred between the Paleozoic, Mesozoic, and Cenozoic eras.

ERA	PERIOD	AGE (millions of years ago)	MAJOR PLANT EVENTS
CENOZOIC			
Climatic Change—Cooler		65	
MESOZOIC	Cretaceous		
	Jurassic		
		200	
	Triassic		
Climate—Warmer and Drier		248	
PALEOZOIC	Permian	300	
	Carboniferous		
	Devonian		
		400	
	Silurian		
	Ordovician		
		500	
	Cambrian		

3. Now indicate on the larger-scale time line in Chapter 24 the point at which plants moved onto land.

4. In the diagram of the life cycle of a pine, label the indicated structures and show where meiosis, pollination, and fertilization take place. Which structures are haploid?

TEST YOUR KNOWLEDGE

MULTIPLE CHOICE: *Choose the one best answer.*

1. Adaptations for terrestrial life seen in all plants are
   a. chlorophylls *a* and *b*.
   b. cell walls of cellulose and vascular tissue.
   c. a waxy cuticle and stomata.
   d. pollen and gametangia.

2. Plants are thought to have originated from chlorophytes
   a. over 425 million years ago.
   b. during the Carboniferous period.
   c. when the climate became warmer and drier at the end of the Paleozoic era.
   d. as the ancient atmosphere became oxygen-rich.

3. Plants are believed to have evolved from green algae because both groups share the following characteristics:
   a. gametangia as sex organs and alternation of generations.
   b. starch as carbohydrate reserve and cytokinesis by cleavage furrow.
   c. thylakoid membranes stacked into grana, chlorophylls *a* and *b*, and cellulose in cell walls.
   d. all of the above.

4. Which of the following lack true roots and leaves?
   a. bryophytes
   b. bryophytes and whiskferns (*Psilotum*)
   c. bryophytes, whiskferns, and lycopods
   d. bryophytes, whiskferns, lycopods, and pterophytes

5. Megaphylls
   a. are gametophyte plants that develop from megaspores.
   b. are leaves specialized for reproduction.
   c. are leaves with branching vascular systems.
   d. are large leaves.

6. Bryophytes differ from the other plant groups because
   a. their gametophyte generation is dominant.
   b. they are lacking cuticle and lignin.
   c. they have flagellated sperm.
   d. all of the above.

7. Which of the following plant groups is incorrectly paired with its gametophyte generation?
   a. angiosperm—a pollen grain
   b. whiskfern (*Psilotum*)—nonphotosynthetic subterranean structure
   c. moss—green matlike plant
   d. fern—frond growing from rhizome

8. Which of the following incorrectly pairs a sporophyte embryo with its food source?
   a. pine embryo — female gametophyte tissue in nucellus
   b. corn embryo — 3N endosperm tissue in seed
   c. moss embryo — archegonium and gametophyte plant
   d. fern embryo — female sporophyte surrounding archegonium

9. A difference between seedless vascular plants and plants with seeds is that
   a. the gametophyte generation is conspicuous in the seedless plants, whereas the sporophyte is dominant in the seeded plants.
   b. the spore is the agent of dispersal in the first, whereas the seed functions in dispersal in the second.

c. the gametophyte is photoautotrophic in the seedless plants but dependent on the sporophyte generation in the other group.
d. the embryo is unprotected in the seedless plants but retained within the female reproductive structure in the other group.

10. As a result of coevolution,
    a. two organisms mutually influence each other's evolution.
    b. a flower may have nectar guides that lead bees to its nectaries.
    c. the red color of ripe fruits may attract birds or mammals that disperse the seeds.
    d. all of the above

11. Which of the following does *not* function in support in gymnosperms?
    a. cuticle.
    b. fibers.
    c. xylem.
    d. lignin.

12. If you could take a time machine back to the Carboniferous period, which of the following scenarios would you most likely confront?
    a. creeping mats of low growing bryophytes
    b. fields of tall grasses swaying in the wind
    c. swamps dominated by large lycopods. horsetails, and ferns
    d. forests of naked-seed trees filling the air with pollen

TRUE OR FALSE: *Indicate T or F, and then correct the false statements.*

1. _____ All photoautotrophic, multicellular eukaryotes are plants.

2. _____ Heteromorphic plants produce male and female gametophytes.

3. _____ Bryophytes are terrestrial nonvascular plants.

4. _____ The club mosses and horsetails are naked seed plants.

5. _____ A sporangium produces spores, no matter what group it is found in.

6. _____ A seed consists of an embryo, nutritive material, and a protective coat.

7. _____ Tracheids are wide, specialized cells arranged end-to-end for water transport and are found in angiosperms.

8. _____ A stamen consists of a filament and anther in which pollen is produced.

9. _____ The female gametophyte in angiosperms consists of haploid cells in which a few archegonia develop.

10. _____ The male gametophyte in angiosperms consists of a pollen grain.

CHAPTER 28

FUNGI

FRAMEWORK

This chapter describes the morphology, life cycles, and ecological and economic importance of the kingdom Fungi. The three divisions of fungi are established on the basis of variations in sexual reproduction. Lichens are symbiotic complexes of fungi and algae. Fungi play an important ecological role, both as decomposers and by their mycorrhizal association with plant roots.

CHAPTER SUMMARY

Characteristics of Fungi

Fungi were once classified with plants, but these eukaryotic, mostly multicellular organisms are now placed in their own kingdom.

Nutrition and Habitats Fungi are heterotrophs that obtain their nutrients by *absorption* of the small organic molecules they form by secreting enzymes into the surrounding food media. Fungi may be saprophytes that decompose nonliving organic material, parasites that absorb nutrients from living hosts, or mutualistic symbionts who feed on but also benefit their hosts. Fungi are associated symbiotically with many organisms and may inhabit land or water.

Structure Most fungi are composed of filaments called *hyphae*, which form a network called a *mycelium*. The hyphae of most fungi are divided into cells by cross walls called *septa*, which usually have pores through which cell organelles and nutrients pass. Cell walls are composed of chitin. Some hyphae are *aseptate* and these *coenocytic* fungi consist of a con-

tinuous cytoplasmic mass with many nuclei. Parasitic fungi may penetrate their host cells with modified hyphae called *haustoria*. None of the organisms considered to be fungi, according to the classification scheme followed in this text, have flagellated cells at any point in their life cycle.

Growth and Reproduction The filamentous hyphae grow rapidly and provide a large surface area for absorption. Although fungi are nonmotile, they rapidly enter into new food territory by the extension of their hyphae.

Two *nuclear-associated organelles* are located outside the nuclear envelope during mitosis. The spindle forms within the nucleus, which later divides after the chromosomes separate. Nuclei of the mycelia are haploid. Genetic heterogeneity may exist when hyphae with different nuclei join. Their haploid nuclei may stay in separate regions of the mycelia or they may engage in a process similar to crossing over.

Fungi produce huge quantities of asexual spores when conditions are favorable. Sexual reproduction may occur when environmental conditions change. In the sexual cycle, syngamy occurs as two separate events. *Plasmogamy*, cytoplasmic fusion, leads to a *dikaryon* stage in which the two nuclei pair up but do not fuse. They may continue to divide in tandem, forming dikaryotic cells. After a period of time, the nuclei fuse (*karyogamy*) and the zygote undergoes immediate meiosis.

Diversity of Fungi

Mycologists classify fungi into three divisions based on variations in sexual reproduction.

Division Zygomycota Zygomycetes, or zygote fungi, are mostly terrestrial fungi, living in soil or on decaying matter. One group forms important mutualistic associations with plant roots. Hyphae are coenocytic. Cell walls are composed of chitosan. Zygote fungi produce large numbers of asexual spores in sporangia at the tips of hyphae. *Zygosporangia*, formed in sexual reproduction, can remain dormant during harsh conditions.

Rhizopus stolonifer, the black bread mold, is a common zygomycete that spreads horizontal hyphae across its food substrate and erects hyphae with bulbous black sporangia containing hundreds of haploid spores. Sexual reproduction involves plasmogamy of hyphal extensions from neighboring mycelia of opposite mating strains. Haploid nuclei pair up and the dikaryotic zygosporangium develops a tough protective coat and may remain dormant until conditions are favorable, when karyogamy between paired nuclei occurs, followed immediately by meiosis to produce haploid spores.

Division Ascomycota The ascomycetes or sack fungi range in complexity from unicellular yeasts to elaborate cup fungi. They are found in a wide variety of habitats and symbiotic associations: some are important plant pathogens; many species are in mutualistic relationships with algae to form lichens; others are in mycorrhizal associations with plant roots; and others are the major non-bacterial saprobes in salt water habitats. The hyphae of multicellular forms are septate, and cell walls are predominantly chitin.

Asexual reproduction by chains or clusters of asexual spores, called *conidia*, which are produced at the ends of hyphae, is very common.

Ascomycetes produce their sexual spores in sacklike *asci*, which may be packed together in *ascocarps* or "fruiting bodies." The dikaryotic stage is associated with the formation of ascocarps. Karyogamy occurs in terminal cells, now called asci, and meiosis yields haploid spores called *ascospores*. In some ascomycetes, the eight ascospores, formed from meiosis and a subsequent mitotic division, are lined up in the ascus in the order in which they were formed, allowing geneticists to study the genetic recombination, crossing over and independent assortment of chromosomes during meiosis.

Division Basidiomycota The basidiomycetes, which include the mushrooms, shelf fungi, puffballs, and rusts, produce a club-shaped *basidium* during a short diploid stage in the life cycle. These club fungi are important saprobes, mycorrhizae-forming mutualists, and plant parasites.

Hyphae are typically dikaryotic and septate, with a donut-like swelling around perforations in the septa in many groups. In response to environmental stim-

uli, an elaborate fruiting body, called a *basidiocarp*, is formed from the dikaryotic hyphae. Karyogamy and meiosis occur in cells at the tip of hyphae lining the gills of a mushroom, forming basidia that produce four haploid basidiospores. Huge quantities of basidiospores are released in a variety of ways by members of this division. Asexual reproduction of spores in conidia is less common than in ascomycetes.

Some Unique Life-Styles of Fungi

Molds Molds are rapidly-growing, asexually reproducing fungi which are saprobes or parasites on a variety of substrates. In later life stages, a mold may reproduce sexually and produce zygosporangia, ascocarps or basidiocarps. Special systems of identification and classification are used for the asexual stages of molds and for those molds that have not been observed to produce sexual spores. If a mold produces asexual spores in sporangia, it is considered a Zygomycota. If it forms conidia, it is classified in either the Ascomycota or Basidiomycota if its sexual structure is known. If its sexual structure is not known, it may be placed in the "form-division," *Deuteromycota* and classified into "form-genera" or "form-species" on the basis of features of their conidia, the *conidiophores* that bear the conidia, and their manner of production. The Deuteromycota are sometimes called *imperfect fungi* in reference to their lack of a sexual stage. Some molds are capable of *parasexuality* in which genetic recombination occurs between different haploid nuclei in a mycelium.

Many imperfect fungi are sources of antibiotics, such as penicillin, and other pharmaceutical products, such as cyclosporine which is used in organ transplants to inhibit immune-system rejection. Some members of the form-genus *Aspergillus* produce *aflatoxins*, which are carcinogenic toxins, in improperly stored grain.

Yeasts Yeasts are unicellular organisms with fungal-like characteristics that grow in liquid or moist habitats. Reproduction is commonly asexual by cell division or budding, but some yeasts produce asci or basidia and are classified accordingly. Others are totally asexual and are placed in the form division Deuteromycota.

The ascomycetous yeast *Saccharomyces cerevisiae* is a domesticated fungus used in baking and brewing. This genus is used extensively in molecular genetic research.

Lichens Lichens are symbiotic associations of millions of algal cells in a lattice of fungal hyphae. The fungus is usually an ascomycete, and the alga is usually a unicellular or filamentous green algae or cyanobacteria. Tissues formed of fungal hyphae give form to the thallus, with the algal component in an inner layer. The alga provides the fungus with food and, in the case of lichens containing cyanobacteria, nitrogen. The fungus creates the thallus that provides for mineral absorption, water retention, and gas exchange. Fungal compounds, including pigments, toxins, and weak acids that bind mineral elements, contribute to the well-being of the association.

Lichens reproduce asexually, either as fragments or by tiny clumps called *soredia*. In addition, it is common for the fungal components to reproduce sexually and for the algal component independently to reproduce asexually.

These associations are given genus and species names. Lichenization may have evolved independently more than ten times. The fungal components do not grow alone in the wild, whereas some lichen algae can. Some lichenologists use the term "controlled parasitism" to describe the symbiosis in lichens, although it appears that the relationships is usually mutualistic.

Lichens can grow on bare rock and withstand desiccation and great cold. In the tundra, herds of caribou and reindeer graze on lichens. Lichens cannot, however, tolerate air pollution, and they are particularly sensitive to sulfur dioxide and other aerial poisons.

Mycorrhizae Mycorrhizae are common and very important mutualistic associations of plant roots and fungi, in which the fungal hyphae provide the plant with minerals absorbed from the soil and the plant provides organic nutrients to the fungus. Basidiomycetes and ascomycetes form sheathing mycorrhizae, which envelop the root with a mantle of hyphae and extend other hyphae between root cortex cells. Zygomycetes form mycorrhizae that penetrate cortex cells with haustoria.

The Ecological Importance of Fungi

Fungi as Decomposers Fungi and bacteria are the principal decomposers of organic matter making possible the essential recycling of chemical elements between living organisms and their abiotic surroundings. Invasive hyphae, enzymes that work on cellulose and lignin, and prolific production of colonizing spores make fungi excellent decomposers of plant material.

Fungi as Spoilers As well as decomposing the organic litter in our ecosystem, fungi work on our food,

clothing, and the wood used in buildings and boats. Keeping materials dry is the best way to protect them from mold.

Pathogenic Fungi Pathogenic fungi cause athlete's foot, ringworm, vaginal yeast infections, and lung infections in humans. Fungal diseases of plants are common. Dutch elm disease, caused by an ascomycete, is eliminating the American elm. Another ascomycete forms ergots on rye that can cause serious symptoms when accidentally milled into flour. Lysergic acid, the raw material of LSD, is one of the toxins in the ergots. Weak doses of some fungal toxins have beneficial medical use.

Edible Fungi Commercially cultivated species of *Agaricus* are eaten, as well as wild mushrooms gathered by knowledgeable collectors. Molds and yeasts are important in cheese, bread, and alcohol production. Truffles, the ascocarps of certain mycorrhizal ascomycetes, develop underground and release strong odors that signal their location. In some cases, the odors released by subterranean ascocarps or basidiocarps mimic sex attractants, which help assure that they are unearthed and their spores dispersed.

Evolution of Fungi

Fungi appear to have had a separate origin from plants and animals, perhaps from a red algae, conjugating green algae, or Chytridiomycota ancestor. Additional rRNA sequencing may help to resolve their origin. The phylogeny of fungal groups is uncertain.

The oldest undisputed fossils of fungi date back 400-440 million years to the Silurian period. The first terrestrial vascular plant fossils have petrified mycorrhizae, indicating that plants and fungi moved onto land together. Terrestrial communities appear to have always been dependent on fungi as decomposers and partners in symbiotic associations.

STRUCTURE YOUR KNOWLEDGE

1. Fill in the following table that summarizes the characteristics of the major groups of fungi.

2. The kingdom Fungi contains members with saprobic, parasitic, and mutualistic modes of nutrition. How do these types of nutrition relate to the ecological and economic importance of this group?

Group	Examples	Morphology	Asexual Reproduction	Sexual Reproduction	Miscellaneous Information
Zygomycota					
Ascomycota					
Basidiomycota					
Deuteromycota					
Lichens					

3. Describe the types of fungal hyphae illustrated by the following three diagrams.

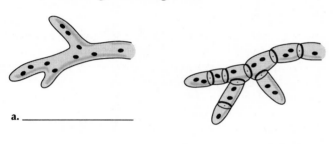

a. _____

b. _____

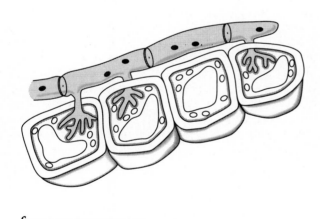

c. _____

4. Indicate on the time line in Chapter 24 the time period during which the first undisputed fossils of fungi appear.

TEST YOUR KNOWLEDGE

FILL IN THE BLANKS

1. _____ division between cells in fungal hyphae

2. _____ hyphal cells with two nuclei

3. _____ component of cell walls in most fungi

4. _____ asexual spores produced in chains at ends of hyphae

5. _____ club-shaped reproductive structure found in mushrooms

6. _____ sacks that contain sexual spores in cup fungi

7. _____ mutualistic associations between plant roots and fungi

8. _____ tough, protective zygote produced by zygomycetes

9. _____ type of hyphae with many nuclei

10. _____ hyphae of parasitic fungi that grows into cells or tissues

MULTIPLE CHOICE: *Choose the one best answer.*

1. The major difference between fungi and plants is that fungi
   a. have an absorptive form of nutrition.
   b. do not have a cell wall.
   c. are not eukaryotic.
   d. are multinucleate but not multicellular.

2. Fungal mitosis
   a. produces spores in dikaryotic cells.
   b. does not involve the formation of a spindle.
   c. takes place within the nucleus.
   d. all of the above

3. A fungus that is both a parasite and a saprobe is one that
   a. digests only the nonliving portions of its host's body.
   b. lives off the sap within its host's body.
   c. first lives as a parasite but then consumes the host after it dies.
   d. lives as a mutualistic symbiont on its host.

4. The fact that karyogamy does not immediately follow plasmogamy
   a. is necessary to create coenocytic hyphae.
   b. allows for the development of more genetic variation.
   c. allows fungi to reproduce asexually most of the time.
   d. creates dikaryotic cells that may benefit from the presence of duplicate copies of alleles.

5. The form division Deuteromycota
   a. is a practical classification that includes molds that usually reproduce asexually by conidia.
   b. includes the fungal components of lichens.
   c. includes the imperfect fungi that have abnormal forms of sexual reproduction.
   d. is home to molds, yeasts, and lichens.

6. Basidiomycetes differ from all other fungi in that they
   a. have dikaryotic cells.
   b. do not commonly reproduce asexually.
   c. form symbiotic relationships with algae.
   d. are coenocytic.

7. In the Ascomycota,
   a. sexual reproduction occurs by conjugation.
   b. spores line up in a sack in the order they were formed by meiosis.
   c. asexual spores form in sporangia on erect hyphae.
   d. most hyphae are dikaryotic.

8. Lichens are symbiotic associations that
   a. usually involve an ascomycete and a green alga or cyanobacterium.
   b. can reproduce asexually by forming soredia.
   c. are able to colonize bare rocks and cold habitats.
   d. are all of the above.

9. The name given to each of the three division of fungi is based on
   a. the structure in which karyogamy occurs during sexual reproduction.
   b. the location of plasmogamy during sexual reproduction.
   c. the location of the dikaryotic stage in the life cycle.
   d. the structure that produces asexual spores.

CHAPTER 29

INVERTEBRATES AND THE ORIGIN OF ANIMAL DIVERSITY

FRAMEWORK

Animals are multicellular eukaryotic heterotrophs whose mode of nutrition is ingestion. Characteristics of animals include sexual reproduction; dominant diploid stage; specialized cells, tissues and organs; embryonic development with a blastula stage; and the presence of muscles and nerves.

This chapter surveys the characteristics and representatives of the major animal phyla. Fossil evidence for Kingdom Animalia's origin and phylogeny is lacking, so comparative anatomy and embryology are used to construct phylogenetic trees. The figure on p. 234 is one version of that phylogeny.

CHAPTER SUMMARY

The evolution of multicellular organisms that eat other organisms opened a new adaptive zone. Animal life began in the Precambrian seas, then radiated, populating first the seas and eventually the land. Animals are grouped into about 35 phyla, with more than a million recognized species and perhaps another million not yet identified. All but one subphylum of these are *invertebrates*, animals that lack backbones, which include over 95% of animal species.

Defining *Animals*

Animals are multicellular heterotrophic eukaryotes that use *ingestion* as their mode of nutrition, either eating other organisms or decomposing organic material, called *detritus*. They store carbohydrates as glycogen. Animal cells lack walls and may have desmosomes, gap junctions and tight junctions, which connect adjacent cells. Muscle and nervous tissues are unique to animals.

The diploid stage is usually dominant and reproduction is primarily sexual, with a flagellated sperm fertilizing a larger, nonmotile egg. The zygote undergoes a series of mitotic divisions, called *cleavage*, usually passing through a *blastula* (hollow ball of cells) stage during embryonic development. The process of *gastrulation* which follows produces embryonic tissues for all of the adult body parts. The life cycle of some animals includes a *larva*—a free-living, sexually immature form. *Metamorphosis* transforms a larva into a sexually mature adult.

The greatest number of animal phyla are found in the stable environment of the oceans. Many animals live in freshwater. Terrestrial habitats have been extensively exploited by only a few animal phyla, notably the vertebrates and arthropods.

Clues to Animal Phylogeny

Because of a fragmented fossil record from the Precambrian and early Cambrian periods, when most animal phyla developed, animal phylogeny depends on comparative anatomy and embryology.

Major Branches of the Animal Kingdom   The animal kingdom may have had its origin from one group of protists. The sponges (phylum Porifera) are called *parazoa* and separated from other animals on the basis of anatomical simplicity. All other animal phyla are grouped into the *eumetozoa*.

The eumetozoa is divided into two groups, partly on the basis of body symmetry: The *radiata*, containing the hydras and jellyfishes, have *radial symmetry*. The *bilateria* include animals with *bilateral symmetry*, having distinct head (*anterior*) and tail (*posterior*) ends and left and right sides. Bilateral animals also have top (*dorsal*) and bottom (*ventral*) sides.

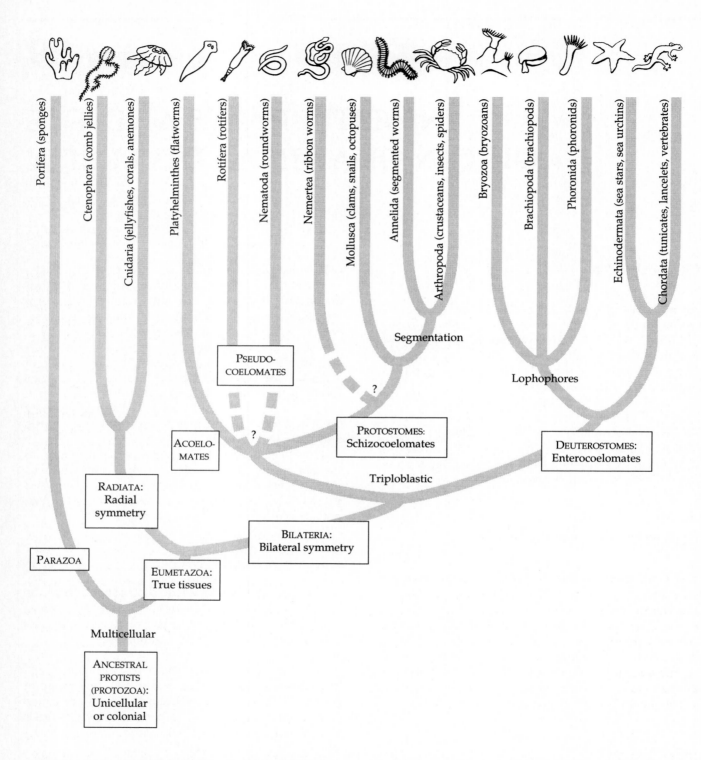

Bilateral symmetry is associated with *cephalization*, the concentration of sensory organs in the head end, which is an adaptation for unidirectional movement. The radial symmetry of a sessile or planktonic animal allows it to confront its environment equally on all sides. Radial symmetry may have evolved secondarily in sedentary eumetozoans.

The basic radiata–bilateria split probably arose early, and some zoologists think the two lines may have come from different protistan ancestors.

Development and Body Plan   During gastrulation a eumetozoan embryo develops layers of cells called the *germ layers*: *Ectoderm* develops into the outer body covering and, in some phyla, into the central nervous system; and *endoderm* lines the primitive gut, or *archenteron*, and gives rise to the lining of the digestive tract and associated organs. The radiata (cnidarians and ctenophores) are *diploblastic*, forming only these two layers. All other eumetozoa are *triploblastic*, producing a middle layer, the *mesoderm*, from which arise muscles and most other organs.

Triploblastic animals that have solid bodies are called *acoelomates*. These include the flatworms (Platyhelminthes) and a few other phyla. A tube-within-in-a-tube body plan with a fluid-filled cavity separating the digestive tract from the outer body wall is found in the other triploblastic, bilateria animals. In *pseudocoelomates*, including the rotifers (Rotifera) and roundworms (Nematoda), the cavity is not completely lined by mesoderm and is called a *pseudocoelom*. *Coelomates* have a true *coelom*, a body cavity completely lined by mesoderm. Mesenteries connecting the inner and outer lining suspend the internal organs. A fluid-filled body cavity cushions internal organs, allows organs to grow and move independently of the outer body wall, and also functions as a hydrostatic skeleton in soft-bodied animals.

The Protostome—Deuterostome Dichotomy Differences in embryology and coelom formation divide the coelomates into two evolutionary lines. The *protostome* line includes annelids, mollusks and arthropods. The *deuterostomes* include echinoderms and chordates.

Most protostomes have *spiral cleavage*, in which the planes of cell division are diagonal to the polar axis of the egg and newly formed cells fit in the grooves between cells of adjacent tiers. The *determinate cleavage* of some protostomes sets the developmental fate of each embryonic cell very early. Many deuterostomes exhibit *radial cleavage*, in which parallel and perpendicular cleavage planes result in aligned tiers of cells in the dividing zygote. *Indeterminate cleavage* in deuterostomes means that early embryonic cells retain the capacity to develop into a complete embryo. Indeterminate cleavage makes possible human identical twins.

The *blastopore* is the opening of the developing archenteron. In typical protostomes ("first mouth"), the blastopore develops into the mouth and a second opening forms at the end of the archenteron to produce an anus. In deuterostomes, the blastopore becomes the anus and the second opening develops into the mouth.

In protostomes, the coelom forms from splits within solid masses of mesoderm, called *schizocoelous* development. In deuterostomes, the mesoderm begins as lateral outpocketings from the archenteron, called *enterocoelous* development.

Parazoa (Phylum Porifera)

Sponges are sessile, mostly marine animals. Water is drawn through pores in the body wall of this saclike animal, into a central cavity, the *spongocoel*, from where it flows out through the *osculum*. Sponges are filter-feeders, collecting food particles from the water circulating through their bodies by the action of collared, flagellated *choanocytes* lining the inside of the body.

In the *mesohyl*, or gelatinous matrix between the two body-wall layers, are *amoebocytes*. These cells take up food from the choanocytes, digest it, and carry nutrients to other cells. Amoebocytes also form skeletal fibers, which may be sharp spicules made of calcium carbonate or silicate, or flexible fibers composed of a protein called spongin.

Most sponges are *hermaphrodites*, producing both eggs and sperm. Sperm, released into the spongocoel and carried out through the osculum, cross-fertilize eggs retained in the mesohyl of neighboring sponges. Swimming larvae disperse through the osculum. After settling on a suitable substrate, the larva inverts, and the flagellated cells line the interior. Sponges are capable of extensive *regeneration*, replacing damaged body parts and reproducing asexually from fragments.

Sponges lack organs, nerves, and muscles, and have relatively unspecialized cells. Ancestors of sponges may have been choanoflagellates, collared protists that resemble choanocytes.

Eumetazoa: Radiata

Phylum Cnidaria The phylum Cnidaria includes hydras, jellyfishes, sea anemones and corals. These mostly marine animals are diploblastic and have radial symmetry, indicating an early origin off the eumetazoan line. Their simple anatomy consists of a sac with a central gastrovascular cavity and a single opening serving as both mouth and anus. This body plan has two forms: polyps, which are sessile, cylindrical forms with mouth and tentacles extending upward, and medusae, which are flattened, mouth-down polyps that move by passive drifting and weak body contractions. Both these body forms occur in the life cycles of some cnidarians.

Cnidarians use their ring of tentacles, armed with *cnidocytes* containing stinging capsules called *nematocysts*, to capture prey. Cells of the epidermis and gastrodermis have bundles of microfilaments arranged into contractile fibers acting as simple muscles. A nerve net coordinates the contraction of cells against the hydrostatic skeleton of the gastrovascular cavity, resulting in movement.

The class Hydrozoa includes animals that alternate

between polyp and medusa forms, as seen in the colonial *Obelia*. Sexual reproduction occurs in the medusa stage, whereas the asexual polyp stage is usually more conspicuous. The common, freshwater *Hydra* exists only in polyp form, reproducing asexually by budding or sexually when environmental conditions deteriorate.

In the class Scyphozoa, the medusa stage is more prevalent. The sessile polyp stage often does not occur in the jellyfishes of the open ocean. The mesohyl between the epidermis and gastrodermis is often thick and jellylike. The class *Anthozoa* includes sea anemones and most corals. They occur only as polyps. Coral polyps secrete calcified external skeletons, and the accumulation of such skeletons produces coral and coral reefs.

Phylum Ctenophora   Comb jellies of the phylum Ctenophora are the largest animals that use cilia for locomotion. Nerves running from a sensory organ to the combs of cilia coordinate movement of these small, transparent marine animals.

Eumetozoa Bilateria: Acoelomates

Phylum Platyhelminthes   The flatworms of the phylum Platyhelminthes are triploblastic and acoelomate, probably diverging from the protostome line prior to the evolution of the schizocoelom. These flattened, bilateral animals have moderate cephalization, and their mesoderm gives rise to organs and muscles. Like cnidarians, typical flatworms have a gastrovascular cavity with only one opening.

The class Turbellaria includes freshwater *planarians* and other, mostly marine, free-living flatworms. Their branching gastrovascular cavity functions for both digestion and circulation. Gas exchange and diffusion of nitrogenous wastes occur across the body wall. Ciliated flame cells function in osmoregulation. Planarians move using cilia to glide on secreted mucus or body undulations to swim.

Eyespots on the head detect light and auricles function for smell. A pair of ganglia leads into a ladderlike nervous system, and planarians can modify their behavior. Planarians reproduce asexually by regeneration or sexually by copulation between hermaphroditic worms.

Flukes are placed in the classes Trematoda and Monogenea. Adapted to live as parasites in or on other animals, flukes have a tough outer covering, suckers, and extensive reproductive organs. The life cycles of flukes are complex, usually including asexual and sexual stages, and intermediate hosts in which larvae develop.

Tapeworms of the class Cestoda are parasites,

mostly of vertebrates. Tapeworms consist of a scolex, with suckers and hooks for attaching to the host's intestinal lining, and a ribbon of proglottids packed with reproductive organs. Absorption of predigested food from the host eliminates the need for a digestive system. The life cycles of tapeworms may include intermediate hosts.

Phylum Nemertea   Most proboscis worms in the phylum Nemertea are marine. Their long, retractable proboscis is used to explore the environment and capture prey. The fluid-filled sac that allows these worms to hydraulically operate the proboscis is considered by some zoologists to be a true coelom, indicating a possible origin from the schizocoelomate line. These worms probably evolved from flatworms, based on the similarity of their excretory, sensory, and nervous systems. They show, however, two new anatomical features: a *complete digestive tract* with separate mouth and anus, and a circulatory system of vessels through which blood flows.

Bilateria: Pseudocoelomates

The evolutionary relationships of the several phyla that have a pseudocoelomate body plan are under debate. The pseudocoelomate condition probably evolved several times. These phyla may be more closely related to protostomes than to deuterostomes.

Phylum Rotifera   Rotifers are smaller than many protists, but have a pseudocoelom that acts as a hydrostatic skeleton and a medium for diffusion, a complete digestive tract, and other organ systems. A crown of cilia draws water and microscopic food organisms into the mouth. Some species reproduce by *parthenogenesis*, by which female offspring develop from unfertilized eggs. In other species, two types of egg develop parthenogenetically: one type forming females and the other developing into degenerate males which produce sperm that fertilize eggs. The resulting resistant zygotes survive harsh conditions in a dormant state.

Phylum Nematoda   Round worms are among the most abundant of all animals, found inhabiting water, soil, and the bodies of plants and animals. These cylindrical worms are covered with tough cuticles and have complete digestive tracts. Their thrashing movement is produced by contraction of longitudinal muscles. The pseudocoelomic fluid produces a hydrostatic skeleton and helps transport nutrients throughout the body. Reproduction is usually sexual; fertilization is internal, and most zygotes form resistant cells.

Numerous nematode species are ecologically important in decomposition and nutrient cycling. The developmental history of every cell of *Caenorhabditis elegans*, an important research organism, has been traced.

Many nematodes are serious agricultural pests and animal parasites. Humans are host to 50 species, including pinworms, hookworms, and *Trichinella spiralis*, the nematode that causes trichinosis.

Bilateria: Protostomes (Schizocoelomates)

Phylum Mollusca Mollusks are mostly soft-bodied marine animals, many of which are protected by a shell. The molluscan body plan has three main parts: a muscular *foot* used for movement, a *visceral mass* containing the internal organs, and a *mantle* that covers the visceral mass and may secrete a shell. The *mantle cavity*, the water-filled chamber formed by the mantle, encloses the gills, anus, and excretory pores. A rasping *radula* is used for feeding by many mollusks. Most mollusks have separate sexes.

Some marine mollusks have a life cycle that includes a ciliated larva called the *trochophore*, also found in some other protostome phyla. Mollusks are not segmented, and many zoologists believe that mollusks did not evolve from a segmented, annelid-like ancestor, but branched earlier on the protostome line. Of the eight molluscan classes, four are discussed in the text.

Chitons, class Polyplacophora, are oval marine animals with shells that are divided into eight dorsal plates. Chitons cling tightly to rocks in the intertidal zone, where they feed on algae.

Gastropoda is the largest molluscan class. Most members are marine, although there are many freshwater species and some snails and slugs are terrestrial. A distinctive feature of this class is *torsion*, the embryonic asymmetric growth of the visceral mass that results in a U-turn of the digestive tract, with the anus and mantle cavity ending up above the head. Most gastropods have single spiraled shells, although slugs and nudibranchs have no shells. Many gastropods have distinct heads with eyes at the tips of tentacles. Land snails lack gills; their vascularized mantle cavity functions as a lung.

The clams, oysters, mussels and scallops of the class Bivalvia have the two halves of their shell hinged at the mid-dorsal line. The shell is kept closed by powerful adductor muscles. Bivalves have neither heads nor radula. They are suspension feeders; water flows into and out of the mantle cavity through siphons, and food particles are trapped in the mucous that coats the gills and then swept to the mouth by cilia. Bivalves are fairly sedentary, using their muscular foot to move and anchor themselves.

The active squids and octopuses of the class Cephalopoda are rapidly moving carnivores. The mouth has beaklike jaws to crush prey and is surrounded by tentacles. The foot has been modified to form the muscular siphon used to jet-propel the squid when water from the mantle cavity is expelled. The shell is reduced and internal in squids, absent in octopuses, and external only in the chambered nautilus. The giant squid is the largest of all invertebrates.

Cephalopods are the only mollusks with *closed circulatory systems*. They have well-developed nervous systems, sense organs, and complex brains—important features for active predators. The ancestors of octopuses and squids were probably shelled, predacious mollusks that eventually evolved and lost their shells. Shelled *ammonites* were the dominant invertebrate predators until their extinction at the end of the Cretaceous period.

Phylum Annelida Annelids are segmented worms, which are found in most marine, freshwater, and soil habitats. The earthworm, a well-known example, has a complete digestive system with specialized regions and a closed circulatory system with some blood vessels functioning as pumping hearts. Respiration occurs across the moist, highly vascularized skin. Septa partition the coelom into segments, in each of which is found a pair of excretory *metanephridia*, which filter metabolic wastes from the blood and coelomic fluid. The nervous system consists of a brain-like pair of cerebral ganglia, a subpharyngeal ganglion, and segmental ganglia along the fused ventral nerve cords. Earthworms contain both male and female reproductive organs; sperm are exchanged between worms during copulation. A mucous cocoon, secreted by the *clitellum*, slides off the worm after picking up its eggs and stored sperm.

Noncompressible coelomic fluid is enclosed by septa within the body segments, serving as a hydrostatic skeleton; circular and longitudinal muscles contract alternately to extend the body and pull it forward along the substrate. Segmentation may first have evolved as an adaptation for this type of burrowing or creeping movement.

The class Oligochaeta includes the earthworms and aquatic species. Earthworm castings improve soil texture. The mostly-marine segmented worms in the class Polychaeta have parapodia on each segment that function in gas exchange and locomotion. Marine polychaetes may be planktonic, bottom burrowers, or tube-dwellers.

Most of the leeches in the class Hirudinea inhabit fresh water. Many feed on small invertebrates, whereas others are parasites that temporarily attach to animals, slit or digest a hole through the skin, and suck the blood of their host. Leeches are still used for some medicinal purposes.

Annelids exhibit two evolutionary innovations. A coelom provides a host of advantages: a hydrostatic skeleton for movement, room for complex organ development, cushion for internal organs, and separation of the action of body wall muscles from internal functions. Segmentation allows for regional specialization of body segments.

Phylum Arthropoda   Arthropods are the most successful phylum of animals if species diversity, distribution, and total numbers are the criteria. Characteristics of arthropods include segmentation, a hard exoskeleton, and jointed appendages. A *cuticle* of chitin and protein completely covers the body as an *exoskeleton*, providing protection and a point of attachment for the muscles that move the appendages. To grow, an arthropod must *molt*, shedding its old exoskeleton and secreting a larger one. Arthropods have extensive cephalization and well-developed sensory organs, including eyes, olfactory receptors, and antennae.

A heart pumps hemolymph through an open circulatory system consisting of short arteries and a network of sinuses known as the hemocoel. Respiratory gas exchange in most aquatic species occurs through gills that extend from the body, whereas terrestrial arthropods have tracheal systems of branching internal ducts.

Arthropods probably evolved from annelids or from a segmented protostome that was a common ancestor of both phyla. Early arthropods may have resembled onychophorans, a phylum of segmented worms with unjointed walking appendages. Arthropods evolved along four evolutionary lines now recognized as subphyla: Trilobitomorpha, Chelicerata, Uniramia, and Crustacea.

Trilobites, with pronounced segmentation and uniform appendages, were common early arthropods throughout the Paleozoic era, but are now extinct. Existing during and beyond this same period were large *eurypterids*, or sea scorpions, marine predators belonging to the chelicerate subphylum.

The chelicerate body is divided into a cephalothorax and abdomen, with the most anterior appendages modified as pincers or fangs, called *chelicerae*. Most marine chelicerates are also now extinct; the horseshoe crab is the one survivor. Most modern chelicerate species are in the terrestrial class Arachnida, which includes spiders, scorpions, ticks, and mites. The cephalothorax has six pairs of appendages: four pairs of walking legs, the chelicerae, and sensing or feeding appendages called pedipalps. Spiders kill prey with poison-equipped, fanglike chelicerae, then, after adding digestive juices, suck up partially digested food. *Book lungs*, consisting of stacked internal plates, function in gas exchange. Many spiders spin characteristic webs of silk from special abdominal glands.

Another evolutionary line produced uniramians and crustaceans with jawlike *mandibles*, one or two pairs of sensory *antennae*, and usually a pair of *compound eyes*. Uniramians have one pair of antennae and unbranched (uniramous) appendages and are believed to have evolved on land. The aquatic crustaceans, with two pairs of antennae and branched appendages, are believed to have evolved in the ocean.

The exoskeleton of early marine arthropods, which probably functioned in protection and anchorage for muscles, helped preadapt arthropods for terrestrial life by providing the support and protection from desiccation needed on land. During the early Devonian period both chelicerates and uniramians spread onto land. The oldest evidence of terrestrial animals are burrows of millipede-like arthropods about 450 million years old.

The millipedes of the class Diplopoda are wormlike, segmented vegetarians with two pairs of walking legs per segment. The centipedes of the class *Chilopoda* are terrestrial carnivores with appendages modified as mouthparts and poison claws for paralyzing prey. Each segment of the trunk has one pair of legs.

The class *Insecta* is divided into about 26 orders and has more species than all other forms of life combined. The study of insects, called *entomology*, is a large field with many subspecialities. The oldest insect fossils date back to the Devonian period, but a major diversification occurred in the Carboniferous and Permian periods with the evolution of flight. Another important radiation occurred as insects evolved along with flowering plants during the Cretaceous period.

Flight is a major key to the success of insects. Many insects have one or two pairs of wings which are cuticle extensions of the dorsal thorax. Insects flap their wings by the action of muscles that change the shape of the thoracic cuticle. Dragonflies were among the first winged insects. Later wing modifications include overlapping or hooking the two pairs together during flight (butterflies and bees) and hardening the anterior wings as a protective cover (beetles).

A generalized insect body has three regions: a head, a thorax, and an abdomen. The head, consisting of several fused segments, has one pair of antennae, a pair of compound eyes, and several pairs of mouth parts modified for various types of ingestion. The segmented thorax has wings and three pairs of walking legs.

The digestive tract has several specialized regions. The circulatory system is open; the heart pumps hemolymph through an artery into the hemocoel. *Malpighian tubules* are outpocketings of the gut that function as excretory organs, removing metabolic wastes from the hemolymph. Gas exchange is accomplished by a *tracheal system* of chitin-lined tubes that ramify throughout the body and open to the outside through spiracles.

The nervous system consists of a pair of ventral nerve cords with segmental ganglia, and a dorsal brain formed from several fused ganglia. The complex behavior of insects appears to be largely innate.

In *incomplete metamorphosis* the young are smaller versions of the adult and pass through several molts. In *complete metamorphosis* the larvae, known as maggots, grubs or caterpillars, look entirely different from the adult. Larvae eat and grow; adults mate and reproduce. Reproduction is usually sexual; fertilization is usually internal.

Insects have a huge impact on humans as pollinators of food crops, vectors of diseases, and competitors for food.

Animals in the subphylum Crustacea have remained mostly in marine and freshwater habitats. Crustaceans have two pairs of antennae, three or more pairs of mouthpart appendages including mandibles, walking legs on the thorax, and appendages on the abdomen. Larger crustaceans have gills. A heart pumps hemolymph through arteries into sinuses that bathe the organs. Nitrogenous wastes pass by diffusion through thin areas of the cuticle and a pair of glands regulates salt balance in the hemolymph. Sexes usually are separate. A swimming larval stage occurs in most aquatic crustaceans.

Lobsters, crayfish, crabs, and shrimp are relatively large crustaceans called decapods. Their cuticle is hardened by calcium carbonate, and the dorsal side of their cephalothorax is covered by a carapace. Isopods are mostly small marine crustaceans, but also include terrestrial sow bugs and pill bugs. The small, very numerous copepods are important members of plankton communities that form the foundation of marine and freshwater food chains. Shrimp-like *krill* form a major food source for many whale species. Barnacles are sessile crustaceans that strain food from the water with their appendages.

Bilateria: Deuterostomes (Enterocoelomates)

The lophophorates, echinoderms, and chordates are grouped together based on their common embryological traits of radial cleavage, coelom formation from the archenteron, and origin of the mouth opposite the blastopore.

The three phyla of *lophophorate animals*, Phoronida, Bryozoa, and Brachiopoda, all have a *lophophore*, a horseshoe-shaped or circular fold bearing ciliated tentacles, which surrounds the mouth and functions in filter-feeding. This complex structure suggests that all three phyla are related. The other characteristics they have in common, a U-shaped digestive tract and lack of a distinct head, may have evolved separately in each group as adaptations to a sessile life. In their embryonic development, the lophophorate animals resemble deuterostomes more closely than protostomes. In the phoronids, however, the mouth forms from the blastopore, suggesting that lophophorate animals may have branched off early from the deuterostome line.

Phoronids are marine tube-dwelling worms that often live buried in sand with their lophophore extended. Bryozoans are tiny, mostly marine animals living in colonies that are often encased in a hard exoskeleton, with pores through which their lophophores extend. Brachiopods, or lamp shells, resemble bivalves. These marine animals attach to the substrate by a stalk and open their shell to allow water to flow through the lophophore. *Lingula* is a brachiopod genus that has changed little in 400 million years.

Phylum Echinodermata Most *echinoderms* are sessile or sedentary marine animals with radial symmetry. They have a thin skin covering an endoskeleton of hard calcareous plates. A *water vascular system* with a network of hydraulic canals controls extensions called tube feet that can function in locomotion, feeding and gas exchange. Sexual reproduction usually involves separate sexes and external fertilization. Bilateral larvae metamorphosize into radial adults.

The sea stars, class Asteroidea, have five arms radiating from a central disc. Sea stars use tube feet lining the undersurfaces of their arms to creep slowly and to open bivalves for food. They evert their stomach through their mouth and slip it into a slightly opened shell, where digestive juices begin to digest the bivalve. Sea stars are capable of regenerating lost arms.

Brittle stars, class Ophiuroidea, have distinct central discs and long and flexible arms. They move by lashing their arms.

Sea urchins and sand dollars, class Echinoidea, have no arms, but are able to move slowly using their five rows of tube feet. Long spines also aid in movement. Their mouth is ringed by complex jawlike structures used to eat seaweeds and other foods.

Sea lilies, class Crinoidea, live attached by stalks to the substrate or crawl, using their long, flexible arms which extend upward from around the mouth and which are used also in suspension feeding. Their form has changed little in 500 million years of evolution.

Sea cucumbers, class Holothuroidea, are elongated animals that bear little resemblance to other echinoderms other than having five rows of tube feet.

The recently discovered sea daisies of the class Concentricycloidea are small animals that live on water-logged wood in the deep sea.

Phylum Chordata   This phylum contains two subphyla of invertebrates plus the subphylum Vertebrata. Echinoderms and chordates share deuterostome developmental characteristics, but have existed as separate phyla for at least 500 million years.

The Origin and Diversification of Animals

Several hypothetical ancestors have been suggested to explain the origin of animals from protists. The *syncytial hypothesis* holds that animals developed from a multinucleate protist, perhaps a ciliate, that became divided into many cells. The *colonial hypothesis* suggests a spherical, heterotrophic, colonial flagellate ancestor, perhaps similar to a present day choanoflagellate colony. Cell layers may have developed from cell proliferation or invagination. If cells proliferated to form an inner mass of digestive and reproductive cells, the protoanimal, called a planuloid, would resemble the planula larva of some cnidarians. If cell layers were formed by invagination, the protoanimal would resemble the embryonic gastrula stage. Another suggested hypothetical ancestor resembles *Trichoplax adhaerens*, a contemporary tiny, flat marine creature with a ciliated epidermis. Lacking a gut, this animal hunches up to form a temporary digestive cavity when it crawls over a food particle. The first archenteron may have originated in this fashion.

There is no fossil record of transitional forms between protists and the first animals. The early diversification of animals is also a phylogenetic mystery. The oldest known fossils come from the end of the Precambrian era, in a span from about 700 to 590 million years ago known as the *Ediacaran period*. These soft-bodied creatures resembled cnidarian medusae, some colonial cnidarians, and annelids.

A second radiation occurred in the Cambrian period at the beginning of the Paleozoic era about 590 million years ago. The evolution of shells and hard skeletons may have opened new adaptive zones. Nearly all modern phyla, along with many extinct phyla, are present in Cambrian fossils. All the basic designs of modern animals evolved in the Cambrian radiation.

STRUCTURE YOUR KNOWLEDGE

1. The diverse phyla of kingdom Animalia are structured into broad groups based on a series of general characteristics (symmetry, coelom, and embryology). Create a concept map or diagram that shows these divisions, the criteria used to determine them, and the major phyla included in each group. Include common examples for each taxon.

2. Why are the origin and phylogeny of kingdom Animalia a mystery? Describe some of the hypotheses concerning animal evolution.

3. Complete the table on p. 241 to show the general characteristics and structures used by members of each phylum for accomplishing basic life functions.

4. Indicate on the timeline in Chapter 24 the date of the oldest known fossils of animals.

TEST YOUR KNOWLEDGE

MATCHING: *Match the following organisms with their classes and phyla. Answers may be used more than once or not at all.*

	PHYLUM	CLASS	ORGANISM
1.	_____	_____	jellyfish
2.	_____	_____	crayfish
3.	_____	_____	snail
4.	_____	_____	leech
5.	_____	_____	tapeworm
6.	_____	_____	cricket
7.	_____	_____	scallop
8.	_____	_____	tick
9.	_____	_____	sea urchin
10.	_____	_____	hydra

PHYLA	CLASSES
a. Annelida	A. Turbellaria
b. Arthropoda	B. Bivalvia
c. Cnidaria	C. Cestoda
d. Echinodermata	D. Crustacea
e. Mollusca	E. Echinoidea

Phylum	Digestion	Excretion	Respiration	Circulation	Nervous Control	Skeleton	Locomotion
Porifera							
Cnidaria							
Platyhelminthes							
Nemertea							
Rotifera							
Nematoda							
Mollusca							
Annelida							
Arthropoda							
Echinodermata							

f. Nematoda
g. Nemertea
h. Platyhelminthes
i. Porifera
j. Rotifera

F. Gastropoda
G. Hirudinea
H. Hydrozoa
I. Insecta
J. Oligochaeta
K. Ophiuroidea
L. Scyphozoa
M. Turbellaria

MULTIPLE CHOICE: *Choose the one best answer.*

1. Invertebrates include
   a. all animals except for the phylum Vertebrata.
   b. all animals without backbones.
   c. only animals that use hydrostatic skeletons.
   d. the radiata and protostomes, but not the deuterostomes.

2. The greatest number of animal phyla are found
   a. in the ocean, where environmental conditions are most stable.
   b. in freshwater habitats, where organisms do not face problems of salt balance.
   c. on land, where gas exchange and locomotion are easiest.
   d. in the tropical rain forests, where temperatures are warm and rainfall is abundant.

3. An insect larva
   a. is a sexually immature organism that does not resemble the adult.
   b. is transformed into an adult by metamorphosis.
   c. is specialized for eating and growth.
   d. is all of the above.

4. Sponges differ from the rest of the Animalia because
   a. they are completely sessile.
   b. they have radial symmetry and are filter feeders.
   c. their simple body structure has no tissues or organs.
   d. they have an unusual feeding structure called a lophophore and their embryology differs from both protostomes and deuterostomes.

5. Cephalization
   a. is the development of bilateral symmetry.

b. is associated with motile animals that concentrate sensory organs in a head region.

c. is a diagnostic characteristic of deuterostomes.

d. is common in radially symmetrical animals.

6. A true coelom
   a. is found in deuterostomes.
   b. is a fluid-filled cavity completely lined by mesoderm.
   c. may be used as a hydrostatic skeleton by soft-bodied coelomates.
   d. is all of the above.

7. Which of the following is descriptive of protostomes?
   a. radial and determinate cleavage, blastopore becomes mouth
   b. spiral and indeterminate cleavage, coelom forms as split in solid mass of mesoderm
   c. spiral and determinate cleavage, blastopore becomes mouth, schizocoelous development
   d. radial and determinate cleavage, enterocoelous development, blastopore becomes anus

8. Hermaphrodites
   a. contain male and female sex organs but usually cross fertilize.
   b. include sponges, earthworms, and most insects.
   c. are characteristically found in rotifers.
   d. are both a and b.

9. A gastrovascular cavity
   a. functions in both digestion and circulation.
   b. has only one opening that serves as mouth and anus.
   c. is found in the phyla Cnidaria and Platyhelminthes.
   d. is all of the above.

10. Which of the following is *not* true of cnidarians?
    a. An alternation of medusa and polyp stage is common in the class Hydrozoa.
    b. They use a ring of tentacles armed with stinging cells to capture prey.
    c. They include hydras, jellyfishes, sea anemones and barnacles.
    d. They have a nerve net that coordinates contraction of microfilaments for movement.

11. Which of the following combinations of phylum and characteristics is incorrect?
    a. Nemertea — proboscis worm, first complete digestive tract
    b. Rotifera — parthenogenesis, crown of cilia, microscopic animals
    c. Nematoda — gastrovascular cavity, tough cuticle, ubiquitous

d. Annelida — segmentation, closed circulation, hydrostatic skeleton

12. Which of the following is either not an excretory structure or is incorrectly matched with its class?
    a. metanephridia — Oligochaeta
    b. Malpighian tubules — Echinoidea
    c. flame cells — Turbellaria
    d. thin region of cuticle — Crustacea

13. The major characteristics of the phylum Mollusca are
    a. radial symmetry, visceral mass, hydrostatic skeleton.
    b. a foot, mantle, visceral mass.
    c. shell, mantle, radula.
    d. tube feet, visceral mass, mantle.

14. Torsion
    a. is embryonic asymmetric growth that results in a U-shaped digestive tract in gastropods.
    b. is characteristic of molluscs.
    c. is responsible for the spiral growth of bivalve shells.
    d. is all of the above.

15. Bivalves differ from other molluscs in that they
    a. are predaceous.
    b. have no heads and are suspension feeders.
    c. have shells.
    d. have open circulatory systems.

16. The oldest fossil evidence of terrestrial animals are
    a. trilobites.
    b. insect castes.
    c. burrows of millipedelike organisms.
    d. eurypterids.

17. The exoskeleton of arthropods
    a. functions in protection and anchorage for muscles.
    b. is composed of chitin and cellulose.
    c. is absent in millipedes and centipedes.
    d. expands at the joints when the arthropod grows.

18. Which of the following is true of the subphylum Uniramia?
    a. The horseshoe crab is the one surviving marine member of this group.
    b. It contains the primarily aquatic crustaceans.
    c. It contains insects, centipedes, and millipedes, characterized by their unbranched appendages.
    d. It is characterized by jawlike mandibles, antennae, compound eyes, and includes all arthropods except the chelicerates.

19. Echinoderms are characterized by
    a. a calcareous exoskeleton.

b. bilateral larvae; radially symmetrical adults with rows of tube feet used for movement and feeding.

c. a water vascular system that serves both digestive and circulatory functions.

d. all of the above.

20. The oldest known animals
a. were soft-bodied creatures from the Ediacaran Period, over 590 million year ago.
b. were from the Cambrian Period at the beginning of the Paleozoic Era.
c. were planuloids—swimming, cigar-shaped, solid masses of cells.
d. were coenocytic ciliates resembling acoel worms.

CHAPTER 30

THE VERTEBRATE GENEALOGY

FRAMEWORK

This chapter focuses on the origin and characteristics of the subphylum Vertebrata and its classes. One version of chordate evolution is illustrated on p. 245. Fossil evidence and speculation about human ancestry is described.

CHAPTER SUMMARY

Phylum Chordata

There are three subphyla of *chordates*: *cephalochordates* and *urochordates*, which are without backbones, and *vertebrates*, which have backbones.

Chordate Characteristics Chordates have four distinguishing anatomical features, which may appear only during the embryo stage: (1) a flexible rod called a *notochord* running the length of the animal as a simple skeleton, (2) a *dorsal, hollow nerve cord* that develops into the brain and spinal cord of the central nervous system, (3) *pharyngeal slits* that open from the anterior region of the digestive tube and function in filter feeding or are modified for gas exchange and other functions, and (4) a *postanal tail* containing skeletal elements and muscles.

Chordates Without Backbones

Subphylum Cephalochordata Cephalochordates, known as lancelets, are tiny marine animals that retain all four chordate characteristics into the adult stage. They filter food particles into a mucous net secreted across the pharyngeal slits. The lancelet swims by coordinated contractions of serial muscle segments that flex the notochord from side to side. Blocks of mesoderm called *somites*, which flank the notochord of chordate embryos, develop into these muscle segments.

Subphylum Urochordata Tunicates or sea squirts are the common names for the mostly sessile, filter-feeding, marine animals in the subphylum Urochordata. Water moves through these saclike, tunic-covered animals from the incurrent siphon, through pharyngeal slits where food is filtered by a mucous net, to the excurrent siphon. The larval traits of notochord, nerve cord, and tail are lost in the adult animal.

The Origin of Vertebrates

Fossils resembling cephalochordates appear in the Burgess Shale of the Cambrian period and have been dated at 550 million years old; vertebrates appear 50 million years later. The fossil record does not show the origin of vertebrates from invertebrate ancestors. Protovertebrates may have been urochordate-like larval forms that became sexually mature and reproduced before undergoing metamorphosis to the sessile adult stage, a phenomenon known as *paedogenesis*. If the larval-like forms became more active swimmers and foragers, natural selection would have favored the development of a head with sensory organs and a brain.

Vertebrate Characteristics

Vertebrate characteristics include cephalization, the development of sense organs and a brain, and *vertebrae*, a series of skeletal units enclosing the nerve cord. The vertebrate skeleton is made of bone and/or cartilage, which is secreted by living cells and, unlike the

244

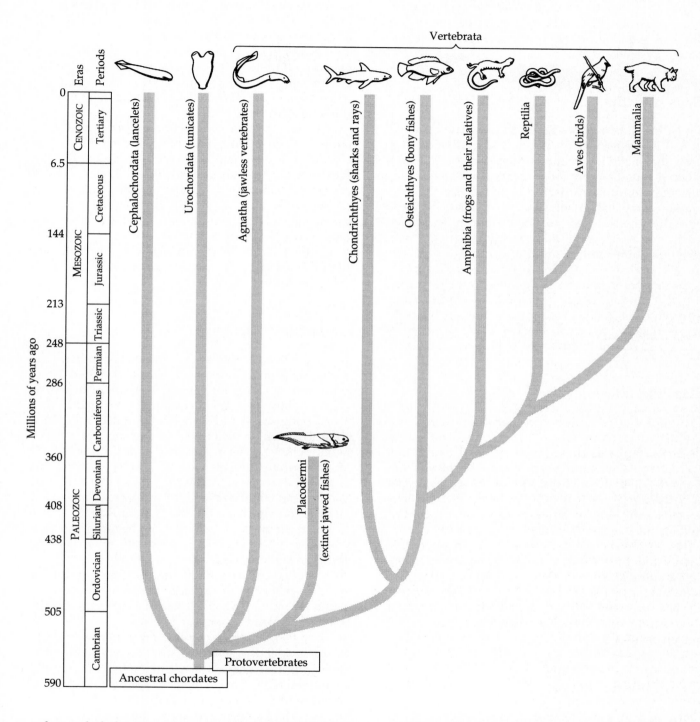

arthropod skeleton, can grow with the animal. The axial skeleton includes the skull enclosing the brain and the vertebral column, as well as, in many vertebrates, ribs and a breastbone. Many vertebrates also have an appendicular skeleton supporting two pairs of limbs.

Vertebrates have a closed circulatory system with a ventral, chambered heart pumping blood through arteries to capillaries and back through veins. The blood, containing hemoglobin in red blood cells, is oxygenated through the skin, gills or lungs. Kidneys remove waste products as the blood passes through them. Reproduction is almost always sexual, with separate male and female sexes. Fertilization may be internal or external. In some species eggs develop by parthenogenesis.

Four classes of the subphylum Vertebrata are called *tetrapods* because they have two pairs of limbs. The other three extant classes and one extinct class are commonly called fishes.

Class Agnatha

The oldest vertebrate fossils are those of jawless creatures called *agnathans*, including the fishlike *ostracoderms* that have an armor of bony plates. The first traces of agnathans are found in the Cambrian strata, but a tremendous radiation occurred by the late Silurian, about 400 million years ago. Agnathans were generally small, lacked paired fins, and probably filtered food by taking in mud or water through their jawless mouths and passing it out through gill slits. The gills probably also functioned in gas exchange.

Ostracoderms and most agnathan groups disappeared during the Devonian period. Jawless fishes are represented today by lampreys, which are suspension-feeders as larvae and blood-sucking parasites as adults, and marine hagfishes, which are scavengers.

Class Placodermi

Armored fishes called *placoderms* replaced the agnathans during the late Silurian and early Devonian periods. These larger fish had paired fins that enhanced swimming and jaws that evolved from the skeletal supports of the anterior pharyngeal slits. Both these traits facilitated a predatory lifestyle. The remaining gill slits, no longer needed for suspension-feeding, maintained their function in gas exchange.

The Devonian period is known as the age of fishes. Sharks and bony fishes diverged from the ancestral line about 425 million years ago. Placoderms and acanthodians, another class of jawed fishes, radiated in both fresh and salt water, and then disappeared almost completely by the Carboniferous period, 350 million years ago.

Class Chondrichthyes

Sharks and rays, the cartilaginous fishes, have skeletons made of cartilage, well-developed jaws, and paired fins. Sharks swim to maintain buoyancy (their paired pectoral and pelvic fins providing lift) and to move water into their mouths and past their gills. Although most sharks are carnivores, the largest sharks and rays suspension-feed on plankton. Shark teeth probably evolved from the jagged scales that cover the skin. A spiral valve in the intestine increases

surface area within the relatively short digestive tract.

Sharks have keen senses: sharp vision, nostrils, auditory organs that can sense percussion, and regions in the head that can detect electric fields generated by muscle contractions of nearby fish. The *lateral line system* is a row of microscopic organs, running along the sides of the body, that are sensitive to water pressure changes.

Following internal fertilization, *oviparous* species of sharks lay eggs that hatch outside the mother; *ovoviviparous* species retain the fertilized eggs until they hatch within the uterus; and, in the few *viviparous* species, developing young are nourished by a placenta until born. The *cloaca* is the common chamber for the openings of the reproductive tract, excretory system, and digestive tract.

Rays, which propel themselves with their enlarged pectoral fins, are primarily bottom-dwellers that feed on mollusks and crustaceans.

Class Osteichthyes

Bony fishes have the largest number of individuals and species of all vertebrate classes. The skeleton of most bony fishes is reinforced with a hard matrix of calcium phosphate. Their skin is often covered by flattened bony scales and coated with mucus to reduce drag. They have a lateral line system similar to that of chondrichthyes. The movements of muscles in the gill chambers and the *operculum*, or flap over the gill chamber, draw water across the gills so that bony fish can breathe while stationary. Bony fishes also have a *swim bladder*, an air sac that controls buoyancy.

The flexible fins of bony fish are better for steering and propulsion than are the stiffer fins of sharks. The common fusiform body shape of fast swimming bony fish, sharks, and aquatic mammals is an example of convergent evolution, an adaptation to reduce drag while swimming. Most species of fish are oviparous and fertilization is external.

Bony fishes probably originated in freshwater. The swim bladder was modified from lungs used to augment gills in stagnant water. Three subclasses evolved by the end of the Devonian: the *ray-finned fishes*, including most of the familiar fish, many of which spread to the seas during their evolution; the *lobe-finned fishes*; and the *lungfishes*, both of which remained in freshwater and continued to use their lungs. Lobefins and some lungfishes had muscular pectoral and pelvic fins that were supported by extensions of the skeleton and may have enabled the fish to "walk" along the bottom or, occasionally, on land. Three present-day genera of lungfishes are found in the Southern Hemisphere. The coelocanth is the only living species of lobe-finned fish.

Class Amphibia

Early Amphibians Lobe-finned fishes with lungs, living in freshwater habitats subject to drought, may have been the ancestors of tetrapods. The amphibians were the first vertebrates on land; the oldest fossils date back to the late Devonian, about 350 million years ago. Amphibians have always been predators, eating the insects and other invertebrates that preceded them onto land. The Carboniferous period, during which a diversity of new forms and sizes evolved, is known as the Age of Amphibians. Amphibians declined during the late Carboniferous, and by the Mesozoic era, the survivors resembled modern amphibians.

Modern Amphibians There are three extant orders of amphibians: Urodeles, or salamanders, Anurans, or frogs, and Apodans, or caecilians. Salamanders, which may be aquatic or terrestrial, swim or walk with a back-and-forth bending of the body. Anurans are more specialized for moving on land, using their powerful hindlegs for hopping. Protective adaptations of frogs include camouflage coloring and distasteful or poisonous skin mucus. Tropical apodans are burrowing, legless, nearly blind, wormlike animals.

Many frogs undergo a metamorphosis from a tadpole, the aquatic, herbivorous larval form with gills and a tail, to the carnivorous, terrestrial adult form with legs and lungs. Many amphibians do not have an aquatic tadpole stage, and some species are exclusively aquatic or terrestrial. Paedogenesis has occurred in some groups of salamanders; the mudpuppy retains gills and other larval features. Amphibians are most abundant in damp habitats, since much of their gas exchange occurs across their moist skin.

In mating, the male frog grasps the female, releasing his sperm as the female sheds her eggs. Fertilization is external. Oviparous species lay their eggs in aquatic or moist environments. Ovoviviparous and viviparous species are also known. Varying amounts of parental care are shown. During breeding season, some frogs communicate with vocalizations. Some species undergo migrations to specific breeding sites, using various forms of communication and navigation.

Class Reptilia

Reptiles, including lizards, snakes, turtles, and crocodilians, exhibit adaptations for living on land not seen in amphibians.

Reptilian Characteristics Skin is covered with keratinized, waterproof scales. Lungs are the main vehicle for gas exchange. Most species lay eggs covered by protective shells, and the embryo develops within a fluid-filled amniotic sac. The evolution of the *amniote egg* allowed vertebrates to complete their life cycle on land. Given this type of egg, fertilization is internal.

Reptiles are *ectothermic*. They absorb external heat rather than generating their own and thus have much lower caloric needs than do mammals. Reptiles regulate their body temperature through behavioral adaptations.

The Age of Reptiles The oldest reptilian fossils date from the upper Carboniferous period, 300 million years ago. *Cotylosaurs,* or "stem reptiles," were the small, insect-eating, lizardlike ancestors of the various reptilian orders that arose in two major adaptive radiations. During the Permian period, the first radiation gave rise to terrestrial predators called *synapsids,* from which one lineage led to the *therapsids.* Mammals are believed to have evolved from the therapsid line. Some stem reptiles returned to water, leading to the plesiosaurs and ichthyosaurs. The *thecodonts* also arose during the Permian radiation and, after surviving the mass extinctions of the Permian period, served as the ancestor to the dinosaurs, crocodilians and birds.

The second great radiation occurred a little over 200 million years ago during the late Triassic period, when diverse groups of dinosaurs evolved from thecodonts. The flying pterosaurs, with wings formed from a membrane of skin stretched between an elongated finger and the body, and the terrestrial dinosaurs were the dominant vertebrates on Earth for millions of years. They included the largest animals ever to live on land. Some scientists postulate that dinosaurs were active, agile, and *endothermic*—able to use metabolic heat to warm the body.

During the Cretaceous period at the end of the Mesozoic era, the climate became cooler and more variable. Probably within 5 to 10 million years, dinosaurs declined and became extinct. A few reptilian groups survived and are successful today.

Reptiles of Today The three largest orders of extant reptiles are Squamata, lizards and snakes; Chelonia, turtles; and Crocodilia, alligators and crocodiles.

Lizards, the most numerous and diverse reptiles, are relatively small. Snakes, apparently descended from lizards that adopted a burrowing lifestyle, are limbless, although primitive species have vestigial pelvic and limb bones. Snakes are carnivorous, with adaptations for locating (heat-detecting and olfactory organs), killing (sharp teeth, often with toxins), and swallowing (loosely articulated jaws) their prey.

Turtles evolved from stem reptiles during the

Mesozoic era and have changed little since then. Crocodiles and alligators are among the largest living reptiles.

Class Aves

Birds evolved from a group of dinosaurs during the reptilian radiation of the Mesozoic era. The amniote egg and scales on the legs are two reptilian traits birds display.

Characteristics of Birds   Bird anatomy has evolved to enhance flight. Weight is reduced; for example, bones are honeycombed and birds are toothless—food is ground in the gizzard. The keratinized beak has assumed a great variety of shapes associated with different diets. Birds are endothermic; feathers and a fat layer help retain metabolic heat. The four-chambered heart segregates oxygenated and oxygen-poor blood, which facilitates a high metabolic rate.

Birds have excellent vision. Their relatively large brains provide for visual processing, motor coordination, and complex behavior, particularly in courtship rituals and parenting. Fertilization is internal. After eggs are laid, they are kept warm during development by the female and/or male bird.

Aerodynamic wings are flapped by large pectoral muscles attached to a keel on the sternum. Feathers made of keratin, a diagnostic characteristic of birds, are extremely light and strong, and shape the wing into an airfoil.

The Origin of Birds   Many dinosaurs, including some *theropods*, built nests, cared for their young, and were most likely endothermic. Some zoologists say that birds should be regarded as reptiles.

Fossils of *Archaeopteryx lithographica*, an ancient feathered creature with clawed forelimbs, teeth, and long tail, have been found from the Jurassic period, 150 million years ago. This weak flyer was probably not a direct ancestor of modern birds. A fossil dating from 125 million years ago seems to represent a later stage of avian evolution. A fossil named *Protoavis* dates from 225 million years ago and shows a mixture of dinosaurian and avian traits.

The evolution of strong flying ability enhanced hunting, foraging, escape from predators, and migration to favorable habitats.

Modern Birds   Flightless birds, called ratites, such as the ostrich, kiwi, and emu, lack a keeled breastbone. Carinate birds have a sternal keel to which attach large breast muscles. Passeriforms, or perching birds, are an order of carinate birds that includes 60 percent of living bird species.

Class Mammalia

Mammalian Characteristics   Mammals have hair made of keratin, which helps to insulate these endothermic animals. Active metabolism is provided by an efficient respiratory system that uses a *diaphragm* to help ventilate the lungs and a four-chambered heart that separates oxygenated and oxygen-poor blood. Milk, produced in mammary glands, is used to nourish mammalian young. Fertilization is internal and the embryo develops in the uterus. In placental mammals, a *placenta*, formed from the uterine lining and embryonic membranes, nourishes the developing embryo.

Mammals have relatively large brains and are capable of learning. Extended parental care of the young provides them time to learn complex skills.

Mammalian teeth come in a diverse assortment of shapes and sizes that are specialized for eating a variety of foods.

Evolution of Mammals   The oldest fossils of mammals date back to the Triassic period, 220 million years ago. Reptilian therapsids were mammalian ancestors. Small Mesozoic mammals, which were probably nocturnal and insectivorous, coexisted with dinosaurs. Mammals underwent an extensive radiation, filling the adaptive zones created by the extinction of dinosaurs and the rise of flowering plants in the late Cretaceous period. There are three major groups of mammals today: *monotremes*, egg-laying mammals; *marsupials*, mammals with pouches; and *placental mammals*.

Monotremes   The platypus and echidnas (spiny anteaters) are the only extant egg-laying mammals. The egg is reptilian in structure, but monotremes have hair and produce milk for their young. Monotremes seem to have descended from a very early mammalian branch.

Marsupials   Marsupials, including opossums, kangaroos, and the koala, complete their embryonic development in a maternal pouch called a *marsupium*. Born very early in development, the neonate crawls into the marsupium, fixes its mouth to a teat and completes its development while nursing.

Australian marsupials have radiated and filled the niches occupied by placental mammals in other parts of the world. Marsupials apparently originated in what is now North America, spreading southward.

After the breakup of Pangaea, South America and Australia became island continents where marsupials diversified in isolation from placental mammals. Placental mammals reached South America in the Cenozoic era, whereas Australia has remained isolated. Geological history helps to explain the evolutionary history and distribution of marsupials.

Placental Mammals Placental mammals complete development attached to a placenta within the maternal uterus. Placentals and marsupials may have diverged from a common ancestor 80 to 100 million years ago. The major orders of placental mammals radiated during the late Cretaceous and early Tertiary, about 70 to 45 million years ago.

There appear to be four main evolutionary lines of placental mammals. The orders Chiroptera and Insectivora, containing bats and shrews, respectively, resemble early mammals. A lineage of medium-sized herbivores gave rise to the lagomorphs (rabbits), perissodactyls (odd-toed ungulates such as horses and rhinos), artiodactyls (even-toed ungulates such as deer and swine), sirenians (sea cows), proboscideans (elephants), and cetaceans (porpoises and whales).

The order Carnivora, the third evolutionary line that developed during the early and middle Cenozoic era, includes cats, dogs, raccoons, and the pinnipeds (seals, sea lions, and walruses). The most extensive radiation resulted in the primate–rodent complex. The order Rodentia includes rats, squirrels and beavers. The order Primates includes monkeys, apes and humans.

The Human Ancestry

Evolutionary Trends in Primates The first primates probably were small, arboreal animals, descended from insectivores late in the Cretaceous. Characteristics that were shaped by this arboreal ancestry include limber shoulder joints that make it possible to *brachiate* (swing from one hold to another), dextrous hands, eye-hand coordination, forward-facing and close-together eyes to provide depth perception, and parental care.

Modern Primates Two suborders of modern primates are Prosimii and Anthropoidea. Prosimians, such as lemurs, lorises, and tarsiers, probably resemble early arboreal primates. The anthropoids include monkeys, apes, and humans. The first fossils of anthropoids are monkeylike primates that probably evolved from prosimian stock about 40 million years ago in Africa or Asia. Somehow ancestral monkeys reached South America. These strictly arboreal New World monkeys have been evolving separately from Old World monkeys for millions of years. Most monkeys in both groups are diurnal and live in social bands.

There are four genera of apes: gibbons, orangutans, gorillas, and chimpanzees, all of which live in the Old World tropics. Modern apes generally are larger than monkeys, with relatively long legs, short arms, and no tails. Ape behavior is more adaptable than that of monkeys.

The Emergence of Humankind Paleoanthropology is the study of human origins and evolution. Chimpanzees and humans represent two divergent branches from a common, less-specialized anthropoid ancestor. Human evolution has not occurred as phyletic change within an unbranched hominid line; there have been times when several different human species coexisted. Different human features have evolved at different rates, known as *mosaic evolution*. Thus, bipedalism or erect posture evolved before an enlarged brain developed.

The oldest known ape fossils are of a cat-sized tree-dweller that lived about 35 million years ago. About 25 million years ago, during the Miocene epoch, ape descendants diversified and spread into Eurasia. About 20 million years ago, the Himalayan range arose, the climate became drier, and the forests contracted. Some of the Miocene or early Pliocene apes living on the edge of the forest may have begun foraging for food on the savanna. Fossil evidence and DNA comparisons indicate that human and chimpanzees diverged from a common ancestor about five million years ago.

In 1924, a British anthropologist found a skull of an early human that he called *Australopithecus africanus*. It appears that *Australopithecus* was an upright hominid, with humanlike hands and teeth, but a small brain, who foraged on the African savannas for nearly two million years, beginning about three million years ago.

In 1974, a petite *Australopithecus* skeleton was discovered in the Afar region of Ethiopia. Lucy, as this 3 million year old fossil was named, and similar fossils have been designated as a separate species, *Australopithecus afarensis*. Fossils have been found that appear to represent the same species, but are a million years older than Lucy. *A. afarensis* thus appears to be the most ancient of the australopithecine group and perhaps the common ancestor of two hominid branches, one leading to *A. africanus* and one leading to *Homo*, our genus. Other anthropologists argue that *A. afarensis* diverged from the lineage leading to *Homo*.

Australopithecus walked erect for over 1 million years with no enlargement of the brain. Larger-brained fossils date back about 2 million years and sometimes have been found with simple stone tools. Some paleoanthropologists place these fossils in the genus *Homo*, calling them *Homo habilis*, or "handy man," whereas other scientists retain them in *Australopithecus*.

Homo habilis and *Australopithecus*, including the more stocky *A. robustus*, coexisted for nearly 1 million years. One theory is that these were distinct lines of hominids, one of which was an evolutionary dead-end, while the other, *Homo habilis*, led to modern humans.

Homo erectus, the first hominid to migrate out of Africa into Asia and Europe, is represented by fossils known as Java Man and Beijing Man, which range in age from 1.6 million years to 300,000 years. They had greater height and brain capacity than *H. habilis* and survived in the colder climates of the north, living in shelters, building fires, wearing animal skins, and using more elaborate stone tools.

The Neanderthals were descendents of *H. erectus*, living from about 130,000 to 30,000 years ago. They had heavier brow ridges, less pronounced chins and larger brains than modern humans. They were skilled tool-makers and left evidence of burials and other rituals. Many paleoanthropologists group post-*H. erectus* fossils dating back 300,000 years with Neanderthals and other Asian and Australasian fossils as the earliest forms of *Homo sapiens*.

According to the *multiregional model* for the emergence of *Homo sapiens*, these archaic *H. sapiens* evolved into modern humans in parallel in different parts of the world. Geographical diversification began a million years ago and accounts for the distinctive features of modern races.

The oldest, fully modern *Homo sapiens* fossils have been found in Africa, dating back 100,000 years. Two groups of fossils found in Israeli caves indicate that Neanderthals may have overlapped with modern humans for about 40,000 years. According to Stringer, this overlap of distinct forms means that Neanderthals and the modern forms did not interbreed, nor were the Neanderthals ancestors to those modern humans. Stringer postulates that Neanderthals and other archaic *H. sapiens* died out and that modern humans evolved from *H. erectus* in Africa, then migrated to other areas. According to this *monogenesis model*, modern humans dispersed from Africa relatively recently, with the diversification of the races occurring in the past 100,000 years.

A group of geneticists recently claimed that mitochondrial DNA comparisons indicated that humans could be traced back to Africa, with the divergence of

mtDNA beginning just 200,000 years ago. Several researchers have challenged the methods used to construct evolutionary trees from mtDNA comparisons. Many paleoanthropologists continue to support the multiregional evolution of modern humans as the best interpretation of the fossil evidence. Perhaps DNA samples recovered from fossils and amplified by polymerase chain reaction will allow molecular biologists and paleontologists to test the predictions made by the two competing models of the origin of modern humans.

Evolution of an erect stance required remodeling the foot, pelvis and vertebral column. Prolonging the period of growth of the brain after birth made possible the enlargement of the human brain. An extended period of parental care helps to make possible the basis of culture—the transmission of knowledge accumulated over the generations. Language, spoken and written, is the major means for this transmission.

Cultural evolution has included several stages: the communal efforts of nomads who hunted and gathered food on African grasslands, the development of agriculture in Eurasia and the Americas about 10,000 to 15,000 years ago, and the Industrial Revolution, which began in the eighteenth century. Since then, the human population and new technology have been growing exponentially.

The biological evolution of life has been linked with environmental changes. Cultural evolution has made *Homo sapiens* a new force in the history of life—a species that no longer needs to adapt to its environment, but simply changes the environment to meet its needs. As agents of rapid environmental change, we are altering the planet faster than many species can adapt. The next and perhaps most devastating crisis in the history of life may be the habitat destruction, pollution, resource decline, and climatic changes resulting from the actions of *H. sapiens*.

STRUCTURE YOUR KNOWLEDGE

1. Place major events in the evolution of phylum Chordata on the geological timetable, p. 251.

2. Describe several examples from vertebrate evolution that illustrate the common evolutionary theme that new adaptations usually evolve from preexisting structures.

3. How did various vertebrate groups meet the challenges of a terrestrial habitat?

ERA	PERIOD	AGE (millions of years ago)	MAJOR ANIMAL EVENTS	PLANT EVENTS
CENOZOIC		5		
CENOZOIC		24		
CENOZOIC	Tertiary			Dominance of angiosperms
Climatic Change—Cooler		65		
MESOZOIC	Cretaceous			Flowering plants appear
MESOZOIC		144		
MESOZOIC	Jurassic			
MESOZOIC		213		
MESOZOIC	Triassic			Conifers and cycads dominant
Climate—Warmer and Drier		248		
PALEOZOIC	Permian	286		Dominance of lycopods, sphenopsids, ferns; great coal forests; first seed plants appear
PALEOZOIC	Carboniferous			
PALEOZOIC		360		Radiation of vascular plants
PALEOZOIC	Devonian	408		
PALEOZOIC	Silurian			Origin of vascular plants
PALEOZOIC		438		
PALEOZOIC	Ordovician			Marine algae abundant
PALEOZOIC		505		
PALEOZOIC	Cambrian			
PALEOZOIC		590		

4. In this idealized version of a chordate, label the four characteristics of phylum Chordata.

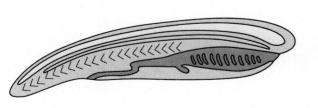

TEST YOUR KNOWLEDGE

FILL IN THE BLANKS

1. _____ Chordate subphylum of sessile, filter-feeding marine animals

2. _____ Chordate subphylum that includes lancelet

3. _____ blocks of mesoderm along notochord that develop into muscles

4. _____ jawless fish with armor of bony plates

5. _____ flap over the gills of bony fishes

6. _____ adaptation that allows reptiles to reproduce on land

7. _____ lizard-like ancestors of the various reptilian orders

8. _____ group from which birds descended

9. _____ egg-laying mammals

10. _____ order in which snakes are placed

11. _____ structure that helps mammals ventilate their lungs

12. _____ mammals that walk on the tips of the toes

13. _____ genus in which the fossil Lucy is placed

14. _____ subgroup that includes monkeys, apes and humans

15. _____ major means for the transmission of human culture

16. _____ theory that modern humans dispersed from Africa relatively recently and races diversified in the past 100,000 years

MULTIPLE CHOICE: *Choose the one best answer.*

1. Pharyngeal slits appear to have functioned first as
   a. suspension feeding devices.
   b. gill slits for respiration.
   c. components of the jaw.
   d. portions of the inner ear.

2. Which of the following characteristics differentiates the vertebrates from the invertebrate chordates?
   a. notochord
   b. extensive cephalization
   c. postanal tail
   d. dorsal hollow nerve cord

3. Modern agnathans are represented by
   a. hagfishes and lampreys.
   b. sharks and rays.
   c. the lobe-fin coelocanth.
   d. lungfishes.

4. Jaws first occurred in the class
   a. Agnatha.
   b. Chondrichthyes.

c. Osteichthyes.
   d. Placodermi.

5. Which of the following is incorrectly paired with its gas exchange mechanism?
   a. amphibians—skin and lungs
   b. lobed-finned fishes—gills and lungs
   c. reptiles—lungs
   d. bony fishes—swim bladder and gills

6. Urodeles include
   a. limbless, burrowing amphibians.
   b. frogs that have a tadpole stage.
   c. frogs and toads that may or may not have a tadpole stage.
   d. salamanders.

7. Reptiles have lower caloric needs than do mammals of comparable size because they
   a. are ectotherms.
   b. have waterproof scales.
   c. have an amniote egg.
   d. move by the bending back and forth of their vertebral column.

8. Which of the following best describes the earliest mammals?
   a. large, herbivorous
   b. large, carnivorous
   c. small, insectivorous
   d. small, herbivorous

9. Oviparity is a reproductive strategy that
   a. allows mammals to bear well-developed young.
   b. is used by both birds and reptiles.
   c. allows vertebrates to reproduce on land.
   d. is a necessity for all flying vertebrates.

10. In Australia, marsupials fill the niches that placental mammals fill in other parts of the world because
    a. they are better adapted and have outcompeted placental mammals.
    b. their offspring complete their development attached to a teat in a marsupium.
    c. after Pangaea broke up, they diversified in isolation from placental mammals.
    d. they evolved from monotremes that migrated to Australia about 12 million years ago.

11. Which of the following characteristics of primates is not associated with an arboreal existence?
    a. shoulder joints that permit brachiation
    b. upright posture
    c. eyes close together on the face to improve depth perception
    d. parental care

12. According to the model of mosaic evolution,
   a. biogeography reflects the adaptive radiation of many groups.
   b. erect posture preceded the enlargement of the brain in human evolution.
   c. modern humans evolved in parallel in different parts of the world, laying the groundwork for racial differences over a million years ago.
   d. modern humans first evolved in Africa and later dispersed to other regions.

CHAPTER 24
EARLY EARTH AND THE ORIGIN OF LIFE

Suggested Answers to Structure Your Knowledge

1. The primitive Earth is believed to have had seas, volcanoes, and large amounts of ultraviolet radiation passing through a thin atmosphere, which probably consisted of H_2O, CO, CO_2, N_2, CH_4, and NH_3. Life was able to evolve in this environment because the reducing (electron-adding) atmosphere and the availability of energy from lightning and UV radiation facilitated the abiotic synthesis of organic molecules, and hot lava rocks may have aided the polymerization of these monomers. Methane or organic fuels may have provided an energy source for early protobionts.

2. (1) The abiotic synthesis of organic molecules provided the amino acids, simple sugars, and nucleotides that served as the building blocks of proteins, energy molecules, and precursors for the information molecules of RNA and DNA.

(2) In order for life to develop, organic monomers had to be linked into polymers that have emergent properties deriving from their structural organization. Early polypeptides may have functioned as simple enzymes, lipids formed bounding membranes, and RNA molecules provided self-replicating structures capable of ordering amino acids into polypeptides.

(3) The aggregation of molecules into discrete packets, separated from the surroundings by selectively permeable membranes, allowed for the concentration of critical molecules and the emerging properties of metabolism, excitability, and primitive reproduction.

(4) The development of a precise mechanism for the recording and replication of genetic material allowed for the reproduction and evolution of successful aggregations of organic molecules.

3.

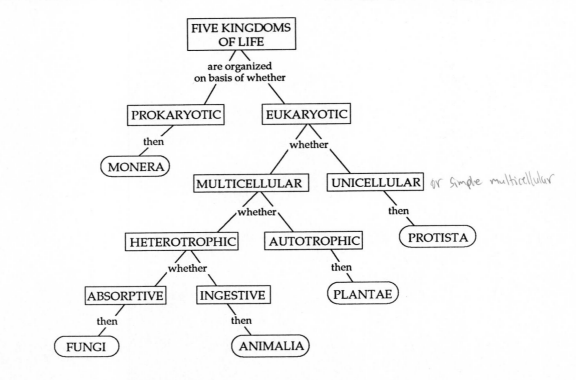

4. The oldest fossils of prokaryotes date from 3.5 billion years ago. The origin of life is presumed to have occurred between 4.1 and 3.5 billion years ago.

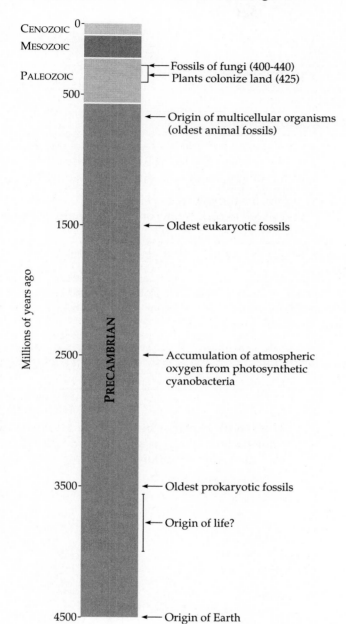

Answers to Test Your Knowledge

Multiple Choice:

1. b	**4.** b	**7.** a	**10.** d
2. c	**5.** c	**8.** c	**11.** d
3. d	**6.** b	**9.** b	**12.** b

Matching:

1. F	**3.** E	**5.** A
2. D	**4.** B	

CHAPTER 25
PROKARYOTES AND THE ORIGINS OF METABOLIC DIVERSITY

Suggested Answers to Structure Your Knowledge

1.

CHARACTERISTIC	DESCRIPTION
Cell shape	May be spheres (cocci), rods (bacilli), or spirals (spirilla).
Cell size	1 - 5 μm in diameter, one-tenth the size of eukaryotic cells
Cell surface	Cell wall made of peptidoglycan, gram-negative bacteria have outer lipopolysaccharide membrane, mycoplasmas are group without cell walls, archaebacteria have no peptidoglycan in walls; sticky capsule for adherence, pili for attachment or conjugation.
Motility	Move with flagella that are either scattered or concentrated at ends of cell; or have filaments that spiral around cell and cause cell to cork-screw; or glide in secreted slime. May show taxis to stimuli.
Internal Membranes	Some infoldings of plasma membrane used in respiration; thylakoid membranes in cyanobacteria for photosynthesis.
Genome	Genophore, the circular DNA molecule with little associate protein, is found in nucleoid region. May have plasmids with antibiotic resistance and other genes and be transferred during conjugation. Have 1/1000 the DNA of eukaryotes.
Growth and Reproduction	Divide by binary fission, geometric growth into colony of progeny. Growth slowed by accumulation of wastes or depletion of nutrients. Some genetic exchange in conjugation; most variation from mutation.

2. The circular structure is an endospore inside a cell of an endospore-forming bacterium. These thick-walled, resistant cells can withstand harsh conditions for long periods of time, breaking dormancy when conditions are favorable.

3. The small sperical cell in this filamentous cyanobacteria is called a heterocyst. It is specialized for nitrogen fixation.

4.

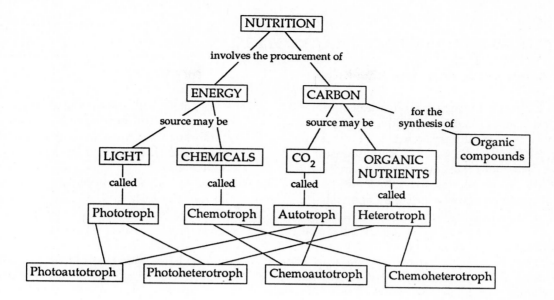

NUTRITION

involves the procurement of

ENERGY   CARBON

source may be   *source may be*   *for the synthesis of*

LIGHT   CHEMICALS CO_2 ORGANIC NUTRIENTS   Organic compounds

called   *called*   *called*   *called*

Phototroph   Chemotroph   Autotroph   Heterotroph

Photoautotroph   Photoheterotroph   Chemoautotroph   Chemoheterotroph

5. • Breakdown of abiotically synthesized organic compounds for energy, led to *glycolysis* and formation of *ATP* by *substrate level phosphorylation*.

• Development of transmembrane *proton pumps* (using ATP) to maintain pH of cell.

• *Electron transport chain* linked oxidation of organic compounds with pumping of H^+ out of cell. Efficient electron transport chain could have formed proton gradient so that H^+ diffused back in, reversing proton pump and generating ATP (*chemiosmosis*).

• Use of light energy absorbed by membrane pigments to drive ATP synthesis. Others may have used light energy to drive electrons from H_2S to $NADP^+$, coopting electron transport chain, to generate reducing power to fix CO_2 (*photosynthesis*).

• Use of water as source of electrons and hydrogen for fixing CO_2 (*photosynthesis evolving oxygen*).

• Use of O_2 to pull electrons from organic molecules down existing electron transport chain (*aerobic respiration*).

6. Decomposers—recycle global nutrients, used for sewage treatment, clean up oil spills

Nitrogen fixers—provide nitrogen to the soil in nodules on legume roots

Symbionts—enteric bacteria, some produce vitamins

Biotechnology—used for production of antibiotics, many useful products

7. Refer to the time line in the answer section for Chapter 24. O_2 began accumulating in the atmosphere between 2.5 and 2 billion years ago.

Answers to Test Your Knowledge

Multiple Choice:

1. c	**4.** b	**7.** d	**10.** d
2. d	**5.** c	**8.** d	**11.** a
3. a	**6.** b	**9.** b	**12.** c

Matching:

1. D	**4.** C	**7.** B	**10.** J
2. I	**5.** H	**8.** G	
3. L	**6.** A	**9.** K	

Fill in the Blanks:

1. cocci	**7.** photoautotroph
2. nucleoid region	**8.** facultative anaerobe
3. gram stain	**9.** saprobes
4. pili	**10.** exotoxins
5. taxis	**11.** nodules
6. chemoheterotroph	**12.** archaebacteria

CHAPTER 26
PROTISTS AND THE ORIGIN OF EUKARYOTES

Suggested Answers to Structure Your Knowledge

1. (a) The kingdom Protista includes unicellular—and a few colonial and multicellular—eukaryotic organisms. Nearly all protists use aerobic respiration and have flagella or cilia at some point in their life. They all reproduce asexually, and some have sexual reproduction as well. They require water or moist habitats (which may be inside the bodies of hosts). Many form cysts to withstand harsh conditions. Protists may be photoautotrophic or heterotrophic or both.

    (b) There is controversy over the boundaries of the kingdom Protista. Originally Whittaker assigned only unicellular eukaryotes to this kingdom. Some colonial and multicellular eukaryotes, however, lack the distinctive traits of either fungi, plants, or animals, and appear to be more closely related to unicellular protists than to these multicellular kingdoms. The expanded kingdom Protista includes these multicellular descendents of unicellular eukaryotes.

2. According to the autogenous hypothesis, eukaryotic cells are thought to have evolved internal membrane structures from the specialization of membranes invaginated from the plasma membrane.

    The endosymbiotic hypothesis claims that at least the mitochondria and chloroplasts of eukaryotic cells developed from photosynthetic and aerobic heterotrophic prokaryotes that became incorporated into larger cells. The lines of evidence for this model include the following similarities: the chloroplasts of red algae to cyanobacteria; the chloroplasts and pigments of green algae to the bacterium *Prochlorothrix*; the size, membrane enzymes and transport systems, division process, circular DNA molecule, ribosomal RNA sequences, and ribosomes of chloroplasts and mitochondria to those found in eubacteria. The existence of endosymbiotic relationships in modern organisms and the recent discovery of transposons, which may have transferred endosymbiont DNA to the host cell nucleus, provide additional evidence for this model.

3.

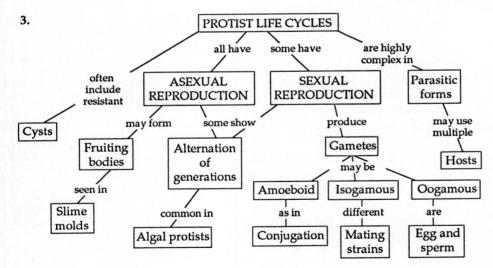

4. *Paramecium* is a member of the Ciliophora.
   a. cilia **d.** contractile vacuole
   b. micronucleus **e.** oral groove
   c. macronucleus

5. Refer to the time line in the answer section of Chapter 24. The oldest known fossils of eukaryotes are acritarchs which date from 1.5 billion years ago.

Answers to Test Your Knowledge

Matching:

1. O	**5.** K	**9.** D	**13.** Q
2. J	**6.** E	**10.** P	**14.** C
3. B	**7.** N	**11.** H	**15.** I
4. L	**8.** F	**12.** M	**16.** A

Multiple Choice:

1. c	**4.** a	**7.** d	**10.** a
2. d	**5.** b	**8.** c	
3. c	**6.** c	**9.** d	

CHAPTER 27
PLANTS AND THE COLONIZATION OF LAND

Suggested Answers to Structure Your Knowledge

1.

PLANT GROUPS	Bryophyta	Lycophyta Sphenophyta	Pterophyta	Coniferophyta	Anthophyta
COMMON NAMES	moss, liverwort, hornwort	club moss, ground pine; horsetail	ferns	pines, conifer	angiosperm, flowering plants
ADAPTIVE CHARACTERISTICS					
Cuticle, stomata	✓	✓	✓	✓	✓
Protected embryo, jacketed reproductive organs	✓	✓	✓	✓	✓
Vascular tissue for transport and support		✓	✓	✓	✓
Protected gametophyte retained on sporophyte plant				✓	✓
Nonswimming sperm, pollen				✓	✓
Fibers for support in vascular tissue				✓	✓
Seed for protection, dispersal, and nourishment of embryo				✓	✓
Flowers and fruit, animal pollinators, protected seeds					✓

2.

ERA	PERIOD	AGE (millions of years ago)	MAJOR PLANT EVENTS
CENOZOIC			Bryophytes Ferns, lycopods, horsetails Conifers Angiosperms
		— 65 —	Dominance of angiosperms
Climatic Change—Cooler			Radiation of flowering plants
MESOZOIC	Cretaceous		
	Jurassic		
		200	Dominance of conifers and cycads
	Triassic		
Climate—Warmer and Drier		— 248 —	
PALEOZOIC	Permian	300	Dominance of seedless plants in great coal forests
	Carboniferous		Seedless vascular plants First seed plants
	Devonian		Bryophytes diverge Early vascular plants
		400	
	Silurian		Origin of plants
	Ordovician		
		500	Algal Ancestors
	Cambrian		

During the Carboniferous period, much of the land was covered by swamps, and giant lycopods, horsetails, and ferns dominated the earth. Formation of the supercontinent Pangaea during the Permian period may have led to a warmer and drier climate, and the lycopods, horsetails, and ferns of the great coal forests were replaced by conifers that were better suited to a dryer climate. With the cooler climates of the Cretaceous period, many groups (including the dinosaurs) disappeared, and some conifers and cycads were replaced by flowering plants.

3. Refer to time line in answer section of Chapter 24.

Plants first colonized land around 425 million years ago.

4. In the life cycle of the pine, meiosis occurs within the sporangia on the microsporophyll to produce microspores that develop into pollen grains and within the nucellus when the megaspore mother cell gives rise to a megaspore that develops into the female gametophyte. Pollination occurs when pollen is drawn in through the micropyle, an opening to the nucellus. Fertilization occurs, usually more than a year later, when a sperm nucleus discharged from the pollen tube fertilizes an egg cell contained in an archegonium. Haploid structures in the pine life cycle include pollen grains and the multicellular female gametophyte enclosed within integuments of the nucellus.

a. pollen cone
b. ovulate cone
c. nucellus
d. pollen grain
e. megaspore
f. female gametophyte
g. egg in archegonium
h. embryo

Answers to Test Your Knowledge

Multiple Choice:

1.	c	5.	c	9.	b
2.	a	6.	a	10.	d
3.	c	7.	d	11.	a
4.	b	8.	d	12.	c

True or False:

1. F—add *or algae.*
2. F—change *heteromorphic* to *heterosporous,* or change *produce . . .* to *have gametophyte and sporophyte that are morphologically distinct.*
3. T
4. F—change *naked seed* to *seedless,* or change *club mosses and horsetails* to *gymnosperms.*
5. T
6. T
7. F—change *tracheids* to *vessel elements*
8. T
9. F—change *angiosperm* to *gymnosperm,* or change *haploid cells…* to *an embryo sac with seven haploid cells.*
10. T

CHAPTER 28
FUNGI

Suggested Answers to Structure Your Knowledge

1.

Group	Examples	Morphology	Asexual Reproduction	Sexual Reproduction	Miscellaneous Information
Zygomycota	zygote fungi, black bread mold	coenocytic hyphae, chitosan cell walls	spores formed in sporangia on aerial hyphae	hyphal extensions fuse, karyogamy, resistant zygosporangia	In soil and decaying matter
Ascomycota	sack fungi, yeasts, cup fungi	septate hyphae, chitin cell walls, some unicellular	abundant spores in chains or clusters, called conidia	dikaryotic hyphae in ascocarp, karyogamy in asci, ascospores	plant parasites, lichens, genetic studies, decomposers
Basidiomycota	club fungi, mushrooms, puffballs, shelf fungi, rusts	dikaryotic hyphae with complex pores, mycelium extensive, chitin	form conidia less commonly than ascomycetes	basidiocarps, karyogamy in basidia, forms basidiospores	edible mushrooms, plant parasites, mycorrhizae
Deuteromycota	*Penicillium,* fungi imperfecti	molds, rapidly growing hyphae	form conidia	no sexual stage, may have parasexuality	nonphylogenetic form division, some produce antibiotics
Lichens	lichens	flat growing thalli	soredia or fragments	fungus may reproduce sexually	mutualistic association, harsh habitats, colonize rocks

2. The role of fungi as saprobes and decomposers of organic material is central to the recycling of chemicals between living organisms and their physical environment. Decomposers also prevent the accumulation of litter, feces, and dead organisms. Parasitic fungi have economic and health effects on humans. Plant pathogens cause extensive loss of food crops and trees (Dutch elm disease, powdery mildews, ergots on rye). Fungal pathogens of humans can cause annoying to serious problems. An example of a mutualistic symbiont is the mycorrhizae found on the roots of most plants. These associations greatly benefit the plant.

3. **a.** diagram of coenocytic hyphae
   b. diagram of septate hyphae with dikaryon cells
   c. diagram of haustoria extending through plant cell walls

4. Refer to the time line in Chapter 24. The first undisputed fossils of fungi date back 400 to 440 million years ago to the Silurian period. The first terrestrial plant fossils have petrified mycorrhizae.

Answers to Test Your Knowledge

Fill in the Blanks:

1. septum	**5.** basidium	**9.** coenocytic
2. dikaryon	**6.** asci	**10.** haustoria
3. chitin	**7.** mycorrhizae	
4. conidia	**8.** zygospore	

Multiple Choice:

1. a	**3.** c	**5.** a	**7.** b	**9.** a
2. c	**4.** d	**6.** b	**8.** d	

CHAPTER 29
INVERTEBRATES AND THE ORIGIN OF ANIMAL DIVERSITY

Suggested Answers to Structure Your Knowledge

1.

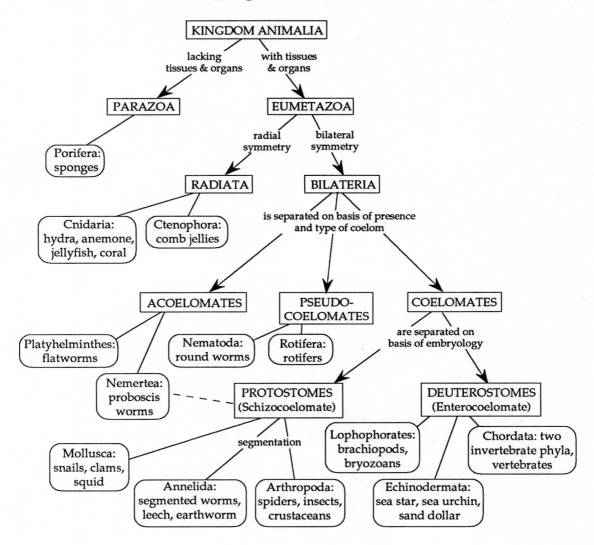

2. Transitional forms between protists and the first animals and between the various animal phyla that would shed light on the origin and phylogeny of animals have not been found in the fossil record. Suggested hypothetical ancestors of the eumetozoa include a multinucleate protist, such as a modern acoel worm (syncytial hypothesis), or a colonial flagellate, such as choanoflagellates (colonial hypothesis). The protoanimal in which cell layers first developed may have been a planuloid, an invaginated colony resembling a gastrula, or an animal that resembled *Trichoplax*, the simplest known animal. The fossil record begins with soft-bodied medusae and annelid-like worms from the Ediacaran period of the Precambrian era (700 to 590 million years ago). The fossils of the next period include nearly all modern phyla. The appearance of so many complex animals in the Cambrian period remains unexplained.

3.

Phylum	Digestion	Excretion	Respiration	Circulation	Nervous Control	Skeleton	Locomotion
Porifera	choano-cytes, amoe-bocytes	diffusion	diffusion	none	none	spicules or spongin fibers	sessile
Cnidaria	gastrovas-cular cavity	diffusion	diffusion	none, g-v cavity distri-butes food	nerve net	hydrostatic	contractile fibers
Platyhelminthes	gastrovas-cular cavity	diffusion, flame cells	diffusion	none, g-v cavity distri-butes food	ganglia, nerve cords, ladderlike	hydrostatic	cilia to glide on mucus or swim
Nemertea	complete digestive tract	diffusion, flame cells	diffusion	simple system with vessels	ganglia, nerve cords, ladderlike	hydrostatic	swim or burrow
Rotifera	complete digestive tract	diffusion in pseudo-coelomic fluid	diffusion	diffusion in pseudo-coelomic fluid	ganglia, nerve cords	hydrostatic, tough cuticle	ring of cilia, muscles
Nematoda	complete digestive tract	diffusion in pseudo-coelomic fluid	diffusion	diffusion in pseudo-coelomic fluid	ganglia, nerve cords	hydrostatic, tough cuticle	thrash with longitudinal muscles
Mollusca	complete digestive tract	nephridia	gills, diffusion	open sys-tem, pump-ing heart	nerve ring, nerve cords	external shell, hydrostatic	muscular foot, jet propulsion
Annelida	digestive tract with specialized regions	metane-phridia in each segment	diffusion, blood vessels in skin	closed, pumping vessels	ganglia, ventral nerve cords	hydrostatic, septa form body segments	longitudinal & circular muscles
Arthropoda	digestive tract with specialized regions	Malphigian tubules	diffusion, gills, tracheal system	open sys-tem pump-ing heart	ganglia, ventral nerve cords	exoskeleton of chitin, jointed	fly, swim, walk, use jointed legs, wings
Echinodermata	digestive tract with specialized regions	diffusion	diffusion, skin, gills, tube feet	diffusion	radial nerve system	calcareous endoskele-ton	tube feet, water vascular system

4. The time line in chapter 24 should show the old-est animal fossils at 700 million years ago.

3. F e   **7.** B e
4. G a   **8.** A b

Answers to Test Your Knowledge

Matching:

1. L	c	**5.** C	h	**9.** E	d
2. D	b	**6.** I	b	**10.** H	c

Multiple Choice:

1. b	**5.** b	**9.** d	**13.** b	**17.** a
2. a	**6.** d	**10.** c	**14.** a	**18.** c
3. d	**7.** c	**11.** c	**15.** b	**19.** b
4. c	**8.** a	**12.** b	**16.** c	**20.** a

CHAPTER 30
THE VERTEBRATE GENEALOGY

Suggested Answers to Structure Your Knowledge

1.

ERA	PERIOD	AGE (millions of years ago)	MAJOR ANIMAL EVENTS	PLANT EVENTS
CENOZOIC		5 24	*Homo sapiens* *Homo erectus, Homo habilis* *Australopithecus*	
CENOZOIC	Tertiary	65	Origin of apes Major radiation and dominance of mammals, birds, pollinating insects	Dominance of angiosperms
	Climatic Change—Cooler			
MESOZOIC	Cretaceous	144	Dinosaurs become extinct Primates probably appear	Flowering plants appear
MESOZOIC	Jurassic	213	Dinosaurs dominant Birds appear	
MESOZOIC	Triassic	248	Second radiation of reptiles first dinosaurs, mammals, and birds	Conifers and cycads dominant
	Climate—Warmer and Drier			
PALEOZOIC	Permian	286	First reptilian radiation Origin of most modern insect orders	Dominance of lycopods, sphenopsids, ferns; great coal forests; first seed plants appear
PALEOZOIC	Carboniferous	360	Age of Amphibians Reptiles appear	
PALEOZOIC	Devonian	408	Age of Fishes	Radiation of vascular plants
PALEOZOIC	Silurian	438	Diversity of jawless fish	Origin of vascular plants
PALEOZOIC	Ordovician	505	First vertebrates	Marine algae abundant
PALEOZOIC	Cambrian	590	First invertebrate Chordate Origin of most invertebrates	

2. The following are examples of new adaptations that evolved from preexisting structures: (1) pharyngeal slits were first used for suspension feeding, then became adapted for gas exchange and the development of gills; (2) skeletal supports for gills became adapted for use as hinged jaws; (3) lungs of bony fish developed for gas exchange in stagnant waters transformed to an air bladder aiding in buoyancy; (4) fleshy fins, supported by skeletal elements, developed into limbs for terrestrial animals; (5) feathers, probably evolved for insulation in endothermic animals, came to be used for flight; (6) dextrous hands, important for an arboreal life, evolved to manipulate tools.

3. Amphibians remained in damp habitats, burrowing in mud during droughts; some secrete foamy protection for eggs laid on land. Reptiles developed scaly, waterproof skin, an amniote egg that provided an aquatic environment for developing embryo, and behavioral adaptations to modulate changing temperatures. All terrestrial groups are tetrapods, using limbs for locomotion on land.

4. The four characteristics of chordates are a dorsal, hollow nerve cord, notochord, postanal tail, and pharyngeal slits.

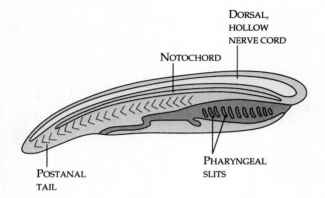

Answers to Test Your Knowledge

Fill in the Blanks:

1. Urochordates	9. monotremes
2. Cephalochordates	10. Squamata
3. somites	11. diaphragm
4. ostracoderms	12. ungulates
5. operculum	13. Australopithecus
6. amniote egg	14. anthropoid
7. cotylosaurs, stem reptiles	15. language
8. theropods, dinosaurs	16. monogenesis model

Multiple Choice:

1. a	5. d	9. b
2. b	6. d	10. c
3. a	7. a	11. b
4. d	8. c	12. b

PLANTS: FORM AND FUNCTION

THE FAR SIDE By GARY LARSON

© 1987 Universal Press Syndicate

Feb. 22, 1946: Botanists create the first artificial flower.

CHAPTER **31**

PLANT STRUCTURE AND GROWTH

FRAMEWORK

A rather large new vocabulary is needed to name the specialized cells and structures in a study of plant anatomy. Focus your attention on how the structures of a plant integrate with the functioning of the entire living organism.

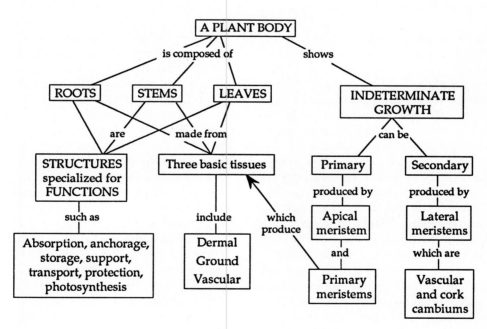

CHAPTER SUMMARY

An Introduction to Plant Biology: The Major Themes

Modern plant biology is using new methods and new research organisms, such as *Arabidopsis thaliana*, to study the genetic control of plant development. The emergence of properties of the whole plant depends on organization and interaction at the molecular and cellular level. Plants show adaptation to their environments on two time scales: the evolutionary interactions that have shaped plants as terrestrial organisms and the physiological and structural responses of individual plants to their environments. As always, the correlation of structure and function will be obvious in our study of plants.

Basic Morphology of Flowering Plants: An Evolutionary Perspective

Plant morphology considers the external structure of plants, whereas *plant anatomy* is the study of internal structure. The most diverse and widespread group of plants is the angiosperms, characterized by flowers and fruits. This group is split into two classes: *monocots*, which have one embryonic seed leaf or cotyledon, and *dicots*, which have two.

To be able to acquire dispersed resources in a terrestrial environment, plants have evolved an underground *root system* for obtaining water and minerals from the soil and an aerial *shoot system* of stems, leaves, and flowers for absorbing light and carbon dioxide for photosynthesis. Vascular tissue transports materials between the interdependent roots and shoots. Xylem carries water and dissolved minerals upward, whereas phloem conducts food made in the leaves to roots and nonphotosynthetic plant parts.

The Root System The structure of roots is adapted for anchorage, absorption, conduction, and storage of food. A *taproot* system is found commonly in dicots, whereas monocots generally have *fibrous root* systems. Most absorption of water and minerals occurs through the tiny *root hairs* that are clustered near the root tips. *Adventitious* roots arise from stems or leaves and may serve as props to help support stems.

The Shoot System The shoot system includes vegetative shoots, which bear leaves, and floral shoots. The stem and attached leaves of a vegetative shoot may be the main shoot or a vegetative branch of the plant. A stem consists of *nodes*, the point at which leaves are attached, and segments between nodes, called *internodes*. An *axillary bud* is found in the angle between a leaf and the stem. A *terminal bud*, consisting of developing leaves, nodes and internodes, is found at the tip or *apex* of a shoot. The terminal bud may exhibit *apical dominance* and inhibit the growth of axillary buds. Axillary buds may develop into flowers or vegetative branches with their own terminal buds, leaves, and axillary buds.

Modifications of stems include stolons, horizontal stems that grow along the surface of the ground; rhizomes, horizontal stems growing underground that may end in enlarged tubers; and bulbs, vertical, underground shoots functioning in food storage.

Leaves, the main photosynthetic organ of most plants, usually consist of a flattened blade and a *petiole*, or stalk. Most monocots lack petioles. Monocot leaves usually have parallel major veins, whereas dicot leaves have networks of branched veins. Leaves emerge from the stem in an alternate, opposite, or whorled arrangement, and they may be simple or compound.

Basic Plant Anatomy: Plant Cells and Tissues

Types of Plant Cells Specialized cells of a multicellular plant have particular adaptations of the cell wall and/or of the *protoplast*, the cell contents without the cell wall. *Parenchyma* cells, relatively unspecialized cells that lack secondary walls and have large central vacuoles, carry on most of the metabolism of the plant, functioning in photosynthesis and food storage.

Collenchyma cells usually lack secondary walls, but have unevenly thickened primary walls. Strands or cylinders of collenchyma cells function in support and elongate with the young parts of the plant.

Schlerenchyma cells have thick secondary walls strengthened with lignin. These specialized supporting elements often lose their protoplasts at maturity. *Fibers* are long, tapered schlerenchyma cells that usually occur in bundles. *Sclereids* are shorter and irregular in shape.

After the cells of xylem form secondary walls and mature, the protoplast disintegrates, leaving behind a conduit through which water flows. In growing parts of a plant, the secondary walls may be deposited in spiral or ring patterns; in areas that are no longer elongating, the secondary walls may have scattered *pits*, which are thinner regions consisting only of primary walls. *Tracheids* are long, thin, tapered xylem cells with lignin-strengthened walls. *Vessel elements* are wider, shorter, thinner-walled, and probably more efficient water conductors. Gymnosperms have only tracheids, whereas angiosperms have both tracheids and vessel elements.

Phloem consists of chains of cells called *sieve-tube members*, which remain alive at functional maturity, but lack nuclei, ribosomes, and vacuoles. In angiosperms, dissolved sucrose flows through pores in the *sieve plates* in the end walls between cells. The nucleus and ribosomes of an adjacent *companion cell*, which is connected to a sieve-tube member by numerous plasmodesmata, may serve both cells.

The Three Tissue Systems of a Plant The *dermal tissue system*, or *epidermis*, is a single layer of cells that covers and protects the young parts of the plant. Root hairs are extensions of epidermal cells that increase the surface area for absorption. The epidermis of leaves and most stems is covered with a *cuticle*, a waxy coating that prevents excess water loss. The *vascular tissue system* consists of xylem and phloem and functions in transport and support. The *ground tissue system*, making up the bulk of a young plant, functions in photosynthesis, support, and storage, and consists predominantly of parenchyma cells.

An Overview of Plant Growth

Most plants exhibit *indeterminate growth*, continuing to grow as long as they live. Animals have *determinate growth* and stop growing after reaching a certain size.

Plants have tissues called *meristems* that remain embryonic and divide to form new cells. Newly formed cells that remain in the meristem are called *initials*, whereas cells that are displaced from the meristem and will specialize into developing tissue are called *derivatives*. *Apical meristems*, located at the tips of roots and in the buds of shoots, produce *primary growth*, resulting in the elongation of roots and shoots. *Secondary growth* is an increase in diameter as new cells are produced by *lateral meristems*, creating a thicker secondary dermal tissue and new layers of vascular tissue. In woody plants, primary growth continues to add to the tips of roots and shoots, while secondary growth thickens and strengthens older regions of the plant.

A plant's lifespan may be genetically programmed or environmentally determined. Plants may be *annuals*, which complete their life cycle in one growing season or year; *biennials*, which have a life cycle spanning two years; or *perennials*, which live many years.

A Closer Look at Primary Growth

The *primary plant body* is produced by primary growth and consists of the primary dermal, vascular, and ground tissues. The youngest parts of a woody plant and herbaceous plants consist of a primary plant body.

Primary Growth of Roots The meristem of the root tip is protected by a *root cap* which secretes a polysaccharide slime to lubricate the growth route. The *zone of cell division* includes the apical meristem and the *primary meristems*. At the center of the apical meristem is the *quiescent center*, a slowly dividing, resistant reserve of meristematic tissue. Derived from the apical meristem are the *protoderm, procambium,* and *ground meristem*, the primary meristems that continue to divide and give rise to the cells of the three tissue systems of the root.

In the *zone of cell elongation*, cells lengthen to more than ten times their original size, which pushes the root tip through the soil. Cells begin to specialize in structure and function as the zone of elongation grades into the *zone of cell differentiation*.

The primary meristems produce the tissue systems

of the root. The protoderm gives rise to the single cell layer of the epidermis. The procambium forms a central vascular cylinder or *stele*. In most dicots, xylem cells radiate from the center of the stele in spokes, with phloem in between. The stele of a monocot may have *pith*, a central core of parenchyma cells, inside the xylem and phloem.

The ground meristem produces the ground tissue system, filling the *cortex* or region of the root between the stele and epidermis. These parenchyma cells store food and are involved in the uptake of minerals from the soil solution entering the root. The innermost cortex layer is the one cell-thick *endodermis*, which regulates the passage of materials into the stele. *Lateral roots* may develop from the *pericycle*, a layer of cells just inside the endodermis with meristematic potential. A lateral root pushes through the cortex to emerge and retains its vascular connection with the central stele.

Primary Growth of Shoots In the shoot, the dome-shaped mass of apical meristem in the terminal bud also gives rise to the primary meristems: protoderm, procambium, and ground meristem. Leaf primordia form as tiny bulges on the sides of the apical dome, and axillary buds develop from clumps of meristematic cells left at the bases of the leaf primordia. Elongation of the shoot occurs by cell division and cell elongation within young internodes. In the grasses and some other plants, internodes have *intercalary meristems* which enable shoots to continue elongation. Axillary buds, with their connection to the vascular tissue located near the stem's surface, may form branches later in development.

Unlike the single central stele of a root, vascular tissue runs through the stem in several *vascular bundles*, each surrounded by ground tissue. In dicots, the vascular bundles may be arranged in a ring, with the pith internal to the ring connected by pith rays to the cortex outside the ring. Xylem is located internal to the phloem in the vascular bundles. In most monocot stems, the vascular bundles are scattered throughout the ground tissue and no definite pith region exists. A layer of collenchyma just beneath the epidermis may strengthen the stem.

The leaf is covered by the wax-coated, tightly-interlocking cells of the epidermis. *Stomata*, tiny pores flanked by *guard cells,* permit both gas exchange and *transpiration*, the evaporation of water from the leaf. To minimize water loss, stomata are more numerous in the lower epidermis of a leaf.

Mesophyll consists of parenchyma ground tissue cells containing chloroplasts. In many dicot leaves, columnar palisade mesophyll is located above spongy mesophyll, which has loosely packed, irregularly shaped cells surrounding many air spaces.

A *leaf trace*, which is a branch of a vascular bundle

in the stem, continues into the petiole as a vein and divides repeatedly within the blade of the leaf, providing support and vascular tissue to the photosynthetic mesophyll.

A shoot has a modular development in which successive nodes and internodes develop from the apical meristem. The developmental phase of the apical meristem can transition from a juvenile vegetative state to a mature vegetative state. The leaves and axillary buds that develop from juvenile nodes retain their juvenile characteristics. The shoot apex may also transition to a reproductive (flower-producing) state, which terminates its primary growth.

A Closer Look at Secondary Growth

The *secondary plant body* consists of the tissues produced by lateral meristems in secondary growth. *Vascular cambium* produces secondary xylem and phloem, and *cork cambium* produces a tough covering for stems and roots. Secondary growth occurs in all gymnosperms and most dicots.

Secondary Growth of Stems   Vascular cambium forms a continuous cylinder from a band of meristematic parenchyma cells located between the xylem and phloem of each vascular bundle and in the pith rays between bundles. *Ray initials*, the cambium cells of the pith rays, produce *xylem rays*, which function in storage and lateral transport of water and nutrients. *Fusiform initials* produce new vascular tissue, with cells on the inside of the vascular cambium differentiating as secondary xylem and cells on the outside becoming secondary phloem. Wood is the accumulation of secondary xylem cells with thick, lignified walls. Annual growth rings result from the seasonal cycle of xylem production in temperate regions, which includes cambium dormancy, spring production of larger xylem cells, and summer wood production. Secondary phloem splits and sloughs off as a tree increases in girth.

The epidermis splits off during secondary growth and is replaced by new protective tissues produced by the cork cambium. Cork cells, with waxy walls impregnated with suberin, are produced externally to the cork cambium. The protective coat formed by the layers of cork and the cork cambium is called the *periderm*. *Bark* refers to phloem and periderm. As secondary growth continually splits the periderm, a new cork cambium develops, eventually forming from parenchyma cells in the secondary phloem. Older secondary phloem, outside the cork cambium, also helps protect the stem as part of the bark.

In older trees, a central supportive column of heart-

wood consists of older xylem with resin filled cell cavities; the sapwood consists of secondary xylem that still functions in transport. *Lenticels* are spongy regions in the bark through which gas exchange occurs.

Secondary Growth of Roots   The two lateral meristems also function in the secondary growth of roots. Vascular cambium produces xylem internal and phloem external to itself, and a cork cambium forms from the pericycle and produces the periderm. Older roots resemble old stems, with annual rings and a thick, tough bark.

STRUCTURE YOUR KNOWLEDGE

1. In the table on p. 273, list the processes or functions of a plant and the specialized cells, tissues, or structures that perform those functions within the roots, stems and/or leaves. Examples of functions would be anchorage, support, transport, gas exchange, and so on.

2. Compare and contrast primary and secondary growth.

3. In the following micrograph of a stem, label the indicated structures. Is this a stem of a monocot of dicot? How can you tell?

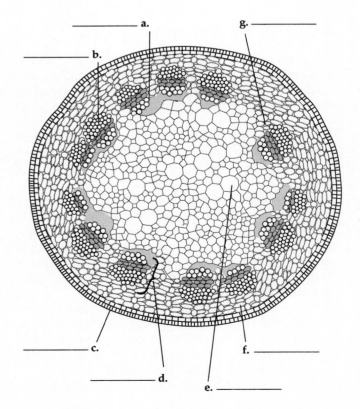

FUNCTION	ROOT	STEM	LEAF
Anchorage			
Support			
Transport			
Gas exchange			

4. Label the indicated structures in this diagram of a leaf.

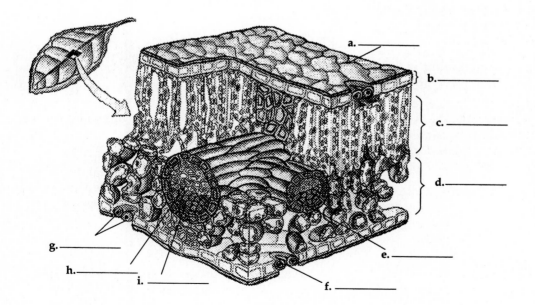

a. _____

b. _____

c. _____

d. _____

e. _____

f. _____

g. _____

h. _____

i. _____

TEST YOUR KNOWLEDGE

MATCHING: *Match the plant structure with its description.*

1. _____ schlerenchyma
2. _____ collenchyma
3. _____ tracheid
4. _____ fibers
5. _____ stele
6. _____ pericycle
7. _____ mesophyll
8. _____ periderm
9. _____ endodermis
10. _____ pith

A. chains of cells that transport food materials

B. tapered xylem cells with lignin in cell walls

C. parenchyma cells with chloroplasts in leaves

D. protective coat of woody stems and roots

E. bundles of long schlerenchyma cells

F. supporting cells with thickened primary walls

G. parenchyma cells inside vascular ring in dicot stem

H. layer from which lateral roots originate

I. central vascular cylinder of root

J. unspecialized cells functioning in food storage or photosynthesis

K. cell layer in root regulating movement into central vascular cylinder

L. supporting cells with thick secondary walls

MULTIPLE CHOICE: *Choose the one best answer.*

1. Monocots differ from dicots in all of the following ways *except* that they
   a. have a taproot rather than a fibrous root system.
   b. have leaves with parallel veins rather than branching venation.
   c. usually have no secondary growth.
   d. have a stem with scattered vascular bundles rather than bundles in a ring.

2. Which of the following is incorrectly paired with its function?
   a. procambium—meristematic tissue that forms protective layer of cork
   b. lenticel—spongy area for gas exchange in woody stem
   c. root hairs—absorption of water and dissolved minerals
   d. root cap—protect root as it pushes through soil

3. Axillary buds
   a. may not develop due to apical dominance of terminal bud.
   b. form at nodes in the angle where leaves join the stem.
   c. have their own apical meristems and leaf primordia.
   d. All of the above are correct.

4. A leaf trace is
   a. a petiole.
   b. the outline of the vascular bundles in a leaf.
   c. a branch from a vascular bundle at a node that extends into a leaf.
   d. a tiny bulge on the flank of the apical dome that grows into a leaf.

5. Ground meristem
   a. produces the root system.
   b. produces the ground tissue system.
   c. produces secondary growth.
   d. is meristematic tissue found at the nodes in monocots.

6. The zone of cell elongation
   a. is responsible for pushing a root through the soil.
   b. has a quiescent center that can undergo mitosis should the apical meristem be destroyed.
   c. produces the protoderm, procambium, and ground meristem tissues.
   d. does all of the above.

7. Perennials differ from annuals in that they
   a. have secondary growth and are woody.
   b. persist through many growing seasons.
   c. have indeterminate growth.
   d. All of the above are correct.

8. The function of stomata is to
   a. allow transport of water vapor into the leaf.
   b. allow transport of sugars out of the leaf.
   c. allow gas exchange between leaf and the atmosphere.
   d. open and close the guard cells.

9. Lateral meristems
   a. are responsible for secondary growth.

b. usually are not present in dicots.

c. produce protoderm, procambium, and ground meristem tissues.

d. produce branch roots and lateral buds.

10. Sieve-tube members

a. are responsible for lateral transport through secondary xylem.

b. control the activities of phloem cells that have no nuclei or ribosomes.

c. are transport cells in which fluid passes through sieve plates in the end walls between cells.

d. have spiral thickenings that allow the cell to grow and function in support.

FILL IN THE BLANKS: *Starting from the outside, place the letter of the tissues in the order in which they are located in a young woody tree trunk.*

___ ___ ___ ___ ___ ___ ___ ___

A. primary phloem
B. secondary phloem
C. primary xylem
D. secondary xylem
E. pith
F. cork cambium
G. vascular cambium
H. cork cells

CHAPTER 32

TRANSPORT IN PLANTS

FRAMEWORK

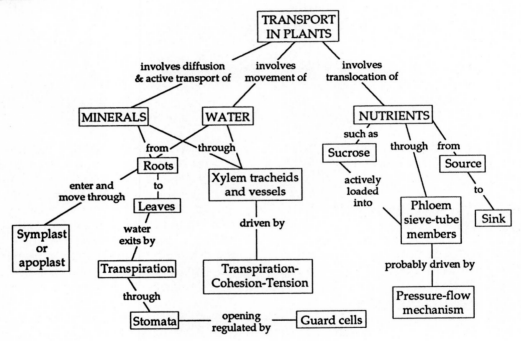

CHAPTER SUMMARY

The move onto land involved the evolutionary specialization of roots for the uptake of water and minerals, shoots for the uptake of CO_2 and exposure to light, and vascular tissue for the transport of materials between roots and shoots.

Basic Transport Processes in Plants: An Overview

Transport in plants involves the absorption of water and solutes by individual cells, the movement of substances from cell to cell, and the long-distance transport of sap in xylem and phloem.

Transport at the Cellular Level Solutes may move across the selectively-permeable plasma membrane by passive transport when they diffuse down their concentration gradients. *Transport proteins*, which speed passive transport, may be specific *carrier proteins*, which selectively bind and transport a solute, or *selective channels*, which allow specific solutes to cross the membrane.

Active transport requires the cell to expend energy to move a solute against its gradient. *Proton pumps* use ATP to pump hydrogen ions out of the cell, generating an energy-storing proton gradient and a membrane potential due to the separation of charges. This potential energy is used by plant cells to drive the transport of many different solutes. The membrane potential helps drive positively-charged potassium ions down their electrochemical gradient and into the negatively-charged cell. In *cotransport*, a solute can be pumped against its concentration gradient when a transport protein links its passage to the "downhill" movement of H^+. *Chemiosmosis* links energy-releasing and energy-consuming processes using a trans-membrane proton gradient like that created by the proton pump.

Water potential, abbreviated by Ψ (psi), is a useful measurement for predicting the direction that water will move when a plant cell is surrounded by a solution. It takes into account both solute concentration and the pressure exerted by the plant cell wall. Water will flow from a region of higher water potential to one of lower water potential. Water potential is measured in *megapascals* (MPa); 1 MPa to about 10 ten atmospheres of pressure.

The water potential of pure water in an open container is defined as 0 MPa. Adding solutes lowers water potential; therefore, a solution at atmospheric pressure has a negative water potential. The addition of pressure increases water potential because it increases the potential for water to move. A *tension*, or negative pressure, will lower Ψ. *Bulk flow* is the movement of water due to a pressure difference. The equation for water potential is $\Psi = P - \Pi$ (P = physical pressure, Π = osmotic pressure). Osmotic pressure measures the tendency of water to enter a solution. Osmotic pressure has a negative effect on Ψ; an increase in solute concentration raises osmotic pressure and thus decreases Ψ.

A plant cell bathed in a more concentrated solution (hyperosmotic) will lose water by osmosis because the solution has a lower Ψ. When bathed in pure water, the cell has the lower Ψ. Water will enter the cell until enough *turgor pressure* builds up so that $P = \Pi$ and $\Psi = 0$ in the cell. Net movement of water will then stop.

The *tonoplast*, the membrane of the central vacuole, also influences transport into the cell. Its proton pumps move H^+ from the cytoplasm into the vacuole, helping to maintain a low cytosolic H^+ concentration and generating a membrane potential that drives transport of certain solutes into the vacuole.

Short-distance or lateral transport of water and solutes within plant tissues can occur by three routes: cell to cell by crossing plasma membranes and cell walls; the *symplast*, the continuum of cytoplasm between cells connected through *plasmodesmata*; and the *apoplast*, the extracellular pathway along cell walls. Solutes and water may change routes during transit. The direction of short-distance transport is largely determined by diffusion due to concentration gradients and bulk flow due to pressure differences.

Long-Distance Transport at the Whole-Plant Level Long-distance transport throughout plants occurs by bulk flow. Water and minerals move through xylem vessels as the result of tension created by transpiration, the evaporation of water from leaves. Sap is forced through phloem sieve tubes by hydrostatic pressure.

Absorption of Water and Minerals by Roots: A Closer Look

Most absorption of water and minerals occurs through root hairs, extensions of epidermal cells located along young root tips. The soil solution soaks into the hydrophilic walls of epidermal cells and along the apoplast into the root cortex, exposing a large surface area of plasma membrane for the uptake of water and minerals. Selective absorption of minerals, using transport proteins in the plasma membrane and tonoplast, allows cells to accumulate essential minerals.

The *endodermis*, the layer of cortex cells surrounding the stele, selectively screens all minerals entering the vascular tissue. A ring of suberin around each endodermal cell, called the *Casparian strip*, prevents water from the apoplast from entering the stele without passing through the selective plasma membrane of an endodermal cell. Water and minerals that had already entered the symplast through a cortex cell pass through plasmodesmata of endodermal cells into the stele.

By a combination of diffusion and active transport, minerals leave the symplast, enter the nonliving xylem tracheids and vessel elements, and move again through the apoplast system.

Ascent of Xylem Sap: A Closer Look

Through *transpiration*, plants lose a tremendous amount of water that must be replaced by water transported up from the roots.

Pushing Xylem Sap: Root Pressure As minerals are actively pumped into the stele and prevented from leaking out by the endodermis, water potential in the stele is lowered. Water flows in by osmosis, resulting in *root pressure*, which forces fluid up the xylem. Root pressure may cause *guttation*, the exudation of water droplets through specialized structures in the leaves when more water is forced up the xylem than is transpired by the plant, as may happen during the night. In most plants, root pressure is not a major mechanism in the movement of xylem sap.

Pulling Xylem Sap: The Transpiration–Cohesion–Tension Mechanism Water vapor from saturated air spaces within a leaf exits to drier air outside the leaf by way of stomata. The thin layer of water that coats the mesophyll cells lining the air spaces then evaporates, forming menisci that increase the surface tension of this water layer. The negative pressure pulls water from the xylem and through the apoplast and symplast of the mesophyll to the cells and surface film lining the air spaces. Water moves along a gradient of decreasing water potential from xylem to neighboring cells to air spaces to the drier air outside the leaf.

The transpirational pull on xylem sap is transmitted from the leaves to the root tips by the cohesiveness of water that results from hydrogen bonding between molecules. Adhesion of water molecules to the hydrophilic walls of the narrow xylem elements and tracheids also helps the upward movement against gravity.

The upward transpirational pull on the cohesive sap creates tension within the xylem, further decreasing water potential, to the extent that water flows passively from the soil, across the cortex, and into the stele.

Cavitation, or a break in the chain of water molecules by the formation of a water vapor pocket in a xylem vessel, breaks the transpirational pull, and the vessel cannot function in transport.

Summary: Long Distance Transport of Water in Plants Occurs by Solar-Powered Bulk Flow The transpiration-cohesion-tension mechanism results in the bulk flow of water from roots to leaves. Water potential differences caused by solute concentration and pressure contribute to the movement of water from cell to cell, but solar-powered tension caused by transpiration is responsible for long distance transport.

The Control of Transpiration: A Closer Look

The Photosynthesis–Transpiration Compromise A plant's tremendous requirement for water is partly a consequence of making food by photosynthesis. To obtain sufficient CO_2 for photosynthesis, leaves must exchange gases through the stomata, which exposes a large surface area—the surfaces of the cells surrounding the air spaces—from which water may evaporate.

A common transpiration-to-photosynthesis ratio is 600 grams of water transpired for each gram of CO_2 incorporated into carbohydrate (600:1). Plants such as corn that use the more efficient C_4 pathway for photosynthesis have ratios of 300:1 or less.

Transpiration also assists in transferring minerals and other substances to the leaves and results in evaporative cooling. When transpiration exceeds the water available, leaves wilt as cells lose turgor pressure.

How Stomata Open and Close Guard cells regulate the size of stomatal openings and control the rate of transpiration. When the kidney-shaped guard cells of dicots become turgid and swell, their radially oriented microfibrils cause them to buckle outward and increase the size of the gap between them. When they become flaccid, they sag and close the space. Monocot guard cells, although different in shape, operate by the same basic mechanism. Guard cells can lower their water potential by actively absorbing potassium ions (K^+), which leads to osmosis and the resulting increase of turgor pressure. The movement of K^+ across the guard cell membrane and into the vacuole is probably coupled with the generation of membrane potentials by proton pumps. A new technique called patch clamping has allowed plant physiologists to study proton pumps and K^+ channels in guard cells. A small patch of membrane is held across the opening of a micropipette while an electrode records ion fluxes across the membrane.

Stomata are usually open during the day and closed at night. The opening of stomata at dawn is related to at least three factors: Light stimulates guard cells to accumulate K^+, perhaps triggered by the illumination of blue-light receptors which activate the proton pumps of the plasma membrane. Light also drives photosynthesis in the guard cell chloroplasts, making ATP available for proton pumping. Second, stomata are stimulated to open when CO_2 is depleted within air spaces of the leaf as photosynthesis begins in the mesophyll. The third factor is a daily rhythm of opening and closing that is endogenous to guard cells. Cycles that have intervals of approximately 24 hours are called *circadian rhythms*.

Environmental stress can cause stomata to close during the day. Guard cells lose turgor when water is in short supply. A hormone called abscisic acid, produced by mesophyll cells in response to a lack of water, signals guard cells to close stomata. High temperatures induce closing, probably because increasing cellular respiration raises CO_2 concentrations in the air spaces in the leaf.

Evolutionary Adaptations that Reduce Transpiration
Many xerophytes, plants adapted to arid climates, have leaves that are small and thick, limiting water loss by reducing their surface to volume ratio. Leaf cuticles may be thick, and the stomata, concentrated on the lower leaf surface, may be located in depressions. Some desert plants lose their leaves in the driest months.

Succulent plants of the family Crassulaceae assimilate CO_2 into organic acids during the night, by a pathway known as CAM (crassulacean acid metabolism), and then release it for photosynthesis during the day. Thus, the stomata are open at night and are closed during the day when water loss would be greatest.

Transport in Phloem: A Closer Look

Translocation is the transport of photosynthetic products throughout the plant through the phloem. Sieve-tube members are the specialized transport cells in angiosperms. Sap flows between these cells end-to-end through porous sieve plates. Phloem sap may have a sucrose concentration as high as 30% and may also contain minerals, amino acids, and hormones.

Source-to-Sink Transport Phloem sap flows from a sugar *source*, where it is produced by photosynthesis or the breakdown of starch, to a sugar *sink*, an organ that consumes or stores sugar. A storage organ, such as a tuber, stores carbohydrates during the summer and is a sugar sink, but when its starch is broken down to sugar in the spring, it becomes a sugar source. The direction of transport in any one sieve tube depends on the location of the source and sink connected by that tube.

Phloem Loading and Unloading In some species, sugar in the leaf moves through the symplast of the mesophyll cells to sieve-tube members. In other species, sugar first moves through the symplast and then into the apoplast in the vicinity of sieve-tube members and companion cells, which actively accumulate sugar. Some companion cells, specialized as *transfer cells*, have ingrowths of their wall that increase surface area for movement of solutes from apoplast to symplast.

Phloem loading requires active transport because sugar concentration in the sieve tubes is two to three times higher than in the mesophyll. Proton pumps and the cotransport of sucrose through membrane proteins along with the returning protons is the chemiosmotic mechanism used for active transport. Sucrose may move by active transport or diffusion out of sieve tubes at the sink end.

Pressure Flow (Bulk Flow) of Phloem Sap The rapid movement of phloem sap from source to sink is due to a pressure flow mechanism. High solute concentration at the source lowers water potential, and the resulting movement of water into the sieve tube produces hydrostatic pressure. At the sink end, the osmotic loss of water following the exodus of sucrose into the surrounding tissue results in a lower pressure. The difference in these pressures causes water to flow from source to sink, transporting sugar. Innovative tests of this model support it as the explanation for the flow of sap in the phloem of angiosperms.

STRUCTURE YOUR KNOWLEDGE

1. Describe the models that explain the movement of xylem and phloem sap throughout the plant.

2. Label the diagram on p. 280 of the cell layers and routes of transport of water and minerals from the soil through the root. **a** and **b** refer to routes of transport; the other letters refer to cell layers or cell structures.

TEST YOUR KNOWLEDGE

MULTIPLE CHOICE: *Choose the one best answer.*

1. Which of the following is *not* a component of the symplast?
   a. sieve-tube members
   b. xylem tracheids
   c. endodermal cells
   d. transfer cells

2. Proton pumps in the plasma membranes of plant cells may
   a. generate a membrane potential that helps drive charged minerals into the cell through their specific carriers.
   b. be coupled to the movement of K^+ into guard cells.
   c. drive the accumulation of sucrose in sieve-tube members.
   d. be involved in all of the above.

3. The Casparian strip prevents water and minerals

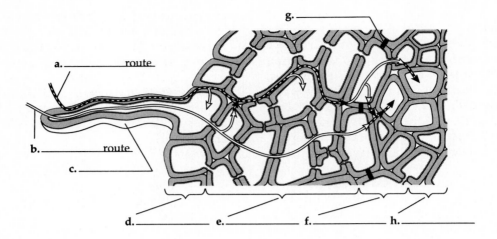

from entering the stele through the
a. plasmodesmata.
b. endodermal cells.
c. symplast.
d. apoplast.

4. The water potential of a plant cell
a. is equal to 0 when the cell is in pure water and is turgid.
b. is greater than that of air.
c. is equal to –0.23 MPa.
d. becomes greater when K^+ ions are actively moved into the cell.

5. Guttation results from
a. the pressure-flow of sap through phloem.
b. a water-vapor break in the column of xylem sap.
c. root pressure causing water to flow up through xylem faster than it can be lost by transpiration.
d. a higher water potential of the leaves than that of the roots.

6. Which of these is *not* a major factor in the movement of xylem sap?
a. transpiration
b. active transport using a proton pump
c. adhesion
d. cohesion

7. Adhesion is a result of
a. hydrogen bonding between water molecules.
b. the pull on the water column as water evaporates from the surface of mesophyll cells.
c. tension within the xylem caused by a lowered water potential.
d. attraction of water molecules to hydrophilic walls of narrow xylem tubes.

8. A plant with a low transpiration-to-photosynthesis ratio
a. would lose less water through transpiration for each gram of CO_2 fixed.
b. could be a C_4 plant.
c. could be a CAM plant.
d. all of the above could be correct.

9. Which of these does *not* stimulate the opening of stomata?
a. a decrease in the turgor of guard cells
b. depletion of CO_2 in the air spaces of the leaf
c. stimulation of proton pumps that result in the movement of K^+ into the guard cells
d. circadian rhythm of guard-cell opening

10. Phloem sap moves
a. from a sugar source to a sugar sink
b. primarily through the symplast.
c. by a pressure-flow mechanism.
d. all of the above are correct.

11. Hydrophytes, plants that are adapted to live in aquatic habitats, are most likely to show which of the following morphologies?
a. very thick cuticle
b. leaves reduced to spines
c. stomata located on the top of the leaves
d. no vascular tissues

12. Your favorite spider plant is wilting. What is the most likely cause and remedy for its declining condition?
a. water potential is too low: apply sugar water
b. the stomata won't open; no remedy available
c. plasmolysis of its cells; water the plant
d. cavitation; perform a phloem-cell bypass

13. A plant cell placed in a solution in an open beaker becomes flaccid.
    a. The water potential of the cell was higher than that of the solution.
    b. The water potential of the cell was lower than that of the solution.
    c. The pressure (P) of the cell was lower than that of the solution.
    d. The cell was hyperosmotic to the solution.

14. What facilitates the movement of K^+ down its concentration gradient into epidermal cells of the root?
    a. cotransport through a membrane protein
    b. bulk flow of water into the root
    c. passage through selective channels, aided by the membrane potential created by proton pumps
    d. active transport through a potassium pump

15. The increase in hydrostatic pressure within sieve-tube members
    a. lowers the water potential at the source end.
    b. raises the water potential at the source end.
    c. lowers the water potential at the sink end.
    d. raises the water potential at the sink end.

16. Metabolic energy of the plant may be required for which of the following transport?
    a. the anti-gravitational movement of water solution through xylem
    b. the movement of soil solution through the apoplast of the root
    c. the anti-gravitational movement of sugars in phloem sap from one sieve-tube member to the next
    d. the movement of sugars from a source into sieve tube members

CHAPTER 33

PLANT NUTRITION

FRAMEWORK

The nutritional requirements of plants includes essential macronutrients and micronutrients. Carbon dioxide enters the plant through the leaves, but the other inorganic raw materials, including water and minerals, must be absorbed through the roots. Soil fertility is influenced by its texture and composition. Nitrogen assimilation by plants is made possible by the decomposition of humus by microbes and nitrogen fixation by bacteria. Mycorrhizae are important associations between fungi and plant roots that increase mineral and water absorption.

CHAPTER SUMMARY

Plants, as photosynthetic autotrophs, make their own organic compounds and are critical to the energy flow and chemical cycling of an ecosystem. Roots, shoots, and leaves are structurally adapted to obtain carbon dioxide, water, and a variety of minerals or inorganic ions from the soil and air.

Nutritional Requirements of Plants

Chemical Composition of Plants Early scientists speculated on whether soil, water, or the air provides the substance for plant growth. Minerals, although essential, make only a small contribution to the mass of a plant. Water supplies most of the hydrogen and some of the oxygen incorporated into organic compounds, but 90% of the water absorbed is lost by transpiration, and most of the water retained is used for growth by cell elongation. By weight, CO_2 from the air is the source of the bulk of organic material of a plant.

Organic substances, most of which are carbohydrates (especially cellulose), make up 95% of the dry weight of plants. Thus, carbon, oxygen and hydrogen are the most abundant elements. Nitrogen, sulfur, and phosphorus—also ingredients of organic compounds—are relatively abundant.

Essential Nutrients *Essential nutrients* are those required for a plant to complete its life cycle from a seed to an adult that produces more seeds. *Hydroponic culture* has been used to determine which of the mineral elements found in plants are essential nutrients. Seventeen elements have been identified as essential in all plants; a few others are necessary for certain plant groups.

Nine *macronutrients* are required by plants in relatively large amounts, including the six major elements of organic compounds, as well as calcium, potassium and magnesium.

Eight *micronutrients* have been identified as needed by plants in very small amounts, functioning mainly as cofactors of enzymatic reactions.

Mineral Deficiencies The symptoms of a mineral deficiency depend on the functions of that nutrient and on its mobility in the plant. For example, a deficiency in magnesium, a component of chlorophyll, causes *chlorosis* or yellowing of the leaves. A mobile nutrient will move to young, growing tissues, so that a deficiency will show up first in older parts of the plant.

Symptoms of a mineral deficiency may be distinctive enough for the cause to be diagnosed by a plant physiologist or farmer. Soil and plant analysis can confirm a specific deficiency. Nitrogen, potassium, and phosphorus deficiencies are most common.

Soil

Plants that grow well in a particular region are adapted to the texture and mineral content of that soil.

Texture and Composition of Soils The formation of soil begins with the weathering of rock and accelerates with the secretion of acids by lichens, fungi, bacteria, and plant roots. The expansion of roots in fissures breaks rocks into smaller pieces. *Topsoil* is a mixture of decomposed rock, living organisms, and *humus* (decomposing organic matter). Several other distinct soil layers, or *horizons*, are found under the topsoil layer.

The size of soil particles varies from coarse sand to fine clay. *Loams*, made up of a mixture of sand, silt, and clay, are often the most fertile soils, having enough fine particles to provide a large surface area for retaining water and minerals but enough coarse particles to provide air spaces with oxygen for respiring roots.

The activities of the numerous soil inhabitants, such as bacteria, fungi, algae, other protists, insects, worms, nematodes, and plant roots, affect the physical and chemical properties of soil.

Humus builds a crumbly soil that retains water, provides good aeration of roots, and supplies mineral nutrients.

Availability of Soil Water and Minerals Water containing dissolved minerals binds to hydrophilic soil particles and is held there in the small spaces between soil particles, available for uptake by plant roots. Positively charged minerals, such as K^+, Ca^{2+}, Mg^{2+}, adhere to the negatively charged surfaces of finely-divided clay particles. Negatively charged minerals, such as nitrate (NO_3^-), phosphate ($H_2PO_4^-$), and sulfate (SO_4^{2-}), tend to leach away more quickly. Roots release acids that facilitate *cation exchange*, in which hydrogen ions displace positively charged mineral ions from the clay particles, making the ions available for absorption. The continuous growth of roots exposes them to new sources of mineral nutrients.

Soil Management Without good soil management, agriculture can quickly destroy the fertility of a soil that has built up over centuries. Agriculture diverts essential elements from the chemical cycles when crops are harvested, thus depleting the mineral content of the soil.

Historically, farmers used manure to fertilize their crops. Today in developed nations, commercially produced fertilizers, usually containing nitrogen, phosphorous and potassium, are used. Manure, fishmeal and compost are called organic fertilizers because they contain organic material that is in the process of

decomposing. These fertilizers slowly decompose to inorganic nutrients, so that they are taken up by the plant in the same form supplied by commercial fertilizers. A disadvantage of commercial fertilizers is that they may be rapidly leached from the soil, polluting streams and lakes.

The acidity of the soil affects cation exchange and can alter the chemical form of minerals and thus their ability to be absorbed by the plant. Managing the pH of soil is an important aspect of maintaining fertility.

Irrigation can make farming possible in arid regions, but it places a huge drain on water resources and raises soil salinity. New methods of irrigation and new varieties of plants that can tolerate less water or more salinity may reduce some of these problems.

In the U.S., topsoil from thousands of acres of farmland is lost to water and wind erosion each year. Agricultural use of cover crops, windbreaks, and terracing can minimize erosion. *Sustainable agriculture* uses a variety of conservation-minded, environmentally safe farming methods.

Nitrogen Assimilation by Plants

Nitrogen is the mineral that usually limits plant growth and crop yields. To be absorbed by plants, nitrogen must be converted to nitrate (NO_3^-) or ammonia (NH_3) by the action of microbes that decompose humus. *Nitrogen-fixing bacteria* convert atmospheric nitrogen into ammonia through the process of *nitrogen fixation*.

Nitrogen Fixation Bacteria capable of nitrogen fixation contain *nitrogenase*, an enzyme that reduces N_2 by adding H^+ and electrons to form ammonia (NH_3). In the soil solution, ammonia forms ammonium (NH_4^+) which plants can absorb. Certain soil bacteria oxidize ammonium, producing nitrate, the form of nitrogen most readily absorbed by roots. Most plants incorporate nitrogen into amino acids or other organic compounds in the roots and then transport it through the xylem in organic form.

Symbiotic Nitrogen Fixation Plants of the legume family have root swellings, called *nodules*, composed of plant cells containing nitrogen-fixing bacteria, in a form called *bacteroids*. The coevolution of this symbiotic relationship between these bacteria of the genus *Rhizobium* and legumes is illustrated by their cooperative synthesis of *leghemoglobin*. This oxygen-binding protein releases oxygen for the respiration needed to supply energy for nitrogen fixation and keeps the free oxygen concentration in the nodules low to prevent inhibition of the enzyme nitrogenase.

The root nodules use most of the symbiotically-fixed nitrogen to make amino acids, which are then transported throughout the plant. Excess ammonium may be secreted into the soil, increasing the fertility of the soil for nonlegumes, which are grown in rotation with the legumes. Rice farmers culture a water fern that has symbiotic cyanobacteria that fix nitrogen, improving the fertility of rice paddies.

Improving the Protein Yield of Crops Protein deficiency is the most common form of human malnutrition. Agricultural research attempts to improve the quality and quantity of proteins in crops. Varieties of corn, wheat, and rice have been developed that are enriched in protein, but they require the addition of large quantities of expensive nitrogen fertilizer. Other strategies for increasing protein yields include improving the productivity of symbiotic nitrogen fixation so that more of the photosynthetic energy goes into legume production, and incorporating into root nodules mutant strains of *Rhizobium* that do not shut off production of nitrogenase when fixed nitrogen accumulates.

The tools of genetic engineering are being used to create varieties of *Rhizobium* that can infect non-legumes and to transplant the genes for nitrogen fixation into other bacteria and, possibly, directly into plant genomes.

Some Nutritional Adaptations of Plants

Parasitic Plants Parasitic plants, such as the mistletoe, produce haustoria that may invade a host plant and siphon xylem sap from its vascular tissue. Epiphytes are plants that grow on the surface of another plant but do not rely on it for nourishment.

Carnivorous Plants Living in acid bogs or other nutrient-poor soils, carnivorous plants obtain nitrogen and minerals by killing and digesting insects that are caught in traps formed from modified leaves.

Mycorrhizae Most plants have modified roots called *mycorrhizae*, which are symbiotic associations between the roots and fungi. The fungus secretes growth hormones that stimulate root growth and branching. The fungus extends hyphae among or in cortex cells, while mycelial projections of the fungus provide the root with a large surface area for the absorption of minerals and water. The plant provides food to the fungus, and the fungus supplies absorbed minerals to the plant and may help to protect the plant from some soil pathogens.

Most plants form mycorrhizae when they grow in their natural habitat. Farmers and foresters now inoculate seeds with the appropriate mycorrhizal fungi to insure optimal plant growth. The fossil record shows that the earliest land plants had mycorrhizae. Plant nutrition reinforces the concept that organisms have adapted to their physical and biological environment.

STRUCTURE YOUR KNOWLEDGE

1. Develop a concept map that organizes your understanding of the basic nutritional requirements of plants.

2. List the key properties of a fertile soil.

3. Explain how nitrogen assimilation by plants depends on bacteria and microbes. Why is nitrogen assimilation by plants so important in agriculture?

TEST YOUR KNOWLEDGE

MULTIPLE CHOICE: *Choose the one best answer.*

1. The inorganic compound that contributes most of the mass to a plant's organic matter is
   a. H_2O.
   b. CO_2.
   c. NO_3^-.
   d. O_2.

2. The effects of mineral deficiencies involving fairly mobile nutrients will first be observed in
   a. older portions of the plant.
   b. new leaves and shoots.
   c. the root system.
   d. the color of the leaves.

3. Most macronutrients are
   a. cofactors in enzymes.
   b. readily available from air and water.
   c. components of organic compounds.
   d. determined by hydroponic culture.

4. The most fertile type of soil is usually
   a. sand because its large particles allow room for air spaces.
   b. loam, which has a mixture of fine and coarse particles.
   c. clay, because the fine particles provide much surface area to which minerals and water adhere.
   d. humus, which is decomposing organic material.

5. Chlorosis is
   a. a symptom of a mineral deficiency indicated by yellowing leaves due to decreased chlorophyll production.
   b. the uptake of the micronutrient chlorine by a plant which is facilitated by symbiotic bacteria.
   c. the production of chlorophyll by a plant within the thylakoid membranes.
   d. a contamination of glassware in hydroponic culture.

6. Negatively charged minerals
   a. are released from clay particles by cation exchange.
   b. are reduced by cation exchange before they can be absorbed.
   c. are leached away by the action of rainwater more easily than positively charged minerals.
   d. are bound when roots release acids into the soil.

7. Nitrogenase
   a. is an enzyme that reduces atmospheric nitrogen to ammonia.
   b. is found in *Rhizobium* and other nitrogen-fixing bacteria.
   c. is inhibited by high concentrations of oxygen.
   d. is all of the above.

8. Epiphytes
   a. have haustoria for anchoring to their host plants and obtaining xylem sap.
   b. are symbiotic relationships between roots and fungi.
   c. live in poor soil and digest insects to obtain nitrogen.

   d. grow on other plants but do not obtain nutrients from their hosts.

9. Which of the following is not an erosion-control measure?
   a. terracing
   b. crop rotation
   c. cover crops
   d. windbreaks such as hedge rows

10. The nitrogen content of the soil may be improved by
    a. the synthesis of leghemoglobin by nitrogen-fixing bacteria.
    b. mycorrhizae on legumes.
    c. water ferns with symbiotic cyanobacteria.
    d. cation exchange.

11. An advantage of organic fertilizers over chemical fertilizers is that they
    a. are more natural.
    b. provide nutrients in the forms most readily absorbed by plants.
    c. release their nutrients over a longer period of time and are less likely to be lost to runoff.
    d. are easier to mass produce and transport.

12. Which of the following is *not* true of mycorrhizae?
    a. They are mutualistic association between roots and fungi.
    b. They are most common on legumes and involve species of *Rhizobium*.
    c. They greatly increase the surface area for absorption of water and minerals.
    d. Most plants have a particular species of fungi with which they form mycorrhizae.

CHAPTER 34

PLANT REPRODUCTION AND DEVELOPMENT

FRAMEWORK

This chapter describes the sexual and asexual reproduction of flowering plants. The flower, with leaves modified for reproduction, produces the haploid gametophyte stages of the life cycle: microspores in the anther develop into pollen grains and a megaspore in the ovule produces an embryo sac. Pollination and the double fertilization of egg and polar nuclei are followed by the development of a seed with a quiescent embryo and endosperm, protected in a seed coat and housed within a fruit. Seed dormancy is broken following proper environmental cues and the imbibition of water.

Vegetative propagation allows successful plants to clone themselves. Agriculture makes extensive use of this type of plant reproduction, by using cuttings, grafts, and test-tube cloning.

Development is the result of the overlapping processes of growth, morphogenesis, and cellular differentiation. Pattern formation is the development of structures in specific locations. This development is linked to position information, which may affect the expression of organ-identity genes.

CHAPTER SUMMARY

Adaptations in reproduction are a key to the spread of plants into a variety of terrestrial habitats. In conifers and angiosperms, pollen dispersed by wind or animals replaced flagellated sperm, and zygotes develop into embryos protected within seeds.

Sexual Reproduction of Flowering Plants

The Angiosperm Life Cycle: An Overview Plants exhibit an *alternation of generations* between haploid

(N) and diploid ($2N$) generations. The diploid plant, the *sporophyte*, produces haploid spores by meiosis. Spores develop into multicellular haploid male or female *gametophytes* which produce gametes by mitosis. Fertilization yields diploid zygotes which grow into new sporophyte plants.

In angiosperms, the gametophytes have become reduced to tiny structures that remain dependent on the sporophyte plant. The flower, the reproductive structure of the sporophyte, evolved from a compressed shoot with four whorls of modified leaves: *sepals, petals, stamens,* and *carpels*. Sepals—the outermost, usually green whorl—enclose and protect a floral bud before it opens. Petals are generally brightly colored and may attract pollinators. Stamens and carpels contain sporangia in which male and female gametophytes, respectively, develop. A stamen consists of a stalk (filament) and a terminal anther, with chambers in which pollen develops. Pollen grains are sperm-containing male gametophytes. A carpel consists of a sticky stigma at the top of a slender neck or style, which leads to an ovary. The ovary encloses one or more ovules in which an embryo sac, the female gametophyte, develops.

Pollination is the arrival of pollen onto the stigma. The pollen grain germinates and grows a tube down the neck of the carpel, releasing its sperm within the embryo sac. Following fertilization, the zygote develops into an embryo as the surrounding ovule develops into a seed. The entire ovary forms a fruit, which aids in seed dispersal.

More About Flowers A *complete flower* has sepals, petals, stamens and carpels. *Incomplete flowers* are missing one or more of these parts. A *perfect flower* has both stamens and carpels; an *imperfect flower* is missing one of these two. Imperfect flowers may be either staminate or carpellate. In *monoecious* plant species, both staminate and carpellate flowers are on the same plant; in *dioecious* species, these flowers are on separate plants. Floral variations include fused carpels, inflorescences, and composite flowers, as well as diverse shapes, colors, and odors adapted to different pollinators.

Development of Pollen: A Closer Look Diploid cells called microsporocytes undergo meiosis to form four haploid *microspores*. The nucleus of a microspore divides once by mitosis to produce a generative cell and a tube cell. The wall surrounding the two cells thickens into the durable coat of the pollen grain, with an elaborate pattern characteristic of the plant species. A pollen grain is an immature male gametophyte.

Development of Ovules: A Closer Look Ovules form within the ovary. The megasporocyte in the sporangium of each ovule grows and undergoes meiosis to form four haploid *megaspores*, only one of which usually survives. This megaspore divides by mitosis three times, forming the female gametophyte, called the *embryo sac*, which typically consists of eight nuclei contained in seven cells. At one end of the embryo sac, an egg cell is lodged between two cells called synergi; three antipodal cells are at the other end; and two nuclei, called polar nuclei, are in a large central cell. Protective layers called integuments form from sporophyte tissue around the embryo sac.

Pollination and Fertilization *Pollination* is the arrival of pollen onto the stigma, carried there by wind or animals. Self-pollination is usually prevented by temporal, structural, or biochemical mechanisms. When pollen does land on a stigma from the same plant, *self-incompatibility*, determined by a single gene with multiple alleles, prevents it from developing because matching alleles are rejected.

The pollen grain grows a tube down the style, and the generative cell divides to form two sperm, the male gametes. The pollen tube probes through the micropyle, an opening through the integuments of the ovule, and releases its two sperm within the embryo sac. By *double fertilization*, one sperm fertilizes the egg to form the zygote and the other combines with the polar nuclei to form a triploid nucleus, which will develop into a food-storing tissue called the *endosperm*. Double fertilization is unique to angiosperms (and one species of gymnosperms). Ovules develop into seeds, and the ovary develops into the enclosing fruit.

The Seed The triploid nucleus divides to form the endosperm, a multicellular mass rich in nutrients, which are provided to the developing embryo and stored for later use when the seed germinates. In many dicots, endosperm is transferred to the cotyledons before the seed matures.

In the zygote the first mitotic division creates a basal cell and a terminal cell. The basal cell divides to produce a thread of cells, called the suspensor, that anchors the embryo and transfers nutrients to it. The terminal cell divides to form a spherical proembryo, on which the cotyledons begin to form as bumps. The embryo elongates and apical meristems develop at the apexes of the embryonic shoot, held between the cotyledons, and the embryonic root, where the suspensor attaches. The primary meristems—protoderm, procambium, and ground meristem—are present in the embryo at this stage.

The root-shoot polarity of the embryo is determined by the first cytoplasmic division of the zygote. Egg organelles and chemicals are unevenly distributed in the cytoplasm, so the two cells differ in cytoplasmic composition. During cellular differentiation, the cytoplasmic environment and position of an embryonic cell influence the selective expression of genes.

As it matures, the seed dehydrates to a water content about 5% to 15% of its weight. The embryo becomes dormant until the seed germinates. The embryo and its food supply, the endosperm and/or enlarged cotyledons, are enclosed in a *seed coat* formed from the ovule integuments.

In a dicot seed, such as a bean, the embryo is an elongate embryonic axis attached to fleshy cotyledons. The axis below the cotyledonary attachment is called the *hypocotyl*; it terminates in the *radicle*, or embryonic root. The upper axis is the *epicotyl*; it terminates as a plumule, a shoot tip with a pair of leaves. In some dicots, the cotyledons remain thin and absorb nutrients from the endosperm when the seed germinates.

A monocot seed, such as a corn kernel, has a single thin cotyledon, called a *scutellum*, which absorbs nutrients from the endosperm during germination. A sheath called a *coleorhiza* covers the root, and a *coleoptile* encloses the embryonic shoot.

Development of the Fruit The ovary of the flower ripens into a *fruit*, which both protects and helps to disperse the seeds. Other floral parts may contribute to what we commonly call a fruit. Hormonal changes following pollination cause the ovary to grow tremendously, its wall becoming the *pericarp*, or thickened wall of the fruit. Fruit usually does not set if a flower has not been pollinated.

A *simple fruit* develops from a single ovary. An *aggregate fruit* results from a single flower that has several separate carpels. A *multiple fruit*, such as a pineapple, develops from an inflorescence, a group of tightly clustered flowers. Fruits usually ripen as the seeds are completing their development. Ripening of fleshy fruits may include softening of the pulp, a change in color, and an increase in sugar content.

Fruits are adapted to disperse seeds, enlisting the aid of wind or animals.

Humans have selectively bred edible fruits. Cereal grains—the wind-dispersed fruits of grasses, which have dry pericarps—are staple foods for humans.

Seed Germination   At germination, the plant resumes the growth and development that was suspended when the seed matured.

Dormancy is an adaptation to terrestrial habitats that increases the chances that the seed will germinate when and where the embryo has a good chance of surviving. The specific cues for breaking dormancy vary with the environment, including heavy rain in deserts, intense heat in areas of frequent fires, and cold in areas with harsh winters. The viability of a dormant seed may vary from a few days to decades or longer.

Imbibition, the absorption of water by the dry seed, causes the seed to expand, rupture its coat, and begin a series of metabolic changes. Stored compounds are digested by enzymes, and nutrients are sent to growing regions. Soon after hydration in cereal seeds, the thin outer layer of the endosperm, called the aleurone, is cued by gibberellic acid, a hormone produced by the embryo, to begin making α-amylase and other enzymes that digest starch stored in the endosperm.

The radicle emerges from the seed first, followed by the shoot tip. In many dicots, a hook that forms in the hypocotyl is pushed up through the ground, pulling the delicate shoot and cotyledons behind it. Light stimulates the straightening of the hook, and the first foliage leaves begin photosynthesis.

Light seems to be the cue that the seedling has broken ground. An etiolated seedling, grown in the dark, fails to straighten its hypocotyl hook. In peas, a hook forms in the epicotyl, lifting the shoot tip out of the soil while the pea cotyledons remain in the ground. In monocots, the coleoptile pushes through the soil, and the shoot tip is protected as it grows up through the tubular sheath.

Germination of a plant seed is a critical and fragile stage in the life cycle. Only a small fraction of seedlings survive in the wild to produce seeds themselves. The enormous production of seeds and fruits provides ample material for natural selection to screen for the most successful genetic combinations, but it is a very expensive means of reproduction.

Asexual Reproduction

Natural Mechanisms of Vegetative Reproduction
Many plant species can clone themselves through asexual or *vegetative reproduction*—an extension of the indeterminate growth of plants, in which meristematic tissues can grow indefinitely and parenchyma cells can divide and differentiate into specialized cells. A common type of vegetative reproduction is *fragmentation*, the separation of a plant into parts that then form whole plants. In some dicot species, the root system gives rise to many adventitious shoots that develop into a clone with separate shoot systems. Some plants, such as dandelions, can produce seeds asexually, a process called *apomixis*.

Vegetative Propagation in Agriculture   Humans have developed many methods to vegetatively propagate trees, crops, and ornamental plants. New plants develop from stem cuttings when a *callus*, or mass of dividing cells, forms at the cut end of the shoot and adventitious roots develop from the callus.

Twigs or buds of one plant can be grafted onto a plant of a different variety or closely related species. The plant that provides the root system is called the *stock*, and the twig is called the *scion*. Grafting can combine the best qualities of different plants.

In test-tube cloning, whole plants can develop from pieces of tissue, called explants, or even from single parenchyma cells. A single plant can be cloned into thousands of plants by subdividing the undifferentiated calluses as they grow in tissue culture. Calluses sprout shoots and roots and develop into plantlets, which can be transferred to soil to develop. Some cultured explants develop into "somatic embryos," which can be packaged along with nutrients into gels to create artificial seeds. Genetic engineering is used to insert foreign DNA into individual plant cells, which then grow into complete plants by test-tube culture.

A technique called *protoplast fusion* is being coupled with tissue culture to create new plant varieties. Protoplasts are screened for agriculturally beneficial mutations and then cultured. In some cases, protoplasts from different species can be fused and cultured to form hybrid plantlets.

Genetic variability in agricultural crops has been eliminated by plant breeders so that plants grow at the same rate and fruits ripen in unison. Vegetative propagation is used to clone exceptional plants, and self-pollinating varieties are used when possible. However, *monoculture*, the cultivation of a single plant variety on large areas of land, produces a fragile ecosystem in which there is little genetic variability and little adaptability. "Gene banks" have been created in which plant breeders maintain seeds of many different plant varieties, so that new varieties can be developed if current ones fail.

A Comparison of Sexual and Asexual Reproduction in Plants: An Evolutionary Perspective

Sexual reproduction in plants generates variation in a population, an advantage when the environment changes. Sexual reproduction also produces seeds, a means of dispersal to new locations and dormancy during harsh conditions. By asexual reproduction, plants well-suited to a stable environment can clone exact copies. The progeny of vegetative propagation are usually not as frail as seedlings. Both modes of reproduction have been useful in the evolutionary adaptation of plant populations to their environments.

Some Cellular Aspects of Plant Development

Development includes all the processes that lead to specialized cells and tissues and a differentiated body.

An Overview of Plant Development Growth, morphogenesis, and cellular differentiation are the overlapping processes involved in development. *Growth* results from cell division and enlargement. *Morphogenesis*, the development of form, accounts for the formation of roots, shoots, and leaves. Unlike animals which move through their environment, plants grow though their environment and must retain embryonic shoot and root tip regions that allow for continuing growth and morphogenesis. *Cellular differentiation* creates a cell's specific structural and functional features.

The Roles of Cell Division and Cell Expansion in Morphogenesis The spatial orientations of cell division and expansion determine the shape of a plant organ. Unlike animal morphogenesis, in which cells migrate, plant morphogenesis depends on the oriented division and expansion of cells that are immobilized by their cell walls. The plane of cell division is determined by the *preprophase band*, a ring of cytoskeletal microtubules that forms during late interphase. The microtubules disperse before metaphase, but leave behind an ordered array of actin microfilaments that first orient the nucleus and later direct the movement of the vesicles that form the cell plate.

About 90% of plant cell growth is due to the uptake of water. The cell wall loosens when acids secreted by the cell break cross-links between cellulose microfibrils, and the hyperosmotic cell swells. The plant cell stops expanding when the wall again becomes rigid enough to offset the osmotic pressure of the cell. Unlike animal cell growth, which requires the synthesis of cytoplasm, this economical means of cell elongation allows for the rapid growth of shoots and roots.

Cells expand perpendicular to the orientation of cellulose microfibrils in the inner layers of the cell wall. The orientation of cellulose microfibrils parallels the orientation of microtubules on the other side of the plasma membrane. These microtubules may control the movement of cellulose-producing enzymes and thus determine the alignment of microfibrils in the wall.

Cellular Differentiation: The Basic Problem Differentiation arises from the synthesis of different proteins in different types of cells. The control of gene expression, and thus the regulation of transcription and translation of specific proteins, allows for the development of different structures and functions of cells, all of which share a common genome.

Pattern Formation and Positional Information *Pattern formation*—the characteristic arrangement of cells within tissues, of tissues within organs, and the development of organs in specific locations—depends on *positional information*, signals that indicate a cell's location and thus direct its differentiation. An embryonic cell may detect its location by gradients of *morphogens*, chemicals that diffuse from specific locations in a developing structure.

Clonal Analysis of the Shoot Apex Using *clonal analysis*, researchers are able to identify cells in the apical meristem and follow their lineages of cells to determine when the developmental fate of a cell becomes determined. Apparently the cells of the meristem are not committed to the formation of specific organs and tissues, but a cell's final position in a developing organ determines what type of cell it will become.

Homeotic Mutations in Flower Development: In Search of the Genetic Basis of Pattern Formation In the transition from a vegetative shoot tip to a floral meristem, growth changes from indeterminate to determinate. Some of the genes that respond to position information and determine what floral organ will develop have been identified. *Homeotic mutations* of these genes result in the placement of one type of organ where another type would normally develop. Wild-type alleles for these genes, called *organ-identity genes*, are presumed to be responsible for the development of normal floral pattern.

Organ-identity genes have been identified and cloned from *Arabidopsis thaliana*, the tiny wild mustard plant with the smallest known angiosperm genome. Similar genes have been isolated in snapdragon, suggesting that these development-controlling genes have been conserved in evolution. These genes code for transcription factors, which are regulatory proteins that bind to specific DNA sites and affect RNA synthesis. Positional information may determine which organ-identity gene is expressed. The transcription factor then influences the transcription of those genes that control development of a specific organ in its proper location.

STRUCTURE YOUR KNOWLEDGE

1. Create a diagram or concept map that sketches out the major events in the life cycle of angiosperms.

2. List the advantages and disadvantages of sexual and asexual reproduction in plants.

3. Identify the flower parts in the diagram shown below.

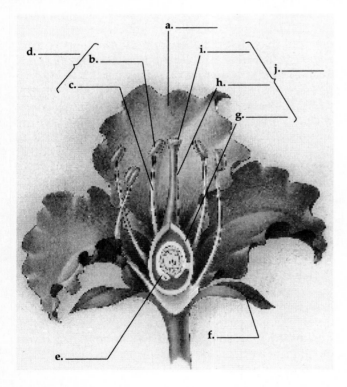

4. In the diagram of pollen germination, label the indicated structures.

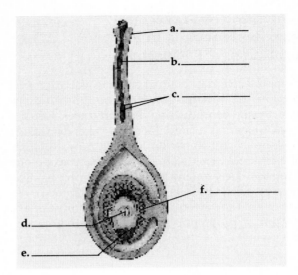

5. Indicate the location within the flower of each of the following events in the angiosperm life cycle.

   a. pollination takes place _____

   b. fertilization takes place _____

   c. meiosis takes place _____

   d. microspore produced _____

   e. megaspore develops _____
      into female gametophyte

TEST YOUR KNOWLEDGE

FILL IN THE BLANKS

1. _____ ring of microtubules that determine future plane of cell division

2. _____ generation that produces spores by meiosis

3. _____ species with male and female flowers on same plant

4. _____ female gametophyte of angiosperms

5. _____ embryonic root

6. _____ embryonic axis above attachment of cotyledon

7. _____ protects dicot shoot as it breaks through the soil

8. _____ twig or stem portion of a graft

9. _____ fruit formed from a flower with several separate carpels

10. _____ mass of dividing cells at cut end of a shoot

MULTIPLE CHOICE: *Choose the one best answer.*

1. A flower on a dioecious plant would be
   a. complete.
   b. perfect.
   c. imperfect.
   d. asexual.

2. Which of the following structures is haploid?
   a. embryo sac
   b. anther
   c. endosperm
   d. both a and b

3. The terminal cell in a zygote
   a. develops into the shoot apex of the embryo.
   b. forms the suspensor that anchors the embryo and transfers nutrients.
   c. develops into the endosperm when fertilized by a sperm nucleus.
   d. divides to form the proembryo.

4. A unique feature of fertilization in angiosperms is that
   a. one sperm fertilizes the egg and the other combines with a cell in the embryo sac and forms the endosperm.
   b. cross fertilization occurs because of mechanisms that prevent self-fertilization.
   c. pollen may be transferred by wind.
   d. a chemical attractant guides the sperm toward the egg.

5. The endosperm
   a. may be absorbed by the cotyledons in the seeds of dicots.
   b. is a triploid tissue.
   c. is digested by enzymes in monocot seeds following hydration.
   d. is all of the above.

6. In an angiosperm, meiotic cell division produces
   a. microspores.
   b. sperm nuclei.
   c. megasporocytes.
   d. all of the above.

7. Which structure protects a monocot shoot as it breaks through the soil?
   a. hypocotyl hook
   b. epicotyl hook
   c. coleoptile
   d. coleorhiza

8. Which of the following is a form of vegetative reproduction?
   a. apomixis
   b. grafting
   c. test-tube cloning
   d. all of the above

9. A disadvantage of monoculture is that
   a. the whole crop ripens at one time.
   b. since there is no genetic variability, the whole crop could be annihilated by a change in conditions, a new pest, or disease.
   c. it predominantly uses vegetative propagation.
   d. all of the above.

10. Protoplast fusion
    a. is used to develop gene banks to preserve genetic variability.
    b. is the method of test-tube cloning.
    c. can be used to form new plant species.
    d. occurs within a callus.

11. In the plant embryo, the suspensor
    a. directs mitotic divisions in the early cells.
    b. connects the early root and shoot apexes.
    c. gives rise to the cotyledons.
    d. is analogous to the umbilical cord in mammals.

12. Flower parts have evolved from modified
    a. leaves.
    b. branches.
    c. sporangia.
    d. sporophytes.

13. Self-incompatibility
    a. prevents cross-pollination.
    b. prevents vegetative reproduction.
    c. controls the timing of pollen maturation.
    d. insures genetic variability of offspring.

14. Morphogenesis in plants is a direct result of
    a. position effects.
    b. the plane of cell divisions and the direction of cell expansion.
    c. morphogens that create gradients in developing organs.
    d. organ-identity genes.

15. Clonal analysis of cells of the shoot apex indicates that
    a. homeotic mutations substitute one body part for another in the mutation's location.

b. a cell's developmental fate is more influenced by position effects than by its meristematic lineage.

c. cellular differentiation results from regulation of gene expression resulting in production of different proteins in different cells.

d. all cells derived from the same parent cell and resulted from vegetative reproduction.

CHAPTER 35

CONTROL SYSTEMS IN PLANTS

FRAMEWORK

Plant hormones—auxin, cytokinins, gibberellins, abscisic acid, and ethylene—control growth, development, movement, flowering, and senescence, as plants respond and adapt to their environments. Plant movements in response to environmental stimuli include phototropism, gravitropism, thigmotropism, and turgor movements. The biological clock of plants controls circadian rhythms, such as stomatal opening and sleep movements, and may use the phytochrome system to time night length in the photoperiodic control of flowering.

CHAPTER SUMMARY

Plants have various mechanisms that enable them to sense and adaptively respond to their environments, generally by altering their patterns of growth and development. These intricate control systems are the product of the evolutionary history of plants interacting with their environments.

The Search for a Plant Hormone: A Case Study of the Scientific Process

Hormones, chemical signals that coordinate the parts of an organism, are produced in one part of the body and translocated to other parts, where minute concentrations are able to trigger responses in target cells and tissues.

The growth of a shoot toward light is called positive *phototropism*. A coleoptile, enclosing the shoot of a grass seedling, grows straight when in the dark but bends toward the light when illuminated from one side. The cells on the darker side elongate faster and produce the bending.

Darwin and his son observed that a grass seedling would not bend toward light if its tip were removed or covered by an opaque cap. They postulated that some signal must be transmitted from the tip down to the elongating region of the coleoptile. Boysen-Jensen demonstrated that the signal was a mobile substance, capable of being transmitted through a block of gelatin separating the tip from the rest of the coleoptile.

In 1926, Went placed coleoptile tips on blocks of agar to extract the chemical messenger. Decapitated coleoptiles kept in the dark elongated and bent away from the side on which one of these agar blocks was placed. Went concluded that the chemical produced in the tip, which he called auxin, promoted growth and that it was in higher concentration on the side away from the light.

Functions of Plant Hormones

Five classes of plant hormones have been identified: *auxin, cytokinins, gibberellins, abscisic acid,* and *ethylene.* These hormones affect cell division, elongation, and differentiation. Depending on the site of action, the developmental stage of the plant, and relative hormone concentrations, the effect of a hormone will vary. Hormones are effective in very small concentrations, indicating that their signal must be amplified in the cell in some way. They may act by affecting the expression of genes, the activity of enzymes, or the properties of membranes.

Auxin Auxins are any substance, including synthetic compounds, that stimulate elongation of coleoptiles. The natural auxin extracted from plants is indoleacetic acid (IAA).

A major site of auxin synthesis is in the apical meristem of a shoot. Within a range of concentrations, auxin stimulates cell elongation. Above that concentration, auxin inhibits growth, probably because it induces the synthesis of ethylene, which inhibits cell elongation.

Auxin appears to be transported through parenchyma tissue in only one direction, from the shoot tip down the shoot. This *polar transport* involves a chemiosmotic mechanism of ATP-driven proton pumps that generate a membrane potential favoring the exit of auxin anions through specific carriers located only at the basal ends of cells. In the more acidic environment outside the cell, also created by proton pumps, auxin picks up a hydrogen ion and the now neutral auxin molecule can move across the plasma membrane into the next parenchyma cell.

According to the *acid-growth hypothesis*, auxin initiates shoot growth in the region of elongation by stimulating proton pumps. The proton pumps lower pH in the cell wall, breaking cross-links between cellulose microfibrils. The turgor pressure of the cell then exceeds the restraining pressure or the wall, and the cell elongates. For continued growth after this relatively fast elongation, the cell must produce more cytoplasm and wall material, processes also stimulated by auxin.

Auxin stimulates cell division in the vascular cambium, differentiation of secondary xylem, and formation of adventitious roots at the cut base of stems. Auxin produced by developing seeds promotes fruit growth; synthetic auxins induce seedless fruit development. The herbicide 2,4-D is a synthetic auxin used to disrupt the normal balance of plant growth in "broad leaf," dicot weeds.

Cytokinins Cytokinins are modified forms of adenine, named because they stimulate cytokinesis. In tissue culture, coconut milk and degraded DNA were found to induce plant cell growth; later cytokinins were identified as the active ingredients. Cytokinins are produced in actively growing roots, embryos, and fruits. Acting with auxin, they stimulate cell division and affect differentiation.

When stem tissue is grown in tissue culture in the absence of cytokinins, the cells grow large but do not divide. If cytokinin and auxin are added, the cells divide. The ratio of the two hormones controls differentiation of the cells: equal concentrations produce undifferentiated cells, more cytokinin results in shoot buds, whereas more auxin leads to root formation. Cytokinins stimulate RNA and protein synthesis, and these proteins may be involved in cell division.

The control of apical dominance involves the interaction between auxin, transported down from the terminal bud, which restrains axillary bud development, and cytokinins, transported up from the roots, which

stimulate bud growth. The interaction between auxin and cytokinins works the opposite way in the development of lateral roots. Together these hormones may coordinate the growth of shoot and root systems. Both hormones may regulate growth indirectly by changing the concentration of ethylene.

Perhaps because they stimulate RNA and protein synthesis, mobilize nutrients, and inhibit protein breakdown, cytokinins can retard aging of some plant organs.

Gibberellins In the 1930s, Japanese scientists determined that the fungus *Gibberella* secreted a chemical that caused the hyperelongation of rice stems or "foolish seedling disease." Over 70 different gibberellins have now been identified, many occurring naturally in plants.

Gibberellins, which are produced by roots and young leaves, stimulate growth in both leaves and stem, affecting cell elongation and cell division. The application of different concentrations of gibberellin to dwarf plants has demonstrated a positive correlation between growth and the concentration of hormone added. Experimental results such as these can be used in a bioassay to determine hormone concentration in a sample of unknown concentration. *Bolting*, the growth of an elongated floral stalk, is caused by a surge of gibberellins.

In some plants, both auxin and gibberellins contribute to fruit set. The commercial spraying of gibberellins is important in the production of Thompson seedless grapes.

In many plants, the release of gibberellins from the embryo signals the seed to break dormancy. The germination of grain is triggered when gibberellins stimulate the synthesis of messenger RNA that codes for the digestive enzymes that break down stored nutrients. Gibberellins are also involved in the breaking of dormancy of apical buds in spring.

Abscisic Acid The hormone abscisic acid (ABA), which is produced in the bud, slows growth, inhibits cell division in the vascular cambium, and induces leaf primordia to develop into scales that protect the dormant bud during winter.

Abscisic acid may act as a growth inhibitor of the embryo when the seed becomes dormant. ABA must be removed or inactivated, or the ratio of gibberellins to ABA must increase, for dormancy to be broken in some seeds. Abscisic acid also acts to help the plant cope with adverse conditions. In a wilting plant, ABA causes stomata in the leaves to close.

Ethylene Plants produce the gas ethylene, which acts as a hormone, promoting fruit ripening, inhibiting growth in roots and axillary buds and contributing to the aging, or *senescence*, of parts of the plant. Ethylene production is induced by a high concentration of auxin. Ethylene initiates or hastens the degradation of cell walls and the decrease in chlorophyll content that is associated with fruit ripening. Many commercial fruits are ripened in huge containers perfused with ethylene gas.

Deciduous leaf loss protects against winter dehydration. Before leaves abscise in the autumn, many of their compounds are stored in the stem awaiting recycling to new leaves. The leaf stops making chlorophyll; fall colors result from a combination of pigments that had been concealed by chlorophyll and new pigments that are made during autumn.

Shortening days and cooler temperatures are the stimuli for leaf abscission. A change in the balance of auxin and ethylene initiates changes in the abscission layer located near the base of the petiole, including the production of enzymes that hydrolyze polysaccharides in cell walls. A layer of cork forms a protective covering on the twig side of the abscission layer.

Plant Movements

Tropisms   Tropisms are growth responses in which a plant organ curves toward or away from a stimulus as a result of differential cell elongation on opposite sides of the organ. These movements may be in response to stimuli such as light, gravity, or touch.

In *phototropism*, cells on the darker side of a stem elongate faster in response to a larger concentration of auxin moving down from the shoot tip. In a mechanism not yet understood, light causes auxin to migrate laterally across the tip toward the dark side. The photoreceptor is believed to be a blue-light sensitive pigment that may also be involved in stomatal opening.

Roots exhibit positive *gravitropism*, whereas shoots show negative gravitropism. The settling of *statoliths*, plastids containing dense starch grains, in cells of the root cap triggers movement of calcium and the lateral transport of auxin. Both of these accumulate on the lower side of the growing root and inhibit cell elongation, causing the root to curve downward.

Most climbing plants have tendrils that grow in coils in response to contact. The directional growth in response to touch is *thigmotropism*. The stunting of growth in height and increase in girth of plants that are exposed to wind or mechanical stimulation is called *thigmomorphogenesis*.

Turgor Movements   In addition to the longer lasting changes in body shape resulting from tropisms, plants can make reversible movements caused by changes in turgor pressure of specialized cells in response to stimuli.

The sensitive plant *Mimosa* folds its leaves after being touched due to the rapid loss of turgor by cells in specialized motor organs called pulvini, located at the joints of the leaf. These cells lose potassium when stimulated, resulting in osmotic water loss. The message travels through the plant from the point of stimulation, perhaps as the result of chemical messengers and electrical impulses, called *action potentials*. These electrical messages may be used in plants as a form of internal communication.

Many members of the legume family exhibit *sleep movements*, the daily raising and lowering of leaves caused by changes in the turgor pressure of motor cells in pulvini. Massive migration of potassium ions from one side of the pulvinus to the other causes these reversible osmotic changes in the motor cells.

Circadian Rhythms and the Biological Clock

Many physiological processes in animals, fungi, protists, and plants fluctuate with the time of day and are controlled by biological clocks—internal oscillators that keep accurate time. A *circadian rhythm* is a physiological cycle with about a 24-hour frequency. These rhythms persist, even when the organism is sheltered from environmental cues. Research indicates that the oscillator for circadian rhythms is endogenous, although the clock is set (entrained) to a precise 24-hour period by daily environmental signals.

Free-running periods, determined in the absence of environmental cues, vary from 21 to 27 hours. The biological clocks in free-running periods still keep perfect time, but they are not synchronized with the outside world. Internal clocks are reset when an organism is once again exposed to normal environment cues.

The nature of the oscillator of the biological clock is still unknown, but it is most likely regulated at the cellular level, perhaps in membranes or protein-synthesizing machinery.

Photoperiodism

Seasonal events in the life cycle of plants usually are cued by photoperiod, the relative length of night and day. A physiological response to day length is called *photoperiodism*.

Photoperiodic Control of Flowering   Garner and Allard discovered that a variety of tobacco plant flowered only when the day length was 14 hours or shorter. They termed it a *short-day plant*, because it seemed to need a light period shorter than a critical length to flower. *Long-day plants* flower when days are longer than a certain number of hours, and the flowering of *day-neutral plants* is unaffected by photoperiod.

In the 1940s, researchers found that it is night length, not day length, that controls flowering and other photoperiod responses. If the dark period is interrupted by even a few minutes of light, a short-day (long-night) plant such as the cocklebur will not flower. Photoperiodic responses thus depend on a critical night length. Short-day plants require a minimum number of hours of uninterrupted darkness, and long-day plants will flower only if they receive less than a maximum number of hours of darkness.

Some plants bloom after a single exposure to the required photoperiod. Others respond to photoperiod only after exposure to another environmental stimulus. The need for pretreatment with cold before flowering is called *vernalization*.

Leaves detect the photoperiod. In some species, exposure of a single leaf to the proper photoperiod will induce flowering. A hormone, named florigen, is believed to be the signal for flowering that travels from leaves to buds, but it has yet to be identified and may be a mixture of several hormones.

Phytochrome   Red light is the most effective in interrupting a plant's perception of night length. A brief exposure to red light breaks a dark period of sufficient length and prevents short-day plants from flowering, whereas a flash of red light during a dark period longer than the critical length will induce flowering in a long-day plant. A subsequent flash of light from the far-red part of the spectrum negates the effect of the red light.

The photoreceptor pigment *phytochrome* alternates between two forms, one of which absorbs red light and the other, far-red light. These two variations of phytochrome are photoreversible; the P_r to P_{fr} interconversion acts as a switch, controlling various events in the life of the plant.

Phytochrome tells the plant that light is present by the conversion of P_r, which the plant synthesizes, to P_{fr} in the presence of sunlight. P_{fr} triggers many plant responses to light, such as breaking seed dormancy. The amount and type of available light, specifically the relative amounts of red and far-red light, is communicated to a plant by the ratio of the two forms of phytochrome.

Role of the Biological Clock in Photoperiodism   In darkness, the phytochrome ratio shifts toward P_r, in part because P_{fr} is converted to P_r and also because P_{fr}

is degraded and new pigment is synthesized as P_r. The conversion to P_r is complete within a few hours of darkness; therefore, the plant cannot use the phytochrome ratio to measure nightlength. It is measured by the biological clock. The role of phytochrome may be to synchronize the clock by signalling when the sun sets and rises.

Signal-Transduction Pathways in Plant Cells

A *signal-transduction pathway*, which links stimuli to cell response, consists of three main steps: reception, transduction, and induction. Reception may be the absorption of a specific wavelength of light by a pigment molecule, as in the photoconversion of phytochrome, or the binding of a hormone to a specific receptor. *Target cells* have receptors for a particular hormone and are thus the only cells able to respond to that hormone message.

The transduction step amplifies the stimulus. A *second messenger* is a substance that increases in concentration in response to the hormonal stimulus and thus serves to amplify the environmental message. Calcium ions (Ca^{2+}) appear to be common second messengers, which increase in concentration in response to a signal reception and bind to the protein *calmodulin*. This complex then activates other cellular proteins and molecules. The specificity of hormone function is due to the types of proteins present in a cell that are activated by second messengers.

The amplified signal now induces a specific cell response. Relatively rapid responses include stomatal closing and cell elongation resulting from cell wall acidification. Slower responses to a signal-transduction pathway involve changes in gene expression, as in the increased transcription of the five genes activated in response to mechanical touch. Thus hormonal and environmental signals are received by specific receptors and then transduced to an amplified chemical message, which then induces specific responses in a cell.

STRUCTURE YOUR KNOWLEDGE

1. The movements of plants allow them to respond and adapt to their environments. Describe examples of tropisms and turgor movements and the mechanisms involved in these plant responses.

2. In the figure on p. 297, indicate whether a short-day plant and a long-day plant would flower under each of the light conditions shown.

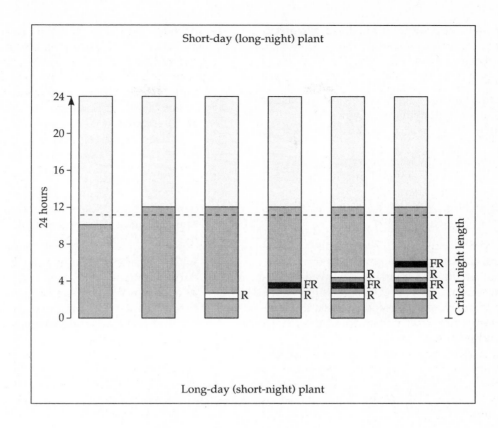

3. Develop a concept map to illustrate your understanding of the photoperiodic control of flowering. Do not forget to include the role of the biological clock.

4. Explain the three steps in the signal-transduction pathway that link stimulus to response in plant cells.

TEST YOUR KNOWLEDGE

TRUE or FALSE: *Indicate T or F and then correct the false statements.*

1. _____ Application of 2,4-D, a synthetic auxin, disrupts the growth of "broad-leaf" dicots.

2. _____ Shoots exhibit positive phototropism and negative gravitropism.

3. _____ Gibberellins, synthesized in the root, counteract apical dominance.

4. _____ Bolting occurs when ethylene stimulates the degradation of cell walls and decreases chlorophyll content.

5. _____ The application of gibberellins may induce the development of seedless fruit.

6. _____ Abscisic acid induces the synthesis of enzymes that hydrolyze polysaccharides in cell walls of the abscission layer.

7. _____ Action potentials are electrical impulses that may be used as internal communication in plants.

8. _____ Thigmotropism is involved in the rapid movements of *Mimosa* leaves.

9. _____ A physiological response to day or night length is called a circadian rhythm.

10. _____ Vernalization is the need for pretreatment with cold before flowering.

MULTIPLE CHOICE: *Choose the one best answer.*

1. The body form of plants within a species may vary more than that of animals within a species because
   a. growth in animals is indeterminate.
   b. plants respond adaptively to their environments by altering their patterns of growth and development.

c. plant growth and development are governed by many hormones.
d. of all of the above.

2. Polar transport of auxin involves
   a. the accumulation of higher concentrations on the side of a shoot away from light.
   b. the reverse movement of auxin from roots to shoot.
   c. movement of auxin ions through carrier proteins located at the basal end of cells, facilitated by the membrane potential.
   d. the unidirectional active transport of auxin into and out of parenchyma cells.

3. According to the acid-growth hypothesis,
   a. auxin stimulates membrane proton pumps.
   b. a lowered pH breaks cross-links between cellulose microfibrils.
   c. the turgor pressure of the cell exceeds the lowered restraining wall pressure, and the cell swells and takes up water.
   d. all of the above are involved in cell elongation.

4. Which of the following situations would stimulate the development of axillary buds?
   a. a large quantity of auxin traveling down from the shoot and a small amount of cytokinin produced by the roots
   b. a small amount of auxin traveling down from the shoot and a large amount of cytokinin traveling up from the roots
   c. an equal ratio of cytokinins to auxin
   d. the absence of cytokinins caused by the removal of the terminal bud

5. The growth inhibitor in seeds is usually
   a. abscisic acid.
   b. ethylene.
   c. gibberellin.
   d. a small amount of ABA combined with a larger ratio of gibberellins.

6. A circadian rhythm
   a. is controlled by an internal oscillator.
   b. is a physiological cycle of approximately a 24-hour frequency.
   c. involves a biological clock that is set by daily environmental signals.
   d. involves all of the above.

7. A bioassay
   a. uses chemical means to measure the activity of biological compounds.
   b. measures the response of living systems to determine the concentration of a biologically active chemical compound.
   c. determines the components present in a mixture of biologically active compounds.
   d. measures the rate of growth of dwarf plants.

8. In order to flower, a short-day plant needs
   a. a burst of red light in the middle of the night.
   b. a night that does not exceed a certain length.
   c. a night that does exceed a minimal length.
   d. a higher ratio of P_r:P_{fr}.

9. A flash of far-red light during a critical-length dark period
   a. will induce flowering in a long-day plant.
   b. will induce flowering in a short-day plant.
   c. will not influence flowering.
   d. will increase the P_{fr} level suddenly.

10. Evidence indicates that
    a. a hormone that travels from leaves to buds induces flowering.
    b. flowering is induced by phytochrome levels in the leaves.
    c. short-day plants need vernalization before photoperiod affects them.
    d. a biological clock in the bud determines photoperiod.

11. The function of calmodulin in a signal-transduction pathway is
    a. to activate other second messengers in the transduction step.
    b. as a steroid hormone that induces the selective activation of genes.
    c. as a membrane-bound hormone receptor that causes an influx of Ca^{2+}.
    d. to form a complex with Ca^{2+} and activate specific proteins and cellular molecules.

12. The specific response to a hormone is a result of
    a. the fact that only target cells have receptors for a particular hormone.
    b. different target proteins within different cells that are activated by a second messenger.
    c. the relative concentration of a hormone in combination with other hormones in a cell.
    d. all of the above.

ANSWER SECTION

CHAPTER 31
PLANT STRUCTURE AND GROWTH

Suggested Answers to Structure Your Knowledge

1.

FUNCTION	ROOT	STEM	LEAF
Anchorage	Tap or fibrous root system Adventitious roots	Adventitious roots	
Absorption	Root hairs on epidermis		
Storage	Parenchyma cells of pith and cortex	Parenchyma cells, may have rhizomes and tubers	Some modified leaves of bulbs
Support	Secondary growth of woody roots	Collenchyma, schlerenchyma, xylem tracheids	Veins, xylem
Protection	Epidermis, root cap, periderm of woody plants	Epidermis with cuticle; bark; periderm (cork and cork cambium) and phloem	Epidermis with cuticle
Transport	Central stele with xylem and phloem	Vascular bundles: xylem and phloem	Veins and xylem and phloem
Photosynthesis		Some parenchyma cells with chloroplasts	Mesophyll (spongy and pallisade)
Growth	Apical meristem forms protoderm, procambium, ground meristem; vascular and cork cambiums	Apical meristem in terminal and axillary buds; intercalary meristem in grasses; vascular and cork cambiums in dicots	Develop from leaf primordia at nodes
Gas exchange	Epidermis; air tubes in swamp trees	Lenticels	Stomata with guard cells

2. Primary growth results in growth at the tips of roots and stems, whereas secondary growth produces greater girth. Apical meristems produce new cells that form three types of primary meristems: protoderm, procambium, and ground meristem. These cells elongate and differentiate into the dermal, vascular, and ground tissue systems.

Secondary growth involves lateral meristems: a vascular cambium, which produces new xylem and phloem cells, and a cork cambium, which produces cork cells that form a waxy protective covering.

3. This micrograph is a dicot stem, indicated by the ring of vascular bundles and the central pith.
a. xylem
b. phloem
c. epidermis
d. vascular bundle
e. pith
f. cortex
g. pith ray

299

4. Structures in the cross section of a leaf:
   a. cuticle
   b. upper epidermis
   c. pallisade parenchyma
   d. spongy mesophyll
   e. vein
   f. stoma
   g. guard cells
   h. xylem
   i. phloem

Answers to Test Your Knowledge

Matching:

1. L	5. I	9. K
2. F	6. H	10. G
3. B	7. C	
4. E	8. D	

Multiple Choice:

1. a	5. b	9. a
2. a	6. a	10. c
3. d	7. b	
4. c	8. c	

Fill in the Blanks:

H   F   A   B   G   D   C   E

CHAPTER 32
TRANSPORT IN PLANTS

Suggested Answers to Structure Your Knowledge

1. The transpiration-cohesion-tension mechanism explains the ascent of xylem sap. The lower water potential of the air surrounding the leaves compared to that of the cortex cells of the root and the soil solution results in the passive movement of water down its water potential gradient, against gravity, in a continuous column of water in xylem vessels. The cohesion of water molecules transmits the pull resulting from transpiration throughout the column, and the adhesion of water to the hydrophilic walls of the xylem vessels aids the flow. Tension created within the xylem by the upward pull also helps the flow of water. Root pressure may contribute to xylem movement in some plants.

A pressure-flow mechanism explains the movement of phloem sap. The active accumulation of sugars in sieve-tube members greatly decreases water potential at the source end, resulting in an inflow of water. The removal of sugar from the sink end of a phloem tube results in the osmotic loss of water from the cell. The difference in hydrostatic pressure between the source and sink end of the phloem tube causes the bulk flow of water with the accompanying transport of sugar.

2.
A. apoplastic	E. cortex
B. symplastic	F. endodermis
C. root hair	G. Casparian strip
D. epidermis	H. stele or xylem vessel

Answers to Test Your Knowledge

1. b	5. c	9. a	13. a
2. d	6. b	10. d	14. c
3. d	7. d	11. c	15. b
4. a	8. d	12. c	16. d

CHAPTER 33
PLANT NUTRITION

Suggested Answers to Structure Your Knowledge

1.

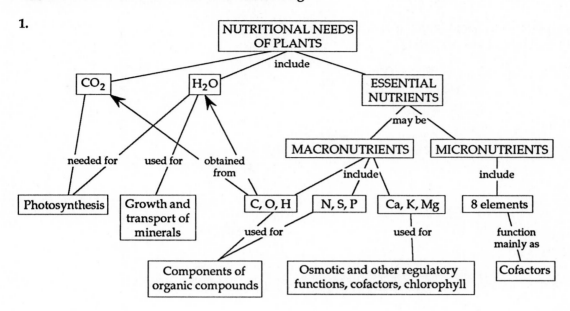

2. The key properties of a fertile soil include:
- *texture*: mixture of sand, silt and clay so that air spaces containing oxygen are provided as well as sufficient surface area for the binding of water and minerals
- *humus*: prevents clay from packing so soil retains water and has air spaces, provides reservoir of mineral nutrients
- *minerals*: adequate supply of nitrogen, phosphorus and potassium as well as other minerals
- *pH*: proper pH level so that minerals are in a form that can be absorbed and cation exchange can take place

3. Nitrogen must be in the form of nitrate or ammonium in order to be absorbed by plants. Nitrogen-fixing bacteria (using the enzyme nitrogenase) reduce N_2 to form ammonia (NH_3), which converts to ammonium (NH^{4+}) in the soil solution.

Other bacteria in the soil oxidize ammonium into nitrate, the form in which plants acquire most of their nitrogen. Microbes decomposing humus also contribute to the ammonium content of the soil. Legumes often have root nodules that house nitrogen-fixing bacteria of the genus *Rhizobium*. Nitrogen is essential to plant growth because it is needed for protein formation. Productive plant growth and high protein content of crops are important in agriculture.

Answers to Test Your Knowledge

1. b	**5.** a	**9.** b
2. a	**6.** c	**10.** c
3. c	**7.** d	**11.** c
4. b	**8.** d	**12.** b

CHAPTER 34
PLANT REPRODUCTION AND DEVELOPMENT

Suggested Answers to Structure Your Knowledge

1. All structures, except the two circled with the dotted line, belong to the sporophyte generation.

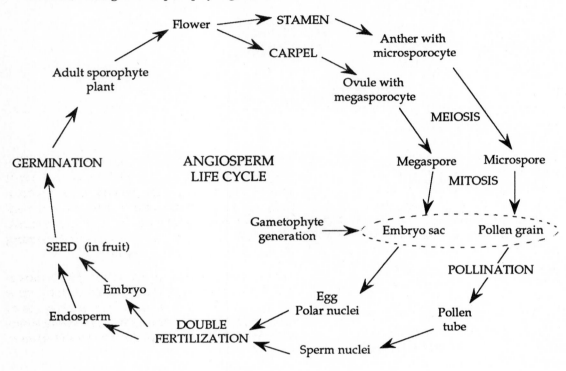

2. *Sexual:* Advantages include increased genetic variability, which provides the potential to adapt to changing conditions, and the dispersal and dormancy capabilities provided by seeds. Disadvantages are that the seedling stage is very vulnerable, sexual reproduction is very energy-intensive, and that genetic recombination may separate adaptive traits.

   Asexual: Advantages include the hardiness of vegetative propagation and the maintenance of well-adapted plants in an environment. Disadvantages relate to the advantages of sexual reproduction: there is no genetic variability from which to choose should conditions change, and there are no seeds for dispersal.

3. a. petal   f. sepal
   b. anther   g. ovary
   c. filament   h. style
   d. stamen   i. stigma
   e. ovule   j. carpel

4. a. stigma   d. polar nuclei
   b. pollen tube   e. egg
   c. two sperm nuclei f. embryo sac

5. a. pollination takes place on the stigma
   b. fertilization takes place within the embryo sac in the ovule
   c. meiosis takes place in the pollen sacs in the anther and in the ovule
   d. microspore produced in the pollen sacs in the anther
   e. megaspore develops into female gametophyte in embryo sac within ovule

Answers to Test Your Knowledge

Fill in the Blanks:

1. preprophase band   4. embryo sac
2. sporophyte   5. radicle
3. monoecious   6. epicotyl

7. hypocotyl or epicotyl hook
8. scion

9. aggregate
10. callus

Multiple Choice:

1. c	5. d	9. b	13. d
2. a	6. a	10. c	14. b
3. d	7. c	11. d	15. b
4. a	8. d	12. a	

CHAPTER 35
CONTROL SYSTEMS IN PLANTS

Suggested Answers to Structure Your Knowledge

1. Tropisms involve growth responses to environmental stimuli such as light, gravity, and touch that are mediated by hormones. In phototropism, auxin moves to the side of a shoot away from light and, according to the acid-growth hypothesis, activates proton pumps which increase the acidity of the cell wall, breaking cross-links in the wall and allowing the cell to take up more water and elongate. Cell elongation on the side away from the light causes the shoot to bend and grow toward light. In gravitropism, specialized plastids called statoliths settle to the lower side of root cap cells and initiate the accumulation of calcium and the lateral movement of auxin to the lower side of a growing root. The growth of these cells is then inhibited, and the root bends and grows toward gravity. In thigmotropism, mechanical stimulation is detected by a plant and responses may include the coiling of tendrils or the thickening of stems exposed to strong winds, called thigmorphogenesis.

Turgor movements are reversible plant movements that result from potassium-mediated turgor changes in specialized cells. Turgor changes of motor cells in pulvini cause the rapid leaf responses of *Mimosa* and the sleep movements of legumes.

2.

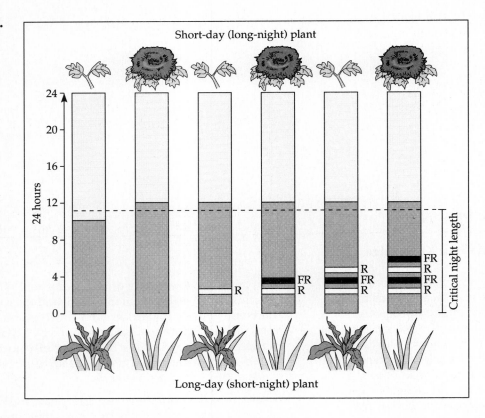

Short-day (long-night) plant

Critical night length

Long-day (short-night) plant

3.

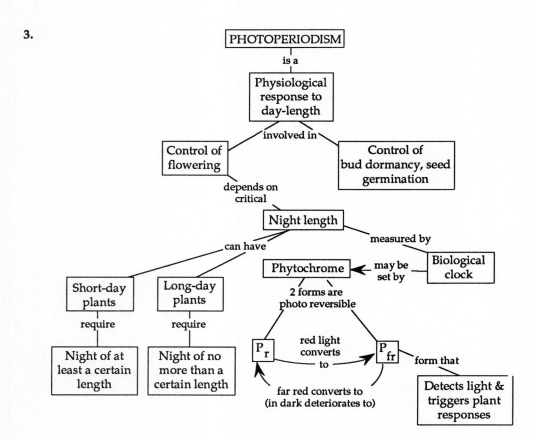

PHOTOPERIODISM

is a

Physiological response to day-length

involved in

Control of flowering

Control of bud dormancy, seed germination

depends on critical

Night length

measured by

Biological clock

can have

Phytochrome

may be set by

Short-day plants

Long-day plants

2 forms are photo reversible

require

require

Night of at least a certain length

Night of no more than a certain length

P_r

red light converts to

P_{fr}

form that

far red converts to (in dark deteriorates to)

Detects light & triggers plant responses

4. In the signal-transduction pathway, a signal is received when a hormone binds to a specific receptor molecule or when a pigment molecule absorbs a certain wavelength of light. The transduction of the signal involves the release of a second messenger, often calcium ions, that serve as chemical amplifiers of the signal. Ca^{2+} binds with calmodulin, and this complex then activates various proteins or other molecules within the cell. These activated molecules then induce specific changes within a cell, which may include fairly rapid responses or alterations of gene expression that may produce longer-term changes. This cascade of events allows a minute concentration of a hormone to produce widespread cellular response.

Answers to Test Your Knowledge

True or False:

1. true
2. true
3. false: change *gibberellins* to *cytokinins*
4. false: change *bolting* to *fruit ripening*; or change to read *Bolting is gibberellin-induced elongation of a flower stalk.*
5. true
6. false: change *ABA* to *ethylene*
7. true
8. false: change *thigmotropism* to *turgor movements*; or change to read *Thigmotropism is the change in the growth of a plant in response to touch.*
9. false: change *circadian rhythm* to *photoperiodism*
10. true

Multiple Choice:

1. b	5. a	9. c
2. c	6. d	10. a
3. d	7. b	11. d
4. b	8. c	12. d

ANIMALS: FORM AND FUNCTION

Testing whether or not animals "kiss."

CHAPTER 36

INTRODUCTION TO ANIMAL STRUCTURE AND FUNCTION

FRAMEWORK

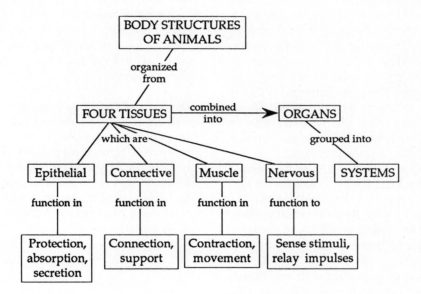

CHAPTER SUMMARY

The comparative study of animals illustrates two general themes: the correlation of structure and function, and the capacity of life to adapt to its environment, both by short-term physiological adjustments and long-term evolutionary changes.

Levels of Structural Organization

A hierarchy of structural order characterizes life. Unicellular protozoa are able to carry out all of their life functions without integration beyond the cellular level, but multicellular organisms have specialized cells grouped into tissues. In most animals, different types of tissues are combined into structural and functional units called organs, and various organs may cooperate in organ systems.

Animal Tissues Tissues are collections of cells with a common structure and function, held together by sticky coatings on the cells or woven together in a fabric of extracellular fibers. Histologists—biologists who study tissues—classify tissues into four categories.

Epithelial tissue lines the outer and inner surfaces of the body in protective sheets of tightly packed cells. Cells at the base of an epithelium are attached to a *basement membrane*, a dense layer of extracellular material. Epithelia are classified based on the number of cell layers and cell shape. *Simple epithelium* has one layer, whereas *stratified epithelium* has multiple layers of cells. The shape of cells at the free surface may be *squamous* (flat), *cuboidal* (boxlike), or *columnar* (pillarlike).

Epithelia may be specialized for absorption or secretion. *Mucous membranes* in the gut and the air passages secrete mucus. Particles are trapped in mucus and moved back up the trachea away from the lungs by cilia on the epithelium.

Connective tissue connects and supports other tissues and is characterized by having relatively few cells suspended in an extracellular *matrix* of fibers, which may be embedded in a liquid, jellylike or solid ground substance.

Loose connective tissue attaches epithelia to underlying tissues and holds organs in place. Its loosely woven protein fibers are of three types. *Collagenous fibers* are made of collagen, the most abundant animal protein. Three collagen molecules coil to form a fibril and several fibrils make up a collagenous fiber, a structure with great tensile strength that resists stretching. *Elastic fibers*, made of the protein elastin, provide loose connective tissue with resilience. Branched *reticular fibers* form a tightly woven connection with adjacent tissues. The most common types of cells enmeshed in loose connective tissue are *fibroblasts*, which secrete the protein of the extracellular fibers, and *macrophages*, amoeboid cells that engulf bacteria and cellular debris by phagocytosis.

Adipose tissue is a special form of loose connective tissue that pads and insulates the body and stores fuel reserves. Adipose cells contain a large fat droplet.

Fibrous connective tissue, with its dense arrangement of parallel collagenous fibers, is found in *tendons*, which attach muscles to bones, and in *ligaments*, which join bones at joints.

Cartilage is composed of collagenous fibers embedded in a rubbery ground substance called *chondrin*. This material is secreted by *chondrocytes* found in scattered lacunae, or spaces, in the ground substance. Cartilage is a strong but somewhat flexible support material, making up the skeleton of sharks and vertebrate embryos. We retain cartilage in some skeletal areas.

Bone is a mineralized connective tissue formed by *osteocytes* that deposit a matrix of collagen and calcium phosphate, which hardens into hydroxyapatite. *Haversian systems* consist of concentric layers of matrix deposited around a central canal containing blood vessels and nerves. Osteocytes are located in lacunae within the matrix and are connected to one another by thin cellular extensions. In long bones, the hard outer region is compact bone built of repeating Haversian systems, whereas the interior is a spongy bone tissue, called marrow. Red marrow, near the ends of long bones, manufactures blood cells.

Blood is a connective tissue that has a liquid extracellular matrix called plasma, containing water, salts, and dissolved proteins. Erythrocytes (red blood cells) carry oxygen, leukocytes (white blood cells) function in defense, and platelets are involved in the clotting of blood.

Muscle tissue consists of long, contractile cells that are packed with microfilaments of actin and myosin. Three types of muscle tissues are found in vertebrate bodies. *Skeletal muscle*—also called striated muscle because it looks striped due to the arrangement of overlapping filaments—is responsible for voluntary body movements and is the only type of muscle under conscious control. *Cardiac muscle*, forming the wall of the heart, is also striated, but its cells are branched, joined at their ends by intercalated discs that rapidly relay electrical impulses. *Visceral muscle*, also called smooth muscle, is composed of spindle-shaped cells lacking striations. It is found in the walls of the digestive tract, arteries, and other internal organs.

Nervous tissue senses stimuli and transmits signals. The *neuron*, or nerve cell, consists of a cell body and two or more nerve processes that conduct impulses toward (dendrites) and away from (axons) the cell body.

Organs and Organ Systems In all animals but sponges and cnidarians, tissues are organized into specialized units of function called *organs*. Organs often consist of a layered arrangement of tissues. Many vertebrate organs are suspended by *mesenteries* in fluid-filled body cavities. Mammals have a *thoracic cavity* separated by a muscular diaphragm from an *abdominal cavity*.

Groups of organs are integrated into *organ systems*, which perform the major functions required for life. The organ systems are coordinated to create the functional integration needed by an organism.

Size, Shape, and the External Environment

Every cell must be in an aqueous medium to allow for oxygen, nutrient, and waste exchange across its plasma membrane. Cell size is limited by the need for a sufficient surface-to-volume ratio. Single-celled organisms, or animals with two-layered saclike bodies or thin flat bodies, can maintain sufficient cellular contact with the environment.

Animals with compact bodies, however, must provide extensive moist internal membranes for exchanging materials with the environment. The surface area of the air chambers of the human lung is about 100 m^2. The millions of tubules forming the internal surface of the kidney are specialized to filter wastes from blood, and the extensive, convoluted lining of the digestive tract absorbs nutrients. The circulatory system connects these exchange surfaces with the aqueous environment bathing the body's cells.

The Animal's Internal Environment

The internal environment of vertebrates is the *interstitial fluid* surrounding the cells, through which oxygen, nutrients, and wastes are exchanged. The ability of an organism to control the "steady state" or constancy of this environment is called *homeostasis*. The mechanisms by which animals maintain homeostasis involve a receptor that detects a change in the internal environment and a control center that processes information and directs an effector to respond. Most homeostatic control mechanisms operate by *negative feedback*. When some variable moves above or below a *set point*, a control mechanism is turned on or off to return the condition to normal. One of the negative feedback mechanisms controlling body temperature in humans involves the sweat glands, which are signaled to increase their activity when the hypothalamus senses a rise in body temperature. When body temperature falls below the set point, the hypothalamus stops sending "sweat" signals.

Positive feedback in a physiological function is a mechanism in which a change in a variable serves to amplify rather than reverse the activity. The stimulation of uterine contractions during childbirth is an example.

Homeostatic systems do not necessarily maintain a constant steady state. Regulated changes in the internal environment allow for adaptive response to changes.

STRUCTURE YOUR KNOWLEDGE

1. Fill in the table below on the structure and function of the four types of animal tissues.

2. Organs usually are composed of layers of several different tissues. Which of the four major animal tissues do you think would be included in all organs? In what types of organs would you predict the remaining tissues would be included?

3. How do compact animals deal with the need for gas, nutrient, and waste exchange between each of their cells and the environment?

4. Identify the following tissues as completely as you can.

a. _____

b. _____

TISSUE	STRUCTURAL CHARACTERISTICS	GENERAL FUNCTIONS	STRUCTURE AND FUNCTION OF SPECIFIC TYPES

c. _____

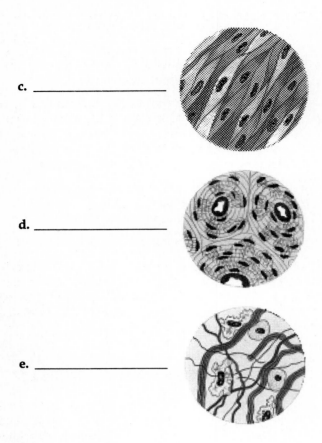

d. _____

e. _____

TEST YOUR KNOWLEDGE

MULTIPLE CHOICE: *Choose the one best answer.*

1. Which of the following is *not* an organ system?
   a. skeletal
   b. connective
   c. digestive
   d. excretory

2. A stratified squamous epithelium would be composed of
   a. several layers of flat cells attached to a basement membrane.
   b. a layer of ciliated, mucus-secreting flattened cells.
   c. a hierarchical arrangement of boxlike cells.
   d. an irregularly arranged layer of pillarlike cells.

3. Which of the following is *not* true of connective tissue?
   a. It consists of few cells surrounded by fibers and a ground substance.
   b. It includes such diverse tissues as bone, cartilage, tendons, adipose, and loose connective tissue.
   c. It connects and supports other tissues.
   d. It forms the internal and external lining of many organs.

4. Which of the following are incorrectly paired?
   a. fibrous connective tissue—chondrocytes embedded in chondrin
   b. bone—osteocytes embedded in hydroxyapatite in Haversian systems
   c. loose connective tissue—collagenous, elastic, reticular fibers
   d. adipose tissue—loose connective tissue with fat-storing cells

5. The best description of visceral muscle is
   a. striated, branching cells, under involuntary control.
   b. spindle-shaped cells, under involuntary control.
   c. spindle-shaped cells connected by intercalated discs.
   d. striated cells containing overlapping filaments, under involuntary control.

6. The diaphragm
   a. is a mesentery.
   b. increases the surface area of the lungs.
   c. is part of the mammalian reproductive system.
   d. separates the thoracic and abdominal cavities in mammals.

7. The interstitial fluid of animals
   a. is the internal environment within cells.
   b. bathes cells and provides for the exchange of nutrients and wastes.
   c. is composed of blood.
   d. is found at the major exchange areas in lungs, kidneys, and intestine.

8. Negative feedback circuits are
   a. mechanisms that maintain homeostasis.
   b. activated when a physiological variable deviates from a set point.
   c. analogous to a thermostat that controls room temperature.
   d. all of the above.

9. Collagen may be the most prevalent protein found in vertebrates because
   a. muscles make up most of the mass of an animal's body.
   b. it is the protein found in most types of connective tissue.
   c. it is the most common protein of blood plasma.
   d. it is an important component of both cartilage and bone.

CHAPTER 37

ANIMAL NUTRITION

FRAMEWORK

Animals eat other organisms to obtain fuel for respiration, organic raw materials, and nutrients such as essential amino acids, vitamins, and minerals. Digestion is the enzymatic hydrolysis of macromolecules into monomers that can be absorbed across cell membranes.

Gastrovascular cavities, found in cnidarians and flatworms, are digestive sacs in which some extracellular digestion takes place before food particles are phagocytized by cells lining the cavity. The more complex animal groups have alimentary canals—one-way tracts with specialized regions for mechanical breakdown of food, storage, digestion, absorption of nutrients, and elimination of wastes.

This chapter details the structures, functions, enzymes, and hormones of the human digestive tract.

CHAPTER SUMMARY

Animals, as heterotrophs, obtain energy and raw materials from the organic compounds in their food. In the long span of evolutionary history, animals have developed diverse mechanisms for obtaining and processing their food.

Food Types and Feeding Mechanisms

Most animals are holotrophs, ingesting plants or animals either whole or in large pieces. *Herbivores* eat plants; *carnivores* eat animals; and *omnivores* consume both plants and animals.

Many aquatic animals are *suspension feeders*, sifting small food particles from the water. *Substrate feeders*

live in or on their food, eating their way through it. *Deposit feeders* salvage pieces of decaying organic matter in detritus. *Fluid feeders* suck fluids from a living plant or animal host. Most animals are *bulk feeders* that eat relatively large pieces of food.

Digestion: A Comparative Introduction

Animals must digest their food. The proteins, fats, and carbohydrates that make up food are too large to pass through cell membranes, and these macromolecules must be reorganized to form the particular macromolecules that make up each heterotroph. *Digestion* breaks apart food particles and splits macromolecules into monomers: polysaccharides into simple sugars, fats into glycerol and fatty acids, proteins into amino acids, and nucleic acids into nucleotides.

Enzymatic Hydrolysis Macromolecules are broken down by *hydrolysis*, the enzymatic addition of a water molecule when the bond between monomers is broken. Hydrogen is added to one side of the bond and a hydroxyl group to the other. Hydrolysis occurs in a specialized compartment to protect the animal's cells from its own hydrolytic enzymes. Following digestion, monomers and small molecules are absorbed into the cells of the animal and the undigested remainder of the food is eliminated.

Intracellular Digestion in Food Vacuoles Large food molecules may be taken into cells by endocytosis and digested in food vacuoles. This *intracellular digestion* is typical of protozoa and sponges. Most animals, however, at least initially break down their food by *extracellular digestion* within a separate compartment of the body.

Digestion in Gastrovascular Cavities The simplest animals have single-opening *gastrovascular cavities*, which function in both digestion and transport of nutrients throughout the body. In the small cnidarian *Hydra*, digestive enzymes secreted into the gastrovascular cavity initiate food breakdown. Gastrodermal cells take in food particles by phagocytosis, and hydrolysis of macromolecules occurs by intracellular digestion within food vacuoles. Undigested materials are expelled through the mouth. Flatworms, such as planaria, also have gastrovascular cavities in which digestion is begun extracellularly and continues within cells that take up food particles by phagocytosis.

Digestion in Alimentary Canals More complex animals have *complete digestive tracts* or *alimentary canals*, with two openings, a mouth and an anus, and specialized regions allowing for the sequential digestion and absorption of nutrients. Food, ingested through the mouth and pharynx, passes through an esophagus that leads to either a crop, stomach, or gizzard—organs specialized for storing or grinding food. In the lengthy intestine, digestive enzymes hydrolyze macromolecules, and nutrients are absorbed across the tube lining. Undigested material exits through the *anus*.

The Mammalian Digestive System

The mammalian digestive tract has a four-layered wall: an inner mucous membrane (the mucosa), a connective-tissue layer, smooth muscle, and a sheath of connective tissue attached to the body cavity membrane. Rhythmic waves of contraction called *peristalsis* push food through the tract. Ringlike valves called *sphincters* regulate the passage of material between some specialized segments.

Accessory organs, including *salivary glands*, the *pancreas*, and the *liver* with its *gall bladder*, deliver digestive enzymes to the alimentary canal through ducts. The following description details the human digestive system.

Oral Cavity Physical and chemical digestion begins in the mouth, where teeth chew and grind food to expose a greater surface area to enzyme action. The presence of food in the oral cavity, as well as the learned associations of food with other stimuli, trigger the release of saliva into the mouth.

Saliva has several components: mucin, a glycoprotein that protects the mouth lining from abrasion and lubricates food for swallowing; buffers to neutralize acidity; antibacterial agents; and *salivary amylase*, which hydrolyzes starch into smaller polysaccharides and prevents a build up of starch between the teeth. The tongue is used to taste, to manipulate food, and to push the food ball or *bolus* into the pharynx for swallowing.

Pharynx The *pharynx* is the intersection leading to both the esophagus and trachea. During swallowing, the windpipe (trachea) moves so that its opening is blocked by the cartilaginous *epiglottis*.

Esophagus Food moves down through the narrow, flexible esophagus to the stomach, squeezed along by a wave of smooth muscle contraction called peristalsis.

Stomach The stomach, with its elastic wall and folds called rugae, can expand to hold about 2 liters of food and fluid. The epithelium lining the lumen (cavity) secretes *gastric juice*, a digestive fluid containing *pepsin*, an enzyme that hydrolyzes specific peptide bonds in proteins, and a high concentration of hydrochloric acid. The acid of the gastric juice breaks down food tissues, kills bacteria, and denatures proteins, increasing the exposure of peptide bonds to the action of pepsin.

Pepsin is synthesized and secreted in an inactive form called *pepsinogen*. It is activated by hydrochloric acid secreted by other stomach epithelia cells and by pepsin itself—an example of positive feedback. Protein-digesting enzymes are often secreted in inactive forms, called *zymogens*. The mucous coating secreted by the epithelium protects the stomach lining from digestion. Lesions called gastric ulcers develop where the lining is eroded faster than it can be regenerated by mitosis.

The sight, smell, or taste of food sends a nervous message from the brain to the stomach that initiates secretion of gastric juice. Substances in the food stimulate the stomach wall to release the hormone *gastrin* into the circulatory system. This hormone stimulates further secretion of gastric juice. If the pH of the stomach contents becomes too low, the release of gastrin is inhibited, which decreases secretion of gastric juice—an example of negative feedback.

Smooth muscles mix the stomach contents about every 20 seconds. *Acid chyme* is the nutrient broth produced by the action of the stomach and its secretions on ingested food. The stomach is usually closed off by two sphincters, a cardiac sphincter that prevents backflow into the esophagus,and a *pyloric sphincter* that regulates passage of acid chyme into the small intestine.

Small Intestine Most enzymatic hydrolysis of macromolecules and nutrient absorption into the blood takes place in the small intestine, the longest section of the alimentary canal.

Several organs contribute to digestion in the small intestine by producing, storing, and secreting digestive juices. The pancreas produces hydrolytic enzymes and a bicarbonate-rich alkaline solution that offsets the acidity of the chyme. The liver produces *bile*, which is stored in the gall bladder. Bile aids in the digestion and absorption of fats and also contains pigments that are byproducts of the breakdown of red blood cells in the liver.

Bile, digestive juices from the pancreas, and secretions from gland cells of the intestinal wall are mixed with the chyme in the *duodenum*, the first section of the small intestine. Regulatory hormones coordinate the release of digestive secretions: *Secretin* is released by cells in the intestinal wall in response to the acidic pH of the chyme and stimulates the pancreas to release bicarbonate. *Cholecystokinin* (CCK), also produced by cells lining the duodenum, stimulates gall bladder contraction and the release of pancreatic enzymes. A fat-rich chyme causes the duodenum to release *enterogastrone*, a hormone that inhibits peristalsis in the stomach, slowing the entry of chyme into the duodenum.

The digestion of starch, begun by salivary amylase in the mouth, is continued by a pancreatic amylase, which hydrolyzes starch into maltose in the small intestine. The enzyme maltase splits maltose into two glucose molecules. It is one of the *disaccharidases*, enzymes specific for hydrolysis of different disaccharides, that are built into the membranes and glycocalyx of epithelial cells, facilitating sugar absorption through the intestinal wall.

Pepsin in the stomach breaks proteins into smaller pieces. Protein digestion is completed in the small intestine by *trypsin* and *chymotrypsin*, enzymes specific for peptide bonds adjacent to certain amino acids; *carboxypeptidase*, which splits amino acids off the free carboxyl end; and *aminopeptidase*, which works from the amino end. *Dipeptidases* are attached to the intestinal lining and split fragments consisting of two or three amino acids. The pancreatic protein-digesting enzymes are secreted as zymogens and activated by the intestinal enzyme *enterokinase*.

Nucleases are a group of enzymes that hydrolyze DNA and RNA into their nucleotide monomers.

The digestion of fats is aided by bile salts, which coat or *emulsify* tiny fat droplets so they do not coalesce, leaving a greater surface area for *lipase* to hydrolyze the fat molecules.

Most digestion is completed while the chyme is still in the duodenum. The *jejunum* and *ileum* are regions of the small intestine specialized for nutrient absorption. The huge surface area of the small intestine is created by large folds covered with fingerlike projections called *villi*, on which the epithelial cells have microscopic extensions called *microvilli*. The core of each villus has a net of capillaries and a lymph vessel called a *lacteal*. Nutrients are absorbed across the epithelium of the villus and then across the single-celled walls of the capillaries or lacteal. Transport may be passive (the nutrient moving down its concentration gradient) or active (the nutrient pumped against a gradient). Active transport of sodium into the intestinal lumen and its passive reentry into epithelial cells seem to drive the uptake of certain nutrients.

Amino acids and sugars enter capillaries and are carried to the liver by the bloodstream. Glycerol and fatty acids are absorbed by epithelial cells where they recombine to form fats and are coated with proteins to make tiny globules called *chylomicrons*. These packages are transported by exocytosis out of the epithelial cells and enter a lacteal. Some fat molecules, bound to specialized proteins, are transported as *lipoproteins* into capillaries.

The nutrient-laden blood from the small intestine is carried directly to the liver by the large *hepatic portal vein*. The liver interconverts and stores molecules, thereby regulating the nutrient content of the blood.

Large Intestine The small intestine leads into the large intestine, or colon, at a junction with a sphincter. A blind pouch, called the *cecum*, with a fingerlike extension, the *appendix*, attaches at this juncture. One of the colon's functions is to finish the reabsorption of the large quantity of water secreted into the digestive tract along with the digestive enzymes. An irritation or infection of the colon lining may result in less water absorption and lead to diarrhea; whereas an excess of reabsorption, occurring when peristalsis moves too slowly, may result in constipation. Digestive wastes are called *feces*.

A rich flora of *Escherichia coli* and other mostly harmless bacteria live on organic material in the feces. Some of these bacteria produce vitamin K, which is absorbed by the host. The feces contain cellulose, other undigested ingredients of food, bile pigments, salts excreted by the colon, and a large proportion of intestinal bacteria. Feces are stored in the *rectum*. A voluntary and involuntary sphincter between the rectum and anus control the elimination of feces, which is initiated by strong contractions of the colon. A meal takes 18 to 36 hours to move through the human digestive system.

Some Evolutionary Adaptations of Vertebrate Digestive Systems

Dentition, the type and arrangement of teeth in the mouth, correlates with diet. Carnivores are characterized by sharp incisors, fanglike canines, and jagged premolars and molars; herbivores have broad molars for grinding plant material; omnivores, such as humans, have a relatively unspecialized dentition.

Herbivores have longer alimentary canals because plant material is more difficult to digest than meat. The extra length also provides more area for absorption of the less concentrated nutrients in vegetation. Specialized structures, such as the spiral valve in a shark's intestine, functionally increase intestinal length by providing additional surface area.

Many herbivorous mammals have special fermentation chambers filled with symbiotic bacteria and protozoa. These microorganisms, often housed in the cecum, digest cellulose to simple sugars and produce a variety of essential nutrients for the animal. Since many of the symbiotic bacteria of rabbits and some rodents live in the large intestine, these animals may ingest their feces so that the nutrients produced by these symbionts may be absorbed as their food goes through the small intestine a second time.

Ruminants have the most elaborate adaptations for a herbivorous diet. The stomach is divided into four chambers, two of which contain symbiotic bacteria that digest cellulose. The cud is regurgitated, rechewed and swallowed into the other chambers, where both the microbially digested cellulose and the microorganisms themselves are digested and their nutrients absorbed.

Nutritional Requirements

Food provides fuel for cellular respiration, organic raw materials for the construction of the animal's molecules, and *essential nutrients*, those which the animal cannot synthesize and must obtain in prefabricated form.

Food as Fuel The monomers of carbohydrates, fats, and proteins can be used as fuel for cellular respiration, although the first two are used preferentially. Energy content of food is measured in *calories*. The Calorie, as used by nutritionists, is actually a *kilocalorie*. Fat supplies about two times as many kcal/g as carbohydrate or protein.

Metabolism must supply sufficient energy to maintain breathing, heart beat, and, in some animals, stable body temperature. The *basal metabolic rate* (*BMR*) is the number of kilocalories a resting organism requires for these processes for a given time. Birds and mammals, which are endothermic and use metabolic energy to maintain a constant body temperature, have higher BMR than do the ectothermic fish, reptiles and amphibians, which absorb their body heat from the environment. In endotherms, body size is inversely related to the calories required per gram of body weight. Smaller animals have greater surface area-to-volume ratios, greater heat loss to the surroundings, and greater energy cost to maintain a stable body temperature. The inverse relationship between body size and metabolic rate also applies to ectothermic vertebrates and even protozoa.

The BMR for humans averages 1600 to 1800 kcal per day for males and 1300 to 1500 kcal for females. BMR is determined by multiplying the O_2 consumption of a resting subject by the 4.83 kcal of energy produced by respiration for each liter of O_2 consumed. BMRs are standardized as kcal/h per kg body weight. Any activity increases the caloric requirement above this level.

When an animal consumes more calories than are needed to meet its energy requirements, excess calories are stored in the liver and muscles as glycogen, a polymer of glucose. When the glycogen stores are full, additional calories are stored in adipose tissue as fat.

An *undernourished* person or other animal has a diet deficient in calories. With severe deficiency, the body breaks down its own proteins for energy, eventually causing irreversible damage. Incidents of undernourishment are usually associated with drought or war, although the condition anorexia nervosa can result in undernourishment.

Overnourishment, or obesity, increases the risk of heart attack, diabetes, and other disorders. "Dieting" is a billion dollar industry in the United States. Successful weight watching requires balancing caloric intake with caloric demand. To lose weight one must eat less, exercise more, or do both.

Food for Fabrication Animals can fabricate most of the organic molecules they need using enzymes to rearrange the carbon skeletons and organic nitrogen acquired from food. In vertebrates, the conversion of nutrients to organic molecules occurs mainly in the liver.

Essential Nutrients Molecules that an animal requires but cannot make are called essential nutrients. These requirements vary from species to species, depending on biosynthetic capabilities. When the diet is missing one or more essential nutrients, the animal is said to be *malnourished*. Malnutrition is more common than undernutrition in human populations.

Eight of the 20 amino acids required to make proteins are essential amino acids in the human diet. Protein deficiency develops from a diet that lacks one or more essential amino acid. In Africa, the resulting syndrome of retarded mental and physical development in children is called *kwashiorkor*, which may arise when a child is weaned from mother's milk to a starchy diet. The trend away from breast-feeding in some less-developed countries has increased the incidence of protein deficiency.

Meat, eggs, and cheese contain complete proteins with all essential amino acids in proportions that meet human requirements. Most plant proteins are incomplete, and diets built on a single staple, such as corn, beans or rice, can result in protein deficiency. The body cannot store amino acids, and a deficiency of a single essential amino acid may prevent protein synthesis and limit the use of other amino acids. A combination of plant foods, complementary in amino acids and consumed at the same meal, can prevent protein deficiencies.

Animals are able to make most of the fatty acids they need. Linoleic acid, an unsaturated fatty acid used to make some membrane phospholipids, is required in the human diet. Deficiencies of essential fatty acids are rare.

Vitamins are essential organic molecules required in small amounts in the diet. Most vitamins are coenzymes or parts of coenzymes. Deficiencies can cause severe syndromes. The first vitamin isolated was thiamine; a deficiency of thiamine results in the disease beriberi.

Thirteen vitamins essential to humans have been identified. Water-soluble vitamins include the B-complex, most of which function as coenzymes in key metabolic processes, and vitamin C, required for production of connective tissue. Excesses of water-soluble vitamins are excreted. The fat-soluble vitamins are A, incorporated into visual pigments; D, aiding in calcium absorption and bone formation; E, seeming to protect phospholipids in membranes from oxidation; and K, required for blood clotting. Excesses of fat-soluble vitamins are deposited in body fat, and overdoses may cause toxic accumulations. There is debate between those individuals who believe that the recommended daily allowances (RDAs) are sufficient and those who believe that optimal intake of certain vitamins is much higher.

A compound that is a vitamin for one species may not be essential for a second species that can synthesize it. Symbiotic intestinal microorganisms may produce vitamins that are used by the host.

Minerals are inorganic nutrients, usually needed in very small amounts. Requirements vary with species. Vertebrates require relatively large quantities of calcium and phosphorus for bone construction. Calcium is also needed for normal nerve and muscle functioning, and phosphorus is needed in ATP and nucleic acids. Iron is a component of the cytochromes and hemoglobin. Other minerals function as cofactors of enzymes. Iodine is needed by vertebrates to make thyroxin, a hormone regulating metabolism. Sodium, potassium, and chlorine are important in nerve function and osmotic balance.

Animal nutrition provides calories for energy needs, organic raw materials for biosynthesis, and essential amino acids, essential fatty acids, vitamins, and minerals.

STRUCTURE YOUR KNOWLEDGE

1. Food provides fuel, organic raw materials, and essential nutrients. Develop a concept map to organize your understanding of these nutritional needs of animals.

2. Label the indicated structures in the diagram of the human digestive system on the following page.

3. Fill in the following chart on the processes that occur in the sections or organs of the human digestive system, and list the glands and digestive fluids associated with these processes.

ORGAN OR SECTION	PROCESSES OCCURRING	ASSOCIATED GLANDS, ORGANS, AND DIGESTIVE FLUIDS
Oral cavity		
Pharynx		
Esophagus		
Stomach		
Small intestine		
Large intestine		

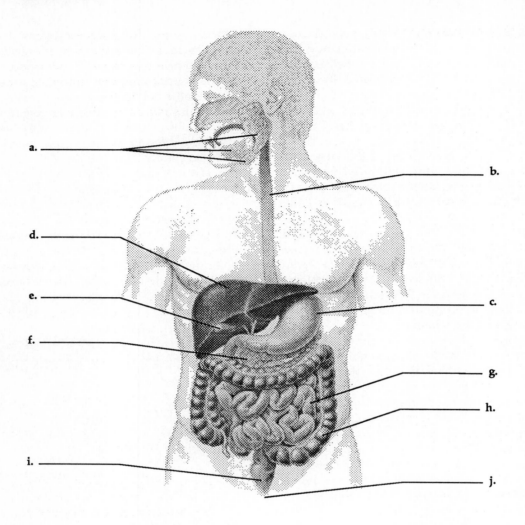

a.

b.

d.

e.

c.

f.

g.

h.

i.

j.

4. Fill in the following table on four major regulatory hormones that control the release of digestive
secretions.

STIMULUS THAT CAUSES RELEASE	HORMONE	EFFECTS OF THE HORMONE
	Gastrin	
	Secretin	
	Cholecysto-kinin (CCK)	
	Enterogas-trone	

TEST YOUR KNOWLEDGE

MATCHING: *Match the description with the correct enzyme or hormone.*

1. _____ enzyme that hydrolyzes peptide bonds, works in the stomach

2. _____ hormone that stimulates secretion of gastric juice

3. _____ hormone that causes the gall bladder to contract and the pancreas to release enzymes

4. _____ enzyme that begins digestion of starch in mouth

5. _____ enzymes specific for hydrolyzing disaccharides

6. _____ enzyme that hydrolyzes peptide bonds at amino end of polypeptide

7. _____ enzyme that hydrolyzes fats

8. _____ intestinal enzyme that activates zymogens

9. _____ enzyme specific for peptide bonds adjacent to certain amino acids, works in duodenum

10. _____ hormone that stimulates secretion of bicarbonate ions from pancreas

A. aminopeptidase

B. bile salts

C. cholecystokinin

D. chymotrypsin

E. disaccharidases

F. enterokinase

G. enterogastrone

H. gastrin

I. hydrochloric acid

J. lipase

K. maltase

L. pepsin

M. amylase

N. secretin

MULTIPLE CHOICE: *Choose the one best answer.*

1. Deposit-feeders
   a. feed mostly on mineral substrates.
   b. filter small organisms from water.
   c. eat plants.
   d. feed on detritus.

2. The energy content of fats
   a. is released by bile salts.
   b. is inversely related to body size.
   c. is approximately two times that of carbohydrate or protein.
   d. can reverse the effects of malnutrition.

3. BMR
   a. stands for bottom metabolic rate.
   b. is higher for ectotherms than for endotherms.
   c. is higher per gram of body weight for smaller animals than for larger ones.
   d. All of the above are correct.

4. Which of the following statements is *not* true?
   a. The average human has enough stored glycogen to supply calories for several weeks.
   b. Eating less and/or exercising more can result in weight loss.
   c. Conversion of glucose and glycogen takes place in the liver.
   d. Excessive calories are stored as fat, regardless of their food source.

5. Kwashiorkor
   a. results from protein deficiency.
   b. is an example of malnutrition.
   c. is a syndrome involving retarded mental and physical development.
   d. All of the above are correct.

6. Incomplete proteins are
   a. lacking in essential vitamins.
   b. a cause of undernourishment.
   c. found in meat, eggs, and cheese.
   d. lacking in one or more essential amino acids.

7. Vitamins
   a. may be produced by intestinal microorganisms.
   b. are consistant from one species to the next.
   c. are stored in the liver.
   d. are inorganic nutrients, needed in small amounts, that usually function as cofactors.

8. Which of the following is most likely to be found in animals with gastrovascular cavities?
   a. absorption of predigested nutrients
   b. extracellular digestion
   c. intracellular digestion
   d. suspension feeding

9. Ruminants
   a. have teeth adapted for an herbivorous diet.
   b. must use microorganisms to digest plant starch.
   c. eat their feces to obtain nutrients digested from cellulose by microorganisms.

d. house symbiotic bacteria and microorganisms in a cecum.

10. The stomach can expand during eating due to its
    a. villi.
    b. rugae.
    c. cecae.
    d. sphincters.

11. The acid pH of the stomach
    a. hydrolyzes proteins.
    b. is regulated by the release of gastrin.
    c. is neutralized by gastric juice.
    d. is produced by pepsin.

12. Acid chyme is the
    a. nutrient broth that moves from the stomach to the duodenum through a sphincter.
    b. nutrient broth that passes through the jejunum and ileum.
    c. gastric secretions of epithelial cells in the stomach.
    d. mixture of macromolecules passing into the stomach.

13. Chylomicrons are
    a. lipoproteins transported by the circulatory system.
    b. small branches of the lymphatic system.
    c. protein-coated fat globules excreted out of epithelial cells into a lacteal.
    d. small peptides acted upon by chymotrypsin.

14. Which of the following is *not* a common component of feces?
    a. intestinal bacteria
    b. cellulose
    c. saturated fats
    d. bile pigments

15. The hepatic portal vein
    a. supplies the capillaries of the intestines.
    b. carries absorbed nutrients to the liver for processing.
    c. carries blood from the liver to the heart.
    d. drains the lacteals of the villi.

16. Peristalsis is controlled by which type of tissue?
    a. squamous epithelial
    b. voluntary muscle
    c. fibrous connecting
    d. none of the above

CHAPTER 38

CIRCULATION AND GAS EXCHANGE

FRAMEWORK

This chapter surveys the basic approaches to circulation and gas exchange found in the animal kingdom, with special attention to the human systems and the problems of cardiovascular disease. The following concept map organizes some of the chapter's key ideas.

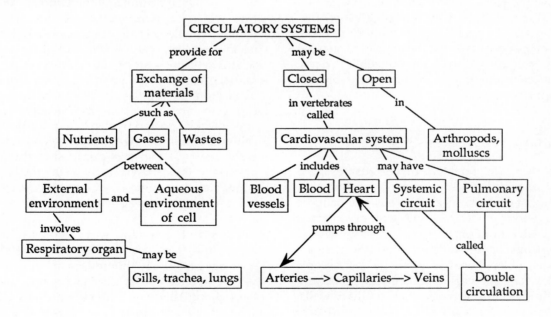

CHAPTER SUMMARY

The exchange of materials is essential for the cells of an organism. Every living cell must reside in an aqueous environment, which provides oxygen and nutrients and permits disposal of carbon dioxide and metabolic wastes.

Diffusion cannot transport substances over the macroscopic distances in animals. All but the simplest animals have internal transport systems for body fluids. The body fluid, usually blood, exchanges materials with the environment across the thin and extensive epithelia of organs specialized for gas exchange, nutrient absorption, and waste removal. The blood then exchanges chemicals with the interstitial fluid that bathes the cells.

Internal Transport in Invertebrates

Gastrovascular Cavities The central gastrovascular cavity inside the two-cell-thick body wall of *Hydra* serves for both digestion and transport of materials. The internal fluid exchanges directly with the water environment through the single opening. Planarians and other flatworms also have gastrovascular cavities that ramify throughout the body.

Open and Closed Circulatory Systems In the *open circulatory system* found in insects and other arthropods, and in most mollusks, *hemolymph* bathes the internal tissues directly. This fluid is circulated as body movements squeeze the *sinuses*, or spaces between organs and by the beating of a heart that is usually part of a dorsal vessel. Hemolymph, drawn into the vessel through ostia when the heart relaxes, is pumped into the interconnected system of sinuses during contraction.

Annelids, some mollusks, and vertebrates have *closed circulatory systems,* in which the blood remains in vessels and exchanges materials with the interstitial fluid bathing the cells. In an earthworm, the dorsal vessel functions as a heart by pumping blood forward. Five pairs of vessels, which loop around the digestive tract and connect the dorsal and ventral main vessels, pulsate and function as auxiliary hearts. Blood circulates more rapidly and efficiently through a closed circulatory system.

Circulation in Vertebrates

The closed circulatory system of vertebrates, also called the *cardiovascular system*, consists of the heart, blood vessels, and blood. The heart has one or more *atria*, which receive blood, and one or more *ventricles*, which pump blood out of the heart. *Arteries*, carrying blood away from the heart, branch into tiny *arterioles* within organs, which then divide into the microscopic *capillaries*. Exchange of substances between blood and interstitial fluid occurs across the thin capillary walls. Capillaries rejoin to form *venules*, which meet to form the *veins* that return blood to the heart.

Vertebrate Circulatory Schemes: An Evolutionary Perspective The ventricle of a fish's two-chambered heart pumps blood first to the capillary beds of the gills, from which the oxygenated blood flows through a vessel to the capillary beds in the rest of the body. Veins return the blood to the atrium. Passage through two capillary beds slows the flow of blood, but body movements help to maintain circulation.

In the three-chambered heart of amphibians, the single ventricle pumps blood through a forked artery into the *pulmonary circuit*, which leads to lungs and skin and then back to the left atrium, and the *systemic circuit*, which carries blood to the rest of the body and back to the right atrium. This *double circulation* repumps blood after it returns from the capillary beds of the lungs, ensuring a strong flow of oxygenated blood to the brain, muscles, and body organs. A ridge in the ventricle diverts most of the oxygenated blood from the left atrium into the systemic circuit and most of the deoxygenated blood from the right atrium into the pulmonary circuit. The three-chambered reptilian heart has a septum that partially divides the single ventricle. In crocodiles, the septum completely divides the ventricle into two chambers.

Delivery of oxygen for cellular respiration is most efficient in birds and mammals, which, as endotherms, have high oxygen demands. The left side of the four-chambered heart handles only oxygenated blood, whereas the right side receives and pumps deoxygenated blood.

In the human circulatory system, oxygenated blood returning from the lungs in the pulmonary veins flows to the left atrium and then into the left ventricle, from which it is pumped through the aorta in the systemic circulation. The aorta gives rise to branch arteries to the heart, brain, limbs, and organs of the body, where arterioles break into thin-walled capillaries, and the blood gives up its oxygen. Deoxygenated blood from the capillary beds returns through venules into veins and is collected by the superior (anterior) and inferior (posterior) vena cavae. These large vessels empty into the right atrium; blood flows into the right ventricle and it is pumped in the pulmonary arteries to the capillary beds of the lungs. The blood completes the pulmonary circuit by returning to the heart in the pulmonary veins.

The Heart The human heart is enclosed in a two-layered, fluid-filled sac just beneath the sternum. The atria have relatively thin muscular walls, whereas the ventricles have thicker muscular walls. The *heart cycle* lasts about 0.8 sec and consists of the *systole*, during which the heart muscle contracts and the chambers pump blood, and the *diastole*, when the heart is relaxed. The atria are filling with blood during the heart cycle except for the first 0.1 sec of the systole when they contract. The slow and powerful contraction of the ventricles lasts 0.3 sec. Blood flows passively into the ventricles during diastole, and atrial contraction finishes filling them.

Atrioventricular valves between each atrium and ventricle are snapped shut when the ventricles contract. Strong fibers prevent these connective tissue flaps from turning inside out. The *semilunar valves* at the exit of the aorta and pulmonary artery are forced open by ventricular contraction and close when the ventricles relax and the elastic walls of the arteries recoil. The heart-beat sounds are caused by the closing of these valves. A heart murmur is the detectable hissing sound of blood leaking back through a defective valve.

The average human heart rate, or *pulse*, is 65 to 75 beats per minute. The inverse relationship between size and pulse is attributable to the higher metabolic rate per gram of tissue of smaller animals.

The *cardiac output*, or volume of blood pumped per minute into the systemic circuit, depends on the heart rate and the *stroke volume*, or quantity of blood pumped by each contraction of the left ventricle.

Cells of cardiac muscle are self-excitable or myogenic; they have an intrinsic ability to contract. The rhythm of contractions is coordinated by the *sinoatrial (SA) node*, or *pacemaker*. When nodal tissue contracts, it generates electrical impulses. The contraction of the SA node, located in the wall of the right atrium near the anterior vena cava, initiates a wave of excitation that signals the cells of the two atria to contract. (Remember that cardiac muscle cells are electrically connected by the intercalated discs.) The *atrioventricular (AV) node*, located at the base of the two atria, relays the impulse after a 0.1 sec delay to the tips of the ventricles. The electrical currents produced during the heart cycle can be detected by electrodes placed on the skin and recorded in an *electrocardiogram* (EKG or ECG). The SA node is controlled by two sets of nerves with antagonistic signals and is influenced by hormones, temperature, and stimulation from increased blood flow during exercise.

Blood Flow The wall of an artery or vein consists of three layers: an outer connective tissue zone with elastic fibers; a middle layer of smooth muscle and more elastic fibers; and an inner lining of *endothelium*, a simple squamous epithelium, and a basement membrane and thin connective-tissue ring. The middle layer is especially thick in arteries, which must be stronger and more elastic. Capillaries have only the endothelial layer.

The flow of blood decelerates after leaving the heart as a result of the increasing cross-sectional area of the branching vessel system. The enormous number of capillaries creates a total diameter much greater than other parts of the system, and friction is also greater in the narrow capillaries. Blood flow is very slow in capillaries, improving opportunity for exchange with interstitial fluid. Blood flow speeds up within veins, due to the decrease in their total cross-sectional area.

Blood pressure, the hydrostatic force exerted against the wall of a blood vessel, is much greater in arteries than in veins and is greatest in systole. *Peripheral resistance*, caused by the narrow openings of the arterioles impeding the exit of blood from arteries, causes the swelling of the arteries during systole. The snapping back of the elastic arteries during diastole maintains a continuous blood flow into arterioles and capillaries. Blood pressure may be measured with a sphygmomanometer. The higher number is the pressure during systole (when blood first spurts through the artery that was closed off by the cuff), and the second is the pressure during diastole (when blood flows smoothly through the artery).

Both cardiac output and peripheral resistance determine blood pressure. Contraction of smooth muscles in arteriole walls increases resistance and thus increases blood pressure, whereas dilation of arterioles lowers blood pressure. Neural and hormonal signals control these muscles.

Blood pressure drops to almost zero after the capillary beds. The one-way valves in veins and the contraction of skeletal muscles between which veins are embedded force blood to flow back to the heart. Pressure changes during breathing also draw blood into the large veins in the thoracic cavity.

Capillaries branch off *thoroughfare channels*, the direct connections between arterioles and venules. Sphincters regulate the passage of blood into capillaries. Only about 5% to 10% of the body's capillaries have blood flowing through them at any one time. The brain, kidneys, liver, and heart are usually heavily supplied with blood; the distribution to the other areas of the body varies with need.

Capillary Exchange The exchange of substances between blood and interstitial fluid may involve bulk transport by endocytosis and exocytosis of the endothelial cells making up the capillary wall, passive diffusion of small substances (water, sugars, salts, oxygen, and urea) through both the cells themselves and the clefts between adjoining cells, and the forcing of fluid out of the capillary by hydrostatic pressure. Blood cells and proteins are too large to pass easily through the endothelium. Blood pressure at the upstream end of a capillary forces fluid out, whereas osmotic pressure at the downstream end tends to draw about 99% of that fluid back in.

The Lymphatic System The fluid that does not return to the capillary, and any proteins that may have leaked through the capillary wall, are returned to the blood through the *lymphatic system*. Fluid diffuses into lymph capillaries intermingled in the capillary net. The fluid, called *lymph*, moves through lymph vessels with one-way valves mainly as a result of the pressure exerted by the movement of skeletal muscles. In *lymph nodes*, the lymph is filtered, and white blood cells attack viruses and bacteria. The lymphatic system also transports fats from the small intestine to the circulatory system.

When the lymphatic system does not return sufficient interstitial fluid to the circulatory system, fluid accumulates in tissues, creating a condition known as edema.

Mammalian Blood

Blood is a type of connective tissue with cells in a liquid matrix called *plasma*. When *whole blood* is centrifuged, the *formed elements* or cells form a dense red pellet that makes up about 45% of the volume of blood.

Plasma   The plasma consists of a large variety of solutes dissolved in water. The collective and individual concentrations of *electrolytes*, or inorganic salts in the form of dissolved ions, are important to osmotic balance between blood and interstitial fluid and to the functioning of muscles and nerves. The kidney is responsible for the homeostatic regulation of ion concentrations. Plasma proteins function as buffers, osmotic components of the blood, antibodies, escorts for lipids, and clotting factors. Blood plasma from which the fibrinogens or clotting factors have been removed is called serum. Nutrients, metabolic wastes, gases, and hormones are also transported in the plasma. Except for its higher protein concentration, plasma is very similar in composition to the interstitial fluid.

Blood Cells   Three classes of cells are found in blood: red blood cells, white blood cells, and platelets. *Red blood cells*, or *erythrocytes*, are the most numerous (about 25 trillion in the body's 5 liters of blood). Mammalian erythrocytes lack nuclei, and all red blood cells lack mitochondria and generate their ATP by anaerobic metabolism. The small size and biconcave shape of red blood cells creates a large surface area of plasma membrane across which oxygen can diffuse. Erythrocytes are packed with *hemoglobin*, an iron-containing protein that binds oxygen.

Erythrocytes are formed from *pluripotent stem cells* in the red marrow of bones, and their production is controlled by a negative feedback mechanism involving the hormone *erythropoietin* secreted by the kidney in response to low oxygen supply in tissues. Red blood cells circulate for about 3 to 4 months before they are phagocytized by cells in the liver and their components recycled.

Leukocytes, or *white blood cells*, fight infections. Some white cells are phagocytes that ingest bacteria and cellular debris; some are lymphocytes that give rise to cells that produce antibodies. Most leukocytes are found in the interstitial fluid and lymph nodes. Leukocytes also arise from stem cells in the bone marrow. Lymphocytes mature in lymphoid organs (spleen, thymus, tonsils, adenoids, and lymph nodes).

Platelets, pinched-off fragments of large cells in the bone marrow, are involved in the blood-clotting mechanism.

Blood Clotting   The blood clotting process usually begins when platelets, clumped together along a damaged endothelium, release clotting factors. By a series of steps, the blood protein *fibrinogen* is converted to *fibrin*. The threads of fibrin weave into a patch. An inherited defect in any step of the complex clotting process causes *hemophilia*, a disease characterized by prolonged bleeding from even minor injuries. Anticlotting factors normally prevent clotting of blood in the absence of injury. A *thrombus* is a clot that occurs within a blood vessel and blocks the flow of blood.

Cardiovascular Disease

More than half of all deaths in the United States are caused by *cardiovascular disease*, usually a result of heart attack or stroke. A heart attack may occur when a thrombus or an embolus, which is a clot formed elsewhere that moves through the circulatory system, blocks a coronary artery. The cardiac muscle that was served by the artery may die, and the conduction of electrical impulses through the cardiac muscle may be interrupted and result in an arrhythmia or cessation of beating. Strokes result from a thrombus or embolus blocking an artery in the brain.

Most heart-attack and stroke victims had been suffering from a chronic disease known as *atherosclerosis*, in which growths called *plaques* develop within arteries and narrow the vessels. Plaques that become hardened by calcium deposits result in *arteriosclerosis*, or "hardening of the arteries." An embolus is more likely to be trapped in narrowed vessels, and plaques are common sites of thrombus formation. Occasional chest pains, known as angina pectoris, may warn that a coronary artery is partially blocked.

Hypertension, or high blood pressure, is thought to damage the endothelium and initiate plaque formation, promoting atherosclerosis and increasing the risk of heart attack and stroke. This condition can be easily diagnosed and controlled by drugs, diet, and exercise. Hypertension and atherosclerosis tend to be inherited. Smoking, not exercising, and eating a fat-rich diet have been correlated with an increased risk of cardio-vascular disease.

The cholesterol level in blood plasma is an indicator of potential atherosclerosis. Most cholesterol is carried as *low-density lipoproteins* (*LDLs*). *High density lipoproteins* (*HDLs*) are another form of cholesterol carriers that appear to reduce cholesterol deposition in plaques. The ratio of LDLs to HDLs is an indication of potential cardiovascular disease.

The death rate from cardiovascular disease has been significantly declining in the United States. Diagnosis and treatment of hypertension, improved intensive care for cardiovascular patients, and healthier life styles may be contributing to this trend.

Gas Exchange in Animals

General Problems of Gas Exchange Gas exchange provides the continuous supply of oxygen needed for cellular respiration and removes the waste product carbon dioxide. The *respiratory medium*, which supplies oxygen, is air for a terrestrial animal and water for an aquatic one. The *respiratory surface*, the portion of an animal's body where gas exchange with the respiratory medium occurs, must be moist and large enough to supply the whole body.

General Structure and Function of Respiratory Organs A localized region of the body surface is usually specialized as a respiratory surface with a thin, moist epithelium separating a rich blood supply and the respiratory medium. When the entire outer skin serves as a respiratory organ, as in an earthworm, the animal must live in damp places and have a relatively small, long and thin body to provide a high ratio of surface area to volume. Most animals use an extensively branched or folded localized region of the body surface for gas exchange, which is usually external in aquatic animals and internal in terrestrial animals. Aquatic animals have *gills*; terrestrial vertebrates use *lungs*; and insects have *tracheae*.

Gills: Respiratory Adaptations of Aquatic Animals Gills are evaginations of the body surface, ranging from the simple bumps on echinoderms, to the complex, finely divided gills of mollusks, crustaceans, fishes, and some amphibians. Delicate gills are usually sheltered by a protective cover. Due to the low oxygen concentration in a water environment, gills usually require *ventilation*, or movement of the respiratory medium across the respiratory surface—often an energy-intensive process.

In the gills of a fish, blood flows through capillaries in a direction opposite to the flow of water. This arrangement sets up a *countercurrent exchange*, in which the diffusion gradient favors the movement of oxygen into the blood throughout the length of the capillary.

Tracheae: Respiratory Adaptations of Insects Air's advantages as a respiratory medium are its higher concentration of oxygen, faster diffusion rate of O_2 and CO_2, and lower density. To prevent the disadvantageous water loss from large, moist surfaces, the respiratory surfaces of most terrestrial organisms are invaginated.

The trachea of insects are tiny air tubes that ramify throughout the body to come into contact with nearly every cell. Openings to the tracheal system are *spiracles*. Some large insects use rhythmic body movements to ventilate their tracheal systems. The efficiency of gas exchange within the tracheal system counteracts the rather slow moving hemolymph of an insect's open circulatory system.

Lungs: Respiratory Adaptations of Terrestrial Vertebrates Lungs are invaginated respiratory surfaces restricted to one location from which oxygen is transported to the rest of the body by the circulatory system. Frog lungs are balloonlike, but mammalian lungs have a spongy texture with a much greater surface area of epithelium for gas exchange.

The lungs of mammals are located in the thoracic cavity and enclosed in a double-walled sac. The two layers adhere tightly to each other, the lungs, and the chest-cavity wall.

Air, entering through the nostrils, is filtered, warmed, humidified, and smelled in the nasal cavity. Air passes through the pharynx and enters the windpipe through the glottis. When food is being swallowed, the glottis is pushed against the epiglottis. The *larynx* functions as a voicebox in humans and many other mammals; exhaled air vibrates a pair of *vocal cords*. The *trachea*, or windpipe, branches into two *bronchi*, which then branch repeatedly into *bronchioles* when they reach the lungs. Ciliated, mucus-coated epithelium lines much of the respiratory tree and removes dust and other particles from the respiratory system.

Multilobed air sacs encased in a web of capillaries are at the tips of the tiniest bronchioles. Gas exchange takes place across the thin moist epithelium of these *alveoli*.

Vertebrate lungs are ventilated by breathing, the alternate inhalation and exhalation of air. A frog breathes by expanding its mouth cavity, which draws air into the mouth, and then raising the jaw, which forces air into the lungs.

Mammals ventilate their lungs by *negative pressure*

breathing. The volume of the lungs and the thoracic cavity is increased by expansion of the rib cage and contraction of the *diaphragm*. Air pressure is reduced within this increased volume, and air flows through the nostrils down to the lungs. Relaxation of the rib muscles and diaphragm compresses the lungs, increases the pressure, and forces air out.

Tidal volume is the normal volume of air inhaled and exhaled by an animal. The maximum volume that can be inhaled and exhaled by forced breathing is called *vital capacity*. The *residual volume* is the air that remains in the alveoli and lungs after forceful exhaling.

Birds have air sacs that penetrate their abdomen, neck, and wings. The air sacs and lungs are ventilated when the bird breathes, through a circuit that includes a one-way passage through tiny channels, called *parabronchi*, in the lungs. Air sacs act as bellows to maintain air flow through the lungs, lower the density of the bird, and help to dissipate heat formed by the metabolism of flight muscles.

Breathing is under automatic control. The *breathing center* in the medulla at the stem of the brain sends nerve impulses to the rib muscles and diaphragm to contract. In a negative feedback loop, stretch sensors in the lungs respond to the expansion of the lungs and send nervous impulses that inhibit the breathing center. When CO_2 concentration increases in the blood, more carbonic acid is formed. The corresponding drop in the pH of the blood, sensed by the breathing center, increases the messages to breathe. Hyperventilation can cause breathing to stop temporarily when excessive deep breathing has rid the blood of most CO_2. Oxygen sensors in key arteries react to severe deficiencies of O_2.

The concentration of gases in air or dissolved in water is measured as *partial pressure*. The partial pressure of oxygen, which makes up 21% of the atmosphere, is 160 mm Hg (0.21 x 760 mm—atmospheric pressure at sea level). The partial pressure of CO_2 is 0.23 mm Hg. Partial pressure is proportional to concentration; a gas will diffuse from a region of higher partial pressure to lower partial pressure.

Blood entering the lungs has a lower P_{O_2} and a higher P_{CO_2} than does the air in the alveoli, and oxygen diffuses into the capillaries and carbon dioxide diffuses out. In the systemic capillaries, pressure differences favor the diffusion of O_2 out of the blood into the interstitial fluid and of CO_2 into the blood.

In most animals, O_2 is carried by *respiratory pigments* in the blood. Hemoglobin, usually located in red blood cells, is the respiratory pigment of almost all vertebrates. Copper is the oxygen-binding component in the blue respiratory protein *hemocyanin*, common in arthropods and many mollusks.

Hemoglobin is composed of four subunits, each of which has a cofactor called a heme group with iron at its center. The binding of O_2 to the iron atom of one subunit induces a change in shape of the other subunits, and their affinity for oxygen increases. Likewise, the unloading of the first O_2 effects a conformational change that lowers the other subunits' affinity for oxygen.

The *dissociation curve* for hemoglobin shows the relative amounts of oxygen bound to hemoglobin under varying oxygen concentrations. At low partial pressures of oxygen, little O_2 is bound, but the percent saturation of hemoglobin rises sharply after the concentration rises high enough for the first oxygen to be bound to a hemoglobin molecule. Thus, in the steep part of this S-shaped curve, a slight change in partial pressure will cause hemoglobin to load or unload a substantial amount of oxygen. A drop in pH lowers the affinity of hemoglobin for O_2. Since a rapidly metabolizing tissue produces more CO_2, which lowers pH, hemoglobin will unload more of its O_2 to that tissue.

Over 20% of CO_2 is transported in blood bound to amino groups of hemoglobin. Seventy percent is transported as bicarbonate ions. Carbon dioxide enters into the red blood cells where it first reacts to form carbonic acid and then dissociates into H^+ and a bicarbonate ion. The hydrogen ions are bound to hemoglobin and other proteins, and the pH value of the blood is not greatly lowered during the transport of carbon dioxide. In the lungs, the diffusion of CO_2 out of the blood shifts the equilibrium in favor of the conversion of bicarbonate back to CO_2, and CO_2 is unloaded from the blood.

Special physiological adaptations have enabled some air-breathing mammals to make long underwater dives. The Weddell seal stores twice the amount of O_2/kg body weight as do humans, mostly by having a larger volume of blood, a huge spleen that stores blood, and a higher concentration of *myoglobin*, an oxygen-storing muscle protein. A diving reflex slows the pulse of diving mammals, reroutes blood to the brain and essential organs, and restricts blood supply to the muscles.

STRUCTURE YOUR KNOWLEDGE

1. Identify the labeled structures in the diagram of a human heart on p. 327.

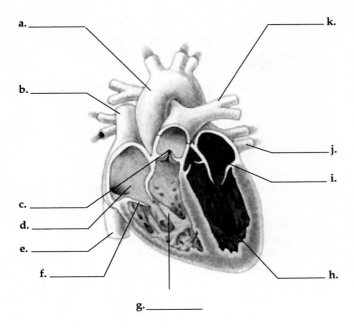

a. _____
b. _____
c. _____
d. _____
e. _____
f. _____
g. _____
h. _____
i. _____
j. _____
k. _____

TEST YOUR KNOWLEDGE

MULTIPLE CHOICE: *Choose the one best answer.*

1. A gastrovascular cavity
   a. is found in cnidarians and annelids.
   b. functions to pump fluids throughout the body.
   c. functions in both digestion and distribution of nutrients.
   d. involves all of the above.

2. Which of the following is *not* a *similarity* between open and closed circulatory systems?
   a. Some sort of pumping device helps to move blood through the body.
   b. Some of the circulation of blood is a result of movements of the body.
   c. The blood and interstitial fluid are indistinguishable from each other.
   d. All tissues come into close contact with the circulating body fluid so that the exchange of nutrients and wastes can take place.

3. In a system with double circulation,
   a. blood is pumped at two locations as it circulates through the body.
   b. there is a countercurrent exchange within the gills.
   c. hemolymph circulates both through a pumping blood vessel and through body sinuses.
   d. blood is repumped after it returns from the capillary beds of the gas-exchange organ.

4. During diastole,
   a. the atria fill with blood.
   b. blood flows passively into the ventricles.
   c. the elastic recoil of the arteries maintains hydrostatic pressure on the blood.
   d. all of the above are occurring.

5. An atrioventricular valve prevents the backflow or leakage of blood
   a. from a ventricle into an atrium.
   b. between ventricles.
   c. from the aorta into the left ventricle.
   d. from the pulmonary vein into the right atrium.

6. During heavy exercise,
   a. stroke volume increases.
   b. heart rate increases.
   c. cardiac output increases.
   d. all of the above occur.

7. Heart beat is initiated by contraction of the
   a. SA node.
   b. AV node.
   c. right atrium.
   d. myogenic cardiac muscle cells.

2. Briefly outline or sketch the circulation of blood through the human heart to the systemic and pulmonary circuits.

3. Develop a concept map or table that shows the components of blood and their functions.

4. Create a concept map that shows your understanding of blood pressure, what it does, what causes it, and where and when it is highest.

5. The following graph shows dissociation curves for hemoglobin at two different pH values. Explain the significance of these two curves.

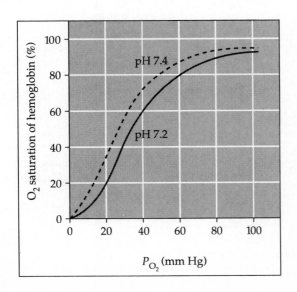

8. Blood flows more slowly in the arterioles than in the arteries because the arterioles
   a. have thoroughfare channels to venules that are often closed off.
   b. collectively have a larger cross-sectional area than do the arteries.
   c. must provide opportunity for exchange with the interstitial fluid.
   d. All of the above are correct.

9. If all the body's capillaries were open at the same time,
   a. blood pressure would fall dramatically.
   b. peripheral resistance would increase.
   c. blood would move too rapidly through the capillary beds.
   d. the amount of blood returning to the heart would increase.

10. Which of the following is *not* a factor in the exchange of substances in capillary beds?
   a. endocytosis and exocytosis
   b. passive diffusion through clefts between endothelial cells
   c. hydrostatic pressure
   d. bulk flow through the apoplast

11. Edema is a result of
   a. swollen lymph glands.
   b. an accumulation of interstitial fluid.
   c. swollen feet.
   d. too high a concentration of blood proteins.

12. A function of the kidney is
   a. control of the lymphatic system.
   b. maintenance of electrolyte balance.
   c. production of lymphocytes.
   d. destruction and recycling of red blood cells.

13. Fibrinogen is
   a. a blood protein that escorts lipids through the circulatory system.
   b. a cell fragment involved in the blood-clotting mechanism.
   c. a blood protein that is converted to fibrin to form a blood clot.
   d. one of the formed elements of blood.

14. Angina pectoris
   a. causes arteriosclerosis when plaques become hardened by calcium deposits.
   b. is a mild form of stroke.
   c. is a serious and hereditary form of hypertension.
   d. may be a warning sign that a coronary artery is partially blocked by plaque.

15. A thrombus
   a. may form at a region of plaque in an artery.
   b. is a traveling embolism.
   c. may cause hypertension.
   d. may calcify and result in arteriosclerosis.

16. The trachea of insects
   a. are stiffened with rings and lead into the lungs.
   b. are filled by positive pressure breathing.
   c. ramify along capillaries of the circulatory system for gas exchange.
   d. are highly branched, coming into contact with almost every cell for gas exchange.

17. Which of the following is *not* involved in speeding up breathing?
   a. a drop in the pH of the blood
   b. stretch receptors in the lungs
   c. impulses from the breathing center in the medulla
   d. severe deficiencies of oxygen

18. The binding of an O_2 to the first iron atom of a hemoglobin molecule
   a. occurs at a very low partial pressure of oxygen.
   b. produces a conformational change that lowers the other subunits' affinity for oxygen.
   c. results in a conformational change and is an example of cooperativity.
   d. occurs more readily at a lower pH value.

19. Most carbon dioxide is transported in the blood
   a. attached to the iron of hemoglobin.
   b. as bicarbonate ions.
   c. at a higher partial pressure than oxygen is.
   d. as carbonic acid.

20. According to the dissociation curve for hemoglobin,
   a. the percent saturation of hemoglobin changes with the amount of oxygen to which it is bound.
   b. there is a range in which a slight change in P_{O_2} will cause hemoglobin to load or unload a large amount of oxygen.
   c. the P_{O_2} in the tissues is lower than that in the alveoli.
   d. a drop in pH value raises the affinity of hemoglobin for O_2.

21. In which animal does blood flow through vessels from the respiratory organs to the rest of the body without returning to the heart?
   a. fish
   b. insect
   c. frog
   d. both a and b

22. Blood leaving the right ventricle of a mammal's heart will pass through how many capillary beds before it returns to the right ventricle?
    a. one
    b. two
    c. one or two, depending on the circuit it takes
    d. at least two, but frequently three

23. Fluid is moved back into the capillaries at the downstream end of a capillary bed by
    a. active transport.
    b. hydrostatic pressure.
    c. osmotic pressure.
    d. bulk transport.

24. In countercurrent exchange,
    a. the flow of fluids or gases in opposite directions maintains a favorable diffusion gradient along the length of an exchange surface.
    b. oxygen is exchanged for carbon dioxide.
    c. a double circulation keeps oxygenated and deoxygenated blood separate.
    d. oxygen moves from a region of high partial pressure to one of low partial pressure, but carbon dioxide moves in the opposite direction.

25. Negative pressure is created in the lungs of mammals by
    a. exhalation of air.
    b. inhalation of air.
    c. relaxation of the diaphragm and rib muscles.
    d. contraction of the diaphragm and rib muscles.

CHAPTER 39

THE BODY'S DEFENSES

FRAMEWORK

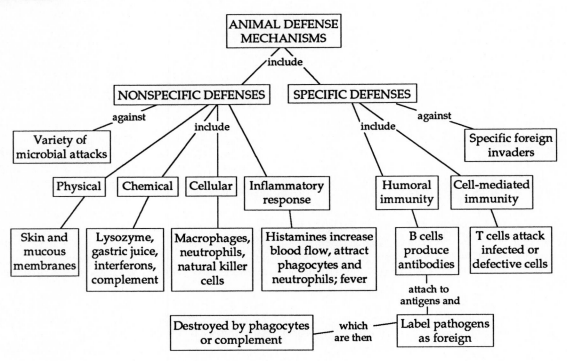

CHAPTER SUMMARY

The defense system protects an animal from external and internal threats such as bacteria, viruses, and early-stage cancer cells. Nonspecific defense mechanisms include the physical barriers of the skin and mucous membranes; the cellular defense by phagocytic white blood cells; and the chemical inflammatory response. The *immune system* consists of cells that react specifically to an invader with the production of antibodies and various cell-mediated responses.

Nonspecific Defense Mechanisms

Nonspecific defense mechanisms, including the skin and mucous membranes, phagocytes, the inflammatory response, and antimicrobial proteins, are effective against a variety of microbial assaults.

The Skin and Mucous Membranes Skin is a physical barrier against microbes, reinforced by oil and sweat secretions with a low pH and containing *lysozyme*, an enzyme that attacks bacterial cell walls. Lysozyme is also present in tears, saliva, and mucus. The ciliated, mucus-coated epithelial lining of the respiratory tract traps and removes particles. The acidity of gastric juice kills most bacteria that reach the stomach.

Phagocytic White Cells and Natural Killer Cells   The major internal nonspecific defense is *phagocytosis* by white blood cells. Short-lived *neutrophils*, making up two-thirds of all white blood cells, leave the blood in response to chemical signals and phagocytize microbes in infected tissue. White blood cells called *monocytes* migrate into tissues and develop into *macrophages*, large, long-lived amoeboid cells that engulf bacteria, viruses, and cellular debris and destroy them with digestive enzymes and reactive forms of oxygen. Some microbes are resistant to the lytic enzymes of macrophages or have capsules that macrophages cannot grasp. Macrophages are found in the interstitial fluid or permanently fixed in organs such as the lungs, liver, lymph nodes and spleen.

Eosinophils are white blood cells that attack larger parasitic invaders, such as worms. *Natural killer cells* destroy the body's infected or aberrant cells by attacking their membranes.

Antimicrobial Proteins   A variety of proteins function in nonspecific defense by attacking microorganisms or interfering with their reproduction. *Complement* is a group of at least 20 proteins that cooperate with other nonspecific and specific defense mechanisms. *Interferons* are proteins produced by virus-infected cells that help other cells resist viral infection. Diffusing from an infected cell to neighboring cells, interferon stimulates production of proteins that inhibit viral replication. Recombinant DNA technology has made it possible to produce large quantities of interferons, which are being tested for their effectiveness in treating viral infections and cancer.

The Inflammatory Response   Physical injury or microorganisms can trigger an *inflammatory response*. Injured cells release chemical signals: basophils in the blood and mast cells in connective tissue release *histamine*, which triggers dilation and increased permeability (leakiness) of blood vessels; white blood cells and damaged tissue cells release prostaglandins and other chemicals, which promote blood flow to the damaged area. The increased blood flow and permeability of capillaries causes redness, swelling and heat, but is healing because blood clotting elements brought to the injured area begin the repair process and help contain the pathogenic microbes in the damaged area, and because phagocytic white blood cells congregate in the interstitial fluid. The pus that may accumulate is a combination of dead cells and fluid that leaked from the capillaries.

An inflammatory response may be localized, as just described, or systemic, including an increase in the number of circulating white blood cells and a fever. Injured cells release compounds that stimulate neutrophil release from bone marrow. Fever may be triggered by toxins produced by pathogens or by *pyrogens*, chemicals released by certain white blood cells. Fevers stimulate phagocytosis and inhibit growth of microorganisms.

The Immune System and Specific Defense: Some Basic Concepts

Basic Features of the Immune System   The *immune system* differs from nonspecific defenses on the basis of its specificity, diversity, self/nonself recognition, and memory. The immune system mounts specific attacks against invaders by recognizing foreign molecules, called *antigens*, and producing special lymphocytes and specific proteins called *antibodies*.

The diversity of the immune system results from the huge variety of lymphocytes that are each keyed to a specific antigenic marker and respond to that antigen by producing the appropriate antibody.

The immune system distinguishes *"self"* from *"nonself"* by recognizing differences in the cell-surface markers of invading bacteria, viruses, fungi, protozoans, pollen, as well as cells of tissue grafts. The "self antigens" on the surface of the body's own cells prevent the body from mounting an immune response against its own cells.

The immune system "remembers" antigens and can react against them more promptly upon re-exposure, a property known as *acquired immunity*.

Active Versus Passive Acquired Immunity   Immunity can be acquired by exposure to an inactive or attenuated pathogen, which initiates the system's long-term capability to respond quickly to the infective agent. Whether by vaccination or exposure and recovery from an infective disease, the body is stimulated to produce antibodies and develop *active immunity*. In *passive immunity*, antibodies supplied through the placenta to a fetus, through milk to a nursing infant, or by an antibody injection provide temporary immunity.

The Immune System's Dual Defense: An Overview of Humoral and Cell-Mediated Responses   *Humoral immunity* involves the production of antibodies that circulate in the blood and lymph and defend against free bacteria and viruses. *Cell-mediated immunity* involves specialized lymphocytes that react against body cells infected by bacteria or viruses and against fungi, protozoans, and worms. The cell-mediated system responds to tissue transplants and may protect the body from its own cancerous cells.

Lymphocytes mediate both humoral and cell-mediated immune responses. *B cells* (or B lymphocytes) produce antibodies; *T cells* (or T lymphocytes) function in cell-mediated immunity.

All blood cells develop from stem cells located in the bone marrow, or in the liver of a fetus. Lymphocytes either migrate to the thymus and differentiate into T cells or continue to develop in the bone marrow as B cells.

B cells and T cells have specific *antigen receptors* on their surfaces that allow them to react to an antigen. *T cell receptors* recognize antigens as specifically as do the bound antibodies that serve as receptors on B cells. When an antigen binds to a lymphocyte with a complementary antigenic receptor on its surface, the lymphocyte is activated to proliferate a clone of *effector cells*. Activated B cells give rise to antibody-producing *plasma cells*. T effector cells include *cytotoxic T cells*, which attack infected cells and cancer cells, and *helper T cells*, which stimulate humoral and cell-mediated immunity.

The Molecular Basis of Antigen-Antibody Specificity
Most antigens are proteins or polysaccharides that are part of the capsule, cell wall, or coat of bacteria or viruses. Antigens also may be part of the surface membrane of transplanted cells. Antibodies recognize a localized region, called an antigenic determinant, or *epitope*, of an antigen. Antigens may have many different epitopes recognized by different antibodies in the immune system.

Antibodies are a class of proteins called *immunoglobulins*, abbreviated *Ig*. The Y-shaped antibody molecule consists of two pairs of polypeptide chains: two identical short light chains and two identical longer heavy chains, linked together by disulfide bridges. Both heavy and light chains have constant regions, located in the stem of the Y, and variable regions, at the ends of the two arms of the Y, which form two antigen-binding sites. The amino acid composition of the variable region creates the unique contours and binding potential of these sites. Weak chemical bonds that form between an epitope and the antigen-binding site of an antibody account for the specificity of antibody–antigen binding.

There are five types of constant regions, creating five classes of antibodies in mammals: IgM, IgG, IgA, IgD, and IgE. Each class plays a different role in the immune response.

Clonal Selection: The Cellular Basis of Immunological Specificity and Diversity   The ability of the immune system to respond specifically to the enormous variety of different antigens depends on the presence of a great diversity of lymphocytes, each able to respond to a particular antigenic epitope. During embryonic development, each lymphocyte develops the ability to respond to a specific antigen. If such an antigen binds to its antigenic receptor, the selected cell then gives rise to a clone of millions of effector cells. *Clonal selection* is the mechanism by which the body mounts an immune response against a specific antigen.

The Cellular Basis of Immunologic Memory   Five to ten days after the first encounter with an antigen, enough B and T effector cells have proliferated to produce a *primary immune response*. If the body reencounters the same antigen, the *secondary immune response* is more rapid, effective, and longer in duration. The ability of the immune system to recognize a previously-encountered antigen is called immunologic memory and is due to the production of *memory cells* along with effector cells. Memory cells are long-lived and able to produce effector cells and more memory cells rapidly when stimulated by the same antigen.

The Development of Self-Tolerance   During embryonic development, lymphocytes carrying antigen receptors for molecules present in the body are destroyed, thus insuring *self-tolerance* and no immune response to the body's cells. The *major histocompatibility complex (MHC)* is a set of glycoproteins embedded in the plasma membrane that enable the immune system to distinguish self from nonself. This biochemical fingerprint, coded for by at least 20 MHC genes, each of which has 50 or more alleles, is unique to each individual (except for identical twins). *Class I MHC* molecules are found on all nucleated cells, whereas *Class II MHC* markers are found on macrophages and B cells, and allow these cells to interact with each other in immune responses.

The Humoral Immune Response: A Closer Look

The Activation of B Cells   The humoral immune response involves the production of antibodies by plasma cells that differentiated from B cells after exposure to an antigen. The selective activation of a B cell occurs when an antigen binds to surface receptors. This process may be assisted by macrophages called *antigen presenting cells (APCs)* that engulf pathogens and then display antigen fragments in a complex with their class II MHC molecules to specialized T cells. These *helper T cells* then stimulate B cells that have also encountered the antigen and are displaying the antigen in combination with a class II MHC molecule to proliferate into plasma cells.

Antigens that require this cooperative approach to antibody production are called *T-dependent antigens*. This sequence of events is illustrated below.

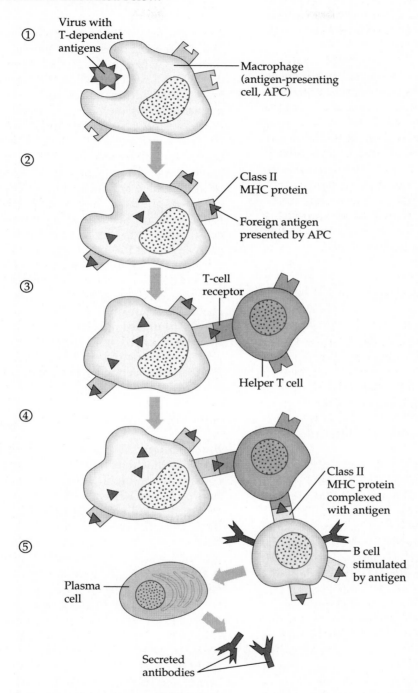

① Virus with T-dependent antigens — Macrophage (antigen-presenting cell, APC)

② Class II MHC protein — Foreign antigen presented by APC

③ T-cell receptor — Helper T cell

④ Class II MHC protein complexed with antigen

⑤ Plasma cell — B cell stimulated by antigen — Secreted antibodies

T-independent antigens trigger antibody production by B cells without the help of macrophages or T cells. Usually found in bacterial capsules and flagella, these antigenic molecules are usually long chains of repeating units, such as protein subunits or polysaccharides, that apparently bind to sufficiently numerous antigen receptors on a B cell to activate the cell.

How Antibodies Work Antibodies label foreign molecules and cells for destruction by one of several effector mechanisms. Antibodies may cover the binding sites of a bacterial toxin or a virus and thereby neutralize it. Because each antibody molecule has at least two antigen-binding sites, the formation of antigen-antibody complexes may cause bacteria to clump (agglutinate) or cause soluble antigen molecules to precipitate. The resulting clumps are then engulfed by phagocytes. Antigen-antibody complexes may activate the complement system, a set of blood proteins that coordinate with other elements of the defense systems.

Monoclonal Antibodies   Techniques for making *monoclonal antibodies* were developed in the late 1970s to supply quantities of identical antibodies for biological research and clinical testing. Such antibodies allow widescale clinical testing for pathogenic microbes and may soon be used to specifically target and destroy cancer cells and microbial toxins. Monoclonal antibodies are produced from a culture of hybrid cells called *hybridomas*, made by fusing tumor cells, which can be cultured indefinitely, with antibody-producing plasma cells, obtained from the spleen of animals that have been exposed to the antigen of interest.

The Cell-Mediated Response: A Closer Look

Whereas the humoral immune response reacts to free pathogens, cell-mediated immunity acts against pathogens that have already entered cells of the body. T cells have membrane-bound T-cell receptors that recognize a self-nonself complex of a specific antigenic epitope and MHC markers displayed by infected cells and respond by proliferating clones of cells to fight the specific pathogen.

Helper T cells (T_H) activate both humoral and cell-mediated responses by releasing signals called *cytokines*, chemicals secreted by one cell that regulate neighboring cells. When a helper T cell recognizes and binds to an antigen-presenting macrophage, the macrophage releases *interleukin-1*, which signals the T_H cell to release *interleukin-2*, a cytokine that stimulates helper T cells to grow more rapidly. Interleukin-2 and other cytokines released by T_H cells also help to activate B cells and the resulting humoral response.

Cytotoxic T (T_C) cells proliferate in response to the interleukin-2 released by helper T cells, attach to infected body cells displaying a complex of a specific antigen and class I MHC markers, and release perforin, a protein that forms a hole in the target cell's membrane. Cytotoxic T cells kill infected cells displaying specific epitopes and probably also attack cancer cells and foreign tissue transplant cells. T_C cells are the only T cells that directly kill other cells.

Another type of lymphocyte, *suppressor T cells (T_S)* probably suppress the immune response when an antigen is no longer present.

The Complement System: A Component of Both Nonspecific and Specific Defenses

Complement proteins cooperate with both specific and nonspecific defense mechanisms. In the "classical pathway," when antibody molecules target an invader, such as a bacterial cell, by attaching to it, a complement protein forms a bridge between two antibody molecules. This molecular association activates complement proteins to form a membrane attack complex that produces a lesion in the cell membrane, causing the foreign cell to lyse.

The several variations of "alternative pathways" do not involve antibodies and are, therefore, nonspecific. Complement may be activated by substances found in some bacteria, yeasts, viruses, virus-infected cells, and protozoans to form membrane attack complexes without the aid of antibodies. Complement can amplify the inflammatory response by stimulating histamine release. One complement attracts phagocytes to infection sites. Some complement proteins coat microorganisms in a process called opsonization, which stimulates phagocytosis. In immune adherence, microbes coated with antibodies and complement proteins adhere to the walls of blood vessels, facilitating their destruction by phagocytic cells.

Self Versus Nonself: Some Applications

Blood Groups   The immune response to the chemical markers that determine *ABO blood groups* must be considered in blood transfusions. Individuals with type A blood have the A antigen of the surface of their red blood cells and produce antibodies against the B antigen. Likewise, individuals with type B blood have B antigens on their blood cells and produce anti-A antibodies. Group AB individuals have both A and B antigens and produce neither antibody; they are called universal recipients since they can receive transfusions of any blood type. Type O individuals are called universal donors because their blood cells carry no antigens and will not be destroyed by complement-mediated lysis when exposed to the antibodies of a blood recipient. They produce A and B antibodies and therefore can only receive type O blood.

An Rh-negative mother may develop antibodies against the *Rh factor*, another red-blood-cell antigen, if fetal blood from an Rh-positive child leaks across the placenta. Should she carry a second Rh-positive fetus, her immunological memory may result in the production of antibodies that cross the placenta and destroy fetal red blood cells. Treatment of the mother with anti-

Rh antibodies just after delivery destroys any fetal cells that may have leaked into her circulation and prevents the mother's immunological response to the antigen.

Tissue Graphs and Organ Transplants Transplanted tissues and organs are rejected because the foreign antigens of the major histocompatibility complex trigger T cell responses. The use of closely related donors, as well as drugs such as cyclosporine, which suppress cell-mediated immunity, help to reduce the immune response to a transplant operation.

Disorders of the Immune System

Autoimmunity Sometimes the immune system turns against self, leading to *autoimmune diseases*, such as lupus erythematosis, rheumatoid arthritis, rheumatic fever, insulin-dependent diabetes, and Grave's disease.

Allergy *Allergies* are hypersensitivities to certain environmental antigens, or allergens. IgE antibodies bound to mast cells (stationary cells in connective tissue) trigger allergic reactions. When antigens bind to these cell-surface antibodies, the mast cells degranulate and release histamines, which create an inflammatory response that may include sneezing, a runny nose, and difficulty in breathing due to smooth muscle contractions. Antihistamines are drugs that counteract histamine. Anaphylactic shock is a severe allergic response in which the abrupt dilation of peripheral blood vessels caused by a rapid release of histamines leads to a life-threatening drop in blood pressure. Severe allergic reactions can be counteracted with injections of the hormone epinephrine.

Immunodeficiency A deficiency may occur in any of the components of the immune system. In the rare congenital disease known as severe combined immunodeficiency, both T and B cells are absent or inactive. Certain cancers, such as Hodgkin's disease, and AIDS depress the immune system, making a person susceptible to infections.

There is growing evidence that general emotional health and immunity are related. Hormones secreted during stress affect the number of white blood cells; nerve fibers penetrate deep into lymphoid tissue and receptors for chemical signals from nerve cells have been found on lymphocytes. Physiologists are now looking at the connections between general health, state of mind and immunity.

AIDS *Acquired immunodeficiency syndrome (AIDS)* was first recognized in the early 1980s, when the increasing incidence of Kaposi's sarcoma and *Pneumocystis* pneumonia was investigated. The infectious agent responsible for AIDS, known as *HIV (human immunodeficiency virus)*, was identified in the United States and France in 1983. This most lethal pathogen ever encountered probably arose by mutation in central Africa. The virus has been identified in preserved blood samples from as early as 1959.

Glycoproteins on the surface of HIV bind to a receptor called CD4 found on the surface of a subclass of helper T cells. Some macrophages and B cells also carry CD4 and may be infected. The virus enters the cell and replicates, and new viruses bud from the host cell. HIV can remain latent as a provirus in the genome of an infected cell.

HIV undergoes rapid antigenic changes during replication, complicating the ability of the immune system to respond to the infection. The number of T helper cells, which, as you have just learned, activate the other T and B lymphocytes, decreases as they are killed by HIV, and the immune system is severely weakened. The collapse of cell-mediated immunity allows for the development of opportunistic infections such as *Pneumocystis* pneumonia and Kaposi's sarcoma. It usually takes about ten years for an HIV infection to progress to the specified reduction of T cells and the appearance of characteristic secondary infections referred to as AIDS.

Most individuals exposed to HIV have circulating antibodies to the virus and will test as *HIV-positive*. This test for HIV antibodies is used to screen blood donations. HIV is transmitted by transfer of body fluids such as blood or semen. Sharing needles between intravenous drug uses, unscreened blood supplies, and unprotected sex—particularly where there is a high incidence of sexually transmitted diseases—are all means by which AIDS is spread.

Newly developed drugs may slow the development of the virus or treat the opportunistic infections, but no cure has been found. The rapid genetic changes of HIV have complicated the development of a vaccine against it.

Immunity in Invertebrates

The ability to distinguish self from nonself is well developed in invertebrates. Many invertebrates have amoeboid cells called coelomocytes that identify and destroy foreign substances. Tissue graft experiments in earthworms have established that their defense systems reject foreign tissue and develop a memory response to these grafts.

STRUCTURE YOUR KNOWLEDGE

This chapter contains a wealth of information that is probably fairly new to you. If you take a little time and pull out the key players of the immune system and organize them first into very basic concept clusters and then develop more interrelated concept maps, you will find that this information is both understandable and fascinating.

1. Fill in the chart below on some of the molecules involved in the immune system.

2. Create a concept map outlining specific defense mechanisms, showing the cells involved in humoral and cell-mediated immunities and their functions.

3 Describe the structure of an antibody molecule and how this structure relates to its function. Briefly explain the clonal selection theory.

4. Briefly describe the illustrated steps of the classical complement pathway shown on p. 337.

TEST YOUR KNOWLEDGE

MULTIPLE CHOICE: *Choose the one best answer.*

1. Which of the following is incorrectly paired with its effect?
   a. gastric juice—kills bacteria in the stomach
   b. fever—stimulates phagocytosis and inhibits microbial growth
   c. histamine—causes blood vessels to dilate
   d. vaccination—creates passive immunity

2. Monoclonal antibodies
   a. are used to treat AIDS.
   b. explain the ability of the immune system to create a large number of identical T or B cells.
   c. are produced in tissue culture by hybridomas.
   d. initiate the secondary immune response.

3. Antigens are
   a. proteins that consist of two light and two heavy polypeptide chains.
   b. proteins or polysaccharides usually found on the cell surface of invading bacteria or viruses.
   c. proteins circulating in the blood that tag foreign blood cells for complement-mediated lysis.
   d. proteins embedded in B-cell membranes.

4. A secondary immune response is more rapid and greater in effect than a primary immune response because
   a. memory cells respond to the pathogen and rapidly clone more effector cells.
   b. the second response is an active immunity, whereas the primary one was a passive immunity.
   c. helper T cells are available to activate other blood cells.
   d. interleukins cause the rapid accumulation of phagocytic cells.

MOLECULE	WHERE PRODUCED OR FOUND	ACTION
Lysozyme		
Histamine		
Interferon		
Complement		
Antibody		
Interleukin-1		
Interleukin-2		
Perforin		

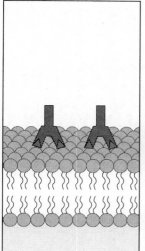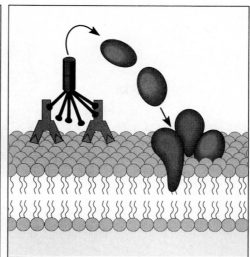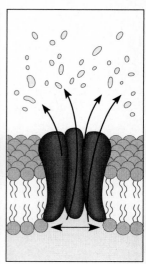

5. Lymphocytes capable of reacting against "self" molecules
   a. are usually not a problem until a woman's second pregnancy.
   b. are destroyed before birth.
   c. are usually kept separate from the immune system.
   d. contribute to immunodeficiency diseases.

6. The major histocompatibility complex
   a. is involved in the ability to distinguish self from nonself.
   b. is a set of several cell-surface antigens.
   c. may trigger T-cell responses after transplant operations.
   d. All of the above are correct.

7. In opsonization,
   a. complement proteins coat microorganisms and help phagocytes bind to and engulf the invading cell.
   b. a set of complement proteins lyses a hole in the foreign cell's membrane.
   c. antibodies cause cells to agglutinate, and the resulting clumps are engulfed by phagocytes.
   d. a flood of histamines is released that may result in anaphylactic shock.

8. Severe combined immunodeficiency
   a. is an autoimmune disease.
   b. is a form of cancer in which the membrane surface of the cell has changed.
   c. is a disease in which both T and B cells are absent or inactive.
   d. is an immune disorder in which the number of helper T cells is greatly reduced.

9. A transfusion of B-type blood given to a person who has A-type blood would result in
   a. the recipient's anti-B antibodies reacting with the donated red blood cells.
   b. the recipient's B antigens reacting with the donated anti-B antibodies.
   c. the recipient forming both anti-A and anti-B antibodies.
   d. no reaction, because B is a universal donor type of blood.

10. Which of the following are incorrectly paired?
   a. variable region—antibody specificity for an epitope
   b. helper T cells—production of plasma cells
   c. cytotoxic T cells—destruction of foreign cells
   d. immunoglobulins—antibodies

11. Which of the following does *not* destroy a target cell by creating a hole in the membrane that causes the cell to lyse?
   a. neutrophil
   b. cytotoxic T cell
   c. natural killer cell
   d. complement proteins

12. A T-independent antigen
   a. binds directly to antibody molecules without the aid of helper T cells.
   b. needs to bind to T_H cells to activate antibody production.
   c. is often a large molecule that simultaneously binds with several antigenic receptors on a B cell at one time.
   d. will result in the production of both plasma cells and memory cells.

CHAPTER **40**

CONTROLLING THE INTERNAL ENVIRONMENT

FRAMEWORK

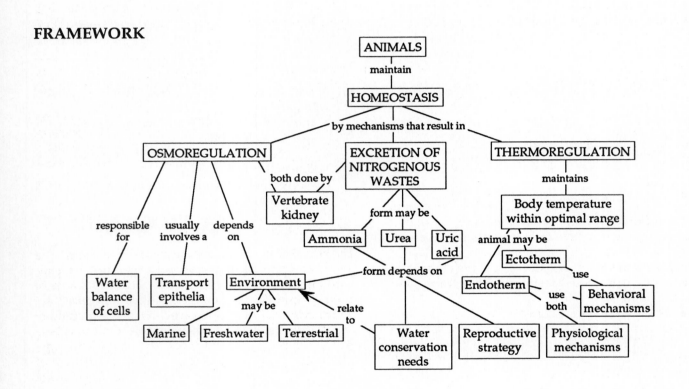

CHAPTER SUMMARY

Animals are able to survive large fluctuations in their external environment by maintaining homeostasis—a relatively constant internal environment. Changes in the internal body fluid that bathes the cells, either hemolymph or interstitial fluid serviced by blood, are tempered by regulatory systems that usually involve feedback mechanisms. The physiological adjustments that maintain homeostasis, such as *osmoregulation*, *excretion*, and *thermoregulation*, enable organisms to cope with short-term environmental changes. These

mechanisms developed from natural selection as populations evolved in specific environments.

Osmoregulation

Whether an animal lives in saltwater, freshwater, or on land, water gain must balance water loss in body cells. There are two basic solutions to the problem of water balance.

Osmoconformers and Osmoregulators Osmosis is the diffusion of water across a selectively permeable mem-

338

brane that separates two solutions differing in *osmolarity* (moles of solute per liter). Osmolarity is expressed in units of milliosmoles per liter (10^{-3} moles/L). Isosmotic solutions are equal in osmolarity, and there is no net osmosis between them. Water flows across a membrane from a hypoosmotic (more dilute) to a hyperosmotic (more concentrated) solution.

Osmoconformers are isosmotic with their aqueous surroundings and do not regulate their osmolarity. *Osmoregulators* must get rid of excess water if they live in a hypoosmotic medium or take in water to offset osmotic loss if they inhabit a hyperosmotic environment.

Problems of Osmoregulation in Different Environments

Most marine invertebrates are osmoconformers, whereas most marine vertebrates are osmoregulators. Sharks have an internal salt concentration lower than that of seawater because their rectal glands pump salt out of the body. They maintain an osmolarity slightly higher than that of seawater, however, by retaining urea (a nitrogenous waste product) and trimethylamine oxide (protects proteins from the damaging effects of urea) within their bodies. They produce a large quantity of urine to balance the osmotic uptake of water.

Many marine bony fishes, having evolved from freshwater ancestors, are hypoosmotic to seawater and must drink large quantities of seawater to replace the water they lose by osmosis. Excess salt is pumped out through salt glands located in the gills. Many marine birds get rid of excess salt through nasal salt glands; marine reptiles also have osmoregulating salt glands.

Freshwater animals constantly take in water by osmosis because their internal fluids are hyperosmotic to their medium. Protozoa use contractile vacuoles to pump out excess water. Freshwater animals excrete large quantities of dilute urine. Salt supplies are replaced from their food, or, in some fish, by active uptake of ions across the gills.

Most animals are *stenohaline*, able to tolerate only small changes in external osmolarity. Animals that move between fresh- and saltwater environments, or that live in brackish water, are *euryhaline*. They can survive in these different osmotic environments by changing their osmoregulatory mechanisms to maintain a constant internal osmolarity.

Some animals are capable of *anhydrobiosis*, or cryptobiosis—surviving dehydration in a dormant state. One mechanism of this adaptation is the production of a large number of disaccharides, whose many hydroxyl groups form hydrogen bonds, replacing water and protecting cellular structures during dehydration.

The most important problem confronting terrestrial animals is the threat of desiccation. Adaptations to prevent dehydration include impervious coverings, drinking and eating food with high water content, nervous and hormonal control of thirst, behavioral adaptations such as nocturnal lifestyles, and water-conserving excretory organs. Only arthropods and vertebrates have successfully colonized land.

Transport Epithelia and Osmoregulation

Osmolarity of the internal environment is usually regulated by transporting salt, followed by the osmotic movement of water, across a *transport epithelium*. Located at the tissue-environment boundary, this single sheet of cells is linked by impermeable tight junctions and is able to regulate the passage of solutes between the extracellular fluid and the environment. Variations in molecular composition of the plasma membrane of the transport epithelium determine its passive permeability to water and salts and its active transport by various membrane proteins.

The transport epithelia of salt glands function exclusively in osmoregulation, whereas the transport epithelia of excretory organs, often arranged in tubular networks, function additionally in metabolic waste excretion.

Excretory Systems of Invertebrates

Protonephridia: Flame-Cell System of Flatworms

The *flame-cell system* of flatworms is a type of protonephridia, consisting of a branched system of tubules embedded in the body tissues. Extracellular fluids pass into the bulbous flame cells at the ends of the smallest tubules. Cilia projecting into the tubule propel the fluid along into excretory ducts that empty by way of nephridiopores. The flame-cell system of freshwater flatworms is primarily osmoregulatory; metabolic wastes are excreted through the gastrovascular cavity.

Protonephridia are simple excretory systems of closed tubules lacking internal openings found in rotifers, some annelids, molluscan larvae, and lancelets, as well as in flatworms.

Metanephridia of Earthworms

Excretory tubules with internal openings that collect body fluids, called *metanephridia*, are found in most annelids. The metanephridia of earthworms occur in pairs in each segment of the worm. An open ciliated funnel called a nephrostome collects coelomic fluid, which then moves through a folded tubule encased in capillaries. The transport epithelium of the tubule pumps salts out of the tubule, and the salts are reabsorbed into the blood. Copious dilute urine, which helps to offset the osmotic uptake of water from moist soil, exits by way of nephridiopores.

Malpighian Tubules of Insects   In insects and other terrestrial arthropods, *Malpighian tubules* remove nitrogenous wastes from the body fluid (hemolymph) and function in osmoregulation. Transport epithelia lining these blind sacs, which open into the digestive tract at the end of the midgut, pump salts and nitrogenous wastes from the hemolymph into the tubule. The fluid passes through the hindgut and into the rectum, where a transport epithelium pumps most of the salt back into the hemolymph, and water follows by osmosis. In this water-conserving system, nitrogenous wastes are eliminated as dry matter along with the feces.

The Vertebrate Kidney

Nephrons, the excretory tubules of vertebrates, are arranged into compact organs, the *kidneys*. The vertebrate excretory system consists of the kidneys, associated blood vessels, and excretory ducts.

Anatomy of the Excretory System   Blood enters the pair of bean-shaped kidneys through the *renal arteries* and leaves by way of the *renal veins*. *Urine* exits through the *ureter* and is temporarily stored in the *urinary bladder*. The body periodically releases urine by micturition (urination) through the *urethra*.

Structure of the Nephron   Each human kidney contains about a million nephrons, each of which consists of a *renal tubule* and associated blood vessels. Water, urea, salts, and small molecules flow from capillaries into the renal tubule, forming the *filtrate*, which is processed by the transport epithelium of the tubule to form urine.

The blind end of the renal tubule is formed into a cuplike *Bowman's capsule* that encloses a ball of capillaries called the *glomerulus*. Filtrate enters the tubule and passes through its three specialized regions: a *proximal convoluted tubule*, the *loop of Henle* with a *descending limb* and an *ascending limb*, and a *distal convoluted tubule*. *Collecting ducts* receive filtrate from many renal tubules, and pass the urine into the renal pelvis, the chamber that drains into the ureter.

Bowman's capsules and the proximal and distal tubules are restricted to the outer *cortex* of the kidney. *Cortical nephrons* have reduced loops of Henle and are located entirely in the cortex. *Juxtamedullary nephrons*, found only in mammals and birds, have long loops of Henle that extend into the *medulla*.

An *afferent arteriole* supplies each nephron, subdi-

viding to form the capillary ball and converging as it leaves the capsule to form an *efferent arteriole*. This arteriole then forms a second capillary network—the *peritubular capillaries*—which surrounds the proximal and distal convoluted tubules. Other capillaries form the *vasa recta*, the descending and ascending capillaries that parallel the loop of Henle. Exchange between the tubules and capillaries is via the interstitial fluid.

General Physiology of the Nephron   *Filtration* occurs when blood pressure forces water and small solutes out of the porous capillaries and through clefts between *podocytes*, which are the cells lining Bowman's capsule, and on into the nephron tubule. The filtrate in the tubule contains salts, nitrogenous wastes, glucose, and other small molecules. Blood cells and plasma proteins remain in the capillary.

Secretion of substances from the interstitial fluid across the tubule epithelium occurs as the filtrate moves through the proximal and distal convoluted tubules. This selective secretion involves both passive and active transport.

Filtration is nonselective, and many necessary molecules pass into the filtrate. Most of the water, sugar, vitamins, and other organic nutrients in the filtrate are *reabsorbed* across the tubule epithelium into the interstitial fluid, from which they are returned to the blood. Selective secretion and reabsorption control the balance of water and various salts in body fluids.

Transport Properties of the Renal Tubule   (1) Reabsorption and secretion both take place in the proximal convoluted tubule. Its transport epithelium selectively secretes ammonia, drugs and poisons processed by the liver, and hydrogen ions into the filtrate and actively transports glucose and amino acids out of the filtrate. Potassium is also reabsorbed. The *brush-border* of the epithelia cells on the interior of the tubule provides a huge surface area for NaCl and water to diffuse into the epithelial cells. The cells actively transport Na^+ across the membrane into the interstitial fluid outside the tubule; Cl^- is transported passively out of the cell to balance the increasing positive charge; and water follows by osmosis. About 75% of the salt and 70% of the water forced into the filtrate is reabsorbed in this area and diffuses into the peritubular capillaries.

(2) The descending limb of the loop of Henle is freely permeable to water but not to salt or other small solutes. The interstitial fluid is increasingly hyperosmotic toward the inner medulla, so water exits by osmosis, leaving a filtrate with high NaCl concentration.

(3) The ascending limb of the loop of Henle is not very permeable to water, but is permeable to salt,

which diffuses out of the lower thin segment of the loop, adding to the high osmolarity of the medulla. Cl⁻ is actively transported out of the thick upper portion of the ascending limb, and Na⁺ follows the negative charge passively. As salt leaves, the filtrate becomes less concentrated.

(4) The distal convoluted tubule is specialized for selective secretion and reabsorption, contributing to homeostasis of the body fluids. It regulates K⁺ concentration by controlling its secretion, and helps to regulate pH by secreting H⁺ and reabsorbing the buffering compound bicarbonate.

(5) As the filtrate moves in the collecting duct back through the increasing osmotic gradient of the medulla, more and more water exits by osmosis. The duct is not permeable to salt, but its lower region is permeable to urea. Urea diffuses out of the duct and helps to form the osmotic gradient that enables the kidney to produce urine that is hyperosmotic to general body fluids.

How the Mammalian Kidney Conserves Water: A Closer Look

The osmolarity gradient of NaCl and urea in the interstitial fluid is produced by the juxtamedullary nephrons and enables the kidney to produce urine up to four times as concentrated as blood and normal interstitial fluid. Filtrate leaving Bowman's capsule has an osmolarity of about 300 mosm/L, the same as blood. Both water and salt are reabsorbed in the proximal convoluted tubule; the volume of filtrate is reduced but the osmolarity remains about the same. In the trip down the descending limb of the loop of Henle, water exits by osmosis, and the filtrate becomes more concentrated. Salt, which is now in high concentration in the filtrate, diffuses out as the filtrate moves up the salt-permeable but water-impermeable ascending limb—helping to create the osmolarity gradient.

The vasa recta has a countercurrent flow to the movement of filtrate in the loop of Henle. As the blood in these capillaries moves down into the inner medulla, water leaves by osmosis and salt enters; as the blood moves back up toward the cortex, water moves back in and salt diffuses out. Thus the capillaries can carry nutrients and other supplies to the medulla without disrupting the osmolarity gradient by carrying away NaCl or watering down the interstitial fluid.

The filtrate makes one final pass through the medulla, this time in the collecting duct, which is permeable to water and urea but not to salt. Water flows out by osmosis, and, as the filtrate becomes more concentrated, urea leaks into the interstitial fluid, significantly adding to the osmolarity of the inner medulla. The resulting concentrated urine is isosmotic to the interstitial fluid of the inner medulla, which can be as high as 1200 mosm/L.

Regulation of the Kidneys

Osmolarity of urine in humans can vary from 70 to 1200 mosm/L, depending on the body's hydration needs. Osmoregulation results from nervous and hormonal control of water and salt reabsorption in the kidneys.

Antidiuretic hormone, or *ADH,* is produced in the hypothalamus and stored in the pituitary gland. When blood osmolarity rises above a set point of 300 mosm/L, *osmoreceptor cells* in the hypothalamus trigger the release of ADH. This hormone increases water permeability of the distal convoluted tubules and collecting ducts, increasing water reabsorption and preventing further increase in blood osmolarity. After consuming water in food or drink, negative feedback decreases the release of ADH. When blood osmolarity is low, the absence of ADH results in the production of large volumes of dilute urine, called diuresis. Alcohol inhibits ADH release and can cause dehydration.

The *juxtaglomerular apparatus (JGA),* located near the afferent arteriole leading to the glomerulus, responds to a drop in blood pressure or a decrease in blood Na⁺ concentration by releasing *renin,* an enzyme that activates the plasma protein *angiotensin.* The active angiotensin II functions as a hormone that constricts arterioles and thus raises blood pressure, and stimulates the adrenal glands to release *aldosterone.* This hormone stimulates Na⁺ reabsorption in the distal convoluted tubules, which leads to the osmotic flow of water from the filtrate and an increase in blood volume and blood pressure. This mechanism prevents water loss should salt concentration and/or blood volume drop. ADH and aldosterone are both hormones that increase water reabsorption, but they respond to different osmoregulatory problems: ADH to an increase in blood osmolarity, and aldosterone to a drop in blood pressure or deficiency of Na⁺.

Atrial natriuretic protein (ANP) counters the renin–angiotensin–aldosterone mechanism. ANP, released by the atrium of the heart in response to increased blood volume and pressure, inhibits the release of renin from the juxtaglomerular apparatus and reduces the release of aldosterone from the adrenals.

Comparative Physiology of the Kidney

Modifications of nephron structure and function in various classes of vertebrates correspond to osmoregulatory and excretory requirements in various habitats. Mammals in dry habitats, needing to conserve water and excrete hyperosmotic urine, have long loops of Henle. Birds also have juxtamedullary nephrons that can produce osmotic gradients and hyperosmotic urine. The kidneys of reptiles have only cortical nephrons and their urine is isosmotic to body fluids. Water is conserved, however, by reabsorption by the epithelium of the cloaca and the production of uric acid, an insoluble nitrogenous waste.

Freshwater fishes are hyperosmotic to their environment and must excrete large quantities of very dilute urine. Salts are conserved by the efficient reabsorption of ions from the filtrate. When in fresh water, frogs and other amphibians actively transport certain salts across their skin and excrete a dilute urine. When on land, water is reabsorbed from the urinary bladder.

The kidneys of marine bony fishes excrete very little urine and function mainly to get rid of divalent cations taken in by drinking sea water. Monovalent ions and nitrogenous wastes in the form of ammonium are excreted by the epithelia of the gills.

The original function of the kidney was osmoregulation; excretion of nitrogenous wastes became a second function over the course of evolutionary time.

Nitrogenous Wastes

Ammonia, a small and toxic molecule, is produced when proteins and nucleic acids are metabolized for energy or converted to carbohydrates or fats. Some animals excrete ammonia; others convert it to less toxic wastes such as urea or uric acid.

Ammonia   Aquatic animals can excrete nitrogenous wastes as ammonia because it is very soluble and easily permeates membranes. Soft-bodied invertebrates lose ammonia across their whole body surface. In freshwater fishes, most of the ammonia is passed as NH_4^+ ions across the epithelia of the gills in exchange for Na^+ taken up from the water.

Urea   Mammals, most adult amphibians, many marine fishes, and turtles produce urea, a much less toxic compound that can be tolerated in more concentrated form and excreted with less loss of water. Ammonia and carbon dioxide are combined in the liver to produce urea, which is then carried by the circulatory system to the kidneys.

Uric Acid   Land snails, insects, birds, and some reptiles produce uric acid, a compound of low solubility in water that can be excreted as a precipitate after nearly all the water has been reabsorbed from the urine.

The mode of reproduction of terrestrial animals seems to have determined whether they excrete uric acid or urea as their nitrogenous waste product. Vertebrates that produce shelled eggs excrete uric acid, which can be stored safely within the egg while

the embryo develops. Mammals and amphibians produce urea, which can be removed by the placental blood or diffuse out of a shell-less egg.

The form of nitrogenous wastes also relates to habitat. Terrestrial reptiles excrete uric acid, whereas crocodiles excrete ammonia and uric acid, and aquatic turtles excrete both urea and ammonia. Some animals actually shift their nitrogenous waste product depending on environmental conditions.

Regulation of Body Temperature

Metabolism is very sensitive to internal changes in temperature. Many animals can maintain their internal temperature within an optimal range even as the environmental temperature fluctuates.

Heat Production and Transfer Between Organisms and their Environments   Heat is exchanged between an organism and the environment by the physical processes of *conduction*, the direct transfer of thermal motion between surfaces in contact; *convection*, the mass flow of air or water past a body; *radiation*, the emission of electromagnetic waves by all objects warmer than absolute zero; and *evaporation*, the conversion of the surface molecules of a liquid to a gas. Water conducts heat much more effectively than air. Convection and radiation account for the bulk of heat loss under relatively mild conditions. Convection and evaporation are the most variable causes of heat loss; wind and the evaporation of sweat can greatly increase the loss of heat.

Ectotherms and Endotherms   Thermoregulation can be classified on the basis of heat source. Animals such as invertebrates, fishes, amphibians, and reptiles absorb their body heat from their surroundings and are called *ectotherms*; whereas mammals and birds, who derive their body heat from their own metabolism, are called *endotherms*.

Endothermy allows terrestrial animals to maintain a constant body temperature in an environment where severe temperature fluctuations may be common. A warm body temperature requires active metabolism, which is also needed to facilitate the energetically expensive movement of animals on land.

Thermoregulation in Terrestrial Mammals   Metabolism generates heat; fat and fur insulation helps retain that heat. Heat production can be increased by muscle contraction (by moving or shivering) and by a rise in metabolic rate. Certain hormones can increase the metabolic rate by *nonshivering thermogenesis*, a form of rapid heat production *Brown fat* in some mammals is specialized for this process. Endothermy is energetically expensive.

The physiological and behavioral processes that enable a land mammal to maintain a constant body temperature include: (1) adjustment of the rate of metabolic heat production; (2) adjustment of the amount of blood flowing to the skin—*vasodilation* increases the rate of heat exchange, whereas *vasoconstriction* decreases it; (3) control of evaporative heat loss by panting or sweating; and (4) changes in behavioral responses, such as moving to a warmer or cooler environment.

In the hypothalamus of the brain are two thermoregulatory areas: a *heating center* that controls vasoconstriction of superficial vessels, shivering, and nonshivering thermogenesis, and a *cooling center* that controls vasodilation and sweating or panting. Both warm and cold temperature-sensing nerve cells are located in the skin, hypothalamus, and some other areas of the nervous system. They stimulate one thermoregulatory area while inhibiting the other. Control of body temperature involves feedback mechanisms, with the hypothalamus functioning as the thermostat.

Some Thermoregulatory Adaptations in Other Animals Birds usually have very high body temperatures. A bird pants to promote evaporative heat loss. Feathers serve as insulation against heat loss. In the legs of water-dwelling birds, the close contact between arteries and veins going to and from the limbs creates a *countercurrent heat exchanger* that reduces heat loss.

Loss of heat to water occurs 50 to 100 times more rapidly than loss to air. Marine mammals maintain their high body temperature by efficient heat-conserving mechanisms, including a thick layer of insulating blubber and countercurrent heat exchange between arterial and venous blood. When in warmer waters, these animals can dissipate metabolic heat by dilating superficial blood vessels.

The metabolic rate of the ectothermic reptiles is very low and contributes little to body temperature. Behavioral adaptations, such as orientating the body to the sun or finding suitable microclimates, allow these animals to regulate their temperature within an optimal range. In diving reptiles, more blood is routed to the body core to conserve heat. Reptiles also can increase thermogenesis by shivering.

Amphibians produce little heat and easily lose heat by evaporation from their moist skin. Behavioral adaptations enable them to maintain their internal temperature within an optimal range.

The body temperature of fishes is determined mainly by the water temperature. Some large, active fish are able to retain the heat produced by their swimming muscles and thus maintain an elevated body-core temperature. A countercurrent heat exchanger, called the *rete mirabile*, keeps the deep swimming muscles warmer than surface tissues. These animals are considered to be partial endotherms.

Some invertebrates adjust their temperature by behavioral or physiological mechanisms. Large flying insects may "warm up" before taking off by contracting their flight muscles. Winter moths have a countercurrent heat exchanger that elevates the temperature of their flight muscles. The social organization of honeybees enables them to regulate body heat in cold temperatures by huddling together. They also can cool their hive by bringing in water and fanning it with their wings to promote evaporation and convection.

Temperature Acclimation Many animals are capable of *acclimation*, a physiological adjustment to a different temperature, such as occurs in seasonal changes. Increased production of enzymes may offset lower enzyme activity at nonoptimal temperatures. Variants of enzymes with different optimal temperatures may also be produced. Changes in the proportions of saturated and unsaturated lipids keep membranes fluid at different temperatures.

Torpor Torpor is a physiological state characterized by decreases in metabolic, heart, and respiratory rates. In *hibernation*, the body temperature is maintained at a lower level. A hibernating animal is able to withstand long periods of cold temperature and decreased food supply. In *aestivation*, animals survive long stretches of elevated temperature and diminished water supply. *Diurnation* is a short torpor that may occur on a daily cycle.

Many ectotherms enter a state of slowed metabolism and inactivity when their food supply decreases. Environmental changes linked to an increase in food supply cause them to come out of torpor. Some hummingbirds and shrews, which have very high metabolic rates due to their small size, enter a daily torpor controlled by the biological clock during the periods when they are not feeding. Hibernation and aestivation may be triggered by changes in the length of daylight.

Interaction of Regulatory Systems

A major challenge of animal physiology is determining how the various regulatory systems are controlled and how they interact to maintain internal homeostasis. The feedback circuits of homeostasis involve nervous communication and hormones.

STRUCTURE YOUR KNOWLEDGE

1. Fill in the table below with a brief description of the osmoregulatory, excretory, and thermoregulatory mechanisms for the animals listed.

ANIMAL	OSMOREGULATION	EXCRETORY SYSTEM	THERMOREGULATION
Marine invertebrate			
Marine bony fish			
Freshwater fish			
Flatworm			
Earthworm			
Insect			
Reptile			
Bird			
Mammal			

2. Look at the table you have completed and draw some generalizations relating the control mechanisms of these animals to their environments and their evolutionary history. In particular, think about the problems of osmoregulation presented by marine, freshwater, and terrestrial habitats, the types of nitrogenous waste products suited to these environments, and the thermoregulatory considerations in an aquatic or terrestrial habitat. Now fill in the following table with your conclusions.

ENVIRONMENT	OSMOREGULATION	NITROGENOUS WASTE	THERMOREGULATION
Marine			
Freshwater			
Terrestrial			

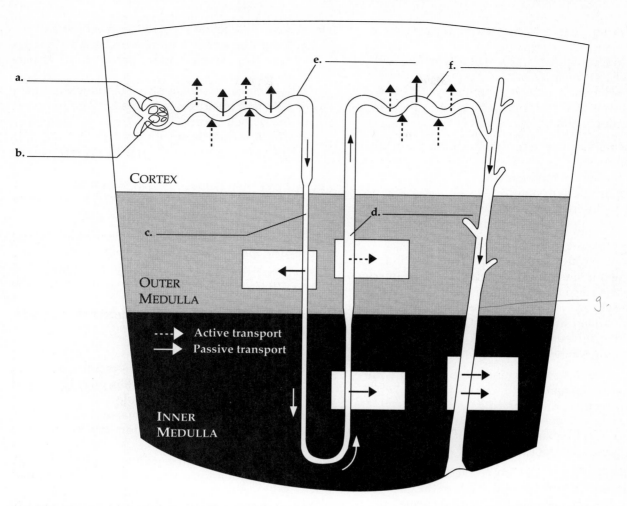

SUBSTANCES FILTERED	SUBSTANCES SECRETED	SUBSTANCES REABSORBED

3. In the diagram of a nephron above, label the parts on the indicated lines. Label the arrows to indicate the movement of salt, water, glucose, K^+, bicarbonate (HCO_3^-), H^+, and urea into or out of the tubule. What enters the tubule through Bowman's capsule? In the table below the diagram, list the substances that are filtered, secreted, and reabsorbed in the production of urine.

4. The osmolarity of urine can vary from 70 to 1200 mosm/L. Develop a concept map to help organize your understanding of the physiological control of water and salt reabsorption in the kidneys. Include the organs or structures involved, the hormones or other proteins released, and their effects.

TEST YOUR KNOWLEDGE

MULTIPLE CHOICE: *Choose the one best answer.*

1. There is a net water flow by osmosis through a membrane from

a. a solution that is hyperosmotic to one that is hypoosmotic.

b. a solution with a lower osmolarity to one with a higher osmolarity.

c. one isosmotic solution to another.

d. cells in a freshwater environment to the surrounding medium.

2. Contractile vacuoles most likely would be found in protists
a. in a freshwater environment.
b. in a marine environment.
c. that are internal parasites.
d. that are hypoosmotic to their environment.

3. Transport epithelia are responsible for
a. pumping water across a membrane.
b. forming an impermeable boundary at an interface with the environment.
c. the exchange of solutes between the extracellular fluid and the environment, followed by the osmotic movement of water.
d. pumping salt into a marine bony fish.

4. Which of the following is incorrectly paired with its excretory system?
a. earthworm—protonephridia
b. flatworm—flame-cell system
c. insect—Malpighian tubules
d. amphibian—kidneys

5. A freshwater fish would be expected to
a. pump salt in through salt glands in the gills.
b. produce copious quantities of dilute urine.
c. diffuse ammonia out across the epithelium of the gills.
d. do all of the above.

6. Which of the following is *not* part of the filtrate entering Bowman's capsule?
a. water, salt, and electrolytes
b. glucose
c. plasma proteins
d. urea

7. Which is the correct pathway for the passage of urine in vertebrates?
a. collecting tubule → ureter → bladder → urethra
b. renal vein → renal ureter → bladder
c. nephron → urethra → bladder → ureter
d. cortex → medulla → bladder → ureter

8. Aldosterone
a. is a hormone that stimulates thirst.
b. is secreted by the adrenal glands in response to a high osmolarity of the blood.
c. stimulates the active reabsorption of Na$^+$ in the nephrons.
d. causes diuresis.

9. Which of the following statements is *incorrect*?
a. Long loops of Henle are associated with steep osmotic gradients and the production of hyperosmotic urine.
b. Ammonia is a toxic nitrogenous waste molecule that invertebrates must actively transport out of their bodies.
c. The form of nitrogenous waste that requires the least amount of water to excrete is uric acid.
d. In the mammalian kidney, urea diffuses out of the collecting duct and contributes to the osmotic gradient within the medulla.

10. The process of secretion in the formation of urine insures that
a. a constant pH is maintained in body fluids.
b. drugs and other poisons are removed from the blood.
c. potassium balance is maintained in body fluids.
d. all of the above

11. The peritubular capillaries
a. form the ball of capillaries inside Bowman's capsule from which filtrate is forced by blood pressure into the renal tubule.
b. intertwine with the proximal and distal convoluted tubules and function in secretion and reabsorption.
c. form a countercurrent flow of blood through the medulla to supply nutrients without interfering with the osmolarity gradient.
d. surround the collecting ducts and reabsorb water, helping to create a hyperosmotic urine.

12. Consumption of alcohol may result in diuresis because
a. the JGA (juxtaglomerular apparatus) releases renin, which activates angiotensin, resulting in an increase in blood pressure.
b. alcohol inhibits the release of ADH, causing excessive water loss in the urine.
c. the atrium of the heart is stimulated to release ANP (atrial natriuretic protein), which inhibits the release of renin and aldosterone and thus lowers blood volume and pressure.
d. the kidneys must work harder to remove alcohol from the system.

13. The diffusion of urea from the collecting tubules
a. helps to maintain the osmotic gradient in the medulla.
b. indirectly contributes to the production of a hyperosmotic urine.
c. must be offset by its active secretion into the distal convoluted tubules.
d. Both a and b are correct.

14. Which of the following is used by terrestrial animals as a mechanism to dissipate heat?
    a. hibernation
    b. countercurrent exchange
    c. evaporation
    d. vasoconstriction

15. The heating center of the hypothalamus
    a. responds to messages from temperature-sensing nerve cells in the skin.
    b. controls shivering and vasoconstriction of superficial vessels.
    c. functions along with a cooling center to thermoregulate by feedback mechanisms.
    d. does all of the above.

16. A countercurrent heat exchange between arterial and venous blood
    a. warms the blood going to the extremities.
    b. warms the blood returning to the body core.
    c. maintains a constant body temperature.
    d. is used by marine mammals when they journey to warmer waters.

17. Physiological adjustments to cooler seasonal temperatures may include
    a. production of enzymes with lower optimal temperatures.
    b. changes in the lipid components of cell membranes.
    c. increased production of enzymes to offset their reduced efficiency.
    d. all of the above.

18. Which of the following sections of the mammalian nephron is *incorrectly* paired with its function?
    a. Bowman's capsule and glomerulus—filtration of blood
    b. proximal convoluted tubule—secretion of ammonia and H^+ into filtrate and transport of glucose and amino acids out of tubule
    c. descending limb of loop of Henle—diffusion of urea out of filtrate
    d. ascending limb of loop of Henle—diffusion and pumping of Na^+ and Cl^- out of filtrate

CHAPTER 41

CHEMICAL SIGNALS IN ANIMALS

FRAMEWORK

This chapter introduces the intricate system of chemical control and communication within animals. The endocrine system produces hormones which regulate homeostasis, growth, and development. Steroid hormones bind with receptors within the nuclei of their target cells and influence gene expression. Peptide hormones bind with cell surface receptors, and second messengers mediate their effects in target cells.

The hypothalamus and pituitary gland play coordinating roles by integrating the nervous and endocrine systems and producing many tropic hormones that control the synthesis and secretion of hormones in other endocrine glands and organs.

CHAPTER SUMMARY

Coordination and communication among the specialized parts of complex animals are achieved by the *nervous system* and the *endocrine system*. The nervous system conveys high-speed messages along neurons, whereas the endocrine system produces chemical messengers called hormones, which travel more slowly and influence development, metabolism, behavior, and homeostasis.

An Overview of Chemical Signals

Hormones are chemical messages that travel between organs of the body; pheromones communicate between different individuals; and some chemical messengers, such as neurotransmitters, carry information between cells.

Hormones An animal *hormone* is a chemical signal that is carried through the circulatory system and elicits a specific response from *target cells*. Endocrine cells produce hormones and are usually assembled into *endocrine glands*, which are ductless secretory organs that release hormones into the circulatory system.

Endocrinology, the study of hormones, is a rapidly growing area of research. Hormones can be grouped into three general classes based on chemical structure: fat-soluble *steroid hormones*, produced from cholesterol; hormones that are amino acids derivatives; and *peptide* hormones, which are chains of amino acids of varying lengths.

Hormonal response is triggered by the binding of a hormone to a complementary *hormone receptor*, which is either within the target cell or built into the target cell membrane. Most homeostatic functions regulated by hormones are controlled by a pair of hormones with antagonistic effects.

Pheromones *Pheromones* are chemical signals that communicate between animals of the same species. Serving as mate attractants, territory markers, or alarm substances, these small and volatile molecules are active in minute amounts.

Local Regulators Local regulators are chemical signals that affect target cells near their points of secretion. The release of neurotransmitter into a synapse between a neuron and its target cell, called synaptic signaling, is a direct form of chemical communication. In paracrine signaling, a cell secretes local regulators, such as histamine and interleukins, into the interstitial fluid, affecting only nearby target cells.

Experiments with culturing mammalian cells on artificial media led to the discovery of *growth factors*, extracellular proteins necessary for certain types of cells to grow and develop. Growth factors bind to surface receptors on target cells, but it is not known how this binding triggers the cell's response. Some onco-genes code for growth-factor-like receptors, but these receptors stimulate growth and division without requiring the binding of a growth factor.

Prostaglandins (*PGs*) are modified fatty acids that are released from most cells and have a wide range of effects on nearby target cells. Small differences in molecular structures can result in profound differences in the effects of prostaglandins. The balance of antagonistic prostaglandins is an important regulatory mechanism. In mammals, PGs stimulate uterine contractions and help to induce labor. Aspirin inhibits the synthesis of prostaglandins and thus reduces their fever-inducing and pain-intensifying actions.

Mechanisms of Hormone Action at the Cellular Level

Hormones act at very low concentrations and can have varying effects on different target cells and in different species. Two *signal transduction pathways* or general mechanisms of hormone action have been identified; both involve the hormone binding to specific receptor molecules. Steroid hormones enter the target cell nucleus and influence gene expression, whereas most nonsteroid hormones attach to the cell surface and stimulate release of intermediaries called second messengers.

Steroid Hormones and Gene Expression   Steroids diffuse through the target cell membrane and, it is thought, into the nucleus, where they bind to receptor proteins. The steroid-receptor complex then attaches to a specific acceptor protein which acts a transcription factor, binding to an enhancer sequence and initiating transcription of mRNA. Thus the signal transduction pathway for steroid hormones involves the activation of genes and the resulting production of protein products.

Two types of cells can respond differently to the same hormone, depending on the particular transcription factor that accepts the hormone-receptor complex and the enhancer sequence it recognizes. Steroid hormones alter gene expression, resulting in the synthesis of proteins and creating a slower, but longer-lasting hormonal response than do peptide hormones.

Peptide Hormones and Second Messengers   Peptide hormones and most hormones derived from amino acids do not enter their target cells, but bind outside the cell to specific receptor proteins embedded in the plasma membrane. This binding results in the cytoplasmic surface of the plasma membrane producing a *second messenger*, which triggers a cascade of biochem-

ical reactions within the cell. Peptide hormones alter the activity of enzymes and other proteins already in the cell, producing a more rapid hormonal response than do steroid hormones.

Through his work with epinephrine and glycogen hydrolysis in liver and muscle cells, Sutherland developed the second messenger model of hormone action, for which he received a Nobel Prize in 1971. He found that the binding of epinephrine to a hormone receptor in the plasma membrane of a liver cell signaled the membrane protein *adenylate cyclase* to convert ATP to cyclic adenosine monophosphate (*cyclic AMP* or *cAMP*). Adenylate cyclase is a membrane enzyme with its active site facing the cytoplasm. Cyclic AMP activates glycogen hydrolysis in the cell. The hormone is the first messenger and cyclic AMP is the second messenger, relaying the message to cytoplasmic enzymes. Enzymatic degradation of cAMP terminates the hormonal response. Many peptide hormones and hormones derived from amino acids use cAMP as their second messenger.

The hormone receptor does not interact directly with adenylate cyclase; rather, a third membrane protein, *G protein*, is the intermediary between them. Binding of a hormone to its receptor activates G protein, which hydrolyzes GTP to GDP, providing the energy to activate adenylate cyclase. The same G protein stimulates cAMP production in response to several different hormone signals. Another G protein inhibits adenylate cyclase in response to the binding of an inhibitory hormone. The balance of these antagonistic hormones allows the cell to precisely regulate its metabolism.

The effect of a hormone is amplified when the second messenger sets off an *enzyme cascade*, in which a sequence of enzymes is activated, with the number of activated products greatly increasing at each step in the sequence. First, cAMP activates cAMP-dependent protein kinase. A *protein kinase* is an enzyme that catalyzes protein phosphorylation, which, depending on the protein, either increases or decreases its activity. In the liver cell's response to epinephrine, cAMP-dependent protein kinase activates phosphorylase kinase, an enzyme that then adds a phosphate group to glycogen phosphorylase—the enzyme that hydrolyzes glycogen. (Or should we say that cAMP-dependent protein kinase phosphorylates phosphorylase kinase, which phosphorylates glycogen phosphorylase!) A few molecules of epinephrine can result in the release of millions of sugar molecules from glycogen.

This basic mechanism is used to mediate responses of many cell types to different hormones. The specificity of hormone action is maintained because different hormone receptors are found on specific target cells, cAMP-dependent protein kinases vary from tissue to tissue, and, in different cell types, cAMP-dependent protein kinases regulate different complements of proteins.

Inositol triphosphate (IP₃) is another second messenger that relays and amplifies a hormonal signal. The binding of a hormone to its receptor activates a G protein that stimulates phospholipase C, a membrane enzyme that cleaves a plasma membrane phospholipid into IP₃ and diacylglycerol. Diacylglycerol activates protein kinase C, which phosphorylates specific proteins. IP₃ causes the release of Ca^{2+} from the endoplasmic reticulum. The calcium either directly alters the activities of certain enzymes or binds to a protein called *calmodulin*, which then binds to and changes the activities of other enzymes and proteins. Neurotransmitters, growth factors, and some hormones use this mechanism of increasing the concentration of cytoplasmic Ca^{2+} to induce changes in their target cells.

Invertebrate Hormones

Invertebrate hormones regulate homeostasis, development, and reproduction.

All groups of arthropods have extensive endocrine systems. Insects have three hormones that interact to control molting and development of adult characteristics: The steroid hormone *ecdysone* stimulates the transcription of specific genes and functions to trigger molts and to promote development of adult characteristics. *Brain hormone* stimulates the prothoracic glands to secrete ecdysone. *Juvenile hormone (JH)* counters the action of ecdysone and promotes retention of larval characteristics during molting. When JH level is high, ecdysone-induced molting produces additional larval stages; only after JH level has decreased does molting result in a pupa.

The Vertebrate Endocrine System

Vertebrate hormones are secreted by more than a dozen tissues and organs, and may target a few or many tissues in the body. *Tropic hormones* target other endocrine glands and are particularly important to chemical coordination.

The Hypothalamus and the Pituitary Gland
The *hypothalamus*, situated in the lower brain, plays a key role in integrating the endocrine and nervous systems. *Neurosecretory cells* of the hypothalamus receive nerve signals and release hormones. One set of neurosecretory cells produces the hormones of the posterior pituitary; another produces *releasing hormones* that regulate the anterior pituitary.

The *pituitary gland*, a small appendage at the base of the hypothalamus, consists of two lobes. A *posterior lobe*, or *neurohypophysis*, stores and secretes two hormones produced by the hypothalamus; the *anterior lobe*, or *adenohypophysis*, produces its own hormones, several of which are tropic hormones whose targets are other endocrine glands.

Oxytocin and *antidiuretic hormone (ADH)* are small peptide hormones synthesized by neurosecretory cells of the hypothalamus. They move down the axons of these cells to the posterior lobe. Oxytocin induces uterine contractions during birth and milk ejection during nursing. ADH functions in osmoregulation. Osmoreceptor cells in the hypothalamus respond to an increase in blood osmolarity by sending impulses to hypothalamus neurosecretory cells. These cells release ADH from their tips, located in the posterior pituitary. ADH binds to receptors on cells lining the collecting ducts in the kidney and sets off a cAMP second-messenger system that results in increased permeability of the tubule epithelia to water. The reabsorption of water from the urine lowers the blood osmolarity and causes the hypothalamus to slow the release of ADH, completing the negative feedback loop. The osmoreceptors also stimulate thirst and initiate a behavioral response that will lower blood osmolarity.

The anterior pituitary produces a number of protein and peptide hormones. *Growth hormone (GH)* affects a variety of target tissues, promoting growth directly and stimulating the release of other hormones, such as *somatomedins*, which are produced by the liver and stimulate bone and cartilage growth. Several human growth disorders are related to abnormal GH production, including gigantism and acromegaly, both related to excessive GH, and hypopituitary dwarfism, caused by GH deficiency. GH deficiency can be treated with GH produced by bacteria genetically engineered to contain the human gene for this hormone.

The hormone *prolactin (PRL)* is a protein very similar to GH, with diverse effects in different vertebrate species, ranging from control of mammary gland growth and milk synthesis in mammals to delay of metamorphosis in amphibians to regulation of salt and water balance in fish.

Three of the tropic hormones produced by the anterior pituitary are glycoproteins. *Thyroid stimulating hormone (TSH)* regulates release of thyroid hormones. *Follicle stimulating hormone (FSH)* and *luteinizing hormone (LH)*, also called *gonadotropins*, stimulate gonad activity.

Several hormones come from fragments of *pro-opiomelanocortin*, a single large protein cleaved into several short pieces inside the pituitary cells. *Adrenocorticotropin (ACTH)* is a tropic hormone that stimulates the adrenal cortex to produce and secrete its steroid hormones. *Melanocyte-stimulating hormone (MSH)* regulates the activity of pigment-containing cells in the skin, as seen in the color changes of amphibians. *Endorphins* (or *enkephalins*) are pro-opiomelanocortin derivatives that are also produced by certain neurons in the brain. These recently-discovered hormones inhibit pain perception and have effects similar to those of morphine and other opiates, which bind to the same receptors in the brain.

Hypothalmic releasing hormones stimulate or inhibit hormone secretion by the anterior lobe of the pituitary. They are produced by neurosecretory cells of the hypothalamus and released into capillaries in an area called the *median eminence*. Veins draining the median eminence lead to capillary beds in the anterior pituitary, creating a portal system that delivers releasing hormones directly to the pituitary.

The Thyroid Gland   The thyroid gland produces two hormones, triiodothyronine (T_3) and thyroxine (T_4), which contain, respectively, three and four iodine atoms. Thyroid hormones, critical to the development and maturation of vertebrates, are required for normal functioning of bone-forming cells and for branching of nerve cells during embryonic development. An inherited deficiency of thyroid hormone in humans results in retarded skeletal and mental development, a condition known as cretinism.

The thyroid gland also regulates metabolism in mammals. Excess thyroid hormone results in a condition known as hyperthyroidism, with symptoms of weight loss, irritability, high body temperature, and high blood pressure. Hypothyroidism can cause cretinism or result in weight gain and lethargy in adults. A goiter is an enlarged thyroid gland, associated with a lack of iodine in the diet.

A negative feedback loop controls secretion of thyroid hormones. Secretion of TSH-releasing hormone, or TRH, by the hypothalamus stimulates the anterior pituitary to secrete thyroid stimulating hormone. TSH binds to receptors on the thyroid gland, generating cAMP and triggering the synthesis and release of thyroid hormones. High thyroid hormone level inhibits the secretion of TSH.

The mammalian thyroid gland also secretes *calcitonin*, a polypeptide hormone that lowers calcium levels in the blood.

The Parathyroid Glands   The four parathyroid glands, located on the thyroid, secrete *parathyroid hormone (PTH)*, which stimulates Ca^{2+} reabsorption in the kidney and release of Ca^{2+} from bone to the blood stream to raise blood calcium levels. A balance between the antagonistic PTH and calcitonin maintains calcium homeostasis. Vitamin D is needed for proper PTH function.

The Pancreas   The pancreas consists mostly of exocrine tissue, but has scattered clusters of endocrine cells known as the *islets of Langerhans*. Within each islet are *beta (ß) cells* that secrete the protein hormone *insulin* and *alpha (α) cells* that secrete the peptide hormone *glucagon*. These antagonistic hormones regulate glucose concentration in the blood. Their secretion is controlled by negative feedback to the glucose receptors on the islet cells which monitor blood sugar level.

Insulin lowers blood sugar levels by promoting the movement of glucose into muscle and other tissues, by slowing the breakdown of glycogen in the liver, and by inhibiting the conversion of amino acids and fatty acids to sugar. Glucagon raises glucose concentrations by stimulating fat and protein breakdown and glycogen hydrolysis.

In diabetes mellitus, the absence of insulin in the bloodstream or the loss of response to insulin in target tissues restricts glucose uptake by cells, and the body must use fats for fuel. Glucose concentration in the blood rises and is excreted in the urine. As more water is excreted with the concentrated urine, dehydration can result.

Type I diabetes mellitus, also known as insulin-dependent or juvenile-onset diabetes, is an autoimmune disorder which destroys pancreatic cells. This type of diabetes is treated by regular injections of insulin. Insulin, previously obtained from animal pancreases, is now produced from human genes inserted into genetically engineered bacteria. More than 90% of diabetics have *type II diabetes mellitus*, or adult onset diabetes, characterized either by insulin deficiency or reduced responsiveness of target cells. Exercise and dietary control are often sufficient to manage this disease.

The Adrenal Glands   In mammals, the *adrenal glands* are located on top of the kidneys and consist of two different glands: the outer *cortex* and the central *medulla*. The adrenal medulla produces *epinephrine* (adrenalin) and *norepinephrine* (noradrenalin)—both of which are *catecholamines* synthesized from the amino acid tyrosine.

Epinephrine, released in response to positive or negative stress, increases the availability of energy sources by stimulating glucose mobilization in skeletal muscle and the liver, and fatty acid release from fat cells. Epinephrine and norepinephrine increase the rate and volume of the heart beat and influence the contraction of smooth muscles to increase blood supply to the heart, brain, and skeletal muscles, while reducing the supply to the skin, gut, and kidneys. The sympathetic division of the autonomic nervous system stimulates the release of epinephrine. Epinephrine and norepinephrine also function as neurotransmitters in the nervous system.

The adrenal cortex responds to endocrine signals of stress rather than to nervous input. The hypothalamus secretes a releasing hormone that causes the anterior pituitary to release ACTH. This tropic hormone stimulates the adrenal cortex to synthesize and secrete *corticosteroids*, a group of steroid hormones.

The two main human corticosteroids are the *glucocorticoids*, such as cortisol, and the *mineralocorticoids*, such as aldosterone. Glucocorticoids promote synthesis of glucose from noncarbohydrates, such as protein, and thus increase energy supplies during stress. Cortisone has been used to treat serious inflammatory conditions such as arthritis, but its immunosuppressive effects can be dangerous.

Mineralocorticoids affect salt and water balance. Aldosterone stimulates kidney cells to reabsorb sodium ions from the filtrate. Aldosterone secretion is regulated by hormones produced in the liver and kidneys in response to plasma ion concentrations.

The Gonads   The testes of males and ovaries of females produce steroids that affect growth, development, and reproductive cycles and behaviors. The three major categories of gonadal steroids—androgens, estrogens, and progestins—are found in different proportions in males and females.

The testes primarily synthesize *androgens*, such as *testosterone*, which determine gender of the developing embryo and stimulate development of the male reproductive system and secondary sex characteristics. *Estrogens*, such as estradiol, are responsible for the development and maintenance of the female reproductive system and secondary sex characteristics. In mammals, *progestins* help prepare and maintain the uterus for the growth of an embryo.

Gonadotropins (follicle-stimulating hormone and luteinizing hormone) from the anterior pituitary gland control estrogen and androgen synthesis. A hypothalamic regulatory hormone, GnRh (gonadotropic releasing hormone) controls secretion of FSH and LH.

Other Endocrine Organs   Many other organs whose main functions are nonendocrine, such as the digestive tract, kidney, and heart, secrete important hormones.

The *pineal gland*, located near the center of the mammalian brain, secretes a modified amino acid hormone, *melatonin*, which regulates functions related to light and changes in daylength. Melatonin is secreted at night; its production may function as a biological clock for daily or seasonal activities such as reproduction.

The *thymus* plays a role in the immune system in young animals. It secretes *thymosin* and other messengers that stimulate development and differentiation of T lymphocytes.

Endocrine Glands and the Nervous System

The endocrine system and nervous system function together to control chemical communication and coordination in animals. They are *structurally* related in that many endocrine glands, such as the vertebrate hypothalamus, posterior pituitary, and the insect brain, are made of nervous tissue. Other endocrine glands, such as the adrenal medulla, have evolved from the nervous system.

The endocrine and nervous systems are *chemically* related in that several vertebrate hormones, such as epinephrine, are used by both systems. And the two systems are *functionally* related. Many physiological processes are coordinated by both nervous and hormonal communications, often in a type of neuroendocrine reflex. The nervous system controls endocrine glands, and hormones affect the development and functioning of the nervous system.

STRUCTURE YOUR KNOWLEDGE

1. Create a concept map that illustrates your understanding of the signal transduction pathways of steroid and peptide hormones.

2. The hypothalamus and pituitary gland produce a number of hormones, tropic hormones, and releasing hormones. Fill in the following table with the major products of each of the regions of these glands.

3. Develop a concept map or diagram of the hormones and mechanisms involved in the body's control of blood sugar levels.

TEST YOUR KNOWLEDGE

MATCHING: *Match the hormone and gland or organ that produces it to the descriptions of hormone action. Choices may be used more than once and not all choices are used.*

HORMONES	GLAND OR ORGAN
A. adrenocorticotropin (ACTH)	**a.** adrenal cortex
B. androgens	**b.** adrenal medulla
C. antidiuretic hormone (ADH)	**c.** hypothalamus

GLAND	HORMONES	MAIN ACTION
Hypothalamus Neurosecretory cells (drain to posterior pituitary)		
(drain to median eminence)		
Pituitary Posterior lobe (neurohypophysis)		
Anterior lobe (adenohypophysis)		

D. calcitonin
E. epinephrine
F. glucagon
G. glucocorticoids
H. insulin
I. melatonin
J. oxytocin
K. thyroxine

d. pancreas
e. parathyroid
f. pineal
g. pituitary
h. testis
i. thymus
j. thyroid

HORMONE GLAND HORMONE ACTION

1. _____ _____ involved in biological clock and seasonal activities

2. _____ _____ involved in synthesis of glucose, suppress inflammatory reaction

3. _____ _____ involved in glycogen breakdown in liver, increase blood sugar

4. _____ _____ stimulate development of male reproductive system

5. _____ _____ stimulate adrenal cortex to synthesize corticosteroids

6. _____ _____ increase available energy supplies, increase heart rate and blood supply to skeletal muscles, heart and brain

7. _____ _____ regulate metabolism, development of bone-forming and nerve cells

8. _____ _____ lower blood calcium levels by inhibiting calcium release from bone

9. _____ _____ influence reabsorption of water from collecting ducts of kidney

10. _____ _____ stimulate contraction of uterine muscles and mammary gland cells

MULTIPLE CHOICE: *Choose the one best answer.*

1. Which of the following is *not* an accurate statement about hormones?
   a. All hormones are secreted by endocrine glands.
   b. Most hormones move through the circulatory system to their destination.

c. Target cells have specific molecular receptors for hormones.

d. Hormones are essential to homeostasis.

2. The best description of the difference between pheromones and hormones is that
   a. pheromones are small, volatile molecules, whereas hormones are steroids.
   b. pheromones are involved in reproduction, whereas hormones are not.
   c. pheromones are a form of olfactory communication; hormones are a form of chemical communication.
   d. pheromones are signals that function between animals, whereas hormones communicate among the parts within an animal.

3. Which one of the following examples is *incorrectly* paired with its chemical structure?
   a. insulin—protein
   b. estrogen—steroid hormone
   c. prostaglandin—modified fatty acid
   d. thyroxin—peptide

4. In the second-messenger model of hormone action,
   a. a tropic hormone signals an endocrine gland to release a hormone.
   b. another molecule relays the message from the membrane-bound hormone to the cytoplasmic enzymes.
   c. negative feedback turns off the production of the hormone.
   d. genes are turned on and new proteins are synthesized.

5. Ecdysone
   a. is a steroid hormone produced in insects that promotes retention of larval characteristics.
   b. is responsible for the color changes in amphibians.
   c. is a peptide hormone secreted from "bag cells" that stimulates egg laying in the sea slug *Aplysia*.
   d. is secreted by prothoracic glands in insects and triggers molts and development of adult characteristics.

6. The role of cAMP in hormone action is to
   a. act as the second messenger and activate other enzymes.
   b. activate adenylate cyclase.
   c. bind a specific hormone to the plasma membrane.
   d. bind to the protein calmodulin and initiate an enzyme cascade.

7. Neurosecretory cells
   a. are found in the hypothalamus.
   b. receive signals from nerve cells and release hormones.
   c. produce releasing hormones that drain into the median eminence.
   d. All of the above are correct.

8. Antidiuretic hormone (ADH)
   a. is produced by cells in the kidney and liver.
   b. stimulates the reabsorption of Na^+ from the urine.
   c. acts through a second messenger system to increase the permeability of kidney collecting tubules to water.
   d. is a steroid hormone produced by the adrenal cortex.

9. The adenohypophysis or anterior lobe of the pituitary
   a. stores oxytocin and ADH produced by the hypothalamus.
   b. is connected by a portal system to the median eminence.
   c. produces several releasing hormones.
   d. does all of the above.

10. Acromegaly is caused by
    a. an excess of thyroxine.
    b. an excess of growth hormone.
    c. an excess of glucocorticoids.
    d. an abnormally high androgen-to-estrogen ratio.

11. Which of the following hormone sequences results in the secretion of estrogens in mammals?
    a. GNRH $\longrightarrow$ FSH $\longrightarrow$ estrogen
    b. LH $\longrightarrow$ FSH $\longrightarrow$ estrogen
    c. CRH $\longrightarrow$ ACTH $\longrightarrow$ FSH $\longrightarrow$ estrogen
    d. gonadotropin $\longrightarrow$ LH $\longrightarrow$ estrogen

12. Pro-opiomelanocortin
    a. is produced by the hypothalamus.
    b. is cleaved into several hormones, including ACTH, melanocyte-stimulating hormone, endorphins, and enkephalins.
    c. produces several releasing hormones.
    d. All of the above are correct.

13. Hyperthyroidism
    a. is responsible for cretinism.
    b. results in the formation of a goiter.
    c. produces weight loss and a high body temperature.
    d. is associated with a deficiency of iodine.

14. Calcium homeostasis is maintained by a balance between
    a. calcitonin, secreted by the thyroid, and PTH, secreted by the parathyroid, which respectively lower and raise blood calcium levels.
    b. PTH and vitamin D.
    c. calmodulin, calcitonin, PTH, and vitamin D.
    d. aldosterone, which stimulates the kidney to reabsorb calcium, and thymosin, which lowers blood calcium levels.

15. Which of the following is *not* true of norepinephrine?
    a. It is secreted by the adrenal medulla.
    b. Its action is to increase the rate and volume of heart beat and constrict selected blood vessels.
    c. Its release is stimulated by ACTH.
    d. It serves as a neurotransmitter in the nervous system.

CHAPTER 42

ANIMAL REPRODUCTION

FRAMEWORK

This chapter covers the patterns and mechanisms of animal reproduction. In sexual reproduction, fertilization may occur externally or internally. Development of the zygote may take place internally within the female, externally in a moist environment, or in a protective, resistant egg. Placental mammals provide nourishment as well as shelter for the embryo and nurse their young.

In the discussion of mammalian reproduction, the human reproductive system is described, including its organs, glands, hormones, gamete formation, and sexual physiology. The chapter also covers pregnancy and birth, contraception, and recently developed reproductive technologies.

CHAPTER SUMMARY

Modes of Reproduction

In *asexual reproduction*, a single individual produces offspring that are genetically identical to itself. In *sexual reproduction*, two individuals contribute genes to the offspring.

Asexual Reproduction Many invertebrates can reproduce asexually by *budding,* in which a new individual grows out from the parent's body, or by *fragmentation,* in which the body is broken into several pieces, each of which develops into a complete animal. Some invertebrates release specialized groups of cells that grow into new individuals, as in the *gemmules* produced by sponges. Regeneration is a means of asexual reproduction when a piece of an animal, such as the detached arm of a starfish, develops into a new organism.

Asexual reproduction allows rapid colonization of new areas and perpetuation of successful genotypes in stable habitats.

Sexual Reproduction In animal sexual reproduction, the two haploid *gametes* fuse to form a diploid *zygote.* The gametes are usually a relatively large, nonmotile female *ovum* and a small, flagellated male *spermatozoon.*

Sexual reproduction produces offspring with varying genotypes and phenotypes, which may enhance reproductive success of parents in fluctuating environments.

Reproductive Cycles and Patterns Most animals show periodic cycles of reproductive activity. A combination of seasonal and hormonal cues control the timing of these cycles, which may be linked to favorable environmental conditions or energy supplies.

The freshwater crustacean *Daphnia,* like aphids and rotifers, produces eggs that develop by *parthenogenesis,* without being fertilized, as well as eggs that are fertilized. Asexual reproduction is associated with stable conditions, whereas sexual reproduction may occur during less favorable environmental conditions. In bees, wasps, and ants, males are produced parthenogenetically, whereas sterile worker females and reproductive females are produced sexually. In a few parthenogenic fishes, amphibians, and lizards, doubling of chromosomes creates diploid "zygotes." In some species of whiptail lizards in which there are no males, females take turns impersonating a male in courtship and mating behavior.

In *hermaphroditism,* found in some sessile, burrowing, and parasitic animals, each individual has functioning male and female reproductive systems. Mating results in fertilization of both individuals. In some fishes and oysters, individuals reverse their sex during their lifetime, a pattern known as *sequential hermaphroditism.* Sex reversal may be related to age or to the relative advantage conferred by size. In a *protandrous* species, an organism is first a male; in a *protogynous* species, it is first a female.

Mechanisms of Sexual Reproduction

Patterns of Fertilization and Development In *external fertilization*, eggs and sperm are shed from the body, and fertilization occurs in the environment. This type of reproduction occurs almost exclusively in moist habitats, where the gametes and developing zygote are not in danger of desiccation. Environmental or pheromone signals may ensure that gamete release is synchronized. Fishes and amphibians that use external fertilization show mating behaviors that trigger the release of gametes and provide a means for mate selection.

Internal fertilization involves the placement of sperm in or near the female reproductive tract so that the egg and sperm unite internally. Behavioral cooperation, as well as copulatory organs and sperm receptacles, are required for internal fertilization.

Developing embryos may receive some type of protection, which may include resistant eggs, development inside the female reproductive tract, and parental protection of eggs. The amniote eggs of birds and reptiles protect the embryo in a terrestrial environment. The embryos of placental mammals develop within the uterus, nourished from the mother's blood supply through the placenta. Parental care of young is widespread among animals.

Diversity in Reproductive Systems Some invertebrates, such as polychaete annelids, do not have distinct *gonads*; eggs and sperm develop from cells lining the coelom. In these mostly marine worms, gametes are stored in the coelom and shed through excretory organs or released by the splitting of the parent.

Insects have separate sexes and complex reproductive systems. Sperm develop in the testes, are stored in the seminal vesicles, and exit through an ejaculatory duct in the penis. Accessory glands may add fluid to the semen. Eggs, produced in the ovaries, pass through the oviducts to the vagina. Females may have a *spermatheca*, or sperm-storing sac.

Flatworms are hermaphrodites with complex reproductive systems. In addition to ovaries, oviducts, and vagina, the female reproductive system includes yolk and shell glands and a uterus, where eggs are fertilized and may begin development. The male system includes a complex copulatory apparatus that may be inserted in the vagina or may inject sperm into the body through hypodermic impregnation.

With the exception of most mammals, vertebrates have a common opening, called the *cloaca*, for the digestive, excretory, and reproductive systems. The uterus of most vertebrates is bicornate—having two separate branches. Nonmammalian vertebrates do not have well-developed penes.

Mammalian Reproduction

Human anatomy is used as the example of a mammalian reproductive system.

Anatomy of Reproduction in Humans The external male *genitalia* include the *scrotum*, a fold of skin enclosing the male gonads or testes, and the *penis*. Sperm are produced in the highly coiled *seminiferous tubules* of the *testes. Interstitial cells*, scattered between the tubules, produce testosterone and other androgens. Sperm production requires a cooler temperature than the internal body temperature of most mammals, so the scrotum suspends the testes below the abdominal cavity.

Sperm pass from the testes into the coiled *epididymis*, in which they mature and are stored. During *ejaculation*, sperm are propelled through the *vas deferens*, into a short *ejaculatory duct*, and out through the *urethra* which runs through the penis.

Three glands add secretions to the *semen*. The *seminal vesicles* contribute a fluid containing mucus; amino acids; fructose, as an energy source for the sperm; and prostaglandins, which stimulate uterine contractions. The *prostate gland* produces an alkaline secretion that balances the acidity of the urethra and the vagina. The *bulbourethral glands* produce a small amount of viscous fluid of uncertain function.

The penis is composed of spongy tissue that engorges with blood during sexual arousal, producing an erection that facilitates insertion of the penis into the vagina. Some mammals have a *baculum*, a bone that stiffens the penis. The head of the penis, called the *glans penis*, is covered by a fold of skin called the *prepuce*, or foreskin.

The female gonads, the *ovaries*, contain many *follicles*, which are sacs of cells that nourish and protect the one egg cell contained within each of them. The follicle produces the primary female sex hormones, the estrogens. Following *ovulation*, the follicle forms a solid mass called the *corpus luteum*, which secretes progesterone and a small amount of estrogen.

The egg cell is expelled into the abdominal cavity and swept by cilia into the *oviduct*, or fallopian tube, through which it is transported to the *uterus*. The *endometrium*, or lining of the uterus, is highly vascularized. The narrow neck of the uterus, the *cervix*, opens into the *vagina*, the thin-walled birth canal and repository for sperm during copulation.

The separate openings of the vagina and urethra are enclosed by two pairs of skin folds, the *labia minora* and the outer *labia majora*, which form a *vestibule*. The *clitoris* is at the top of the vestibule. Both the labia minora and the clitoris are composed of erectile tissue. The vaginal orifice is initially covered by a thin membrane called the *hymen*.

The *mammary gland*, or breast, is composed of deposits of fatty tissue and a series of *alveoli*, small milk-secreting sacs that drain into ducts that join together and open at the nipple.

The external genitalia of both sexes arise from common *primordia*, or undifferentiated embryonic tissue, whose development into male or female structures is dependent on the presence or absence of androgens.

Hormonal Control of Mammalian Reproduction

Androgens are responsible for the male primary (associated with reproduction) and secondary (associated with the voice, hair distribution, and muscle growth traits of men) sex characteristics. The most important androgen is *testosterone*. Androgens are also determinants of sexual and other behaviors. Androgen secretion and sperm production are controlled by hormones from the anterior pituitary and hypothalamus.

Female humans and many primates have *menstrual cycles*, during which the endometrium lining thickens to prepare for the implantation of the embryo and then is shed if fertilization does not occur. This bleeding, called *menstruation*, occurs on a cycle of approximately 28 days in humans. Other mammals have *estrous cycles*, during which the endometrium thickens, but, if fertilization does not occur, it is reabsorbed and there is no bleeding. Estrous cycles are often coordinated with season or climate, and females are receptive to sexual activity only during *estrus*, or heat, the period surrounding ovulation.

The human menstrual cycle consists of the *menstrual phase*, the few days of menstrual bleeding; the *proliferative phase*, the week or two during which the endometrium begins to thicken; and the *secretory phase*, about two weeks during which the endometrium becomes more vascularized and secretes a glycogen-rich fluid.

The *ovarian cycle* begins with the *follicular phase*, during which an egg cell enlarges, its follicle cells become multilayered and enclose a fluid-filled cavity, and the large, mature follicle forms a bulge near the ovary surface. The *luteal phase* begins with ovulation and the rupture of the follicle and adjacent ovary wall. The remaining follicular tissue develops into the hormone-secreting corpus luteum.

Five hormones coordinate the menstrual and ovarian cycles using positive and negative feedback. During the follicular phase, the hypothalamus secretes GnRH (gonadotropin-releasing hormone), which stimulates the anterior pituitary to secrete FSH (follicle-stimulating hormone) and LH (luteinizing hormone). FSH stimulates follicular growth, and the cells of the follicle secrete estrogen. When the level of estrogen secretion rises sharply, the hypothalamus is stimulated to increase GnRH output, which results in a rise in LH and FSH release.

By positive feedback, the increase in LH, caused by increased estrogen secretion from the follicle, induces maturation of the follicle, and ovulation occurs. In the luteal phase, LH stimulates the transformation of the ruptured follicle and maintains the corpus luteum, which secretes estrogen and progesterone. The rising level of these two hormones exerts negative feedback on the hypothalamus and pituitary, inhibiting secretion of LH and FSH. The corpus luteum, deprived of its LH stimulation, disintegrates, thereby dropping the levels of estrogen and progesterone. This drop releases the inhibition of the hypothalamus and pituitary, and FSH and LH secretion begins again, stimulating growth of new follicles and the start of the next follicular phase.

The hormones of the ovarian cycle synchronize the menstrual cycle and the preparation of the uterus for possible implantation of an embryo. Estrogen causes the endometrium to begin to thicken in the proliferative phase. After ovulation, estrogen and progesterone stimulate increased vascularization and gland development of the endometrium. Thus, the luteal phase corresponds with the secretory phase of the menstrual cycle. The rapid drop of ovarian hormones caused by the disintegration of the corpus luteum reduces blood supply to the endometrium and begins its degeneration, leading to the menstrual phase of the next cycle.

Estrogens are also responsible for female secondary sex characteristics. They induce fat deposition in the breasts and hips, affect water retention and calcium metabolism, and influence sexual behavior.

Gamete Formation (Gametogenesis)

Spermatogenesis, the production of mature sperm cells, occurs continuously in the seminiferous tubules of the testes. The haploid nucleus of a sperm is contained in a thick head, tipped with an *acrosome*, which contains enzymes that help the sperm penetrate the egg. Mitochondria in the neck of the spermatazoon provide ATP for movement of the flagellum, or tail.

In *oogenesis*, meiotic cytokinesis is unequal, producing one large ovum and three small haploid polar bodies that disintegrate. In spermatogenesis, all four meiotic products develop into mature sperm. At birth, an ovary already contains all its presumptive egg cells, whereas the precursor cells of sperm continue to divide mitotically throughout a male's reproductive years. Oogenesis happens in stages: the first meiotic division occurs within maturing follicles, whereas the second division occurs after ovulation. In humans, this second division is triggered by penetration of the egg cell by the sperm.

Sexual Maturation In humans, the onset of reproductive ability, termed puberty, is a gradual process that is completed around the age of 12 to 14 and includes the onset of menstruation or the production of viable sperm, as well as the development of secondary sex characteristics.

Human Sexual Physiology The human sexual response cycle includes two types of physiological reactions: *vasocongestion*, or increased blood flow to a tissue, and *myotonia*, or increased muscle tension. During the *excitement* phase, vasocongestion of the penis, testes, labia, vagina, and breasts occurs; myotonia results in nipple erection and tension of the arms and legs. The excitement phase prepares the vagina and penis for *coitus*, or sexual intercourse. The *plateau* phase continues vasocongestion and myotonia, and breathing rate and heart rate increase. In both sexes, *orgasm* is characterized by rhythmic, involuntary contractions of the reproductive system. In males, emission deposits semen in the urethra, and ejaculation occurs when the urethra contracts and semen is expelled. In the *resolution* phase, vasocongested organs return to normal size, muscles relax, and nonreproductive reactions (breathing and heart rates) return to normal.

Conception, Pregnancy, and Birth The development of an embryo in the uterus, preceded by *conception* and ending with birth, is called *pregnancy* or *gestation*. The gestation period correlates with body size and the degree of development of the young at birth. Human pregnancy averages 266 days (38 weeks).

Human gestation can be divided into three *trimesters*. Following fertilization in the oviduct, the *zygote* travels to the uterus in about 3 to 5 days. After about a week of *cleavage*, or cell division, the zygote develops into a hollow ball of cells called a *blastocyst* and implants in the endometrium. Tissues grow out of the developing embryo to meet with the endometrium and form the *placenta*, a disk-shaped organ in which gas and nutrient exchange and waste removal take place between the maternal and fetal circulations. Differentiation leads to *organogenesis*, and, by the eighth week, the embryo has all the rudimentary structures of the adult and is called a *fetus*.

The embryo secretes *human chorionic gonadotropin (HCG)*, which maintains the corpus luteum's secretion of progesterone and estrogen through the first trimester. High progesterone levels initiate growth of the maternal part of the placenta, enlargement of the uterus, and cessation of ovulation and menstrual cycling.

During the second trimester, HCG declines, the corpus luteum degenerates, and the placenta secretes its own progesterone, which maintains the pregnancy. The fetus grows rapidly and is quite active. The third trimester is a period of rapid fetal growth.

Labor consists of a series of strong contractions of the uterus and results in birth, or *parturition*. During the first stage of labor, the cervix dilates. The second stage consists of the contractions that force the fetus out of the uterus and through the vagina. The placenta is expelled in the final stage of labor.

Lactation is unique to mammals. Prolactin secretion stimulates milk production, and oxytocin controls the release of milk during nursing.

Several hypotheses attempt to explain why a mother does not reject an embryo, which has paternal as well as maternal chemical markers. There is evidence that the trophoblast, a protective layer that develops from the blastocyst and surrounds the embryo, induces the development of white blood cells that suppress other white cells from mounting an immune attack. One hypothesis suggests that this suppression can occur only after an initial immune response to the trophoblast. Some researchers suggest that if the initial response is too weak, then suppression may not occur, and the continued immunological attack results in spontaneous abortion of the embryo.

Contraception, the deliberate prevention of pregnancy, can be accomplished by one of several methods: preventing release of egg or sperm, preventing fertilization, or preventing implantation. The *rhythm method* is based on refraining from intercourse during the period that conception is most likely, the few days before and after ovulation. The failure rate of the rhythm method is 10% to 20%, meaning that 10 to 20 women become pregnant during a year out of every hundred using the method.

Several *barrier methods* of contraception have failure rates of less than 10%. *Condoms* and diaphragms, used in conjunction with spermicidal foam or jelly, present a physical and chemical barrier to fertilization. *Coitus interruptus* is an undependable method of contraception.

The intrauterine device (IUD) probably prevents implantation by irritating the endometrium. IUDs have a low failure rate but have been associated with harmful side effects.

The release of gametes may be prevented by chemical contraception, as in birth control pills, and by sterilization, as in *tubal ligation* in women or *vasectomy* in men. Birth control pills are combinations of synthetic estrogen and progestin, which act by negative feedback to stop the release of GnRH by the hypothalamus and FSH and LH by the pituitary. The absence of these hormones results in a cessation of ovulation and follicle development. Cardiovascular problems are a potential harmful effect of birth control pills. In 1990, the FDA approved a progestin patch that will release progestin for up to 5 years when inserted under the skin.

Abortion is the termination of a pregnancy. Spontaneous abortion, or miscarriage, occurs in as many as one-third of all pregnancies. The drug RU486 is available to women in France who choose to terminate pregnancy in the first few weeks. As an analog of progesterone, RU486 blocks progesterone receptors in the uterus.

Condoms are the only form of birth control that offer some protection against sexually transmitted diseases.

Reproductive Technology Some genetic diseases and congenital defects can be detected while the fetus is in the uterus. The use of *ultrasound* produces an image of the fetus from the echoes of high-frequency sound waves. In *amniocentesis*, a sample of amniotic fluid is withdrawn with a long needle, and fetal cells are cultured and analyzed for chromosomal defects. In *chorionic villi sampling*, a small sample of tissue is removed from the fetal part of the placenta. This procedure can be done earlier and produces faster results than amniocentesis, but carries a higher risk of complications.

The new technique of *in vitro fertilization* involves the removal of ova from a woman whose oviducts are blocked, fertilization within a petri dish, and implantation of the developing embryo in the uterus.

The development of male chemical contraception is currently under research.

STRUCTURE YOUR KNOWLEDGE

1. Trace the path of a human sperm from the point of production to the point of fertilization, briefly commenting on the functions of both the structures it passes through and the associated glands.

2. Describe the path of a human ovum that does not become fertilized.

3. In the diagram of the human menstrual cycle on p. 361, label the lines indicating the levels of the gonadotropic and the ovarian hormones, and the phases of the ovarian and menstrual cycle.

4. Answer the following questions concerning the human menstrual cycle.

a. What does GnRH do?

b. What does FSH stimulate?

c. What causes the spike in LH level?

d. What does this high level of LH induce?

e. What does LH maintain during the luteal phase?

f. What inhibits secretion of LH and FSH?

g. What allows LH and FSH secretion to begin again?

5. List the three general types, along with examples, of birth control methods. Which examples are most likely to prevent pregnancy? Which are least likely to do so?

6. Describe how birth control pills work. How does the new French drug RU486 function as a contraceptive?

TEST YOUR KNOWLEDGE

FILL IN THE BLANKS

1. _____ type of asexual reproduction in which a new individual grows while attached to the parent's body

2. _____ sequential hermaphroditic species in which an organism is first a male and then becomes a female

3. _____ development of egg without fertilization

4. _____ individual with functioning male and female reproductive systems

5. _____ common opening of digestive, excretory, and reproductive systems in nonmammalian vertebrates

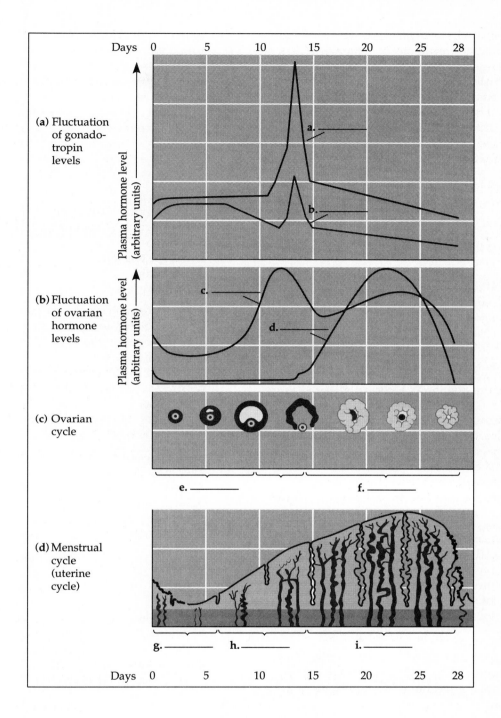

(a) Fluctuation of gonado-tropin levels

Plasma hormone level (arbitrary units)

Days 0   5   10   15   20   25   28

a.

b.

(b) Fluctuation of ovarian hormone levels

Plasma hormone level (arbitrary units)

c.

d.

(c) Ovarian cycle

e. _____   f. _____

(d) Menstrual cycle (uterine cycle)

g. _____   h. _____   i. _____

Days 0   5   10   15   20   25   28

6. _____ type of reproductive cycle in which thickened endometrium is reabsorbed

7. _____ hormone that prepares the uterus for pregnancy

8. _____ common duct for urine and semen in males

9. _____ filling of a tissue with blood due to increased blood flow

10. _____ period when reproductive ability begins in humans

MULTIPLE CHOICE: *Choose the one best answer.*

1. Which of the following is an explanation for the periodicity of reproductive cycles in animals?
   a. Reproduction may correspond with periods of

increased food supply, during which energy can be invested in gamete formation.

    b. Seasonal cycles may allow offspring to be produced during favorable environmental conditions when chances of survival are highest.

    c. Hormonal control of reproduction may be tied to biological clocks and seasonal cues.

    d. All of these may contribute to periodic reproductive activity.

2. Which of the following is *not* required for internal fertilization?
   a. internal development of the embryo
   b. copulatory organ
   c. sperm receptacle
   d. behavioral interaction

3. Insect reproductive systems may include all of the following *except*
   a. seminal vesicles.
   b. an ejaculatory duct.
   c. a spermatheca.
   d. an endometrium.

4. Which of the following is incorrectly paired with its function?
   a. seminiferous tubules—add fluid containing mucus, fructose, and prostaglandins to semen
   b. scrotum—encase testes and hold below abdominal cavity
   c. epididymis—store sperm
   d. prostrate gland—add alkaline secretion to semen

5. The function of the corpus luteum is to
   a. nourish and protect the egg cell.
   b. produce prolactin in the alveoli.
   c. produce progesterone and estrogen.
   d. convert into a hormone-producing follicle following ovulation.

6. In an estrous cycle,
   a. the endometrium does not thicken unless fertilization occurs.
   b. females are not receptive to sexual activity during estrus.
   c. the endometrial lining is reabsorbed if fertilization does not occur.
   d. the thickened endometrium lining is shed and the cycle is coordinated with season or day length.

7. Which of the following hormones is incorrectly paired with its function?
   a. GnRH—control release of FSH and LH
   b. prolactin—stimulate production of milk in alveoli of mammary gland
   c. estrogen—responsible for primary and secondary female sex characteristics
   d. human chorionic gonadotropin—secreted by corpus luteum and stimulates growth of placenta

8. Myotonia is
   a. a congenital birth defect.
   b. the hormone responsible for breast development.
   c. muscle tension.
   d. rhythmic contractions of reproductive organs during orgasm.

9. The secretory phase of the menstrual cycle
   a. is associated with dropping levels of estrogen and progesterone.
   b. is when the endometrium begins to degenerate and menstrual flow occurs.
   c. corresponds with the luteal phase of the ovarian cycle.
   d. corresponds with the follicular phase of the ovarian cycle.

10. Examples of birth-control methods that prevent release of gametes from the gonads are
    a. sterilization and chemical contraception.
    b. birth control pills and IUDs.
    c. condoms and diaphragms.
    d. abstinence and chemical contraception.

11. The ability of a pregnant woman not to reject her "foreign" fetus may be due to
    a. the fact that fetal and maternal blood never mix.
    b. the suppression of her immune response to the paternal chemical markers on fetal tissue.
    c. the protection of the fetus in the placenta.
    d. the production of human chorionic gonadotropin that maintains the pregnancy.

12. Which of the following is the least invasive technique of checking for birth defects?
    a. ultrasound
    b. amniocentesis
    c. chorionic villi sampling
    d. in vitro fertilization

CHAPTER 43

ANIMAL DEVELOPMENT

FRAMEWORK

Fertilization initiates physical and molecular changes in the egg cell. Early embryonic development includes cleavage of the fertilized egg, gastrulation, and organogenesis. The amount of yolk in the egg and the taxonomic placement of the animal determine how these processes occur in any particular animal group.

Development may be mosaic, in which the fate of blastomeres is fixed by the first cleavage divisions, or regulative, in which blastomeres remain totipotent for a longer time. Determination and differentiation of cells are the result of the control of gene expression by both cytoplasmic determinants and the positional information a cell receives within a developing structure. Developmental biologists, using transplant experiments and other techniques, are gradually unraveling some of the mechanisms underlying the complex processes of animal development.

CHAPTER SUMMARY

Animal development charts the course from a single fertilized egg to an organism made of many differentiated cells organized into specialized tissues and organs. An animal's form and function develop throughout its lifetime and include such events as embryonic development, postembryonic growth and maturation, metamorphosis, regeneration, wound-healing, and aging.

An Overview of Developmental Processes

Preformation, or the belief that the egg or sperm contains a preformed, miniature embryo, was once the favored theory of animal development. *Epigenesis*, first proposed by Aristotle, is the view that the form of an embryo gradually develops from an egg. With evidence seen through the microscope, epigenesis replaced preformation as the theory of embryology. The developmental plan of an embryo is predetermined by the zygote genome and by the distribution of material in the egg cytoplasm. As cell division separates cytoplasmic components, nuclei are exposed to different environments that affect which genes are expressed by different cells.

The three key processes of embryonic development are cell division, the production of large numbers of cells; *differentiation*, the formation of specialized cells that are then arranged in tissues and organs; and *morphogenesis*, the movement of cells and tissues that produce body shape and form.

Fertilization

Fertilization, the union of egg and sperm, combines the haploid sets of chromosomes of these specialized cells and activates the egg by initiating metabolic reactions that trigger embryonic development.

The Acrosomal Reaction In sea urchin fertilization, the acrosome at the tip of the sperm discharges hydrolytic enzymes when it comes in contact with the jelly coat of an egg, allowing the *acrosomal process* to penetrate the jelly coat. Fertilization within the same species is assured when a protein called *bindin*, on the surface of the acrosomal process, attaches to specific receptor molecules on the *vitelline layer*, which is external to the plasma membrane of the egg.

Enzymes probably digest a hole through the vitelline layer, and the sperm's plasma membrane fuses with that of the egg, allowing the sperm nucleus to enter the egg. Fusion of the membranes opens membrane ion channels in the egg membrane, and Na^+ ions flow into the egg. The resulting depolarization of the membrane prevents other sperm cells from fusing with the egg, providing a *fast block to polyspermy*.

The Cortical Reaction   Membrane depolarization initiates the release of calcium ions into the egg cytoplasm, causing *cortical granules* located in the gelatinous outer zone of cytoplasm, called the cortex, to fuse with the plasma membrane. These vesicles release their contents by exocytosis, causing the vitelline layer to loosen from the plasma membrane and the resulting space to swell by osmotic uptake of water. The elevated vitelline layer hardens to form the *fertilization membrane*, which functions as a *slow block to polyspermy*.

Activation of the Egg   The release of calcium ions also *activates* the egg by increasing the rates of cellular respiration and protein synthesis. It stimulates the extrusion of H^+ from the cell, and the increase in pH may be what triggers activation.

Parthenogenic development can be initiated by injecting calcium into an egg. Even an enucleated egg can be activated to begin protein synthesis, showing that inactive mRNA has been stockpiled in the egg.

After the sperm nucleus fuses with the egg nucleus, DNA replication begins in preparation for the cleavage division that begins the development of the embryo.

Early Stages of Embryonic Development

The early development of a sea urchin and the frog *Xenopus* are used to illustrate the steps of cleavage, gastrulation, and organogenesis.

Cleavage   *Cleavage* is a succession of rapid cell divisions during which the embryo becomes partitioned into many small cells, called *blastomeres*. *Yolk*, stored nutrients in the egg, provides the nourishment for these early divisions. Cleavage separates different domains of the cytoplasm, containing different cytoplasmic components, into cells and sets the stage for cellular differentiation.

The axis of a *Xenopus* egg is defined by the *animal pole*, which is the point at which the polar body budded from the egg during meiosis, and the opposite end called the *vegetal pole*. The animal hemisphere has melanin granules in the cortex, whereas the vegetal hemisphere contains the yellow yolk. The two hemispheres of a sea urchin egg are not differentiated by color. The first two cleavage divisions are vertical, or

polar; the third division is equatorial. Deuterostomes—echinoderms and chordates—exhibit radial cleavage, in which each tier of cells aligns directly with the cells of the lower tier. Most protostomes—annelids, arthropods, and mollusks—follow a pattern known as spiral cleavage.

Further cleavage in sea urchins produces a *morula*, a solid ball of cells, followed by the arrangement of cells into a hollow ball, the *blastula*, surrounding a fluid-filled cavity called the *blastocoel*. The frog blastula has its blastocoel enclosed in a wall that is several cells thick and restricted to the animal hemisphere.

Gastrulation   Changes in cell motility, shape, and adhesion are part of the rearrangement of cells in *gastrulation* that result in a multi-layered embryo called the gastrula. In a sea urchin, gastrulation occurs when cells at the vegetal pole flatten into a plate that then buckles inward by a process known as *invagination*. Some cells near the plate detach and move into the blastocoel as migratory mesenchyme cells. The invagination deepens to form a narrow pouch called the *archenteron*, a cavity that will become the digestive tract. The archenteron opening is the *blastopore*, which develops into the anus. Other mesenchyme cells detach from the tip of the archenteron. Cytoplasmic extensions (filopodia) from cells at the tip of the archenteron adhere to the blastocoel wall and pull the archenteron to the animal pole. The tip fuses with the outer wall to form the mouth.

The gastrula consists of the *primary germ layers*—the outer *ectoderm*, the middle *mesoderm* formed from migrating mesenchyme cells, and the inner *endoderm*. The organ systems develop from this triploblastic body plan, characteristic of most animals.

Gastrulation in frog development is more complicated because of the multilayered blastula wall and the large, yolk-filled cells of the vegetal hemisphere. A change in the shape of "bottle cells" begins gastrulation, forming a tuck that will become the *dorsal lip* of the blastopore. In a process called *involution*, surface cells roll over the lip and migrate along the roof of the blastocoel. The blastopore is first crescent shaped and eventually forms a circle surrounding a *yolk plug* of large, yolk-filled cells. Gastrulation produces an endoderm-lined archenteron, surrounded by mesoderm, with the outer layer of the gastrula composed of ectoderm.

Organogenesis   Rudiments of organs develop from the primary germ layers in a process known as *organogenesis*. The neural tube and notochord form first in chordates. The *notochord* forms from dorsal mesoderm along the roof of the archenteron. Ectoderm above the developing notochord thickens to form a neural plate, which sinks and rolls into a tube. This hollow *neural tube* develops into the central nervous system. The ectoderm also gives rise to the epidermis, inner ear, and eye lens.

Mesoderm along the sides of the notochord separate into blocks, called *somites*, which give rise to the segmented vertebrae and muscles associated with the axial skeleton. Lateral to the somites, the mesoderm forms the lining of the coelom. It also forms muscles, skeleton, gonads, kidneys, and most of the circulatory system. The digestive tract lining and the organs that form as outpocketings of the archenteron (lungs, liver, pancreas) arise from endoderm.

A band of cells called the *neural crest* forms above the neural tube. These cells later migrate to form pigment cells in the skin; bones and muscles of the skull; teeth; the adrenal medulla; and components of the peripheral nervous system.

Embryology of Amniotes

Reptiles, birds, and mammals are *amniotes,* meaning that they create the aqueous environment necessary for embryonic development with a fluid-filled sac surrounded by a membrane called the amnion. Amnions within the shelled egg of reptiles and birds and the uterus of placental mammals make reproduction on land possible.

Avian Development In the large egg cell of a bird, the yolk is so dense that cleavage of the fertilized egg occurs in only a small disc of cytoplasm at the animal pole, producing a cap of cells called the *blastodisc. Meroblastic cleavage* is this incomplete division of a yolk-filled egg, whereas *holoblastic cleavage* is the complete division of eggs with little to modest amounts of yolk.

The blastodisc separates into an upper epiblast and a lower hypoblast layer, forming a cavity between them comparable to a blastocoel. A straight invagination, called the *primitive streak*, marks the anterior–posterior axis of the embryo and is the site of gastrulation: Cells of the epiblast move toward the primitive streak, detach and ingress, forming the mesoderm and contributing to the endoderm formed from the hypoblast. Cells remaining in the epiblast become ectoderm. The borders of the embryonic disc fold down and join, forming a triple-layered tube attached by a stalk to the yolk. Organogenesis proceeds in a fashion similar to the frog embryo.

The primary germ layers also give rise to four *extraembryonic membranes*, essential to development within the avian egg shell. The *yolk sac* grows to enclose the yolk and develops blood vessels to carry

nutrients to the embryo. The *amnion* encloses a fluid-filled sac, which provides an aqueous environment for development and acts as a shock absorber. The *allantois*, growing out of the hindgut, serves as a receptacle for the nitrogenous waste, uric acid, and expands to press the *chorion* against the lining of the eggshell. The vascularized allantois, in conjunction with the chorion, provides gas exchange for the developing embryo.

Mammalian Development Fertilization occurs in the oviduct and embryonic development begins on the journey to the uterus. The egg of placental mammals has little yolk, thus cleavage of a mammalian zygote is holoblastic. Gastrulation and early organogenesis, however, are similar in pattern to those of birds and reptiles.

The *blastocyst* consists of an outer epithelium, the *trophoblast*, surrounding a cavity into which protrudes a cluster of cells, called the *inner cell mass*. These cells develop into the embryo and some of the extraembryonic membranes. The trophoblast secretes enzymes that enable the blastocyst to embed in the uterine lining (endometrium) and extends projections that develop into the fetal portion of the placenta. The human embryo reaches the uterus as a blastocyst and implants about a week after fertilization.

The inner cell mass forms a flat *embryonic disc*, which resembles the two-layered blastodisc of birds and reptiles. In gastrulation, cells from the upper layer move through a primitive streak to form the mesoderm and endoderm. Organogenesis begins with the formation of the notochord, neural tube, and somites. In the human embryo, all major organs have begun development by the end of the first trimester.

The four extraembryonic membranes are homologous to those of reptiles and birds. The chorion surrounds the embryo and the other membranes. The amnion encloses the embryo in a fluid-filled amniotic cavity. The yolk-sac membrane, enclosing a small fluid-filled cavity, is the site of early formation of blood cells. The allantois forms blood vessels that connect the embryo with the placenta through the umbilical cord.

Mechanisms of Development

The mechanisms underlying animal development are organized into two basic concepts: (1) heterogeneous cytoplasm in the egg is partitioned through cleavage to different blastomeres and (2) interactions between cells affect a cell's developmental fate. Both of these phenomena influence gene expression, which leads to the differentiation of cells.

Polarity of the Embryo   A bilaterally symmetrical animal has a left and right side, an anterior–posterior axis, and a dorsal–ventral axis, polarities that may be established in the process of fertilization.

In the amphibian egg, an animal–vegetal axis exists due to the yolk platelets concentrated in the vegetal pole and cortical pigment granules concentrated near the animal pole. During fertilization, the cortex of the egg moves toward the point of sperm entry, and the edge of the pigmented layer opposite that point is pulled toward the animal pole, exposing a lighter-colored cytoplasm. This half-moon shaped mark is the *gray crescent*, which will be bisected by the first cleavage division and marks the future location of the dorsal lip of the blastopore. The gray crescent determines the dorsal–ventral and the left–right axis of the embryo, and the animal–vegetal axis of the egg becomes the anterior–posterior axis of the embryo.

Mammalian eggs, in contrast, do not have an apparent polarity, and early cleavage divisions are randomly oriented.

Localized Cytoplasmic Determinants   Certain unevenly distributed cytoplasmic components, called *cytoplasmic determinants*, fix the developmental fates of cells. In mollusks, the polar lobe that forms on one of the first two blastomeres is necessary for subsequent formation of mesodermal structures in the embryo. If separated, the blastomere without the polar lobe does not develop normally.

Early blastomeres that can be separated and still produce complete embryos are said to be *totipotent*. Even though cytoplasmic determinants are asymmetrically distributed in the frog egg, the first cleavage division bisects the gray crescent and equally separates these determinants. A zygote's pattern of cleavage influences the developmental fate of cells.

Mosaic and Regulative Development   *Mosaic* development is that in which the early blastomeres are not totipotent. In *regulative* development, cells are totipotent longer. Mammalian embryos remain regulative longer than do those of most other animals.

Fate Maps and the Analysis of Cell Lineages   In the 1920s, Vogt developed a *fate map* for the amphibian embryo by labeling regions of the blastula with dye to determine where specific cells showed up in later developmental stages. Cell lineage analysis follows the mitotic descendants of individual blastomeres. The developmental history of every cell has been determined for the nematode *Caenorhabditis elegans*.

Morphogenetic Movements   Extension, contraction, and adhesion create morphogenetic movements of cells. Changes in cell shape usually involve reorganization of the cytoskeleton. Formation of a wedge shape leads to invaginations and evaginations. Some morphogenesis involves amoeboid movement. Cells at the leading edge of migrating tissue may extend filopodia and then drag along neighboring cells attached to them by intercellular junctions. Amoeboid cells wander individually from the neural crest to various parts of the embryo.

An extracellular matrix of adhesive substances and fibers may help to guide the movement of cells. The orientation of *fibronectin* fibrils, extracellular glycoproteins, corresponds to the arrangement of contractile microfilaments in the cytoskeleton of migrating cells. *Cell adhesion molecules (CAMS)* on the surface of cells hold the cells of specific tissues and organs together during morphogenesis.

Induction   In *induction*, one group of cells influences the development of another. Induction by the chordamesoderm that forms the notochord causes neural plate formation. Using transplant experiments, Mangold and Spemann established that the dorsal lip is the primary organizer of the embryo because of its influence in early organogenesis. A growth factor called *activin* has been identified as the chemical signal involved in the formation of chordamesoderm. Development of the vertebrate eye involves a series of reciprocal inductions.

Differentiation   Cells differentiate, developing characteristic structures and functions, because they express different portions of their common genome. Differentiation is marked by the production of *tissue-specific proteins*. Following induction by retinal cells in the vertebrate eye, immature lens cells begin producing mRNA for the protein crystallin. Even after cells begin expressing different genes, nearly all cells in an organism still possess the same genes and have *genomic equivalence*.

To determine whether genes are irreversibly inactivated during differentiation, Briggs and King transplanted nuclei from embryonic and tadpole cells into eunucleated frog and toad egg cells. The ability of the transplanted nuclei to direct normal development was inversely related to its developmental age. These studies indicate that the nucleus of a differentiated cell still contains all the genes necessary for development, but that the genome of a cell undergoes some change during differentiation.

Once a cell has become restricted to a particular path of development, it is said to be *determined*. Transplant experiments show whether tissues at various stages of development have been determined. *Imaginal disks* are groups of undifferentiated, yet determined, cells in insect larvae that are destined to form specific organs during metamorphosis. Determination appears to develop progressively; the developmental options of cells become more and more narrow.

The cytoplasmic environment of a cell apparently controls genome expression. Cytoplasmic determinants may affect which genes will be expressed when the cell differentiates later. Morphogenetic movements also contribute to determination and differentiation by exposing cells to differing chemical and physical environments.

Pattern Formation *Pattern formation* is the ordering of cells and tissues into three-dimensional structures that make up the parts of an animal. *Homeotic genes*, first identified in fruit flies, are regulatory genes that control the developmental fate of groups of cells. A sequence of 180 nucleotides, called a *homeobox*, was found in each of the fruit-fly homeotic genes. The same or similar homeotic nucleotide sequences have now been identified in every eukaryotic organism examined, and homeoboxes appear to have a large role in controlling patterns of development.

These nucleotide sequences are translated into amino acid sequences, called homeodomains, which are able to bind to specific DNA sequences. The proteins containing homeodomains have turned out to be regulatory proteins that control the transcription of other genes by binding to DNA. Thus, homeotic genes code for proteins that may coordinate the transcription of developmental genes.

The similarity of eukaryotic homeodomain amino acid sequences and their relationship to DNA-binding domains of prokaryotic regulatory proteins indicate that these sequences must have arisen early and been conserved through evolution as important regulators of gene expression and development.

But more than genetic information guides the formation of body parts. The cells of a rudimentary organ differentiate based on *positional information* that indicates their position within a field of cells.

Wolpert has been studying positional information in chick limb development by doing transplant experiments. The "zone of polarizing activity" (ZPA) appears to communicate position along the anterior–posterior axis in a developing limb bud. Another region of mesoderm at the tip of the wing bud may assign position along the proximal–distal axis. Position along the dorsal-ventral axis may reflect relative distance from the dorsal and ventral ectoderm of the limb. The substances producing the chemical gradients to which developing cells respond are called *morphogens*. Experiments have isolated *retinoic acid* as the possible morphogen produced by the ZPA.

Other experiments indicate that positional informa-

tion contributes to overall body form, not just the pattern of individual parts. Undifferentiated mesoderm from the thigh region developed into a toe when transplanted to the tip of a wing bud. The mesoderm had already been determined to form a part of the leg, but it still responded to its location at the tip of a developing limb.

Installments of positional information are provided to cells throughout their development. Researchers have identified a hierarchy of pattern formation in *Drosophila*. The anterior-posterior and dorsal-ventral polarity of the fruit fly is determined by the unequal distribution of substances in the unfertilized egg. Nurse cells at one end of the egg secrete bicoid mRNA into the egg. After fertilization this mRNA is translated, and a gradient of bicoid protein indicates the head end of the animal. Posterior, dorsal, and ventral morphogens also affect this first stage of pattern formation. These morphogens lead to regional differences in the expression of segmentation genes which determine the number of segments in the embryo. Products of the segmentation genes then influence which homeotic genes will be expressed in each segment. Pattern formation results from a sequence of gene activations that continue to refine body form.

STRUCTURE YOUR KNOWLEDGE

1. Create a flow chart that shows the sequence of key events in the fertilization of a sea urchin egg and the functions of these events.

2. Label below the indicated structures of early organogenesis in the frog and bird embryo.

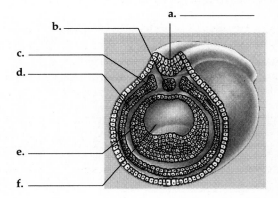

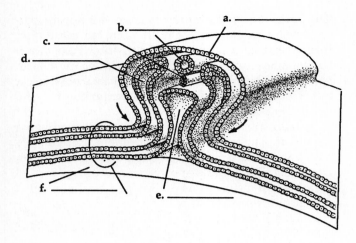

3. Fill in the table below, briefly describing the early stages of development for sea urchin, frog, bird, and mammalian embryos.

4. Use the following terms to create a concept map that illustrates the relationships among these concepts. Add additional concepts as needed.

development
determination
cytoplasmic determinants
positional information
differentiation
cell division
pattern formation
morphogenesis

TEST YOUR KNOWLEDGE

MULTIPLE CHOICE: *Choose the one best answer.*

1. The vitelline layer
   a. is inside the fertilization membrane.
   b. releases calcium, which initiates the cortical reaction.
   c. has receptor molecules that are specific for bindin proteins on the acrosomal process of sperm.
   d. All of the above are correct.

2. The fast block to polyspermy
   a. prevents sperm from other species from fertilizing the egg.
   b. is produced by the depolarization of the membrane.
   c. is a result of the formation of the fertilization membrane.
   d. is caused by the exocytosis of cortical granules.

3. The activation of an enucleated egg by pricking shows that
   a. calcium ions initiate activation.
   b. inactive mRNA had been stockpiled in the egg cell.
   c. parthenogenic development may replace fertilization in sea urchins.
   d. the nucleus is not needed for the development of a sea urchin embryo.

4. Which of the following groups does not have eggs with definite animal and vegetal poles?
   a. sea urchin
   b. frog
   c. bird
   d. mammal

5. The archenteron develops into the
   a. anus in deuterostomes.
   b. blastocoel.
   c. neural tube.
   d. digestive tract.

6. Organogenesis
   a. produces the gastrula by morphogenesis.
   b. follows a different pattern in the embryos of frogs and birds.
   c. is the rudimentary development of organs from the primary germ layers.
   d. begins with the formation of the neural plate.

	SEA URCHIN	FROG	BIRD	MAMMAL
Cleavage				
Blastula				
Gastrulation				

7. Which of the following is incorrectly paired with its primary germ layer?
   a. muscles—mesoderm
   b. central nervous system—ectoderm
   c. lens of the eye—mesoderm
   d. liver—endoderm

8. In a frog embryo, gastrulation
   a. is impossible because of the large amount of yolk.
   b. proceeds by involution as cells roll over the dorsal lip of the blastopore.
   c. produces a blastocoel displaced into the animal hemisphere.
   d. occurs along the primitive streak.

9. The embryonic disc of mammalian embryos
   a. develops from the inner cell mass.
   b. resembles the blastodisc of birds.
   c. develops into the embryo and some extraembryonic membranes.
   d. All of the above are correct.

10. The function of the allantois in birds is to
    a. provide for nutrient exchange between the embryo and yolk sac.
    b. store nitrogenous wastes in the form of urea.
    c. form a respiratory organ in conjunction with the chorion.
    d. provide an aqueous environment for the developing embryo.

11. Which of the following is a *correct* statement of a *difference* between avian and mammalian early embryology?
    a. The bird egg has meroblastic cleavage whereas the mammalian egg has holoblastic cleavage.
    b. The bird embryo is surrounded by an amnion membrane, whereas the mammalian embryo is enclosed by the placenta.
    c. Gastrulation occurs through the primitive streak in birds but by way of the blastopore in mammals.
    d. The bird egg has four extraembryonic membranes, whereas the mammalian embryo develops only two.

12. Somites are
    a. blocks of mesoderm circling the archenteron.
    b. cells from which the notochord arises.
    c. serially arranged mesoderm blocks that develop into vertebrae and skeletal muscles.
    d. mesodermal derivatives of the notochord that line the coelom.

13. Which of the following is an example of induction?
    a. a gradient of retinoic acid that signals the anterior-posterior axis in a limb bud
    b. the extension of filopodia which then pull the archenteron toward the animal pole
    c. movement of cells over the dorsal lip of the gray crescent and into the interior of frog embryo
    d. interactions between optic vesicle and ectoderm to form optic cup and lens

14. The imaginal disks of an insect larva
    a. are cells that are determined but not differentiated and will form various organs of the adult following metamorphosis.
    b. establish the polarity of the embryo as the result of a gradient of bicoid mRNA.
    c. contain all three primary germ layers as a result of invagination along the primitive streak.
    d. are like somites that develop into the segmented structure of the adult insect.

15. The gray crescent
    a. is the future location of the dorsal lip of the blastopore.
    b. is created when the pigmented layer is pulled toward the animal pole at fertilization.
    c. is bisected by the first cleavage division, creating an equal division of cytoplasmic determinants.
    d. is all of the above.

16. In regulative development,
    a. cells remain totipotent until the gastrula stage.
    b. early blastomeres that are separated produce complete embryos.
    c. blastomeres without the polar lobe do not develop normally.
    d. one group of cells regulates or influences the development of another.

17. Cytoplasmic determinants
    a. are unevenly distributed cytoplasmic components that fix the developmental fates of cells.
    b. are involved in the regulation of gene expression.
    c. are involved in the differentiation of cells and tissues.
    d. are all of the above.

18. Cells of specific tissues and organs are held together during morphogenesis by
    a. fibronectin fibrils.
    b. cell adhesion molecules.
    c. induction.
    d. morphogens.

19. Pattern formation appears to be determined by
    a. positional information a cell receives from gradients of chemical signals called morphogens.
    b. differentiation of cells, which then migrate into developing organs.
    c. the movement of cells along fibronectin fibrils.
    d. the induction of cells from chordamesoderm cells found in the center of organs.

20. Homeodomains are
    a. coded for by homeobox DNA sequences found in homeotic genes.
    b. part of regulatory proteins that bind to DNA and influence transcription.
    c. similar to DNA-binding domains of prokaryotic regulatory proteins.
    d. All of the above.

CHAPTER 44

NERVOUS SYSTEMS

FRAMEWORK

This chapter describes the structural components of nervous systems and how they functionally integrate, coordinate, transmit and respond to information from the internal and external environment.

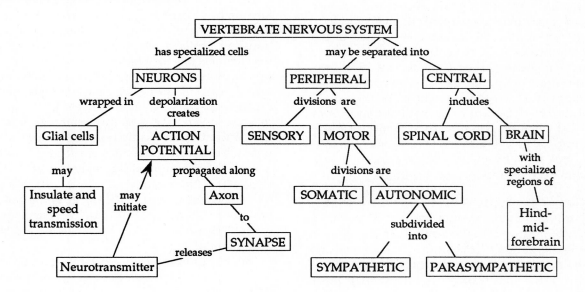

CHAPTER SUMMARY

The nervous system enables an animal to respond rapidly and appropriately to environmental stimuli. The coordination of behavior and physiology is a product of both the nervous and endocrine systems. Three characteristics distinguish the nervous system from the endocrine system: rapid electrical transmission of information along specialized cells called *neurons*; pinpoint control, with messages traveling direct-

ly to individual target cells; and a greater structural complexity enabling the integration of more kinds of information and responses.

The three main functions of the nervous system are sensory input, integration, and motor output. Information from sensory receptors is conducted to integration centers in the brain and spinal cord (*central nervous system*) where it is interpreted. Motor output is the conducting of signals to *effector cells*, such as muscle or gland cells. The *peripheral nervous system* carries sensory and motor information between the body and the central nervous system in the form of electrical and chemical signals.

Cells of the Nervous System

The cells of the nervous system include neurons, which transmit signals, and supporting cells, which strengthen, insulate, and protect neurons.

Neurons A neuron consists of a cell body, containing a nucleus and organelles, and long fiberlike processes, *axons* and *dendrites*, which conduct impulses. A neuron usually has one axon which transmits signals away from the cell body. Dendrites are the more numerous, shorter, and branched processes that carry signals toward the cell body.

Axons originate from a region of the cell body called the axon hillock. In many vertebrates, axons are wrapped in *Schwann cells* that collectively form the insulating *myelin sheath*. Branched axons may terminate in hundreds or thousands of specialized endings called *synaptic terminals*, which release neurotransmitters that relay nervous signals across the *synapse* between the axon and another neuron or effector.

The cell bodies of most neurons are located in the central nervous system. Others are located elsewhere in the body, in clusters called ganglia. *Sensory neurons* transmit information about the internal and external environment from sensory receptors to the central nervous system. *Motor neurons* carry information from the central nervous system to effector organs. *Interneurons,* located within the central nervous system, transmit signals between neurons and integrate information from sensory to motor neurons.

Supporting Cells Supporting cells give structural integrity to the nervous system and are involved in neural function as well. In the central nervous system they are known as *glial cells,* and include *astrocytes,* which line capillaries in the brain, contributing to the *blood-brain barrier* that restricts the passage of most substances into the brain; and *oligodendrocytes,* which insulate axons in a myelin sheath. Oligodendrocytes and the supporting cells in the peripheral nervous systems, called Schwann cells, wrap to form concentric membrane layers around axons. The resulting myelin sheaths provide electrical insulation between neurons.

Transmission of Electrical Signals Along a Neuron

Electrical signals, created by ion fluxes across the plasma membrane, transmit signals along the length of neurons.

The Origin of Electrical Membrane Potential An electrical voltage gradient across the membrane, called the *membrane potential,* results from the difference in charge between cytoplasm and extracellular fluid, and is a fundamental feature of all cells. Electrophysiologists can measure the magnitude of the charge separation by using microelectrodes placed inside and outside a cell and connected to a voltmeter. The *resting potential,* or membrane potential of a nontransmitting neuron, is about −70 mV (millivolts). Neurons and muscle cells are *excitable cells*; they are able to respond to a stimulus by changing their membrane potential to conduct signals.

Membrane potential depends on the different concentrations of ions on either side of the membrane and the selective permeability of the membrane to different ions. The principle cation outside cells is sodium (Na^+) and the principle anion is chloride (Cl^-). Inside cells, potassium (K^+) is the principal cation and a group of negatively-charged, mostly large organic molecules (proteins, negative amino acids, phosphate, etc.) are considered the principal anion (A^-).

Due to the phospholipid bilayer, ions must cross membranes by way of ion channels through specific transmembrane protein molecules or be carried by transport proteins. The number and type of selective ion channels in a membrane determines its permeability to different ions. Cells are usually much more permeable to K^+ than to Na^+. The internal anions (A^-) contribute a constant internal negative charge because they are too large to cross the membrane. The potassium concentration gradient drives K^+ out of the cell until the electrical gradient, created by the nonmoving A^- and the exiting positive charge, starts to drive K^+ back across the membrane. The *equilibrium potential* for potassium is about −85 mV. It is the potential at which the concentration and electrical gradients would balance each other, and there would be no net movement of K^+ ions.

The higher external concentration of Na^+, as well as the negative charge inside the cell, tend to move Na^+ into the cell, increasing the internal positive charge and raising the resting potential of a cell to −70mV. Sodium–potassium pumps maintain these concentration gradients using energy from ATP to move sodium ions back out of the cell and potassium ions into the cell.

The Action Potential When a neuron receives a stimulus, *gated ion channels* open, causing the cell to change its membrane potential. Should the stimulus open potassium channels, the efflux of K^+ will cause the membrane potential to become more negative, resulting in *hyperpolarization.* When sodium channels open and Na^+ is allowed to flow in, the membrane potential is reduced, *depolarizing* the membrane. With this type of *graded potential*, the magnitude of a voltage change is proportional to the strength of the stimulus; the stronger the stimulus, the more gated ion channels that open and the greater the hyperpolarization or depolarization.

Once depolarization of a neuron reaches a particular level, called the *threshold*, an *action potential* is triggered. The threshold potential in a typical neuron is –55 to –50 mV. The action potential operates on the *all or none* principle, always creating the same voltage spike, regardless of the intensity of the stimulus. An action potential causes the membrane potential to first reverse polarity in the depolarizing phase, and then return to the normal negative resting potential during a repolarizing phase. The membrane potential may temporarily become more negative than the resting potential in a period called the undershoot.

The *voltage-gated channels* or *voltage-sensitive channels* of excitable cells open and close in response to changes in membrane potential. The sodium activation gate opens rapidly in response to depolarization, allowing Na^+ to enter the cell, which further depolarizes the membrane and opens more voltage-gated sodium channels. The slower acting sodium-channel inactivation gate closes in response to depolarization, greatly reducing sodium permeability during the repolarizing phase. Voltage-sensitive potassium channels also open in response to depolarization, allowing K^+ to flow from the cell. This response is slower than that of the Na^+ channels and helps return the cell to its resting-stage negativity. During the undershoot, the sodium-channel inactivation gates have returned sodium permeability to the resting level, but the potassium channels have not yet closed in response to repolarization, and the increased permeability to K^+ may temporarily hyperpolarize the membrane and produce a voltage more negative than the resting potential. During the *refractory period*, which occurs at the end of an action potential before the inactivation gates have reopened, the neuron cannot respond to another stimulus.

Though the amplitude of an action potential is always the same, its frequency varies with the intensity of a stimulus. A neuron will "fire" repeatedly after each refractory period in response to a strong stimulus.

Propagation of the Action Potential Stimuli are usually received at the dendrites or cell body of a neuron, and the action potential travels along the axon. As sodium ions move into the cell during the action potential, they depolarize adjacent sections of the membrane, bringing them to the threshold potential. Voltage-sensitive channels open and an action potential is generated farther along the membrane, in turn

inciting another action potential. Thus local depolarizations and action potentials across the membrane result in the propagation of serial action potentials along the length of the neuron. Because of the brief refractory period, the action potential is usually propagated in only one direction.

Speed of Propagation of the Action Potential Resistance to current flow is inversely proportional to the cross-sectional area of the conducting "wire." The greater the axon diameter, the lower the resistance and the faster the action potential is conducted. Some invertebrates have giant axons which conduct impulses rapidly.

In vertebrates, supporting cells form a thick myelin sheath around many axons, with small gaps, called *nodes of Ranvier*, where these cells abut. Voltage-sensitive ion channels are concentrated in the node regions, which are the only place where the neuron membrane has contact with extracellular fluid. Action potentials can only be generated at these nodes, and a nerve impulse "jumps" from node to node, resulting in a faster mode of transmission known as *saltatory conduction*.

The Synapse: Transmission Between Cells

Synapses between neurons conduct impulses from the axon terminal of a *presynaptic cell* to a dendrite or cell body of the *postsynaptic cell*.

Electrical Synapses Electrical synapses allow action potentials to flow directly from presynaptic to postsynaptic cell via cytoplasmic connections called gap junctions. Electrical synapses are found in the giant neurons of some crustaceans, but are less common than chemical synapses in vertebrates and most invertebrates.

Chemical Synapses At a chemical synapse, the electrical message is converted to a chemical message that travels from the presynaptic cell across a synaptic cleft to the postsynaptic cell. The synaptic terminal, at the tip of a branch of the presynaptic axon, contains numerous *synaptic vesicles*, in which thousands of molecules of the chemical messenger, called a *neurotransmitter*, are stored. The depolarization of the *presynaptic membrane* at the axon terminal opens voltage-sensitive calcium channels in the membrane. The influx of Ca^{2+} causes the synaptic vesicles to fuse with the presynaptic membrane and release neurotransmitter into the cleft.

The *postsynaptic membrane* of the postsynaptic cell

contains receptor proteins for different neurotransmitter molecules, which are associated with particular ion channels. These chemically sensitive gates allow ions, such as Na⁺, K⁺, or Cl⁻, to cross the membrane, altering the membrane potential by either bringing it closer to the threshold potential or hyperpolarizing it. Enzymes rapidly break down the neurotransmitter.

Summation: Neural Integration at the Cellular Level

A neuron must integrate the information it receives from numerous neighboring neurons at thousands of its excitatory and inhibitory synapses. At an excitatory synapse, binding of the neurotransmitter to receptors opens a gated channel that allows Na⁺ to flow into and K⁺ to flow out of the cell. A net flow of positive charge into the cell depolarizes the membrane, creating an *excitatory postsynaptic potential (EPSP)* and bringing the membrane potential closer to an action potential threshold. At an inhibitory synapse, binding of the neurotransmitter opens ion gates to allow K⁺ to flow out and/or Cl⁻ to move into the cell, hyperpolarizing the membrane and producing an *inhibitory postsynaptic potential (IPSP)*. The response to a neurotransmitter depends on the type of receptor and the gated ion channels associated with it on the postsynaptic membrane.

EPSPs and IPSPs are graded potentials. Their magnitude depends on the number of neurotransmitter molecules that bind to receptors. *Summation* of several postsynaptic potentials is usually necessary to bring the axon hillock to threshold potential. *Temporal summation* occurs with repeated release of neurotransmitters from one or more synaptic terminals before the postsynaptic potential returns to its resting potential. *Spatial summation* occurs when several different presynaptic terminals, usually from different neurons, release neurotransmitter simultaneously. Several EPSPs arriving simultaneously may depolarize the axon hillock to the threshold potential and generate an action potential. Summated IPSPs will hyperpolarize the membrane. Concurrent EPSPs and IPSPs also summate, countering their electrical effects. The membrane potential of the axon hillock at any given time is determined by the sum of all EPSPs and IPSPs.

Neurotransmitters and Receptors

About ten molecules have been identified as neurotransmitters, and more may be added to the list. The criteria for neurotransmitters are as follows: (1) the compound must be contained in synaptic vesicles and discharged when the presynaptic cell is stimulated, and it must alter the membrane potential of the postsynaptic membrane; (2) when experimentally injected into the synapse, it must cause an EPSP or IPSP; and (3) the compound must be rapidly degraded or removed from the synapse.

Acetylcholine is a common neurotransmitter in invertebrates and vertebrates. In vertebrate neuromuscular junctions, acetylcholine released from the motor axon depolarizes the postsynaptic muscle cells. It can also be inhibitory, for example, slowing down the heart rate of vertebrates and mollusks.

Biogenic amines, neurotransmitters derived from amino acids, usually function within the central nervous system. *Epinephrine, norepinephrine,* and *dopamine* are biogenic amines derived from the amino acid tyrosine. *Serotonin*, synthesized from tryptophan, and dopamine affect sleep, mood, attention, and learning. Imbalances of these transmitters have been associated with mental illness.

The amino acids *glycine, glutamate,* and *gamma aminobutyric acid (GABA)* function as neurotransmitters in the central nervous system. GABA is the most common inhibitory transmitter in the brain.

Endorphins and *enkephalins* are *neuropeptides* produced in the brain during physical or emotional stress that function as natural pain killers. An endorphin that is released from the anterior pituitary also functions as a hormone.

Acetylcholine and the amino acid transmitters bind to receptors on the postsynaptic membrane, affect ion permeability, and create either EPSPs or IPSPs. Biogenic amines and neuropeptides have a longer lasting effect because, when they bind to their receptors, they activate a second messenger that affects the metabolism of the postsynaptic cell.

Neural Circuits and Clusters

Groups of neurons that interact and carry information along specific pathways are called circuits. In *convergent circuits*, several neurons come together and feed information into a single postsynaptic neuron. *Divergent circuits* spread out information from one neuron to several postsynaptic neurons. *Reverberating circuits* are circular paths in which signals return to their source.

Nerve cell bodies are arranged in functional groups called *ganglia*. A ganglion in the brain is often called a *nucleus*.

Invertebrate Nervous Systems

The cnidarian *nerve net* is a loosely organized system of nerves, connected by electrical synapses, in which impulses are conducted in both directions. Some centralization is seen in jellyfish, in which clusters of nerve cells around the margin of the bell coordinate swimming movements. Modified nerve nets are also found in echinoderms.

Bilateral animals with more active lifestyles show the evolutionary trend of *cephalization*—the concentration of sense organs and feeding structures in the head, along with the enlargement of anterior ganglia that process sensory information and control feeding responses.

Flatworms have a simple brain and two or more nerve trunks, with ladderlike transverse nerves. Annelids and arthropods have a prominent brain and a ventral nerve cord, which may have ganglia within each segment of the body. Sessile mollusks have little cephalization and only simple sense organs. Cephalopod mollusks, such as the octopus, have large brains, image-forming eyes, and giant axons, all of which contribute to their active predatory life and their ability to learn and remember.

The Vertebrate Nervous System

Peripheral Nervous System The peripheral nervous system consists of the *sensory* or *afferent nervous system*, which brings information from sensory receptors to the central nervous system (CNS), and the *motor* or *efferent nervous system*, which carries signals away from the CNS to effector cells. The human peripheral nervous system contains 12 pairs of cranial nerves and 31 pairs of spinal nerves. A few of the cranial nerves are sensory only; the other cranial and spinal nerves contain both sensory and motor neurons.

The sensory nervous system brings in messages from both the external and internal environments. The motor nervous system has two divisions, one of which governs responses to the external environment, while the other coordinates functions of the internal organs. The *somatic nervous system* carries signals to skeletal muscles, the movement of which is largely under conscious control. Many skeletal muscle movements, however, are determined by *reflexes*, which are automatic, subconscious reactions to stimuli mediated by the spinal cord or lower brain. The *autonomic nervous system* maintains involuntary control over smooth and cardiac muscles and various organs.

The autonomic nervous system is subdivided into the *parasympathetic* and the *sympathetic nervous systems*. In general, the parasympathetic division slows the heart, stimulates digestion, and enhances activities that gain and conserve energy. The sympathetic division accelerates the heart rate and increases the metabolic rate, preparing an organism for action. Parasympathetic nerves originate at the top and bottom of the CNS, whereas sympathetic nerves come from the upper and central spinal cord. In the sympathetic nervous system, the first of the two neurons synapse with cell bodies of the second neurons in

prominent ganglia near the spinal cord. Acetylcholine is released in this synapse, whereas norepinephrine is released by sympathetic neurons at the target organs. Acetylcholine is released by both the first and second parasympathetic neurons in the chain.

Central Nervous System The central nervous system, which bridges the sensory and motor components of the peripheral nervous system, consists of the *spinal cord* and the *brain*. Protective layers of connective tissue called the *meninges* cover these bilaterally symmetrical organs. Axons are in bundles or tracts; *white matter* is named for the white color of their myelin sheaths. Neuron cell bodies make up the *gray matter*, which is outside the white matter in the brain, but inside it in the spinal cord.

Vertebrate nervous systems are hollow. Spaces in the brain called *ventricles* are continuous with the narrow *central canal* of the spinal cord and filled with *cerebrospinal fluid*. This fluid, formed by filtration of the blood, cushions the brain and carries out circulatory functions.

The spinal cord integrates simple responses to some stimuli, usually in the form of a reflex, and carries information to and from the brain. The knee-jerk reflex involves a stretch receptor in the knee; a sensory neuron, with its cell body outside the spinal cord in a dorsal root ganglia; and a motor neuron, with its cell body in the gray matter. Most reflexes have interneurons located between sensory and motor neurons.

The vertebrate brain evolved as bulges in the anterior end of the spinal cord. Three regions—the *rhombencephalon* or *hindbrain*, the *mesencephalon* or *midbrain*, and the *prosencephalon* or *forebrain*—are present in all vertebrates, although these regions may be subdivided to provide a greater capacity for integration of complex activities.

Three evolutionary trends in the vertebrate brain include an increase in relative size, subdivision into areas with specific functions, and additional complexity of the forebrain. More sophisticated behaviors are correlated with increased size of one region of the forebrain, the *cerebrum*. Folding or convolution of the cerebral cortex increases surface area and capacity of the cerebrum.

The human brain develops from the three primary regions that differentiate into specialized structures. The *brainstem* forms from the hindbrain and midbrain and extends from the spinal cord to the middle of the brain.

The hindbrain has three parts: the *medulla oblongata*, which contains control centers for homeostatic functions including respiration, heart and blood-vessel actions, and digestion; the *pons*, which functions with the medulla in some of these activities and in conducting information between the rest of the brain and the spinal cord; and the *cerebellum*, which coordinates movements. The tracts of motor neurons from the mid- and forebrain cross in the medulla, so that the right side of the brain controls much of the movement of the left side of the body and vice versa. The cerebellum integrates information from the auditory and visual systems with sensory input from the joints and muscles as well as motor pathways from the cerebrum to provide unconscious coordination of movements and balance.

The upper portion of the brainstem, or midbrain, receives and integrates sensory information and sends this information to specific regions of the forebrain. Fibers involved in hearing pass through or terminate in the *inferior colliculi*. The *superior colliculi* are important visual centers, forming prominent optic lobes in nonmammalian vertebrates and coordinating visual reflexes in mammals. The *reticular formation*, which includes a major group of nuclei (ganglia) in the midbrain, regulates states of arousal.

The most intricate neural processing occurs in the forebrain. The *diencephalon* contains the thalamus and hypothalamus, and the *telencephalon* contains the cerebrum.

The *thalamus* is a major relay area for sensory information going to the cerebral cortex. It filters information sent to the cerebrum and receives input from the cerebrum and from parts of the brain controlling emotion and arousal.

The *hypothalamus* is the major site for homeostatic regulation. It produces the posterior pituitary hormones and the releasing hormones that control much of the hormone secretion of the anterior pituitary. The hypothalamus contains the regulating centers for many autonomic functions and also plays a role in sexual response, mating behaviors, the alarm response, and pleasure.

The basal ganglia, a cluster of nuclei below the cerebral cortex, are important in relaying motor impulses and coordinating motor responses.

The *cerebral cortex* has changed most during vertebrate evolution. The human cortex is divided into five lobes. The two hemispheres of the cortex are connected by the *corpus callosum*, a thick band of fibers. The cortex contains bilateral primary sensory and motor areas, which directly process information, and association areas, which integrate information from several sources.

The primary sensory area receives impulses from touch, pressure, and pain receptors throughout the body. The primary motor area sends impulses to the skeletal muscles. The proportion of the primary sensory and motor areas devoted to controlling each part of the body is correlated with the importance of that area. The special senses of vision, hearing, smell, and taste are controlled by other regions of the cortex.

Integration and Higher Brain Functions Nerve impulses are integrated on all levels of the nervous system, from the spinal reflex to the intellectual creations of the cerebral cortex.

Arousal is a state in which an individual is aware of the external world, whereas sleep is a state in which the individual is not conscious of external stimuli. Different patterns in the electrical activity of the brain can be recorded by an *electroencephalograph*, or *EEG*. Slow, synchronous alpha waves are produced by a person lying quietly with eyes closed. Faster beta waves are associated with opened eyes or thinking about a complex problem. Quite slow and highly synchronized delta waves occur during sleep. Periods of delta waves alternate with periods of a desynchronized EEG and rapid eye movements, REM sleep, when most dreaming occurs.

The reticular formation filters the sensory information reaching the cortex. The level of arousal relates to the amount of input the cortex receives. Sleep-producing centers are located in the pons and medulla, and serotonin may be the neurotransmitter involved in these areas. A center that causes arousal is found in the midbrain.

Human emotions have been tied to interactions between the cerebral cortex and a group of nuclei in the lower forebrain called the *limbic system*. Destruction of this portion of the brain may result in docility.

The association areas of the cerebral cortex are not bilaterally symmetrical. The left hemisphere controls speech, language, and calculation, whereas the right hemisphere controls artistic ability and spatial perception. Much of the information on this *lateralization* of the brain comes from Sperry's work with "split-brain" patients.

Language and speech are controlled by areas on the left cerebral hemisphere. *Wernicke's area* stores information required for speech content and construction, whereas *Broca's area* involves information required for speech production by the lips, tongue, and other muscles. Different types of aphasia, the inability to speak coherently, occur if one or the other of these regions is damaged.

Human memory consists of *short-term memory*, the immediate sensory perception of an object or idea, and *long-term memory*, the recall of these perceptions after the passage of time. Transferring information from short-term to long-term memory is facilitated by rehearsal, a favorable emotional state, and associations with previously learned and stored information. Fact memories can be consciously recalled from your memory bank. Skill memories are formed by repetition of motor activities and do not require conscious recall of specific information.

Neuroscientists have determined that, in what appears to be a memory circuit, sensory information is transmitted from sensory regions of the cerebral cortex to the *hippocampus* and *amygdala*, components of the limbic system, from which impulses are relayed to other regions of the forebrain. An integrating area called the basal forebrain then transmits impulses back to the same sensory area of the cortex that perceived the sensory input. Chemical or structural changes in the sensory cortex may store the information as a memory. Some neuroscientists suggest that changes in dendrite structure enhance future transmission across a synapse, thus facilitating learning.

Simpler organisms are often used to study the mechanisms involved in memory and learning. The habituation of the sea hare *Aplysia* to mild touches, as well as its sensitization to harmful stimuli, seem to involve changes in ion channels at the synapses of sensory and motor neurons.

STRUCTURE YOUR KNOWLEDGE

1. Label the indicated structures on this diagram of a neuron.

2. Develop a flow chart, diagram, or description of the sequence of events in the creation and propagation of an action potential, and the transmission of this potential across a chemical synapse.

3. The vertebrate nervous system can be separated into a sequence of divisions and subdivisions. Create a concept map that shows these subsets and their functions.

4. Fill in the table on p. 378 by briefly describing the functions of the parts of the human brain and identifying the section (hind-, mid-, or forebrain) in which they are found.

TEST YOUR KNOWLEDGE

MULTIPLE CHOICE: *Choose the one best answer.*

1. Which of the following does *not* distinguish the nervous system from the endocrine system?
   a. Nervous communication is more rapid.
   b. Nervous system uses chemical communication.
   c. Nervous system messages are delivered directly to target cells or organs.

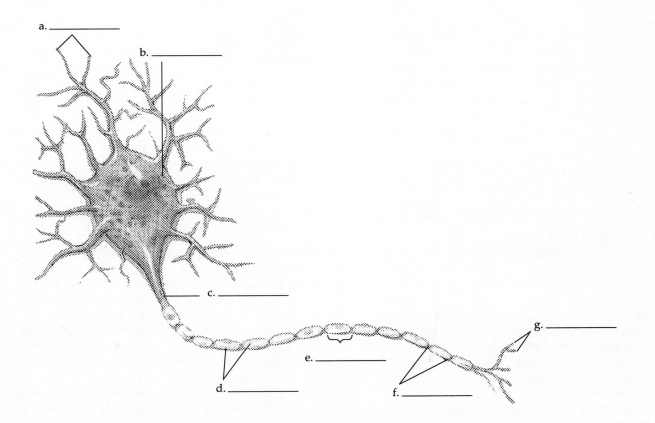

a. _____
b. _____
c. _____
d. _____
e. _____
f. _____
g. _____

PART OF BRAIN	SECTION	FUNCTION
Medulla		
Pons		
Cerebellum		
Inferior, superior colliculi		
Reticular formation		
Thalamus		
Hypothalamus		
Cerebral cortex		

  d. The structural complexity of the nervous system allows for integration of more information and responses.

2. Interneurons
   a. may connect sensory and motor neurons.
   b. are confined to the peripheral nervous system.
   c. do not have cell bodies.
   d. are electrical synapses between neurons.

3. Nodes of Ranvier are
   a. gaps where Schwann cells abut at which action potentials are generated.
   b. neurotransmitter-containing vesicles located in the synaptic terminals.
   c. the major components of the blood-brain barrier that restrict the passage of substances into the brain.
   d. clusters of receptor proteins located on the postsynaptic membrane.

4. Which of the following is *not* true of the resting potential of a neuron?
   a. The inside of the cell is more negative than is the outside.
   b. There are concentration gradients with more sodium outside the cell and a higher potassium concentration inside the cell.
   c. It is about –70 mV and can be measured by using microelectrodes placed inside and outside the cell.
   d. It is formed by the opening of voltage-sensitive channels.

5. After an action potential, the resting potential is restored by
   a. the opening of voltage-sensitive potassium channels and the closing of sodium activation gates.
   b. the opening of sodium activation gates.
   c. the refractory period in which the membrane is hyperpolarized.
   d. the delay in the action of the sodium–potassium pump.

6. The threshold potential of a membrane
   a. is a positive value, often equal to about 35mV.
   b. opens voltage-sensitive channels and permits the rapid outflow of sodium ions.
   c. is the depolarization that is needed to generate an action potential.
   d. is a graded potential that is proportional to the strength of a stimulus.

7. Which of the following is *not* true of chemical synapses?
   a. Synaptic terminals at the ends of branching dendrites contain synaptic vesicles, which enclose the neurotransmitter.
   b. The influx of calcium when an action potential reaches the presynaptic membrane causes synaptic vesicles to release their neurotransmitter into the cleft.
   c. The binding of neurotransmitter to receptors on the postsynaptic membrane changes the membrane's permeability to certain ions.

d. An excitatory postsynaptic potential forms when sodium channels open and the membrane potential moves closer to an action potential threshold.

8. In spatial summation,
   a. the sum of simultaneously arriving neurotransmitters from different presynaptic nerve cells determines whether the postsynaptic cell fires.
   b. several action potentials arrive in fast succession from the same presynaptic cell.
   c. several IPSPs arrive concurrently, bringing the presynaptic cell closer to its threshold.
   d. the voltage spike of the action potential that is initiated is higher than normal.

9. An inhibitory postsynaptic potential occurs when
   a. sodium flows into the postsynaptic cell.
   b. binding of the neurotransmitter opens ion gates that result in the membrane becoming hyperpolarized.
   c. enzymes do not break down the neurotransmitter in the synaptic cleft.
   d. acetylcholine is the neurotransmitter.

10. Which of the following does *not* function as a neurotransmitter?
    a. acetylcholine
    b. neuropeptides
    c. biogenic amines
    d. steroids

11. Groups of neurons that interact and carry information along pathways are called
    a. ganglia.
    b. nuclei.
    c. circuits.
    d. nerve nets.

12. Which of the following animals is mismatched with its nervous system?
    a. sea star (echinoderm)—modified nerve net, central nerve ring with radial nerves
    b. hydra (cnidarian)—ring of ganglia, paired ventral nerve cords
    c. annelid worm—brain, ventral nerve cord with segmental ganglia
    d. vertebrate—dorsal central nervous system of brain and spinal cord

13. Which of the following is *not* true of the autonomic nervous system?
    a. It is a subdivision of the somatic nervous system.
    b. It consists of the sympathetic and parasympathetic divisions.
    c. It is part of the peripheral nervous system.
    d. It controls smooth and cardiac muscles.

14. If you needed to obtain nuclei from neurons for an experiment, you would want to make preparations of
    a. astrocytes of the brain.
    b. the gray matter of the brain.
    c. the inner portion of the spinal cord.
    d. both b and c.

15. Cerebrospinal fluid
    a. cushions the brain.
    b. supplies the brain with nutrients and oxygen and removes wastes.
    c. is a filtrate of blood.
    d. is or does all of the above.

16. Which of the following structures is incorrectly paired with its function?
    a. pons—conducts information between spinal cord and brain
    b. cerebellum—contains tracts that cross motor neurons from one side of the brain to the other side of the body
    c. thalamus—screens and relays incoming impulses to the cerebrum
    d. corpus callosum—band of fibers connecting left and right hemispheres

17. During REM sleep,
    a. a desynchronized EEG is produced.
    b. dreaming occurs.
    c. rapid eye movement occurs.
    d. all of the above occur.

18. The reticular formation
    a. filters sensory information and produces arousal.
    b. relays motor impulses and coordinates motor responses.
    c. is responsible for sleep and uses serotonin as its neurotransmitter.
    d. is severed in "split brain" patients.

19. The limbic system
    a. controls speech patterns and prevents aphasia.
    b. is responsible for lateralization of the brain.
    c. is a group of nuclei in the lower forebrain associated with emotions.
    d. regulates sexual response, mating behaviors, and pleasure.

20. A simple reflex, such as that of rapidly pulling a finger from a scalding tea kettle, includes
    a. the transfer of information to and from the white matter of the spinal cord.
    b. an exchange of information between interneurons.
    c. a sensory receptor, a sensory neuron, and a motor neuron.
    d. an interneuron and a stretch receptor in the skin.

CHAPTER **45**

SENSORY AND MOTOR MECHANISMS

FRAMEWORK

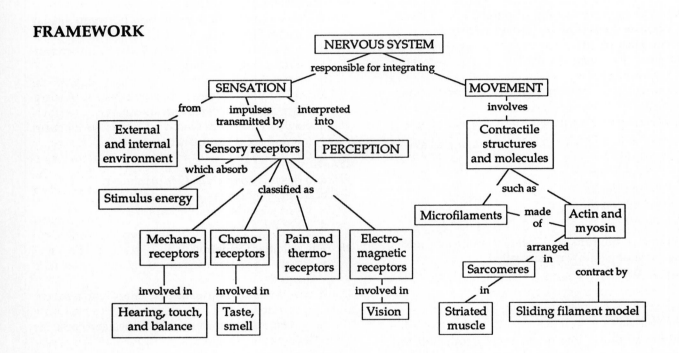

CHAPTER SUMMARY

Sensory Receptors

Sensation and Perception Information is transmitted in the nervous system as action potentials. *Sensations*, or impulses traveling along sensory neurons, are routed to different parts of the brain that interpret them as *perceptions*. The destination of the impulses in the brain determines what is perceived.

General Function of Sensory Receptors *Sensory receptors* are usually modified neurons that occur singly or in groups within sensory organs and that collect and

transmit information from environmental stimuli. *Reception* is the absorption of energy from a particular stimulus. The conversion of that energy into the electrochemical energy of action potentials is called *transduction*. The stimulus energy may need to undergo *amplification*, either by accessory structures of sense organs or as part of the transduction process, in order to enter the nervous system.

Transmission of the changes in the sensory cell membrane potential to the nervous system may occur either as an action potential, if the receptor is a sensory neuron, or by the release of neurotransmitter into a synapse with a sensory neuron, which then translates it into an action potential in the neuron. The intensity of the *receptor potential*, as this sensory cell membrane potential is called, is graded and correlates with the frequency of action potentials or the quantity of neurotransmitter released. Changes in the spontaneous firing rate of sensory cells provides information on the presence or absence of a stimulus, as well as on its intensity.

Integration of sensory information begins with the summation of graded potentials by receptors and continues through *sensory adaptation* of the receptor cell to ongoing stimulation and changes in the sensitivity of receptors under varying conditions.

Types of Receptors Sensory receptors may be *exteroreceptors*, which receive information from the outside environment, or *interoreceptors*, which provide information from the internal environment. Receptors may also be categorized based on the type of energy stimulus to which they respond.

Mechanoreceptors respond to the mechanical energy of pressure, touch, motion, and sound. Bending or stretching of the mechanoreceptor cell membrane increases its permeability to sodium and potassium ions, creating a receptor potential. In humans, *Pacinian corpuscles*, which are modified dendrites of sensory neurons found in deep skin layers, respond to strong pressure, whereas *Meissner's corpuscles* and *Merkel's discs* are closer to the surface and detect light touch.

The position of body parts is monitored by *muscle spindles*, stretch receptors which are stimulated by stretching of the muscles. *Hair cells*, mechanoreceptors that detect motion, are found in the vertebrate ear, the lateral line organs of fishes and amphibians, and the balance organs of arthropods. When motion produces bending in the cilia or microvilli projecting upward from a hair cell, the membrane stretches, ion permeabilities change, and the rate of action potential firing is affected.

Chemoreceptors include both general receptors that monitor the total solute concentration and specific receptors that respond to individual kinds of molecules. Stimulus molecules bind to membrane sites on the receptor cell and initiate changes in membrane permeability. *Gustatory* (taste) and *olfactory* (smell) *receptors* respond to groups of related chemicals.

Electromagnetic receptors respond to the wavelengths of electromagnetic radiation. *Photoreceptors* detect visible light and are often organized into eyes. Some animals have receptors that detect infrared rays (snakes), electric currents (some fish), and magnetic fields (some homing or migrating animals).

Thermoreceptors respond to heat or cold and help to regulate body temperature. *Ruffini's end organs* and *Krause's end bulbs*, receptors consisting of encapsulated, branched dendrites in mammalian skin, have been proposed as heat and cold receptors, respectively.

Naked dendrites called *nociceptors* detect pain. Different groups of receptors respond to excess heat, pressure, or chemicals released from damaged tissues. Histamines and acids trigger pain receptors, and prostaglandins increase pain by sensitizing receptors.

Vision

Light Receptors and Vision of Invertebrates Receptors containing light-absorbing pigments are used by most invertebrates to detect light. The *eye cup* of planaria, which detects light intensity and direction, consists of receptor cells within a darkly pigmented cup open only to one side. The brain compares impulses from the left and right eye cups to help the animal navigate a direct path away from a light source.

Two types of image-forming eyes have evolved in invertebrates. The *compound eye* of insects and crustaceans contains up to thousands of light detectors called *ommatidia*, each of which has a cornea and lens. The different intensities of light entering the many ommatidia produce a mosaic image. Compound eyes are adept at detecting movement, partly due to the rapid recovery of the photoreceptors, and may detect color and ultraviolet radiation.

Some jellyfish, spiders, and many mollusks, have a *single-lens eye*, in which light is focused through the single lens onto the retina, consisting of a bilayer of photosensitive receptor cells.

Vertebrate Vision The eyeball consists of a tough, outer, connective tissue layer called the *sclera*, and a thin, pigmented, inner layer called the *choroid*. At the front of the eye, the sclera becomes the transparent *cornea* and the choroid forms the colored *iris* which regulates light entering the *pupil*. The *retina*, a layer inside the choroid, contains the photoreceptor cells. The optic nerve attaches to the eye at the optic disc.

The transparent *lens* focuses an image onto the retina. The *ciliary body* produces the *aqueous humor* that fills the anterior eye cavity. Glaucoma is a disease in which an accumulation of aqueous humor increases pressure in the eye. Jellylike *vitreous humor* fills the posterior cavity of the eye. Many fishes focus by moving the lens backward or forward. Mammals focus by *accommodation*, in which ciliary muscles change the shape of the lens.

Rod cells and *cone cells* are the photoreceptors in the retina. The relative proportion of each of these receptors correlates with the activity pattern of the animal: Rods are more light sensitive and enable night vision, whereas cones distinguish colors. In the human eye, rods are concentrated toward the edge of the retina, whereas the center of the visual field, the *fovea*, is filled with cones. Birds of prey, such as hawks, have a high density of cones and keen eyesight.

Both rods and cones have visual pigments embedded in an outer stack of folded membranes. *Retinal* is the light-absorbing molecule and is bonded to a membrane protein called an *opsin*. Rods contain the molecule *rhodopsin*, which dissociates or "bleaches" when it absorbs light. Enzymes recombine the retinal and opsin in the dark. In bright light, the rhodopsin remains bleached and cones are responsible for vision. The three subclasses of red, green, and blue cones, each with its own type of opsin that binds with retinal to form a *photopsin*, are named for the color of light they are best at absorbing.

When retinal is stimulated by light, it initiates metabolic changes that decrease the permeability of the cell membrane to Na^+ ions, creating a graded receptor potential that hyperpolarizes the cells and reduces the release of neurotransmitter. Rods and cones synapse with *bipolar cells* in the retina, which in turn synapse with *ganglion cells*, whose axons transmit action potentials to the brain. *Horizontal cells* and *amacrine cells* are neurons in the retina that help to integrate visual information.

Signals from rods and cones may follow the vertical pathway directly from receptor cells to bipolar cells and then to ganglion cells. Several receptor cells synapse with each bipolar cell, and several bipolar cells pass information to one ganglion cell. Alternately, signals may follow a lateral pathway; lateral integration involves horizontal cells, which carry signals from one receptor cell to others and to several bipolar cells, and amacrine cells, which relay information from one bipolar cell to several ganglion cells. *Lateral inhibition* of nonilluminated receptors and bipolar cells by the horizontal cells and of ganglion cells by amacrine cells enhances the contrast between a spot of light and its surroundings.

The receptive field of a ganglion cell (the photoreceptors supplying information to the cell—information that is integrated by bipolar, horizontal, and amacrine cells) is either an on-center field that responds to light spots surrounded by darkness, or else an off-center field that responds to spots of darkness surrounded by light. Both of these receptive fields enhance the patterns of bright and dark spots that constitute much of the visual information transmitted to the brain.

Axons of the ganglion cells form the optic nerve. The left and right optic nerves meet at the *optic chiasma* at the base of the cerebral cortex. Information from the left sides of both eyes travels to the right side of the brain, whereas what is sensed in the right fields of

view goes to the left side of the brain. Most axons of the ganglion cells go to the *lateral geniculate nuclei* of the thalamus, where neurons lead to the *primary visual cortex* in the occipital lobe of the cerebrum. Other interneurons carry information to visual processing and integrating centers in the cortex.

Somehow the coded spots, lines, and movements that are projected onto the visual cortex are integrated into our perception and recognition of objects.

Hearing and Balance

The Mammalian Ear In mammals and most terrestrial vertebrates, the mechanoreceptors for hearing and balance are located within the ear. The ear consists of three regions: The external *pinna* and the *auditory canal* make up the outer ear. The *tympanic membrane* (eardrum) of the *middle ear* transmits sound waves to three small bones—the *malleus* (hammer), *incus* (anvil), and *stapes* (stirrup)—which conduct the waves to the *inner ear* by way of a membrane beneath the stapes, called the *oval window*. The *auditory (Eustachian) tube*, connecting the pharynx and the middle ear, equalizes pressure within the middle ear. The inner ear is a fluid-filled labyrinth of channels within the temporal bone.

The coiled *cochlea* of the inner ear has two large fluid-filled chambers—an upper vestibular canal and a lower tympanic canal—separated by the smaller cochlear duct, filled with endolymph. The *organ of Corti*, located on the floor of the cochlear duct, contains the receptor hair cells. The tips of the hair cells extend into the cochlear duct, and some attach to the tectorial membrane, which overhangs the organ of Corti. The following diagram shows a cross section of part of the cochlea.

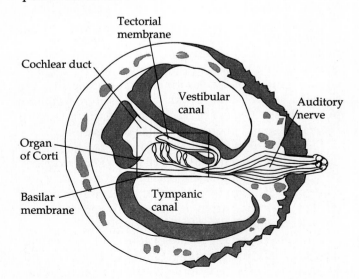

Sound waves—which were transmitted and amplified by the tympanic membrane, the three bones of the middle ear, the oval window, and pressure waves in the cochlear fluid—are transduced into action potentials in the cochlea. Pressure waves, traveling from the vestibular canal through the tympanic canal and dissipating when they strike the *round window*, vibrate the basilar membrane on which the organ of Corti lies. The bending of the hairs against the tectorial membrane stretches the plasma membrane of the receptor cell, making it more permeable to sodium. The resulting depolarization increases neurotransmitter release from the hair cell and increases the frequency of action potentials generated in the sensory neuron and carried through the auditory nerve to the brain.

Volume is a result of the *amplitude*, or height, of the sound wave; a stronger wave bends the hair cells more and results in more action potentials. *Pitch* is related to the *frequency* of sound waves, usually expressed in hertz (Hz), or vibrations per second. Fibers of different lengths span the basilar membrane and vibrate in response to different frequencies, transmitting their impulses to specific regions of the cerebral cortex where the sensation is perceived as a particular pitch.

Within the inner ear is the vestibular apparatus responsible for balance and equilibrium: two chambers, the *utricle* and *saccule*, and three *semicircular canals*. Hair cells in the utricle and saccule project into a gelatinous material containing calcium carbonate particles called otoliths. These heavy particles exert a constant pull on the hairs in the direction of gravity; movements relative to gravity alter the output of neurotransmitters by the receptor hair cells. Other hair cells in the swellings at the base of the semicircular canals (the ampulla) project into a gelatinous mass called the cupula and are able to sense rotational movements by the pressure of the endolymph in the semicircular canals against the cupula.

Hearing and Equilibrium in Other Vertebrates Mechanoreceptor units called *neuromasts*—clusters of hair cells with sensory hairs embedded in a gelatinous cap or cupula—are contained in the *lateral line system* of fish and aquatic amphibians. Water moving in a tube past these mechanoreceptors bends the cupula and stimulates the hair cells, enabling the fish to perceive its movement, water currents, sounds of low vibration, and the pressure waves generated by other moving objects.

Fishes also have inner ears, consisting of a saccule, utricle, and semicircular canals, within which sensory hairs are stimulated by the movement of otoliths or granules. Sound wave vibrations pass through the skeleton of the head to the inner ears. Some fishes have a *Weberian apparatus*, a series of bones that transmits vibrations from the swim bladder to the inner ear.

In amphibians and reptiles, sound vibrations are conducted by a tympanic membrane and single bone to the inner ear. A cochlea has evolved in birds.

Sensory Organs for Hearing and Balance in Invertebrates Many arthropods sense sounds with body hairs that vibrate in response to sound waves of specific frequencies. Localized "ears" are found on the legs of many insects, consisting of a tympanic membrane with attached receptor cells stretched over an internal air chamber.

Invertebrates have mechanoreceptors, called *statocysts*, which often consist of a layer of hair cells around a chamber containing statoliths, dense granules or grains of sand. The stimulation of the hair cells under the statocysts provides positional information to the animal.

Taste and Smell

The sense of taste (gustation) detects chemicals in a solution, whereas smell (olfaction) detects airborne chemicals. Chemical senses are used by animals to find mates, navigate, communicate, and feed. *Setae*, or sensory hairs containing chemoreceptor taste cells, are found on the feet and mouthparts of insects. Insects also have olfactory setae, usually located on their antennae.

In mammals, the chemical senses of taste and smell are produced when a molecule binds to a receptor protein in a receptor cell membrane and triggers a membrane depolarization and the release of neurotransmitter. *Taste buds*, which contain groups of receptor cells, are scattered on the tongue and mouth. The four primary taste sensations—sweet, sour, salty, and bitter—are detected in distinct regions of the tongue. The brain integrates the input from the taste buds as they differentially respond to different chemicals and creates the perception of a complex flavor.

Olfactory receptor cells line the upper part of the nasal cavity; their axons go to the olfactory bulb of the brain. Chemicals bind to specific receptor molecules on the cilia of receptor cells that extend into the mucous layer of the nasal cavity.

Introduction to Animal Movement

Animals move to obtain food, escape from danger, and find mates. The evolution of various skeletal and muscle designs and body shapes reflect adaptations to overcome the challenges of friction and gravity in a water, land, or air environment.

On a celular level, all animal movement depends on protein strands moving past each other, either in microtubules, which are responsible for beating of cilia and flagella, or in microfilaments, which are involved in amoeboid movement and muscle contraction.

Skeletons and Their Roles in Movement

Skeletons function in support of the body, protection of soft tissues, and movement.

Hydrostatic Skeletons Fluid under pressure in a closed body compartment creates a *hydrostatic skeleton*. As muscles contract and change the shape of the fluid-filled compartment, the animal moves. Hydrostatic skeletons are found in most cnidarians, flatworms, nematodes, and annelids.

Exoskeletons Typical of mollusks and arthropods, *exoskeletons* are hard coverings deposited on the surface of animals. The *cuticle* of an arthropod contains fibrils of *chitin*, a polysaccharide similar to cellulose, embedded in a protein matrix. Where protection is needed, this flexible support is hardened by cross-linking proteins and addition of calcium salts, but it remains pliable at the joints of the skeleton.

Endoskeletons The supporting elements of an *endoskeleton* are embedded in the soft tissues of the animal. Sponges have endoskeletons composed of spicules or fibers, and echinoderms have ossicles or hard plates beneath the skin. The endoskeletons of chordates are composed of cartilage and/or bone. The vertebrate axial skeleton consists of skull, vertebral column, and rib cage; the appendicular skeleton contains the pectoral and pelvic girdles and the limb bones.

Muscles

Movement involves the contraction of antagonistic pairs of muscles working against some type of skeleton.

Structure and Physiology of Vertebrate Skeletal Muscle Vertebrate *skeletal muscle* consists of a bundle of multinucleated muscle cells, called fibers, running the length of the muscle. Each fiber is a bundle of *myofibrils*, each composed of two kinds of *myofilaments*: *thin filaments*, consisting of two strands of actin coiled with a strand of regulatory protein, and *thick filaments*, made of myosin molecules.

The regular arrangement of myofilaments produces repeating light and dark bands; thus skeletal muscle is also called striated muscle. The repeating units are called *sarcomeres* and are delineated by *Z lines*. Thin filaments attach to the Z line and project toward the center of the sarcomere. Thick filaments lie free in the center, forming the *A band*, which is broken by an *H zone* in the center where the thin filaments do not reach. At the edges of the sarcomere, where thick filaments do not extend when the muscle is at rest, is an *I band* of only thin filaments.

According to the *sliding filament model* of muscle contraction, the thick and thin filaments do not change length, but slide past each other, increasing their degree of overlap. This overlap pulls the Z lines together, shortens the I bands, and eliminates the H zone.

The mechanism for the sliding of filaments is the hydrolyzing of ATP by the globular head of a myosin molecule, which then changes to a high-energy configuration that binds to actin and forms a *cross-bridge*. When the energy is released, the myosin head relaxes to its low-energy configuration, which bends the head toward the molecule's fibrous tail and pulls the attached thin filament toward the center of the sarcomere. When a new molecule of ATP binds to the myosin head, it breaks its bond to actin, and the cycle repeats with the high-energy configuration attaching to an actin molecule farther along on the thin filament. Most of the energy for muscle contraction comes from *phosphagens*, such as *creatine phosphate* in vertebrates, which supply phosphate to ADP and make new ATP.

When the muscle is at rest, the myosin-binding sites of the actin molecules are blocked by a strand of the regulatory protein *tropomyosin*, whose position is controlled by another set of regulatory proteins, the *troponin complex*. When troponin binds to calcium ions, the tropomyosin-troponin complex changes shape, and the actin filament is exposed.

Calcium ions are actively transported into the *sarcoplasmic reticulum* of a muscle cell. When an action potential of a motor neuron causes the release of acetylcholine into the neuromuscular junction, an action potential spreads across the plasma membrane of the muscle cell and into *transverse tubules*, infoldings of the plasma membrane. The action potential carried by the transverse tubules changes the permeability of the sarcoplasmic reticulum, and calcium ions are released into the cytoplasm. Calcium binding with troponin exposes the actin sites and contraction begins. Contraction ends when the sarcoplasmic reticulum pumps calcium back out of the cytoplasm, and the tropomyosin–troponin complex again blocks the actin sites.

A contraction of a single muscle fiber is an all-or-none action, producing a twitch. More graded contraction of muscles occurs by temporal summation when action potentials arrive so rapidly that the muscle does not relax between them, and each contraction is stronger than the last. Motor neurons usually deliver a volley of action potentials, producing a state of smooth and sustained contraction called *tetanus*.

Graded muscular contractions also result from involving additional fibers. A single motor neuron and all the muscle fibers it innervates make up a *motor unit*. The strength of a muscle contraction depends on the size and number of motor units that are activated. Muscle tension can be increased by the stimulation—*recruitment* of additional motor units.

The muscle fatigue of extended tetanus is caused by depletion of ATP, loss of ion gradients required for depolarization, and accumulation of lactic acid. Some support muscles are always in a state of partial contraction, in which different motor units take turns contracting.

Slow muscle fibers have less sarcoplasmic reticulum; thus calcium remains in the cytoplasm longer and twitches last longer. Slow fibers are common in muscles that must sustain long contractions. They have many mitochondria, a good blood supply, and *myoglobin*, which extracts and stores oxygen from the blood. Fast muscle fibers are used for rapid and powerful contractions.

Other Types of Muscles Vertebrate *cardiac muscle* cells are connected by *intercalated discs*, across which action potentials spread to all the cells of the heart. *Pacemaker channels* in the plasma membrane of a cardiac muscle cell generate action potentials by rhythmic depolarizations. The action potentials of cardiac muscle cells last up to 20 times longer and have a role in controlling the duration of contraction.

The spiral arrangement of actin and myosin filaments in *smooth muscle* accounts for its nonstriated appearance. Smooth muscle lacks a transverse tubule system and a well-developed sarcoplasmic reticulum. Its relatively slow contractions can occur over a greater range of lengths.

Invertebrates have muscles similar to the skeletal and smooth muscles of vertebrates. The flight muscles of insects are capable of independent and rapid contraction.

Animal behavior is a function of the integration and interpretation of inputs from sensory receptors by a nervous system which then signals muscles to move and respond.

STRUCTURE YOUR KNOWLEDGE

1. Label the parts of the eye on the diagram below.

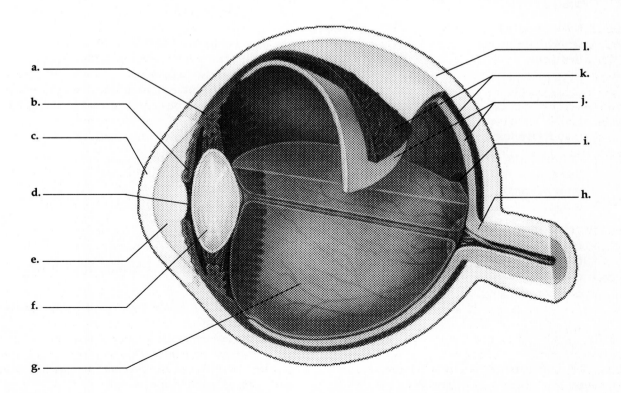

2. Describe the path of a light stimulus from where it enters the eye to its transmission as an impulse through the optic nerve.

3. Arrange the following structures in an order that enables you to trace the passage of sound waves from the external environment to the brain.

round window	basilar membrane
oval window	cerebral cortex
organ of Corti	vestibular and tympanic canal
auditory nerve	tectorial membrane
cochlea	tympanic membrane
auditory canal	malleus, incus, stapes

4. Trace the sequence of events in muscle contraction from an action potential in a motor neuron to the relaxation of the muscle.

TEST YOUR KNOWLEDGE

MATCHING: *Match the term with its description.*

1. _____ oxygen-storing compound in muscles

2. _____ stretch receptor on muscle, monitors position

3. _____ stimulus energy converted into electrochemical energy

4. _____ deep pressure receptors in humans

5. _____ naked dendrites that detect pain

6. _____ sensory hairs on feet and mouthparts of insects

7. _____ invertebrate mechanoreceptors for position; chamber of hair cells

8. _____ energy-storing compound in muscles

9. _____ changing shape of lens to focus

A. accommodation

B. adaptation

C. ampulla

D. creatine phosphate

E. cupula

F. ganglion cells

G. muscle spindle

H. myofibrils

I. myoglobin

J. nociceptors

K. ommatidia

L. Pacinian corpuscles

10. _____ numerous light detectors of compound eye

11. _____ decrease in sensitivity of receptor during constant stimulation

12. _____ blocks myosin-binding sites on actin

M. setae

N. statocyst

O. transduction

P. transmission

Q. transverse tubules

R. tropomyosin

MULTIPLE CHOICE: *Choose the one best answer.*

1. The absorption of energy from a particular stimulus is called
   a. reception.
   b. transduction.
   c. integration.
   d. transmission.

2. Prostaglandins
   a. increase perception of pain by sensitizing pain receptors.
   b. act like histamines to trigger pain receptors.
   c. are released by nociceptors in response to a pain stimulus.
   d. are stimulated by aspirin and decrease pain.

3. The compound eye found in insects and crustaceans
   a. is composed of thousands of prisms that focus light onto a few photoreceptor cells.
   b. cannot sense colors.
   c. contains many ommatidia, each of which consists of a cornea and lens that focus light from a tiny portion of the field of view.
   d. All of the above are correct.

4. Which of the following statements is *not* true?
   a. The iris regulates the amount of light entering the pupil.
   b. The choroid is the thin pigmented inner layer of the eye that contains the photoreceptor cells.
   c. The ciliary muscle changes the shape of the lens.
   d. The thin aqueous humor fills the anterior eye cavity.

5. Which of the following is incorrectly paired with its function?
   a. cones—respond to different light wavelengths producing color vision
   b. horizontal cells—produce lateral inhibition of receptor and bipolar cells

c. bipolar cells—relay impulses between rods or cones and ganglion cells

d. ganglion cells—synapse with the optic nerve at the optic disc or blind spot

6. The fovea is
   a. the blind spot.
   b. the center of the visual field, containing only cones.
   c. the layer that contains photoreceptor cells.
   d. the optic disc, where the optic nerve attaches to the eye.

7. The molecule rhodopsin
   a. is involved with color vision.
   b. dissociates when struck by light and is only functional in complete dark.
   c. is recombined by enzymes from retinal and opsin in the light.
   d. is more sensitive to light than are photopsins and is involved in black and white vision in dim light.

8. Rotation of the head is sensed by
   a. the Weberian apparatus located in the utricle and saccule.
   b. changes in the action potentials of hair cells in response to the pull on them by calcium carbonate particles.
   c. bending of hair cells in the semicircular canals in response to the inertia of the endolymph.
   d. hair cells in response to movement of fluid in the coiled cochlea.

9. The tympanic membrane
   a. conducts sound waves to the fluid in the vestibular canal.
   b. sets up vibrations in the basilar membrane.
   c. transmits sound waves to the malleus, incus, and stapes.
   d. equalizes pressure between the middle ear and the outside.

10. Fish can perceive pressure waves from moving objects with their
    a. lateral line system containing mechanoreceptor units called neuromasts.
    b. tectorial membrane as it is stimulated by the bending of sensory hairs embedded in a gelatinous cap.
    c. inner ears, which consist of a saccule, utricle, and semicircular canals, within which sensory hairs are stimulated by the movement of otoliths.
    d. Weberian apparatus, a series of bones that transmits vibrations from the swim bladder to the inner ear.

11. The binding of a molecule to a receptor protein on a receptor cell membrane
    a. triggers a membrane depolarization and release of neurotransmitter.
    b. is the mechanism of reception and transduction in chemoreceptors.
    c. is involved in both gustatory and olfactory sensation.
    d. involves all of the above.

12. Hydrostatic skeletons are used for movement by all of the following *except*
    a. cnidarians.
    b. echinoderms.
    c. nematodes.
    d. annelids.

13. When striated muscle fibers contract,
    a. the Z lines are pulled closer together.
    b. the I band remains the same.
    c. the A band shortens.
    d. all of the above occur.

14. Tetanus
    a. is muscle fatigue resulting from a build up of lactic acid during anaerobic respiration.
    b. is the all-or-none contraction of a single muscle fiber.
    c. is the result of stimulation of additional motor neurons.
    d. is the result of a volley of action potentials, whose temporal summation produces increased and sustained contraction of a muscle.

15. What is the role of ATP in muscle contraction?
    a. to form cross bridges between thick and thin filaments
    b. to break the cross bridge when it binds to myosin and provide energy to myosin to form its high-energy configuration that binds to actin
    c. to remove the tropomyosin–troponin complex from blocking the binding site on the actin filament
    d. to bend the cross bridge and pull the thin filaments toward the center of the sarcomere

16. When an action potential spreads into the transverse tubules,
    a. calcium is released from the tubules into the sarcoplasmic reticulum.
    b. acetylcholine is released into the neuromuscular junction.
    c. the sarcoplasmic reticulum membrane becomes permeable and calcium ions are released into the cell.
    d. calcium binds with troponin and the actin sites are exposed.

17. Most of the readily available energy for muscle contraction in vertebrates comes from
    a. phosphagens.
    b. myoglobin.
    c. glycogen.
    d. both a and b.

18. A motor unit is
    a. a sarcomere that extends from one Z line to the next Z line.
    b. a motor neuron and the muscle fibers it innervates.
    c. the myofibrils of a fiber and the transverse tubules that connect them.
    d. the muscle fibers that make up a muscle.

19. Which of the following is *not* a characteristic of cardiac muscle?
    a. spiral arrangement of actin and myosin filaments
    b. intercalated discs that spread action potentials between cells
    c. action potentials that last a long time
    d. ability to generate action potentials without nervous input

20. Smooth muscle contracts relatively slowly because
    a. the only ATP available is supplied by fermentation.
    b. its contraction is stimulated by hormones, not motor neurons.
    c. it does not have a well-developed sarcoplasmic reticulum, and calcium enters the cell through the plasma membrane during an action potential.
    d. it is not striated.

FILL IN THE BLANKS: *Use this diagram of a section of a skeletal muscle fiber to answer the following.*

1. _____ Identify a.

2. _____ Identify b.

3. _____ Identify c.

4. _____ Identify d.

5. _____ Identify e.

6. _____ Identify f.

7. _____ Identify g.

8. _____ In which of these structures is troponin found?

9. _____ Which band stays the same length during contraction?

10. _____ Which structure forms cross bridges?

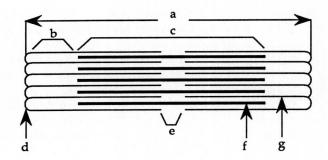

CHAPTER 36
INTRODUCTION TO ANIMAL STRUCTURE AND FUNCTION

Suggested Answers to Structure Your Knowledge

1.

TISSUE	STRUCTURAL CHARACTERISTICS	GENERAL FUNCTIONS	STRUCTURE AND FUNCTION OF SPECIFIC TYPES
Epithelial	Tightly packed cells; basement membrane; cuboidal, columnar, squamous shapes; simplified, stratified layers	Protection, absorption, secretion; lines body surfaces	Mucous membrane: lubricate surfaces, may have cilia that sweep mucus; Skin: multiple layers of cells that form protective, waterproof covering
Connective	Few cells that secrete extracellular matrix of protein fibers in liquid, gel or solid ground substance	Connect and support other tissues	Loose connective: loose weave of collagenous or elastic fibers, hold organs in place. Adipose: adipose cells store fat. Fibrous connective: collagenous fibers form tendons, ligaments. Cartilage: chondrocytes secrete chondrin, rubbery matrix, flexible support. Bone: osteocytes in Haversian system, mineralized matrix, strong support. Blood: erythrocytes, leukocytes, and platelets in plasma matrix.
Muscle	Long cells with many microfilaments of actin and myosin	Contraction, movement	Skeletal: striated, voluntary movement. Visceral: smooth, spindle-shaped cells, in organs, involuntary. Cardiac: striated, branched cells, intercalated discs, in heart.
Nervous	Neurons with cell body, axons, dendrites	Sense stimuli, conduct impulses	

2. All organs are covered with epithelial tissue, and internal lumens, ducts, and blood vessels will be lined with epithelium. Some types of connective tissue would be found in most organs, usually serving to connect other tissues and supply blood to the nonvascularized epithelium. Muscle tissue would be present if the organ must contract. Nervous tissue innervates most organs. The small intestine is an example of an organ composed of many tissues. An epithelial mucosa lines the lumen; the submucosa contains blood vessels and loose connective tissue; a layer of circular and longitudinal smooth muscle comes next; and the outer layer is an epithelium.

3. A compact animal must provide gas exchange, nutrients, and waste removal for each cell. Cells of a large animal are bathed in an intersitial fluid that keeps cell surfaces moist and allows for exchange of gases, nutrients, and wastes. Large internal surface areas are specialized for obtaining oxygen, for digestion and absorption, and for removal of waste from the blood. The circulatory system usually provides these resources to all the cells of the body through the interstitial fluid.

4. **a.** stratified columnar epithelium
   b. simple squamous epithelium
   c. smooth muscle
   d. loose connective
   e. bone

Answers to Test Your Knowledge

Multiple Choice:

1. b	3. d	5. b	7. b	9. b
2. a	4. a	6. d	8. d	

CHAPTER 37
ANIMAL NUTRITION

Suggested Answers to Structure Your Knowledge

1.

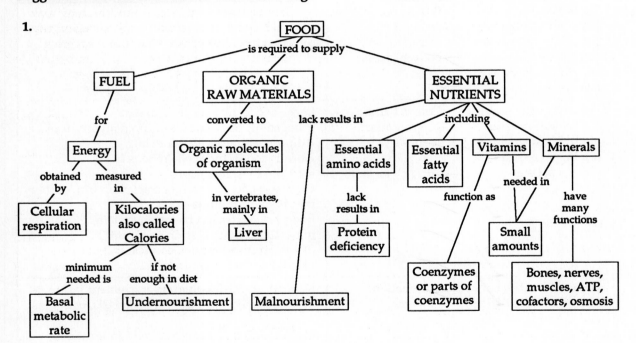

2. **a.** salivary glands   **f.** pancreas
   b. esophagus   **g.** small intestine
   c. stomach   **h.** large intestine
   d. liver   **i.** rectum
   e. gall bladder   **j.** anus

3.

ORGAN OR SECTION	PROCESSES OCCURRING	ASSOCIATED GLANDS, ORGANS, AND DIGESTIVE FLUIDS
Oral cavity	Mechanical breakup of food, mix with saliva, begin breakdown of starch, lubricate with mucin	Salivary glands secrete salivary amylase, mucin, buffers, antibacterial agents
Pharynx	Swallowing of bolus, epiglottis closes trachea	
Esophagus	Bolus through esophageal sphincter, peristalsis moves bolus to stomach through cardiac sphincter	
Stomach	Food tissues broken down by low pH, protein denatured and hydrolysis begun, churning mixes acid chyme, exits through pyloric sphincter	Epithelial glands secrete gastric juice with pepsin and HCL; hormone gastrin regulates gastric-juice release
Small intestine	Complete digestion of starch and proteins, fats emulsified and hydrolyzed, absorption of nutrients across epithelial lining into capillaries and lacteals, large surface area due to villi and microvilli	Pancreatic release of bicarbonate neutralizes acid chyme; pancreatic enzymes: amylase, trypsin, and other protein-hydrolyzing enzymes; liver produces bile salts, stored in gall bladder, emulsify fats; lipase hydrolyzes fats; brush border of epithelium contains disaccharidases and dipeptidases; nucleases hydrolyze DNA and RNA. Hormones: secretin, CCK, enterogastrone control secretions and peristalsis
Large intestine	Reabsorption of water, feces stored in rectum; microorganisms may produce vitamin K	

4.

STIMULUS THAT CAUSES RELEASE	HORMONE	EFFECTS OF THE HORMONE
Presence of food in stomach, stomach distension, previous gastrin secretion	Gastrin	Stimulates secretion of gastric juices, too low pH in stomach inhibits gastrin release
Acidic pH of chyme entering duodenum	Secretin	Stimulates release of bicarbonate from pancreas
Amino acids or fatty acids in duodenum	Cholecysto-kinin (CCK)	Stimulates release of digestive enzymes from pancreas and bile from gall bladder
Chyme rich in fats in duodenum	Enterogas-trone	Inhibits stomach peristalsis and acid secretion

Answers to Test Your Knowledge

Matching:

1. L	5. E	9. D	
2. H	6. A	10. N	
3. C	7. J		
4. M	8. F		

Multiple Choice:

1. d	5. d	9. a	13. c
2. c	6. d	10. b	14. c
3. c	7. a	11. b	15. b
4. a	8. c	12. a	16. d

CHAPTER 38
CIRCULATION AND GAS EXCHANGE

Suggested Answers to Structure Your Knowledge

1. **a.** aorta
   b. superior vena cava
   c. semilunar valve
   d. right atrium
   e. inferior vena cava
   f. atrioventricular valve
   g. right ventricle
   h. left ventricle
   i. left atrium
   j. pulmonary vein
   k. pulmonary artery

2.

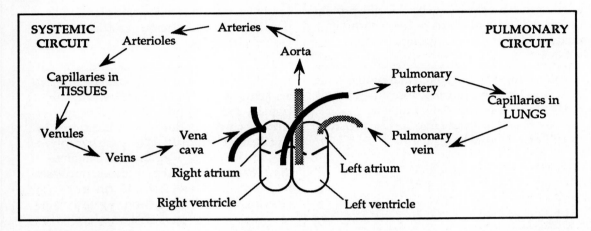

3.

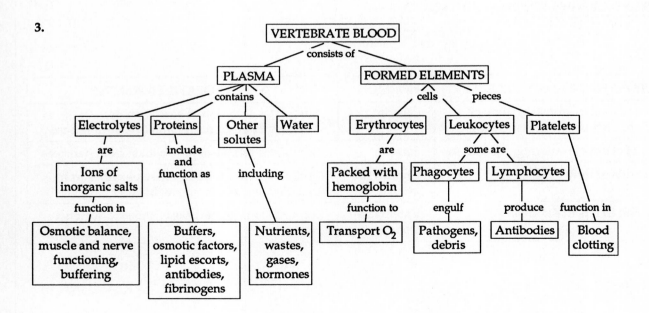

4.

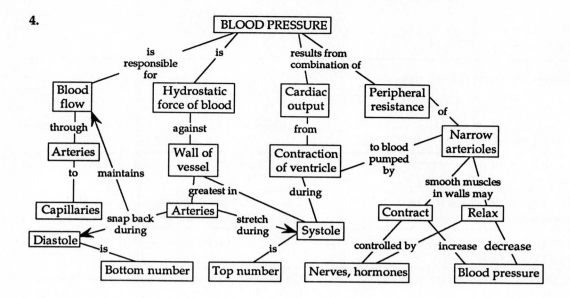

5. The dissociation curve for hemoglobin is shifted to the right at a lower pH, meaning that at any given partial pressure of O_2, hemoglobin is less saturated with oxygen. A rapidly respiring tissue produces more CO_2, which reacts with water to form carbonic acid. The resulting lower pH of the tissue produces a conformational change in hemoglobin that causes it to more readily release the O_2 it is carrying, thus supplying more O_2 to rapidly respiring cells.

Answers to Test Your Knowledge

Multiple Choice:

1. c	6. d	11. b	16. d	21. a
2. c	7. a	12. b	17. b	22. b
3. d	8. b	13. c	18. c	23. c
4. d	9. a	14. d	19. b	24. a
5. a	10. d	15. a	20. b	25. d

CHAPTER 39
THE BODY'S DEFENSES

Suggested Answers to Structure Your Knowledge

1.

MOLECULE	WHERE PRODUCED OR FOUND	ACTION
Lysozyme	Enzyme in perspiration, tears, mucus, and saliva	Attacks cell wall of many bacteria
Histamine	Released by injured cells and by degranulation of mast cells	Causes dilation and leakiness of small blood vessels, initiates inflammatory response
Interferon	Produced by cell infected by virus and diffuses to neighboring cells	Inhibits viral replication in cells that it enters
Complement	Set of proteins that circulate in blood	Cooperates with specific defense mechanisms in classical pathway to form membrane attack complex which lyses cells; cooperates with non-specific defenses in various alternate pathways, including opsonization and histamine release
Antibody	Protein (I_g) made by plasma cells derived from B lymphocytes, circulate in blood and lymph	Binds to specific antigen, marks foreign cells and molecules for destruction, neutralizes virus or toxin
Interleukin-1	Cytokine secreted by antigen-presenting macrophage (APC) after binding with helper T cell	Stimulate helper T cell to release interleukin-2
Interleukin-2	Cytokine released by activated T_H cell	Stimulate helper T and cytotoxic T cells to reproduce, help activate B cells
Perforin	Released by cytotoxic T cells when attached to cells displaying antigen-MHC complexes	Forms hole in target cell's membrane, causing it to lyse

2.

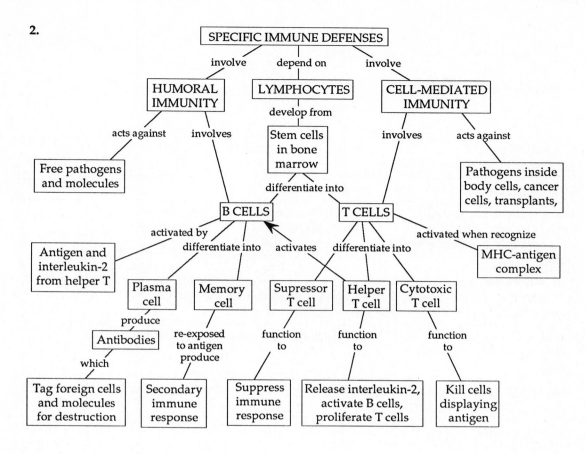

3. An antibody is a protein that typically consists of two identical light polypeptide chains and two identical heavy chains, held together by disulfide bonds in a Y-shaped molecule. The amino acid sequences in the variable sections of the light and heavy chains in the arms of the Y account for the specificity in binding between antibodies and antigenic determinants. The constant region of the antibody determines its effector function: five types of these constant regions correspond to five classes of antibodies. As an example, IgE immunoglobulins are attached to mast cells, which create allergic responses, and IgA is found in body secretions and is the major antibody in colostrum, the first breast milk produced.

According to the clonal selection theory, T and B cells become fixed early in their development to produce a specific antigen receptor or antibody. The body's ability to respond to a great variety of antigens depends on a lymphocyte population with a huge diversity of receptor specificities. When a T or B cell encounters its antigen, it is selectively activated to proliferate and produce a clone of lymphocytes, all having the same antigenic specificity.

4. (1) Antibodies attach to antigens on pathogen's plasma membrane.

(2) Complement proteins bridge between two antibodies.

(3) Complement proteins form a membrane attack complex and attach to pathogen's membrane.

(4) An area of the target cell's membrane is lysed and the pathogen is destroyed.

Answers to Test Your Knowledge

Multiple Choice:

1. d	**5.** b	**9.** a
2. c	**6.** d	**10.** b
3. b	**7.** a	**11.** a
4. a	**8.** c	**12.** c

CHAPTER 40
CONTROLLING THE INTERNAL ENVIRONMENT

Suggested Answers to Structure Your Knowledge

1.

ANIMAL	OSMOREGULATION	EXCRETORY SYSTEM	THERMOREGULATION
Marine invertebrate	Isosmotic to seawater, osmoconformer	Diffusion of ammonia through body wall	Ectothermic
Marine bony fish	Drink water to compensate for loss to hyperosmotic seawater, salt glands in gills pump out salt	Little urine excreted, ammonia diffuses through gills	Ectothermic, may have heat exchanger
Freshwater fish	Dilute urine to compensate for osmotic gain, may pump salts in through gills	Kidneys produce copious, dilute urine, ammonia lost across epithelium of gills	Ectothermic
Flatworm	Flame cell system (protonephridia), dilute fluid excreted	Most wastes excreted into gastrovascular cavity	Ectothermic
Earthworm	Metanephridia, excrete dilute urine to offset osmosis from habitat	Metanephridia	Ectothermic
Insect	Malpighian tubules, salt and most water reabsorbed	Malpighian tubules, uric acid conserves water	Ectothermic, behavioral responses
Reptile	Nephron in kidney, uric acid conserves water	Nephron in kidney, uric acid in most, related to shelled egg	Ectothermic, behavioral responses
Bird	Nephron in kidney, nasal salt glands in sea birds, uric acid conserves water	Nephron in kidney, juxtamedullary nephron, uric acid related to shelled egg	Endothermic, high metabolism, pant, countercurrent heat exchanger in legs
Mammal	Nephron in kidney, concentration of urine related to habitat, hormonal feedback control	Nephron in kidney, juxtamedullary nephron, urea	Endothermic, vasodilation and constriction, heat production, behavioral responses

2.

ENVIRONMENT	OSMOREGULATION	NITROGENOUS WASTE	THERMOREGULATION
Marine	Most invertebrates are osmoconformers; hypoosmotic fish lose water, drink sea water, secrete salt	Ammonia diffuses out easily into water, excreted through gills in fish	Stable environment, mostly ectotherms; mammals have blubber & countercurrent exchange
Freshwater	Animals are hyperosmotic, gain water, excrete large quantities of urine; ion uptake across gills in fish	Ammonia diffuses easily into water, exchanged across gills for Na^+ in fish	Fairly stable environment, mostly ectotherms
Terrestrial	Lose water by evaporation, drink, regulate water loss through excretion	Urea less toxic, less water needed to excrete; uric acid most concentrated, found in animals with shelled egg	Variable environment, behavioral adaptations if ectotherms; endotherms can maintain temperature and move rapidly

3. a. Bowman's capsule
  b. glomerulus
  c. descending limb, loop of Henle
  d. ascending limb, loop of Henle
  e. proximal convoluted tubule
  f. distal convoluted tubule
  g. collecting tubule

- Filtrate enters through the Bowman's capsule.
- NaCl, H_2O, glucose, and K^+ are reabsorbed from the proximal convoluted tubule. H^+ and NH_3 are secreted into the tubule.
- Water is reabsorbed from the descending limb of the loop of Henle.
- NaCl diffuses from the thin segment of the ascending limb and is pumped out of the thick segment.
- Na^+, H_2O, and HCO_3^- is reabsorbed from the distal convoluted tubule. K^+ and H^+ are secreted into the tubule.
- Urea and H_2O diffuse out of the collecting tubule.

SUBSTANCES FILTERED	SUBSTANCES SECRETED	SUBSTANCES REABSORBED
water salt glucose amino acids vitamins urea other small molecules	ammonia hydrogen ions potassium (maintain balance) drugs and poisons	water glucose amino acids vitamins potassium NaCl bicarbonate

4.

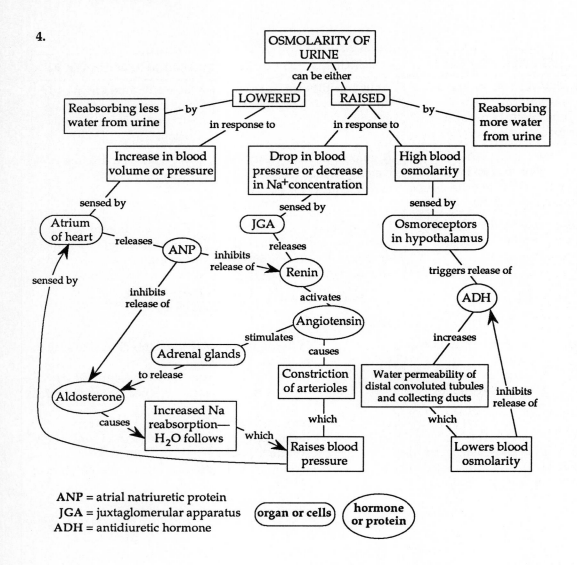

ANP = atrial natriuretic protein
JGA = juxtaglomerular apparatus
ADH = antidiuretic hormone

organ or cells   hormone or protein

Answers to Test Your Knowledge

Multiple Choice:

1. b	**5.** d	**9.** b	**13.** d	**17.** d
2. a	**6.** c	**10.** d	**14.** c	**18.** c
3. c	**7.** a	**11.** b	**15.** d	
4. a	**8.** c	**12.** b	**16.** b	

CHAPTER 41
CHEMICAL SIGNALS IN ANIMALS

Suggested Answers to Structure Your Knowledge

1.

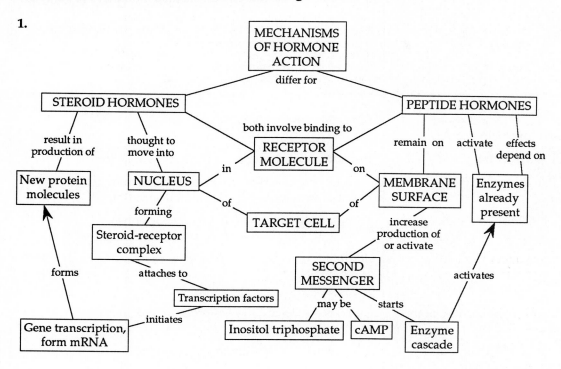

2.

GLAND	HORMONES	MAIN ACTION
Hypothalamus Neurosecretory cells (drain to posterior pituitary)	Oxytocin Antidiuretic hormone (ADH)	Stimulate uterine muscle contraction, mammary glands Increase water permeability of collecting ducts, promote reabsorption of water from urine
(drain to median eminence)	Releasing hormones (e.g., TRH and GnRH)	Carried by portal system to anterior pituitary, control release of hormones from anterior pituitary
Pituitary Posterior lobe (neurohypophysis)	Stores and secretes oxytocin and ADH	See actions listed above.
Anterior lobe (adenohypophysis)	Growth hormone (GH) Prolactin (PRL) Thyroid-stimulating hormone (TSH) Folicle-stimulating, lutenizing hormones (FSH and LH) Adrenocorticotropin (ACTH) Endorphins and enkaphalins	Promote growth, stimulate other growth factors Various effects depending on species: mammary gland growth, milk production and secretion Stimulate thyroid gland to secrete hormones Stimulate activities of gonads, ovulation and sperm production Stimulate adrenal cortex to produce and secrete glucocorticoids Reduce pain, effects similar to morphine

3.

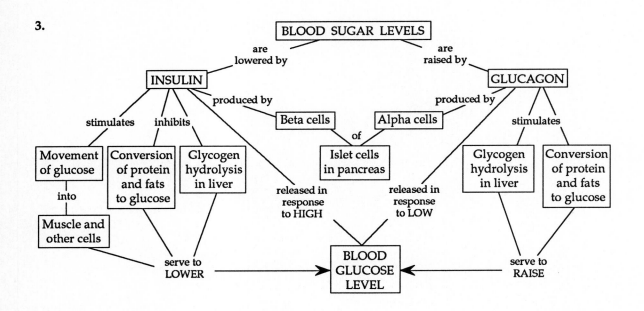

Answers to Test Your Knowledge

Matching:

1. I	f	**5.** A	g	**9.** C		c, released from g
2. G	a	**6.** E	b	**10.** J		c, released from g
3. F	d	**7.** K	j			
4. B	h	**8.** D	j			

Multiple Choice:

1. a	**4.** b	**7.** d	**10.** b	**13.** c
2. d	**5.** d	**8.** c	**11.** a	**14.** a
3. d	**6.** a	**9.** b	**12.** b	**15.** c

CHAPTER 42
ANIMAL REPRODUCTION

Suggested Answers to Structure Your Knowledge

1. PATH OF SPERM FROM FORMATION TO FERTILIZATION

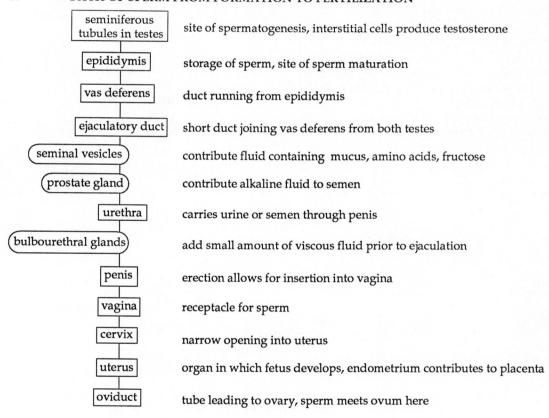

seminiferous tubules in testes	site of spermatogenesis, interstitial cells produce testosterone
epididymis	storage of sperm, site of sperm maturation
vas deferens	duct running from epididymis
ejaculatory duct	short duct joining vas deferens from both testes
seminal vesicles	contribute fluid containing mucus, amino acids, fructose
prostate gland	contribute alkaline fluid to semen
urethra	carries urine or semen through penis
bulbourethral glands	add small amount of viscous fluid prior to ejaculation
penis	erection allows for insertion into vagina
vagina	receptacle for sperm
cervix	narrow opening into uterus
uterus	organ in which fetus develops, endometrium contributes to placenta
oviduct	tube leading to ovary, sperm meets ovum here

2. A mature egg cell within a follicle is released during ovulation and swept into the fallopian tube or oviduct. It travels to the uterus, where it flows out through the vagina with the bleeding associated with menstruation.

3. a. LH
b. FSH
c. estrogen
d. progesterone
e. follicular phase
f. luteal phase
g. menstrual phase
h. proliferative phase
i. secretory phase

4. a. GnRH is a releasing hormone of the hypothalamus that stimulates the pituitary to secrete FSH and LH.
b. FSH stimulates growth of follicles.
c. The spike in LH level is caused by an increase in GnRH production that resulted from

increasing estrogen levels produced by the developing follicle.

d. The high level of LH induces the maturation of the follicle and ovulation, and transforms ruptured follicle to corpus luteum.

e. LH maintains the corpus luteum during the luteal phase, which secretes estrogen and progesterone.

f. High levels of estrogen and progesterone act on the hypothalamus and pituitary to inhibit the secretion of LH and FSH.

g. Following the inhibition of LH and FSH, the corpus luteum disintegrates without LH to maintain it. Estrogen and progesterone production ceases, which allows LH and FSH secretion to begin again.

5. Birth control methods (a) prevent fertilization—abstinence, rhythm method, condom, diaphragm, *coitus interruptus*; (b) prevent implantation of embryo—IUD, RU486; (c) prevent release of gametes from gonads—sterilization, chemical contraception such as birth control pills, progestin patch. The most effective methods are abstinence, sterilization, chemical contraception. Least effective are the rhythm method and coitus interruptus.

6. Birth control pills, which contain a combination of synthetic estrogen and progestin, act by negative feedback to prevent the release of GnRH by the hypothalamus and FSH and LH by the pituitary, thus preventing the development of follicles and ovulation. RU486 is an analog of progesterone which blocks progesterone receptors in the uterus, thus preventing progesterone from maintaining pregnancy.

Answers to Test Your Knowledge

Fill in the Blanks:

1.	budding	6.	estrous cycle
2.	protandrous	7.	progesterone
3.	parthenogenesis	8.	urethra
4.	hermaphrodite	9.	vasocongestion
5.	cloaca	10.	puberty

Multiple Choice:

1.	d	5.	c	9.	c
2.	a	6.	c	10.	a
3.	d	7.	d	11.	b
4.	a	8.	c	12.	a

CHAPTER 43
ANIMAL DEVELOPMENT

Suggested Answers to Structure Your Knowledge

1.

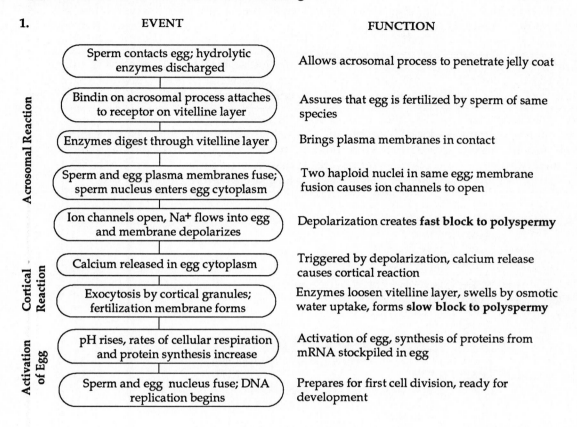

EVENT	FUNCTION
Acrosomal Reaction — Sperm contacts egg; hydrolytic enzymes discharged	Allows acrosomal process to penetrate jelly coat
Bindin on acrosomal process attaches to receptor on vitelline layer	Assures that egg is fertilized by sperm of same species
Enzymes digest through vitelline layer	Brings plasma membranes in contact
Sperm and egg plasma membranes fuse; sperm nucleus enters egg cytoplasm	Two haploid nuclei in same egg; membrane fusion causes ion channels to open
Ion channels open, Na+ flows into egg and membrane depolarizes	Depolarization creates **fast block to polyspermy**
Cortical Reaction — Calcium released in egg cytoplasm	Triggered by depolarization, calcium release causes cortical reaction
Exocytosis by cortical granules; fertilization membrane forms	Enzymes loosen vitelline layer, swells by osmotic water uptake, forms **slow block to polyspermy**
Activation of Egg — pH rises, rates of cellular respiration and protein synthesis increase	Activation of egg, synthesis of proteins from mRNA stockpiled in egg
Sperm and egg nucleus fuse; DNA replication begins	Prepares for first cell division, ready for development

2. Frog Embryo
   a. neural groove   **d.** coelom
   b. notochord   **e.** endoderm
   c. somite (mesoderm)   **f.** archenteron

   Bird Embryo
   a. ectoderm   **d.** coelom
   b. neural tube   **e.** archenteron
   c. notochord   **f.** yolk

3.

	SEA URCHIN	FROG	BIRD	MAMMAL
Cleavage	Holoblastic, animal pole at end where polar body formed, radial cleavage	Holoblastic, but yolk impedes divisions, creates unequal cells	Meroblastic, only in cytoplasmic disc on top of yolk	Holoblastic, no polarity to egg, blastomeres equal
Blastula	Hollow ball of cells, one cell thick, enclose blastocoel	Blastocoel in animal hemisphere, walls several cells thick	Blastodisc, blastocoel cavity between epiblast and hypoblast	Embryonic disc, resembles blastodisc of birds and reptiles
Gastrulation	Invagination to form archenteron, opening is blastopore, migrating mesenchyme cells form mesoderm	Involution of cells at dorsal lip of blastopore, extends around yolk plug, forms 3 cell layers and archenteron	Ingression of epiblast cells at primitive streak, 3-cell layered disc folds under to form archenteron	Gastrulation through primitive streak, similar to birds

4.

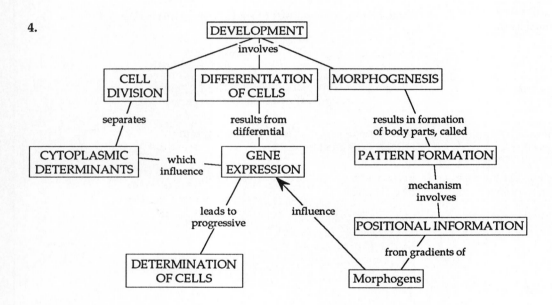

Answers to Test Your Knowledge

Multiple Choice:

1. c	**5.** d	**9.** d	**13.** d	**17.** d
2. b	**6.** c	**10.** c	**14.** a	**18.** b
3. b	**7.** c	**11.** a	**15.** d	**19.** a
4. d	**8.** b	**12.** c	**16.** b	**20.** d

CHAPTER 44
NERVOUS SYSTEMS

Suggested Answers to Structure Your Knowledge

1. **a.** dendrite   **e.** Schwann cell
   b. cell body   **f.** Nodes of Ranvier
   c. axon   **g.** synaptic terminal
   d. myelin sheath

2.

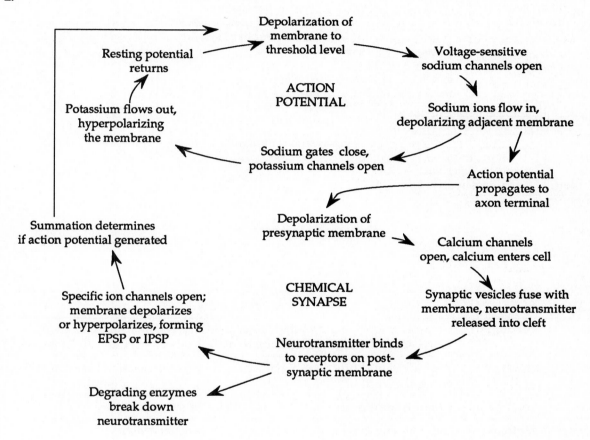

3.

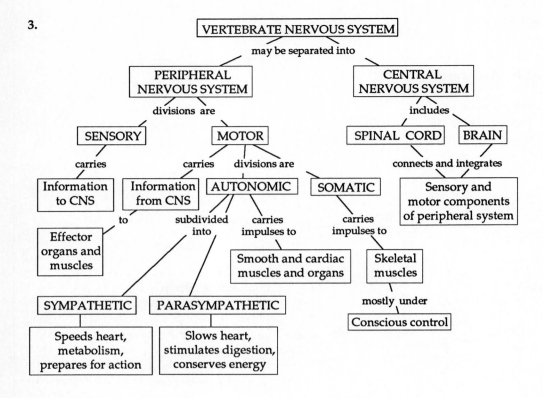

4.

PART OF BRAIN	SECTION	FUNCTION
Medulla	Hindbrain	Control centers for homeostatic functions–respiration, heart and blood vessel actions, digestion; conducts information between brain and spinal cord
Pons	Hindbrain	Aids in some of functions of medulla; conducts information between brain and spinal cord
Cerebellum	Hindbrain	Controls unconscious coordination of movement and balance
Inferior, superior colliculi	Midbrain	Inferior—involved with hearing; superior—visual center
Reticular formation	Midbrain	Regulates state of arousal; filters all sensory information going to cerebral cortex
Thalamus	Forebrain	Major screening area; relays input to cerebral cortex; contains nuclei of reticular formation
Hypothalamus	Forebrain	Produces hormones; regulates pleasure, alarm response, sexual response and mating behaviors; autonomic functions
Cerebral cortex	Forebrain	Processes sensory and motor information; integrates information; responsible for thinking, memory, emotions, speech

Answers to Test Your Knowledge

Multiple Choice:

1. b	**5.** a	**9.** b	**13.** a	**17.** d
2. a	**6.** c	**10.** d	**14.** d	**18.** a
3. a	**7.** a	**11.** c	**15.** d	**19.** c
4. d	**8.** a	**12.** b	**16.** b	**20.** c

CHAPTER 45
SENSORY AND MOTOR MECHANISMS

Suggested Answers to Structure Your Knowledge

1. **a.** ciliary body **g.** vitreous humor
    b. iris **h.** optic nerve
    c. cornea **i.** fovea
    d. pupil **j.** retina
    e. aqueous humor **k.** choroid
    f. lens **l.** sclera

2. Light enters the eye through the pupil and is focused by the lens onto the retina. When rhodopsin or photopsin, the photopigments in rods and cones, absorbs light energy, it isomerizes, decreasing the receptor cell's permeability to sodium and hyperpolarizing the membrane. The reduction in the release of neurotransmitter by rods and cones serves to depolarize or hyperpolarize connected bipolar cells, which in turn affect action potentials in ganglion cells—the sensory neurons that form the optic nerve. Horizontal and amacrine cells provide lateral integration of visual information.

3. Sound waves travel down the *auditory canal* to the *tympanic membrane*, where they are amplified and transmitted to the *malleus, incus, and stapes.* Vibration of the stapes against the *oval window* sets up pressure waves in the fluid in the *vestibular canal* within the *cochlea*, from which they are transmitted to the *tympanic canal* and then dissipated when they strike the *round window.* The pressure waves vibrate the *basilar membrane*, on which the *organ of Corti* is located. Tips of the hair cells arising from the organ of Corti are embedded in the *tectorial membrane.* When vibrated, they bend, triggering a depolarization, release of neurotransmitter, and initiation of action potentials in the sensory neurons of the *auditory nerve*, which leads to the *cerebral cortex.*

4. A motor neuron releases acetylcholine into the neuromuscular junction, initiating an action potential within the muscle fiber, which spreads into the cell through the transverse tubules. The action potential changes the permeability of the sarcoplasmic reticulum membrane which releases calcium ions into the cytoplasm. The calcium binds with troponin, changes the shape of the tropomyosin–troponin complex, and exposes the myosin-binding sites of the actin molecules. Heads of myosin molecules in their high-energy configuration bind to these sites, forming cross bridges. Myosin relaxes and its head bends, pulling the thin filament toward the center of the sarcomere. When myosin binds to an ATP, the cross bridge breaks. Hydrolysis of the ATP returns the myosin to its high-energy configuration, and it binds further down the actin molecule. This sequence continues as long as there is ATP and until calcium is pumped back into the sarcoplasmic reticulum, when the tropomyosin-troponin complex again blocks the actin sites.

Answers to Test Your Knowledge

Matching:

1. I	**4.** L	**7.** N	**10.** K
2. G	**5.** J	**8.** D	**11.** B
3. O	**6.** M	**9.** A	**12.** R

Multiple Choice:

1. a	**6.** b	**11.** d	**16.** c
2. a	**7.** d	**12.** b	**17.** a
3. c	**8.** c	**13.** a	**18.** b
4. b	**9.** c	**14.** d	**19.** a
5. d	**10.** a	**15.** b	**20.** c

Fill in the Blanks:

1. sarcomere
2. I band
3. A band
4. Z line
5. H zone
6. thick filament, myosin
7. thin filament, actin and tropomyosin
8. in g, in thin filament
9. A band, letter c
10. head of myosin of thick filament (letter f), attaches to actin of thin filament (letter g)

CHAPTER 46

ECOLOGY: DISTRIBUTION AND ADAPTATIONS OF ORGANISMS

FRAMEWORK

Ecology deals with the distribution and abundance of organisms and the interrelationships between organisms and both the abiotic and biotic factors in the environment. This chapter describes the organizational levels at which ecological questions are asked, the abiotic factors to which organisms have adapted in both an ecological and an evolutionary time frame, and the major world communities, or biomes, in which adaptations to climate and abiotic factors have produced similar and characteristic life forms.

CHAPTER SUMMARY

Ecology is the scientific study of the interactions of organisms with their environments. Using scientific observation and experimentation to test hypotheses is often challenging because of the scope and complexity of ecological questions. Ecology is a multidisciplinary science, incorporating aspects of genetics, evolution, physiology, and behavior. Ecology focuses on interactions because organisms are affected by, and, in turn, affect both the *abiotic* and *biotic* factors of their environments.

The science of ecology is not synonymous with the growing "ecological" consciousness about current environmental problems, but it helps us to understand these problems and their possible solutions.

The Scope and Development of Ecology

The Questions of Ecology Ecology focuses on the factors that determine the distribution and abundance of organisms. *Organismal ecology* considers behavioral,

morphological, and physiological responses of an organism to its physiochemical environment. Population ecology is concerned with the factors that control the size of *populations*, which are groups of individuals of the same species in an area. The *community* includes all the populations of organisms in an area; ecology on this level looks at interactions such as predation and competition. *Ecosystem ecology* considers abiotic factors as well as the biological community, and ecological questions include the flow of energy and chemical cycling.

Ecology as an Experimental Science Historically, ecology was a descriptive science, but currently an experimental approach in both the laboratory and the field is being used as well to investigate ecological questions. Tremendous creativity is often required to control variables and manipulate communities in field experiments. Some ecologists develop mathematical models and computer simulations to study the interactions of variables and gain insight into ecological questions.

Ecology and Evolution Interactions between organisms and their environments occur within ecological time. The cumulative effects of these interactions are realized on the scale of evolutionary time. The distribution and abundance of organisms are the result of both evolutionary history and present interactions with the environment.

Terrestrial Biomes

The *biosphere* is the relatively thin layer of Earth inhabited by life, including the waters of seas, lakes, and streams, the land to a few meters below the soil surface, and to a few kilometers into the air. *Biomes* are characteristic communities found in broad geographical regions. The general appearance of a biome may be similar over large areas, even though species composition varies locally, because in many places, convergent evolution has produced superficial resemblance of unrelated "ecological equivalents." Ecologists organize biomes by various schemes; the textbook shows a map of eight terrestrial biomes, most of which are named for their predominant vegetation.

The predominant ecological communities in a biome result from natural succession; patchiness in biomes results from natural or human disturbances. Human disturbances have resulted in biomes that ecologists classify as "urban" and "agricultural."

The geographic distribution of the world's major biomes is related to the prevailing climate, particularly temperature and rainfall. Geology of the area may affect nutrient availability and soil structure.

The following survey of major terrestrial biomes travels in a general direction from the equator to the poles.

Tropical Forest *Tropical forests*, within 23° latitude of the equator, vary little in temperature (averaging around 23°C) or day length. Tremendous variations in rainfall result in *tropical thorn forests*, lowlands where rainfall is scarce; *tropical deciduous forests*, where trees and shrubs drop their leaves during a long dry season that alternates with a shorter monsoon; and *tropical rain forests*, where rainfall is abundant.

The tropical rain forest, with its dense canopy formed by tall trees, lianas (woody vines), and epiphytes, has the largest species diversity of all communities. The soil is generally poor and thin because high temperatures and humidity lead to rapid decomposition and recycling of nutrients back into plant material. The animals, typically tree-dwellers, have mutualistic interactions with plants as they disperse pollen, fruits, and seeds while foraging. Human destruction of the tropical rain forest by mining, lumbering, and farming may greatly reduce species diversity and cause large-scale climate changes.

Savanna Tropical and subtropical grasslands with only scattered trees, *savannas* typically have three distinct seasons: cool and dry, hot and dry, and warm and wet. Frequent fires and large grazing mammals restrict vegetation to grasses and *forbs*, small broadleaf plants. Large herbivores and burrowing animals are most common. Areas where forest and grassland intergrade are also referred to as savanna.

Desert Characterized by low precipitation (less than 30 cm/yr), *desert* temperatures may be hot or cold, depending on location. Perennial vegetation, in deserts where there is enough precipitation to sustain it, con-

sists of widely scattered shrubs or succulents. Rainy periods are marked by rapid blooms of annual plants. Seed eaters, such as ants, birds and rodents, and their reptilian predators are common animals. Desert animals have physiological and behavioral adaptations to dry conditions and extreme temperatures.

Chaparral The *chaparral*, or shrubland, is common along coastlines in mid-latitudes that have mild, rainy winters and hot, dry summers. The dominant vegetation, dense, spiny evergreen shrubs and annuals, is maintained by and adapted to periodic fires; many species have large storage roots that permit quick regeneration or seeds that germinate only after a fire. Animals are typically browsers, fruit-eating birds, seed-eating rodents, and reptiles.

Temperate Grassland Somewhat similar to savannas, but found in regions with relatively cold winters, *temperate grasslands* are maintained by fire, drought, and grazing, which prevent the invasion of woody shrubs and trees. Soils are deep and rich in nutrients. Large grazing mammals, their large carnivore predators, and burrowing rodents are typical animals.

Temperate Deciduous Forest Characterized by broad-leaved deciduous trees, *temperate deciduous forests* grow in midlatitude regions that have adequate moisture to support the growth of large trees. Hot summers alternate with cold winters, during which trees drop their leaves and become dormant. Forbs, shrubs, and one or two strata of trees are typical; a rich diversity of animal life is associated with the variety of food and habitat. Human activity has greatly reduced these forests worldwide.

Taiga Found in northern latitudes and high elevations at lower latitudes, the *taiga*, or coniferous or boreal forest, is characterized by harsh winters and short summers. Soil is usually thin and acidic. Coniferous trees grow in dense, uniform stands, shading out undergrowth. Heavy snowfall insulates the soil and protects small mammals. Animals include seed-eaters, insect herbivores, larger browsers, and predators.

Tundra The northernmost limit of plant growth is in the *arctic tundra* and is characterized by low shrubby or matlike vegetation. The *alpine tundra*, found at all latitudes on high mountains above the tree line, has similar flora and fauna, but otherwise quite a different environment. The arctic tundra has long periods of very little light, interspersed with a brief warm summer marked by long days and rapid plant growth. The *permafrost* of the arctic tundra prevents roots from penetrating very far into the continually-saturated soil. Animals adapt to the cold by living in burrows, having large, well-insulated bodies, or migrating. Some large herbivores are found. In the lower latitude alpine tundra, day length is more even and plant growth is slow, but steadier.

Freshwater Biomes

Aquatic habitats, in which life arose and evolved for almost 3 billion years before moving onto land, still make up the largest portion of the biosphere. As in terrestrial biomes, similar aquatic environments have similar communities of organisms that show similar adaptations. Freshwater biomes, characterized by a salt concentration less than one percent, are closely linked with and shaped by the surrounding terrestrial biomes.

Ponds and Lakes Small bodies of standing fresh water are called *ponds*; larger ones are *lakes*. Except where shallow, ponds and lakes are stratified in availability of light, temperature, and community structure. The *photic zone* receives sufficient light for photosynthesis, whereas little light penetrates into the lower *aphotic zone*. A narrow *thermocline* separates surface waters, which warm during the summer, from the cold bottom layer.

The shallow and warm waters close to shore, called the *littoral zone*, support a diverse community of rooted and floating aquatic plants, attached algae, molluscs, insects, crustaceans, fish, and amphibians. Reptiles, waterfowl, and some mammals feed on the abundant plants and animals of the littoral zone. The open surface waters of the *limnetic zone* are occupied by a variety of phytoplankton, which are fed upon by zooplankton, leading to a food chain of larger organisms. *Detritus*, dead organic matter, settles into the deep, aphotic *profundal zone* where it is decomposed by microbes and other organisms, causing these cold, dark waters to become oxygen depleted but nutrient rich. Seasonal mixing of water layers in temperate zones brings oxygen to the profundal zone and nutrients to the limnetic zone.

Oligotrophic lakes are deep, clear, nutrient poor, and fairly nonproductive, with little detritus to reduce oxygen levels. The shallower, nutrient rich waters of *eutrophic* lakes support large, productive phytoplankton communities. Runoff carrying nutrients and sediment may gradually convert oligotrophic lakes into eutrophic lakes. Municipal wastes and enriched runoff from fertilized land contribute excessive nitrogen and phosphorus and may result in explosive algal growth, high production of detritus, and depletion of oxygen.

Streams and Rivers Streams and rivers are freshwater, flowing habitats, whose physical and chemical characteristics vary from their source to their point of entry into oceans or lakes. Clear, fast-flowing, and rocky-bottomed headwaters gradually change to a more turbid, slower-flowing, and nutrient-rich river. *Riffles* form where shallow water flows rapidly over a rough bottom; *pools* are found where deep water flows slowly over a smooth bottom; and smooth *runs* are deep water flowing rapidly over a flat bottom. Overhanging vegetation contributes to nutrient content. Oxygen levels are high in turbulently flowing water and low in murky, warm waters. Temperature varies with altitude, latitude, and season.

Biotic communities reflect these abiotic factors. Plankton communities are largely absent in rapidly flowing waters; instead food chains are built on the photosynthesis of attached algae and rooted plants, or the addition of organic matter and nutrients from the land. The composition of animal communities, including *benthic* (bottom-dwelling) communities, varies with changes in the physical environment along the length of streams and rivers. Organisms show evolutionary adaptations that allow them to exploit the variety of stream and river habitats.

Marine Biomes

Three-fourths of Earth is covered by marine biomes. Algae produce a large portion of the world's oxygen. The average salt concentration of seawater is three per cent. The marine environment also can be classified into a photic zone, in which phytoplankton, zooplankton, and many fishes are found, and an aphotic zone, into which light does not penetrate. The *intertidal zone* is the shallow area along the shore. The *neritic zone* is the shallow region over the continental shelf. The *oceanic zone*, past the continental shelf, reaches great depths. Open water of any depth is the *pelagic zone*, the bottom surface of which is the *benthic zone*.

Estuaries Where a freshwater river or stream meets the ocean, an *estuary* is formed—highly productive areas often bordered by intertidal mudflats or saltmarshes. Within an estuary, salinity varies both spatially and daily with the rise and fall of tides. Saltmarsh grasses, algae, and phytoplankton are the major producers. Estuaries serve as spawning grounds for many marine organisms and feeding and breeding areas for crustaceans, mollusks, fish, and waterfowl. Little undisturbed estuary habitat remains, as these areas have been developed for commercial or residential use and destroyed by the pollution of their streams or rivers.

The Intertidal   The daily cycle of tides exposes the shoreline to variations in water, nutrients, and temperature, and to the mechanical force of wave action. Rocky intertidal communities are vertically stratified, with organisms adapted to firmly attach to the hard substrate and to withstand variations in exposure and salinity. A diverse array of algae, seaweeds, invertebrates, and fishes is usually present. Sandy or mudflat intertidal zones are less stratified, have few large algae or plants, and are home to burrowing worms, clams, and crustaceans. Shorebirds and surface-dwelling crabs are predators or scavengers of this often buried community.

Coral Reefs   Found in tropical waters in the neritic zone, *coral reefs* are highly diverse and productive biomes. The structure of the reef is produced by the calcium carbonate skeletons of the coral and serves as a substrate for other corals and algae. Coral feed on microscopic organisms and organic debris and are nourished by symbiotic photosynthetic dinoflagellates. The communities support a huge variety of herbivores and carnivores. Coral reefs are easily damaged by pollution, development, native and introduced predators, and souvenir hunters.

The Oceanic Pelagic Biome   The generally cold water of the *oceanic pelagic biome* is typically nutrient poor. Photosynthetic plankton flourish in the 100 meter deep photic region, responsible for nearly half of the photosynthetic activity on earth, and are grazed on by numerous types of zooplankton. Morphological adaptations of plankton help them maintain buoyancy. Free-swimming animals, called necton, include squid, fishes, turtles, and marine mammals. Many pelagic birds feed on fish of the surface waters.

Benthos   The communities that occupy the sea bottom of the neritic and pelagic zones are collectively called *benthos*. Nutrients reach the fairly nutrient-poor benthic zone as detritus falling from the waters above. Neritic benthic communities are very diverse and productive. Organisms in the *abyssal zone*, the deep benthic region where light does not penetrate, are adapted to darkness, cold, and high water pressure. A fairly diverse community of invertebrates and fishes inhabit this region. Chemosynthetic bacteria form the basis of a collection of organisms adapted to the hot, anoxic environment surrounding deep sea hydrothermal vents.

Environmental Diversity of the Biosphere

The patchiness of the biosphere is a result of spatial and temporal differences in abiotic factors, such as temperature, rainfall, and light, which create a variety of habitats.

Important Abiotic Factors   Temperature is an important environmental factor because of its effects on metabolism and enzyme activity. Most organisms cannot maintain body temperatures that vary more than a few degrees from ambient temperature.

Depending on their habitat, organisms have adaptations to regulate their osmolarity in aquatic environments or to obtain water and reduce water loss in terrestrial habitats.

Light energy drives almost all ecosystems. The intensity and quality of light are important abiotic factors in all environments and limiting factors in aquatic environments. Most photosynthesis occurs relatively near the surface of the water, as about 45% of red light and 2% of blue light is absorbed in each meter of water. Many plants and animals are sensitive to photoperiod, which serves as an indicator of seasonal changes.

Wind increases the rate of heat and water loss in organisms. Wind may also affect plant morphology. Soils, which vary in their physical structure, pH, and mineral composition, affect the distribution of plants, and, in turn, the distribution of animals. Substrate composition in aquatic environments influences the types of organisms that can inhabit intertidal and benthic zones.

Fire, hurricanes, and other periodic disturbances can drastically affect biological communities. Many plants have evolved adaptations to periodic fires.

Climate and the Distribution of Biomes   The climate, or prevailing weather conditions of a locality, is determined by temperature, water, light, and wind. A climograph plots annual mean temperature and rainfall for a region; generally these values correlate reasonably well with the distribution of various biomes. Overlaps of biomes on a climograph indicate that seasonal patterns of variation, such as the distribution of rainfall or the range of temperatures, influence the types of communities that can develop. Geology also affects the soil structure, nutrient availability, and types of vegetation.

Global Climate Patterns   The absorption of solar radiation heats the atmosphere, land, and water, setting patterns for temperature variations, air circulation, and water evaporation that cause latitudinal variations in climate. The spherical shape of the Earth and the tilting of its axis creates seasonal variation in the intensity of solar radiation. Only the *tropics* (from 23.5° north latitude to 23.5° south latitude) receive sunlight from directly overhead, making it the region with the greatest amount of and least variation in solar radiation. Seasonal variations in daylength and mean temperature increase with latitude.

The global circulation of air begins as intense solar radiation near the equator causes warm, moist air to rise, producing the characteristic wet tropical climate, the arid conditions around 30° north and south, the fairly wet though cool climate about 60° latitude, and the cold and rainless climates of the arctic and antarctic. Air flowing in the lower levels of the three major air circulation cells on either side of the equator produces predictable global wind patterns.

Local and Seasonal Effects of Climate Regional climatic patchiness is influenced by proximity to water and topographical features. Coastal areas are generally more moist, and ocean currents cool or warm overlying air masses that may then pass over the land. Large bodies of water generally moderate daily climate. Mountains also influence solar radiation, temperature, and rainfall. South facing slopes in the northern hemisphere are warmer and drier. Air temperature drops approximately 6°C with every 1000 m increase in elevation. The windward side of a mountain range receives much more rainfall than the leeward side, as the warm, moist air rising over the mountain releases moisture and the drier cooler air absorbs moisture as it descends down the other side.

Seasonal changes affect local climate. Seasonal changes in wind patterns affect ocean currents, sometimes causing upwellings of cold, nutrient rich water. Seasonal temperature changes produce the biannual mixing of waters in ponds and lakes.

Climate also varies on a very small scale. Microclimates within larger areas have slight differences in physical features that affect the distributions of organisms.

Responses of Organisms to Environmental Variation

The geographical distribution of a species is largely determined by its ability to tolerate a combination of abiotic factors. Organisms must cope with the entire set of environmental factors within their habitats, using adaptations that may represent evolutionary compromises to conflicting needs.

Homeostasis and the Principle of Allocation Plants and animals that use behavioral and physiological mechanisms to maintain a steady internal environment are called *regulators*. *Conformers*, organisms whose internal conditions vary with environmental changes, are often those that live in relatively stable environments.

Regulating mechanisms require an expenditure of energy, and the costs and benefits of homeostasis have been selective pressures affecting the evolution of

organisms facing various environmental challenges. The *principle of allocation* is used to analyze how an organism divides its limited energy among various processes, such as reproduction, eating, growing, escaping from predators, and dealing with environmental fluctuations. The maintenance costs of endotherms are significantly higher than those of ectotherms.

The distribution of organisms is related to different systems of energy allocation. In stable environments, organisms can allot more energy for growth and reproduction, but their resulting intolerance of environmental change restricts them to such environments. Organisms that put more of their energy into dealing with environmental fluctuation can successfully live in a wider range of habitats.

Environmental Grain Organisms of different sizes and activity levels relate to spatial variations within an environment, called *environmental grain*, in different ways. A coarse-grained environment has patches so large in relationship to the size and activity of the organism that it can choose from different environmental patches. Patches in a fine-grained environment are so small relative to the size and activity of an organism, that the organism does not choose between patches. Temporal environmental variations can also be described as fine- or coarse-grained, with seasonal shifts in climate considered coarse-grained for even large organisms. Adaptations to spatial and temporal variations vary in their activation time and reversibility.

Behavioral Responses Behavioral responses may include relocation to a more favorable environment, locally or more distantly, as in seasonal migrations. Cooperative social behavior, such as huddling, can help organisms respond to unfavorable conditions.

Physiological Responses Physiological responses to environmental change are usually slower than are behavioral responses. Optimal environmental conditions and tolerance limits for an organism can be determined experimentally by varying abiotic factors. Tolerance limits help to determine the spatial distribution of organisms. Physiological adjustments to environmental change that shift the tolerance limits is called *acclimation*.

Morphological Responses Morphological responses to environmental changes, such as changes in growth that result in altered body form, are most common in plants and provide these rooted organisms with a means of adapting to changing environments.

Adaptation over Evolutionary Time Behavioral, physiological, and morphological responses occur within an ecological time scale, but are based on adaptations that have developed by natural selection over an evolutionary time span. The adaptation of organisms to localized environments may restrict their geographic distribution and often their ability to respond to environmental change. In order for a species to exist in a particular location, it must first have reached that location, and second, the conditions there must be within its range of toleration.

STRUCTURE YOUR KNOWLEDGE

1. • Define *ecology*.

   • What are the levels of ecological study, and what is the focus of inquiry at each of these levels?

   • What methods are used to answer ecological questions?

   • What theory guides the interpretation of data?

2. • What are biomes?

   • What accounts for the similarities in life forms throughout a biome?

   • Temperature and precipitation are two of the key factors that influence the vegetation found in a biome. On the climograph shown below, label the North American biomes (arctic and alpine tundra, coniferous forest, desert, grassland, temperate forest, and tropical forest) represented by each area of temperature and precipitation.

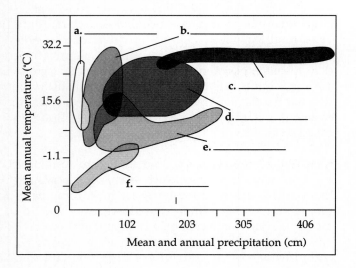

TEST YOUR KNOWLEDGE

MULTIPLE CHOICE: *Choose the one best answer.*

1. Which level of ecology considers energy flow and chemical cycling?
   a. community
   b. ecosystem
   c. organismal
   d. population

2. Evolutionary history
   a. influences the distribution of organisms.
   b. involves the cumulative effects of the interactions between organisms and their environments.
   c. has produced the adaptations of organisms to different environments.
   d. involves all of the above.

3. Ecologists use mathematical models and computer simulations because
   a. ecological experiments cannot be performed.
   b. most of them are mathematicians.
   c. ecology is becoming a more descriptive science.
   d. these approaches allow them to study the interactions of variables and simulate large-scale experiments.

4. An animal that expends most of its energy in growth and reproduction probably lives in
   a. a stable environment.
   b. an unstable environment.
   c. a terrestrial environment.
   d. an aquatic environment.

5. Acclimation
   a. is a morphological response to environmental change.
   b. can extend the tolerance limit of an organism.
   c. involves behavioral responses to environmental change.
   d. occurs more rapidly in plants than in animals.

6. Which of the following is incorrectly paired with its description?
   a. neritic zone—shallow area over continental shelf
   b. abyssal zone—benthic region where light does not penetrate
   c. littoral zone—area of open water
   d. intertidal zone—shallow area at edge of water

7. A conformer is most likely to be successful in the
   a. desert.
   b. tropical rain forest.
   c. taiga.
   d. savanna.

8. Two communities have the same mean temperature and rainfall but very different compositions and characteristics. The best explanation for this phenomenon is that the two
   a. have a different range of temperatures and pattern of rainfall throughout the year.
   b. are composed of species that have very low dispersal rates.
   c. are found on different continents.
   d. receive different amounts of sunlight.

9. Phytoplankton are the basis of the food chain in
   a. streams.
   b. estuaries.
   c. the oceanic pelagic biome.
   d. rocky intertidal zones.

10. A biome that is characterized by three seasons—cool and dry, hot and dry, and warm and wet—is a
    a. tropical forest.
    b. savanna.
    c. chaparral.
    d. temperate forest.

11. The permafrost of the arctic tundra
    a. prevents plants from getting established and growing.
    b. helps to keep the soil from getting wet since water cannot soak in.
    c. prevents plant roots from penetrating very far into the soil, probably restricting plants to low growing forms.
    d. both b and c.

12. Many plant species have adaptations for dealing with the periodic fires typical of a
    a. savanna.
    b. chaparral.
    c. temperate grassland.
    d. All of the above may be correct.

13. A fast running stream with a rocky bottom would be a coarse-grained environment to
    a. benthic insect nymphs.
    b. trout.
    c. phytoplankton.
    d. rooted plants.

14. The ample rainfall of the tropics and the arid areas around 30° north and south latitudes are caused by

a. ocean currents that flow clockwise in the northern hemisphere and counterclockwise in the southern.
b. the global circulation of air initiated by intense solar radiation near the equator producing wet and warm air.
c. the tilting of the earth on its axis and the resulting seasonal changes in climate.
d. the heavier rain on the windward side of mountain ranges and the "rainshadow" on the leeward side.

15. Upwellings in the ocean
    a. are locations of reef communities.
    b. occur over deep sea hydrothermal vents.
    c. bring nutrient rich water to the surface.
    d. are responsible for ocean currents.

MATCHING: Match the biotic description with its biome.

BIOME

1. _____ chaparral
2. _____ desert
3. _____ savanna
4. _____ taiga
5. _____ temperate forest
6. _____ temperate grassland
7. _____ tropical rain forest
8. _____ tundra

BIOTIC DESCRIPTION

A. broad-leaved deciduous trees

B. lush growth, trees, lianas, epiphytes

C. dense shrubs, fire-adapted vegetation

D. tropical grasses and forbs

E. coniferous forests

F. low shrubby or matlike vegetation

G. grasses in relatively cool regions

H. widely scattered shrubs, cacti, succulents

CHAPTER 47

POPULATION ECOLOGY

FRAMEWORK

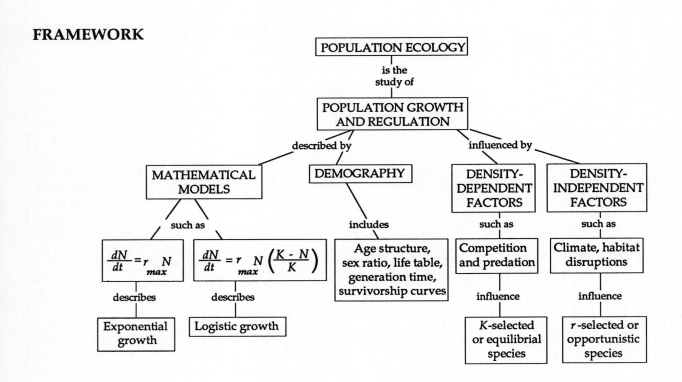

CHAPTER SUMMARY

The continuing growth of the human population in the face of limited and even dwindling resources is a critical biological phenomenon. Population ecology is the study of the fluctuations in and regulation of population size. This branch of ecology overlaps with organismal and community ecology. The interactions of individuals with the environment on both ecological and evolutionary time scales determine the characteristics of a population.

Density and Dispersion

Every population of organisms has geographic boundaries; ecologists determine the geographic boundaries of the population they are studying based upon the type of organism and the research question being asked. The number of individuals per unit area or volume is the population *density*; the spacing of those individuals within the boundaries of the population is referred to as *dispersion*.

Measuring Density Only rarely is it possible to measure population size by counting all individuals. More often, density is measured by using one of a variety of sampling techniques. Indirect indicators of organisms, such as burrows or nests, also may be counted. In the *mark–recapture method*, animals are trapped, marked, released, and trapped again. The proportion of total animals trapped the second time to marked animals recaptured is multiplied by the initial number marked to estimate population size.

Patterns of Dispersion Individuals may be dispersed in the population's *geographical range* in several patterns. *Clumping* may indicate a heterogeneous environment, with organisms congregating in suitable microenvironments. Clumping also may be related to social interactions between individuals. *Uniform* distribution is usually related to competition for resources, resulting in antagonistic interactions between organisms or the establishment of territories. *Random* spacing, indicating the absence of strong attractions or repulsions between individuals, is not very common.

Demography

Population size changes in response to the relative rates of birth and immigration versus death and emigration. *Demography* is the study of the vital statistics that affect population size, such as sex ratio and age structure, which affect the birth and death rates.

Age Structure and Sex Ratio The *age structure* of a population is the relative number of individuals of each age. Each age group in a population has a characteristic fertility and death rate. Often young and old members are most likely to die and intermediate-age individuals have the highest birth rate. Populations with the greatest proportion of members of reproductive age or slightly younger grow the fastest. A shorter *generation time*, the average span between the birth of individuals and birth of their offspring, results in faster population growth.

Sex ratio is the proportion of each sex in a population. The number of females in the population may have a greater effect on birth rate, if males of the species typically mate with many females.

Life Tables and Survivorship Curves A *life table* summarizes the mortality rate for each age group and can be constructed either by following a *cohort* of newborn organisms throughout their lives or by calculating from the age at death of a sample of individuals. Age-specific *fecundity* is the average number of offspring produced by a female of a particular age.

Age-specific mortality can be graphed on a *sur-*

vivorship curve, which shows the number of members of a cohort still alive at each age. Three types of survivorship curves are shown below.

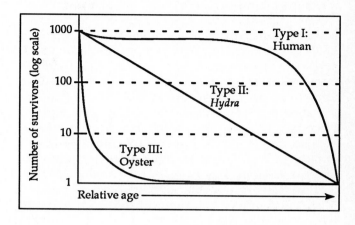

Type I, typical of modern human populations and other large mammals, shows low mortality during early and middle age and a rapid increase with old age. Populations that produce relatively few offspring and provide parental care usually have Type I curves. In a Type II curve, death rate is relatively constant throughout the life span. A Type III curve is typical of populations that produce many offspring, most of which die off rapidly. The few that survive are likely to reach adulthood, as is the case with many fishes and marine invertebrates. Many species show intermediate or more complex survivorship patterns.

Evolution of Life Histories

The *life history* of an organism—its pattern of birth, reproduction, and death—affects population growth in ecological time, but evolved as a result of natural selection operating in evolutionary time.

Three factors of a life history influence the number of offspring a female will produce: (1) clutch size; (2) the number of reproductive episodes in a lifetime; and (3) age at first reproduction. Clutch size usually is inversely related to the size of offspring. Organisms with a Type III survivorship curve usually produce large numbers of small eggs or offspring. Some organisms reproduce only once, investing all the organism's energy budget into the production of offspring, whereas other organisms reproduce more than once in a lifetime, dividing their energy into maintenance, growth, and reproduction. Delaying first reproduction can result in a maximum lifetime reproductive output if older females produce larger clutches and if survivorship to an older age is likely. Darwinian fitness is measured ultimately by how many offspring survive to produce their own offspring. Because organisms have a finite energy budget, they cannot maximize all three of these factors simultaneously.

Ecologists developed the concepts of opportunistic and equilibrial life histories to explain variations in life histories in relation to evolutionary selective pressures. An *opportunistic life history* is often found in populations of generally small species that produce large numbers of offspring at one time and live in variable environments. Such rapidly growing and reproducing species are able to take advantage of new or disturbed habitats, but survivorship is low and population size fluctuates dramatically. Populations of larger species may have an *equilibrial life history*, with repeated production of fewer, better-endowed offspring that are each more likely to survive into adulthood. Slow maturation and parental care are typical of such populations; they tend to be less affected by environmental fluctuations due to better homeostatic capabilities and thus able to maintain a more constant population size.

The concepts of opportunistic and equilibrial life histories, while now seen as oversimplifications, have led to many field and laboratory studies that show a relationship between environmental stability and life histories. Some populations, such as those of dandelions, show different life history characteristics in response to the severity of habitat disruptions. The concept of stress, the general difficulty of environmental conditions, has been suggested as an additional environmental factor that influences life history, such that stressed organisms grow slowly, producing their few offspring infrequently.

Models of Population Growth

Mathematical modeling is a common tool that ecologists use to describe rates of population growth, to study variables affecting growth, and to predict population sizes.

Exponential Population Growth Growth of a small population in a very favorable environment will be restricted only by the physiological ability of that species to reproduce. Ignoring immigration and emigration, the change in population size during a specific time period is equal to the number of births minus deaths for that time and may be signified as $\Delta N / \Delta t = B - D$. Births and deaths also can be expressed in terms of a per capita birth and death rate (bN and dN); the change in population size per unit time can be signified as $\Delta N / \Delta t = bN - dN$. The difference in per capita birth and death rates is the per capi-

ta population growth rate, symbolized by r; $r = b - d$. *Zero population growth* (ZPG) occurs when $r = 0$. The formula describing change in the population at any one instant (instantaneous change) uses the notation of differential calculus and is written as $dN/dt = rN$.

The *intrinsic rate of increase* (r_{max}) is the fastest growth rate possible for a population reproducing under ideal conditions. Under these conditions, there is *exponential population growth*, expressed as $dN/dt = r_{max} N$, which produces a J-shaped growth curve when graphed. The larger the population (N) becomes, the faster the population grows. Periods of exponential growth may occur for short times in opportunistic species. Such species are sometimes described as *r-selected*, because their growth rate may be close to r_{max}.

Logistic Population Growth A population may grow exponentially for only a short time before its increased density limits the resources available within its geographic range. Growth rate slows as the population approaches its *carrying capacity (K)*, which is the maximum stable population size that an environment can support over a relatively long period of time. Crowding and resource limitation may lead to decreased per capita birth rates and increased per capita death rates. The *logistic equation* ($dN/dt = r_{max} N(K-N)/K$) includes the term $((K-N)/K)$ to reflect the impact of the increasing population size (N) on r as the population approaches the carrying capacity.

In the logistic equation, when N is small the $(K-N)/K$ term is close to 1 and growth is approximately exponential ($r_{max}N$). As population size (N) approaches the carrying capacity (K), the $(K-N/K)$ term becomes a small fraction, and exponential growth ($r_{max}N$) is reduced by that fraction. When N reaches K, the term $(K-N)/K$ is 0, and the growth rate (r) is 0. At this point, birth rate equals death rate, and the size of the population does not increase. Maximum increase in population numbers occurs when N is intermediate—when the population is at about one-half the carrying capacity. The formulas and graphs for exponential and logistic growth are shown on the next page.

Species with equilibrial life histories may show logistic growth and are sometimes referred to as *K-selected* because their populations stabilize near carrying capacity.

According to the logistic model, an increasing population ultimately leads to resource limitations that curb population growth. *Intraspecific competition* by members of the same species for the same resources increases with population size, causing r to decline. Food and other resources may be protected in defend-

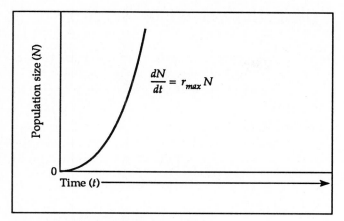

(a) Exponential population growth.

$$\frac{dN}{dt} = r_{max} N$$

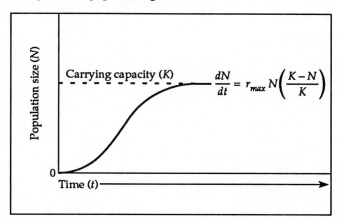

(b) Logistic population growth.

ed territories by animals that exhibit *territoriality*, but then the availability of territorial space may become the limiting resource. The accumulation of toxic metabolic wastes and the availability of nesting sites or shelters may also be limiting factors.

The logistic equation has been found to describe the growth of some populations raised experimentally under constant environmental conditions and without predators. The logistic model makes the assumption that any increase in population numbers will have a negative effect on population growth. For some populations, however, too low a population size is detrimental. According to the *Allee effect*, individuals may benefit as the population grows, from physical support, as in plants, or from social interactions important to reproduction or protection from predators.

The logistic model also assumes that populations approach their carrying capacity smoothly, but many populations overshoot because of a lag time before the negative effects of increasing numbers are realized. Thus populations may oscillate above and below a general carrying capacity.

There are also populations to which the concept of a carrying capacity does not apply because population

density is not an improtant factor. These opportunistic populations are often reduced by environmental conditions before resources have a chance to become limiting.

Regulation of Population Size

The factors regulating population size may differ for opportunistic and equilibrial species and may be a combination of density-dependent and density-independent factors.

Density-Dependent Factors As a population grows, *density-dependent factors*, such as dwindling resources or increasing predation, may affect a larger proportion of individuals and act to slow population growth. Resource limitation may reduce the number of offspring produced by each individual and affect health and survivorship.

Predation may be a density-dependent factor when a predator shows *switching behavior* and feeds preferentially on a prey population that has reached a high density. Intrinsic factors may also regulate population size. Studies of mice have shown that even when food or shelter is not limiting, population size stabilizes at high densities as mice undergo physiological changes that inhibit reproduction. The fluctuation of field and laboratory populations about a carrying capacity provides circumstantial evidence for density-dependent population regulation, but cause and effect relationships are difficult to prove.

Density-Independent Factors Density-independent factors, such as weather or habitat disruption, affect the same proportion of the population whether it is large or small. Density-independent factors may reduce population size before resources or other density-dependent factors become limiting.

Interaction of Regulating Factors Many populations remain fairly stable in size, apparently influenced by density-dependent factors that maintain the populations close to their carrying capacity. Short-term or seasonal fluctuations in population size, however, may be due to density-independent factors. For example, survival of bobwhite quail through the winter varies from 20% to 80% with snow depth, but population size at the end of each summer is quite constant. When only a few adults survive the winter, their reproductive rate increases, probably reflecting higher per capita abundance of resources. Most populations are probably regulated by some mix of density-dependent and density-independent factors.

Population Cycles Some populations of birds, mammals, and insects show regular density fluctuations called population cycles. One hypothesis is that crowding may affect the endocrine systems of organisms, reducing fertility, increasing aggressiveness, or inducing mass migrations. The time lag in a population's response to density-dependent factors is another explanation for the regular oscillation about the carrying capacity. Cycles in the snowshoe hare population were once thought to be caused by the predation of lynx on the increasing density of prey, but these cycles have now been found in hare populations where lynx are absent. The current thinking is that an increasing herbivore population deteriorates its food supply, which may regulate population size more than increasing predator pressure.

Human Population Growth

The human population has been growing exponentially for centuries. It is unlikely that any other population of large animal has ever sustained exponential growth for so long. It took 200 years for the population to double to 1 billion. In the 80 years between 1850 and 1930, it doubled to 2 billion; it doubled again in the next 45 years and is projected to reach 8 billion by the year 2017 if the present growth rate is maintained. In 1992, the world population stood at about 5.5 billion. The growth since the Industrial Revolution has resulted mainly from a drop in death rates, especially among infants. Severe environmental problems are caused by this explosive population growth.

The development of agricultural and then industrial technology has increased the carrying capacity of Earth for humans, and ecologists cannot agree on its ultimate carrying capacity.

Worldwide population growth rates vary by country. High birth rates are more common in underdeveloped and developing countries. The age structure of a population influences growth rate; a large proportion of individuals of reproductive age or younger results in more rapid growth.

Human population growth is unique in that it can be consciously controlled by voluntary contraception or government sanctions. In many cultures, women are delaying marriage and reproduction, thus slowing population growth. Remember that generation time is inversely related to population growth rate.

When and how human population growth will level off is an issue of great concern—whether *r* will reach zero by a reduction in birth rate or by increased mortality caused by resource limitation and environmental degradation.

STRUCTURE YOUR KNOWLEDGE

1. Create a concept map to organize your understanding of the exponential and logistic equations—the mathematical models of population growth.

2. How do the demographic factors of age structure, age specific fecundity, sex ratio, generation time, and survivorship influence population growth rates.

TEST YOUR KNOWLEDGE

MULTIPLE CHOICE: *Choose the one best answer.*

1. In a range with a heterogeneous distribution of suitable habitats, the dispersion pattern of a population probably would be
   a. clumped.
   b. uniform.
   c. random.
   d. unpredictable.

2. In a mark–recapture study, an ecologist traps, marks and releases 25 voles in a small wooded plot. Three nights later she resets her traps and captures 30 voles, 10 of which were marked. What is her estimate of population density of voles in that plot?
   a. 45
   b. 75
   c. 99
   d. 150

3. The age structure of a population influences population growth because
   a. younger females have more offspring than do older females.
   b. populations with shorter generation times grow more rapidly.
   c. different age groups have different reproductive capabilities.
   d. life tables show that mortality rates change with age.

4. A Type I survivorship curve is level at first with a rapid increase in mortality in old age. This type of curve is
   a. typical of many invertebrates that produce large numbers of offspring.
   b. typical of humans and other large mammals.
   c. found most often in *r*-selected species.
   d. almost never found in nature.

5. The middle of the S growth curve in the logistic growth model
   a. is the period when the population is increasing the fastest.
   b. is best described by the term $r_{max}N$.
   c. shows that reproduction ceases when the population size reaches K and dN/dt becomes 0.
   d. is the period when competition for resources is highest.

6. A Type III survivorship curve is most likely to be found in
   a. an opportunistic species.
   b. an equilibrial species.
   c. a species that undergoes periodic molting.
   d. a species that is territorial.

7. A few members of a population have reached a very favorable habitat with few predators and unlimited resources, but their population growth rate is slower than that of the parent population. What is a possible explanation for this situation?
   a. The genetic makeup of these founders may be less favorable than that of the parent population.
   b. The parent population may still be in an exponential part of its growth curve and not yet limited by density-dependent factors.
   c. The Allee effect may be operating; there are not enough population members present for successful reproduction.
   d. All of the above may apply.

8. The term $(K-N)/K$
   a. is the carrying capacity for a population.
   b. is greatest when K is very large.
   c. is zero when population size equals carrying capacity.
   d. increases in value as N approaches K.

9. Density-independent factors
   a. tend to maintain a population around the carrying capacity.
   b. are involved in the population cycles seen in some mammals.
   c. are important in the regulation of K-selected populations.
   d. include climatic events and habitat disruptions.

10. Which of the following is *not* true? A population with a large r_{max} value
    a. probably has a short generation time.
    b. probably has large clutch sizes.
    c. is most likely found in variable environments.
    d. is most likely to be regulated by density-dependent factors.

11. The age structure of a population
    a. can be inferred from a survivorship curve.
    b. is a result of age-specific fertility.
    c. greatly influences the population growth rate.
    d. is most uniform in populations with high growth rates.

12. Which of the following is *not* a characteristic of an equilibrial species?
    a. usually one reproductive episode per lifetime with little parental care
    b. extensive homeostatic capability to deal with environmental fluctuations
    c. long maturation time and long generation time
    d. large body size and large offspring or eggs

13. The human population is growing at such an alarmingly fast rate because
    a. technology has increased our carrying capacity and, thus, density-dependent factors have not slowed reproduction.
    b. the death rate has greatly decreased since the Industrial Revolution.
    c. the age structure of underdeveloped countries is highly skewed toward younger ages.
    d. all of the above are correct.

Questions 14–20. Use the following choices to indicate how these life history characteristics would be affected by the described changes.
   a. increase
   b. decrease
   c. stay the same
   d. no apparent relationship

14. How would r_{max} be expected to vary with an increase in generation time?

15. How would generation time be expected to vary with an increase in body size?

16. For a population regulated by density-dependent factors, how might clutch size change with increased population density?

17. How would clutch size be expected to vary with an increase in parental care?

18. For a population in a particular environment, how would K be expected to change with an increase in N?

19. In a population showing exponential growth, how would *dN/dt* be expected to change with an increase in *N*?

20. How would generation time be expected to vary with an increase in *K*?

CHAPTER 48

COMMUNITY ECOLOGY

FRAMEWORK

Communities are composed of populations of various species that may interact through competition, predation, and symbiosis. The structure of a community—its species diversity, the prevalent forms of vegetation, trophic relations, and stability—is determined by these interactions, the past history of the community, and its successional stage. Disturbances and chance events contribute to the dynamic equilibrium (or non-equilibrium) of a community.

Biogeography studies the distribution of species throughout the world. Island biogeography permits the study of community structure and equilibrium within a more simple system.

CHAPTER SUMMARY

An assemblage of populations of different species living close enough to allow for interaction is called a *community*. Community ecology studies the factors involved in structuring a community—in determining species composition and the absolute and relative abundance of species in a community.

Two General Views of Communities

Ecologists in the 1920s and 1930s developed two divergent views on community composition based upon plant distributions. Gleason advanced the "individualistic" concept of communities as chance groupings of species found in the same area because of similar abiotic requirements. Clements advocated the "interactive" concept of a community functioning as an integrated unit.

The individualistic view predicts that plant species have independent distributions along environmental gradients and that boundaries between communities are indistinct. The interactive hypothesis, in contrast, predicts that plant species are clustered into discrete communities. In most plant communities, species appear to vary on a continuum, with each species independently distributed along environmental gradients according to its tolerance range. In places where some key environmental factor changes abruptly, there may be sharp delineations in species composition between communities, but this does not imply that the plant species are locked into prescribed relationships. The distribution of animal species may be more dependent on the presence of other species. Community composition is most likely determined by both abiotic gradients and biological interactions.

Population Interactions

Interspecific interactions occur among populations of species living in the same community.

Coevolution Both abiotic and biotic components of the environment act as selective factors for evolution. *Coevolution* refers to reciprocal evolutionary adaptations of two species, in which a change in one acts as a selective agent on the other species, whose adaptations in turn act as selective forces on the first. The adaptations of flowers and their exclusive pollinators are examples of coevolution.

A complex example of coevolution that illustrates the difficulty in sorting out the importance of various selective forces is found in passion-flower vines, which have chemicals toxic to most herbivorous insects, and the butterfly *Heliconius*, whose larvae are specialized feeders on these vines. Female butterflies tend to avoid depositing their bright yellow eggs on leaves that already have eggs. Some species of *Passiflora* have leaves with yellow dots that appear to mimic these eggs and, as a result, protect the vine from a heavy infestation of *Heliconius* larvae. The yellow dots are nectaries that attract ants and wasps, which prey on the *Heliconius* eggs and larvae. The presence of ants on a leaf appears to discourage butterflies from laying eggs.

Coevolution is marked by reciprocity and specificity. Whereas it is often difficult to determine whether intertwined adaptations among several species have served as selective factors for each other, it is generally accepted that interactions between species in ecological time are linked to adaptations over evolutionary time.

Predation   Species that eat other organisms are *predators*; *prey*, whether plant or animal, are the species that are eaten. Adaptations to increase success in predation include acute senses, speed and agility, and physical structures such as claws, fangs, teeth, and stingers.

Plants, which cannot run away from their predators, may defend themselves with mechanical devices, such as thorns or microscopic spines, or chemical compounds. Distasteful or toxic chemicals, called *secondary compounds* because they form as metabolic byproducts of biochemical pathways, include such well-known compounds as strychnine, morphine, nicotine, mescaline, spices, and tannins. Counter-adaptations often enable herbivore populations to overcome plant defenses.

Animals can defend against predation by hiding, fleeing, or defending. Potential prey may use camouflage in the form of *cryptic coloration* or body form to blend in with the background. Deceptive markings may confuse predators. The porcupine's mechanical defenses and the skunk's chemical defenses discourage predation. From the food they eat, some animals passively accumulate compounds that would be toxic to their predators. Bright, conspicuous, *aposematic coloration* warns predators not to eat animals with these defenses.

Mimicry, in which a *mimic* species resembles a *model* species, may be used by prey or predators. In *Batesian mimicry*, a harmless species resembles a poisonous or distasteful species. *Müllerian mimicry* is mutual imitation by two or more distasteful species. Predators may use mimicry to "bait" their prey.

Interspecific Competition   Intraspecific competition (among individuals of the same species) for limited resources limits population growth. If two species use the same resource, *interspecific competition* may affect the density of both species. Competition for resources

that involves fighting is called *interference competition*, whereas the use of the same resources is called *exploitative competition*.

Lotka and Volterra modified the logistic equation to include the effects of interspecific competition and predicted that two species that used the same resources could not coexist in the same community. Gause's laboratory experiments with *Paramecium* supported this *competitive exclusion principle*.

An organism's *ecological niche* is described as its place in an ecosystem—its habitat and use of biotic and abiotic resources. The *fundamental niche* includes resources an organism theoretically could use; the *realized niche* is the resources it actually uses as determined by constraints such as competition, predation, or the absence of a potential resource.

The competitive exclusion principle holds that two species with the identical niche cannot coexist in a community. It is difficult to evaluate the role of competition in nature, since it is presumed not to be able to operate very long before a species either becomes extinct or adapts to a slightly different niche.

Slight variations in niche that allow ecologically similar species to coexist provide circumstantial evidence that competition has been a major factor in the development of ecological relationships. *Resource partitioning* is observed in many sympatric species, including the several species of *Anolis* lizard with characteristically different perching sites that appear to minimize competition among them. *Character displacement* of some morphological or behavior trait allows closely related sympatric species to avoid competition. When these species are allopatric, the differences between them may be much less. In addition to the accumulated circumstantial evidence, controlled field studies have documented the competitive exclusion of species in areas of overlap in fundamental niche.

Symbiotic Interactions   Close association between two species is *symbiosis*. Interaction between a *symbiont* and its *host* influence community composition and structure. In *parasitism*, the host is harmed; in *commensalism*, one individual is benefitted and the other unaffected; and in *mutualism*, both symbionts benefit from the relationship. Parasites that live within a host are *endoparasites*; those that feed briefly on the surface of a host are *ectoparasites*.

Parasites best adapted to find and feed on hosts and hosts best able to defend against parasites are favored by natural selection. Examples of host defenses are secondary plant chemicals and the immune system of vertebrates. Many parasites have evolved adaptations that allow them to feed without killing their host. Coevolution tends to stabilize host–parasite relationships.

It is difficult to establish that a relationship is truly commensal, that one member is completely unaffected by the other. Cattle egrets and the cattle that flush the insects on which the egrets feed are an example of commensalism. Evolutionary adaptations in a commensal relationship are likely to involve only the species that benefits.

Coevolution in mutualistic relationships affects both species, as changes in one alter the fitness of the other. Examples of mutualism include the nitrogen-fixing bacteria within legumes, mycorrhizae, flowering plants and their pollinators, and the ants on acacia trees. Mutualistic interactions may have evolved when organisms became able to derive some benefit from their predator or parasite.

Community Structure: Patterns and Processes

Community Characteristics The community level of organization has emergent properties based on the following characteristics: (1) *vegetation structure and associated animals*, the vertical profile or the most common vegetative growth forms influence which animals are associated with the community; (2) *trophic structure*, the feeding relationships (plant–herbivore, host-parasite, and predator–prey interactions) and accompanying energy flow and nutrient cycling; (3) *species richness*, the number of species found in a community and *relative abundance*, the relative numbers of individuals in each species are both components of *species diversity*; and (4) *stability*, the ability of a community to resist change and return to its original makeup following a disturbance. Diversity indexes are mathematical formulas that combine data on species richness and relative abundance. Relative abundance of a species, however, may not reflect the importance of that species to community interactions and structure.

Determinants of Community Characteristics Ecologists attempt to identify which factors determine community characteristics such as species diversity and stability. In order to show that competition was a major factor limiting species diversity, many researchers conducted field experiments documenting the importance of competition. This research showed that interspecific competition is important, but does not always lead to competitive exclusion. These studies may have overstated the influence of competition because subjects expected to show competitive interactions were often the species chosen for research. Indeed, competition is not thought to influence herbivorous populations, which tend not to be limited by

their food supply but to be regulated by density-independent factors. Competition is seen more frequently among carnivorous and plant species, but is difficult to quantify and even more difficult to know what occurred in evolutionary history.

The role of predation in structuring communities may be to stabilize species diversity. As one prey species dwindles in number, most predators will feed on another species. This switching behavior may serve to maintain a variety of species within a community. Paine's study of a predatory sea star in an intertidal community helped to develop the concept of a *keystone predator* and its role in maintaining species diversity by reducing the density of a highly competitive prey species.

Environmental patchiness helps to maintain species diversity because different species are best adapted to particular local conditions.

Competition, predation, symbiosis, environmental patchiness, or an interaction of these factors may all affect community structure and diversity. Which factors are most important may vary from community to community and even in different components of the same community. Determining the evolutionary factors that led to today's ecological relationships is extremely difficult.

Succession

A transition in species composition in a community, usually following some disturbance, is known as *ecological succession*. In traditional views, this succession follows a definite sequence of predictable transitional stages called *seres*, leading to a characteristic *climax community* that is the final, stable stage of succession. If no organisms had been present previously, due to the absence of soil, the process is called *primary succession*. *Secondary succession* occurs when an existing community is disrupted by fire, logging, or farming. Both the primary glacial-till succession and the secondary old field succession take about 200 years to reach a climax community.

Causes of Succession Early successional seres are dominated by *r*-selected opportunistic species that disperse readily and grow rapidly. Species whose tolerances best fit the specific abiotic environment of an area are the most successful colonizers.

Successional changes may be *autogenic*, attributed to conditions induced by the biotic community. Early colonizers may use *inhibition* and retard the growth of other species by exploitative or interference competition. Autogenic succession also may occur because of *facilitation*, in which one stage of vegetation alters the environment, perhaps by changing the pH or increasing organic matter in the soil, and makes it more suitable for species in the next sere.

Allogenic factors that periodically disturb communities may prevent succession to a typical climax community. For example, grasslands may be maintained because periodic fires prevent the invasion of trees.

Human Disturbance   Human activities have altered the structure and succession of communities all over the world. Logging and agriculture disrupt mature communities and restart successional growth. The disappearance of the tropical rain forests and the creation of vast barren areas in Africa are the results of human disturbances.

Community Equilibrium and Species Diversity   During succession, *K*-selected species tend to replace *r*-selected species and the rate at which new species enter the community slows down. Species diversity generally increases during succession, and predation, competition, and symbiosis become more extensive. According to the equilibrium model, which stresses population interactions, succession reaches a climax when interactions are so intricate that no new niches are available for additional species.

The nonequilibrium model views communities as being in continual flux, with the number and identity of species changing even in the so-called climax stage. A mature community is described as a *polyclimax*, an unpredictable combination of patches at different successional stages. Chance events such as dispersal and disturbance are thought to have major roles in the process of succession, which depends, therefore, on the species that happen to colonize an area and on environmental disturbances that prevent species diversity from becoming constant. Severe and frequent disturbances may restrict the community to good colonizers, whereas mild and infrequent disturbances may allow late-successional, highly competitive species to become dominant. According to the *intermediate disturbance hypothesis*, species diversity is greatest when disturbances are intermediate in severity and frequency, allowing organisms from different successional stages to be present.

Diversity and Community Stability   Ecological theory once held that increased diversity led to increased stability, to communities either more resistant to change or more resilient following disturbance. Some mathematical models developed in the 1970s predict the opposite—that increased diversity leads to decreased stability, but these models were rejected by other ecologists. Ecologists continue to debate the relationship between species diversity and community stability.

Biogeographical Aspects of Diversity

Biogeography is the study of past and present distributions of species. Biogeographic realms, roughly correlated with the patterns of continental drift after the breakup of Pangaea and separated in places by deserts or mountain ranges, may have unique taxonomic grouping reflecting both past history and present interactions of biotic and abiotic components of the environment.

Limits of Species Ranges   A species may be limited to a particular range for one of three reasons: it never dispersed beyond that range; pioneers to areas outside the range failed to tolerate the physical environment or to compete successfully with resident species; or it retracted from a former larger range in the course of evolutionary time. Transplant experiments can determine which of the first two reasons has operated to limit a species' range. Paleontology and historical biogeography can provide evidence of species for which the third explanation applies.

Global Clines in Species Diversity   Many taxonomic groups increase in diversity as one moves from northern and southern latitudes toward the equator along a global climatic cline. Terrestrial organisms often decrease in diversity with an increase in altitude, and diversity of benthic animals generally increases with depth. Many hypotheses try to explain the high diversity of the tropics: the tropics rarely experience natural disturbances and their age has allowed complex interactions to coevolve; they generally experience intermediate levels of disturbance and have more environmental patchiness; the stability of the climate may allow organisms to specialize which results in a finer level of resource partitioning; increased solar radiation results in increased photosynthesis and a large resource base; the structural complexity provides a great variety of microhabitats; and the existing diversity is self-propagating, preventing any populations from establishing dominance. A complex combination of these factors may be involved in latitudinal gradients and the high diversity of the tropics. The shallow-to-deep water cline is not correlated with increased productivity, however, but may relate to the long-term stability of deep sea benthic communities. All of these explanations are difficult to address experimentally.

Island Biogeography   Any habitat surrounded by a significantly different habitat is considered an island and allows ecologists to study interactions that are difficult to isolate in more complex systems. In the 1960s, MacArthur and Wilson developed a general theory of island biogeography, stating that the size of the island and its closeness to the mainland (or source of dispersing species) are important variables directly correlated with species diversity. Species diversity is greater on larger islands, on which more niches are available, and on closer islands, on which a higher immigration rate accounts for higher diversity. Following a period of initial immigration and colonization, fewer new species can become established and the extinction rate of present species increases. When immigration and extinction rates are equal, an equilibrium in species diversity develops, although species composition may continue to change.

Observations and experiments, including the work of Simberloff and Wilson with arthropod diversity, provide support for the island biogeography theory. These experiments have shown that, whereas species diversity relates to size and closeness of islands, chance events such as dispersal affect the actual species composition of communities. Ecologists now find that community structure and dynamics are much less predictable and deterministic than once believed.

Conservation Biology: Lessons from Community Ecology and Biogeography

Efforts to conserve biological diversity in the face of increasing human encroachment on natural communities are based on our esthetic regard for other life forms, the awareness that products useful to humans may be derived from newly discovered species, and the recognition that natural environments maintain the proper functioning of the biosphere. Resistance to conservation comes from religious and cultural opposition to birth control and widespread economic difficulties.

Information from community ecology and biogeography can help to direct efforts to preserve biodiversity. Island biogeography can be applied to help determine the size of nature preserves necessary to provide for the requirements of different species. The prefer-

ability of one large preserve versus a collection of smaller preserves depends on the heterogeneity of the environments involved. Several smaller preserves connected by passageways of suitable habitat might allow for recolonization should a local species become extinct and for the spread of protected species between preserves. The design of nature preserves is an area of active research.

STRUCTURE YOUR KNOWLEDGE

1. Develop a concept map organizing your understanding of the important factors that structure a community.

2. Describe succession and the factors that contribute to this process.

3. Many biogeographical studies have found that large islands have greater species richness than small islands. Label the lines on the following graph that show how immigration rate and extinction rate vary with the number of species on large and small islands. Indicate the equilibrium number on the x axis for a small and large island.

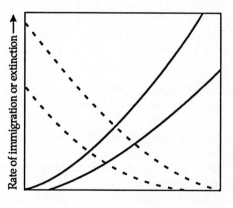

TEST YOUR KNOWLEDGE

MULTIPLE CHOICE: *Choose the one best answer.*

1. Which of the following is *not* part of Gleason's individualistic concept of communities?
   a. Communities are chance collections of species that are in the same area because of similar environmental requirements.
   b. There should be no distinct boundaries between communities.
   c. The consistent composition of a community is based on interactions that cause it to function as an integrated unit.
   d. Species are distributed independently along environmental gradients.

2. Two species, A and B, occupy adjoining environmental patches that differ in several abiotic factors. When species A is experimentally removed from a portion of its patch, species B colonizes the vacated area and thrives. When species B is experimentally removed from a portion of its patch, species A does not successfully colonize the area. From this one might conclude that
   a. both species A and species B are limited to their range by abiotic factors.
   b. species A is limited to its range by competition and species B is limited by abiotic factors.
   c. both species are limited to their range by competition.
   d. species A is limited to its range by abiotic factors and species B is limited to its range because it cannot compete with species A.

3. The species richness of a community refers to
   a. the relative numbers of individuals in each species.
   b. the number of different species found in a community.
   c. the feeding relationships or trophic structure within the community.
   d. the species diversity that is characteristic of that community.

4. Which of the following is *not* an example of coevolution?
   a. adaptations of flowers and their exclusive pollinators
   b. passion-flower vines and the butterfly *Heliconius*
   c. a parasite that is specific for one host
   d. aposematic coloration of monarch butterfly and predators that learn not to eat them

5. Through resource partitioning,
   a. two species can compete for the same prey item.
   b. slight variations in niche allow closely related species to coexist in the same habitat.
   c. two species can share the same realized niche in a habitat.
   d. competitive exclusion results in the success of the superior species.

6. Which of the following is an example of a predator using mimicry?
   a. a snapping turtle whose tongue resembles a wriggling worm
   b. the porcupine's quills
   c. a caterpillar's body shape and color resembling its background
   d. wing markings of io moth that look like large eyes

7. The ability of some herbivores to eat plants that have toxic secondary compounds may be an example of
   a. coevolution.
   b. mutualism.
   c. commensalism.
   d. niche partitioning.

8. A good-tasting prey species may defend against predation by
   a. Müllerian mimicry.
   b. Batesian mimicry.
   c. secondary compounds.
   d. aposematic coloration.

9. A role predation may play within a community is to
   a. maintain species diversity by switching between prey species as each becomes more numerous.
   b. increase the relative abundance of the dominant species.
   c. stabilize host–parasite relationships.
   d. encourage resource partitioning and coevolution.

10. A highly successful parasite
    a. will not harm its host.
    b. may benefit its host.
    c. will be able to feed without killing its host.
    d. will kill its host fairly rapidly.

11. The most important factor in determining community structure
    a. may change from one community to another.
    b. is predation.
    c. is competition.
    d. is history.

12. Primary succession
    a. involves the first colonists that arrive in an area.

b. occurs where no soil is present in an area.
c. is a result of facilitation.
d. occurs following a disruption of a community.

13. Which of the following would *not* be an autogenic factor in succession?
 a. early colonizers that shade other species
 b. accumulation of humus from decomposition
 c. periodic fires that restart secondary succession
 d. changes in soil pH from accumulated tannins that make soil unsuitable for present inhabitants

14. Inhibition
 a. may prevent the achievement of a climax community.
 b. is evidence for the equilibrium theory of succession.
 c. is one of the factors that determines the most tolerant species in an area.
 d. interferes with the successful colonization of other species.

15. According to the nonequilibrium model of succession,
 a. chance events such as dispersal and disturbance play major roles in succession, and species composition remains in flux.
 b. species diversity is greatest in the climax community.
 c. when succession reaches a climax community, only extinctions make room for new colonists.
 d. the communities with the greatest diversity have the greatest resiliency and resistance to change .

16. The taxonomic similarity of species within biogeographic realms
 a. is due to the similarity in environmental characteristics.
 b. is mostly a reflection of the past history of species' distributions.
 c. is greatest on larger islands.
 d. reflects the inability of species to live outside their range.

17. A species may be restricted to a particular range because
 a. it cannot tolerate environmental conditions outside that range.
 b. it has never dispersed beyond that range.
 c. it would outcompete native species if it were transplanted to their habitat.
 d. both a and b.

18. The increasing diversity of the benthic community with depth is an example of
 a. the diversity of an area correlating with its productivity.
 b. a cline that may be related to increasing environmental stability.
 c. the equilibrium model of community structure.
 d. a community of intermediate disturbance reaching a higher species diversity.

19. An island that is small and far from the mainland, as compared to a large island close to the mainland,
 a. would be expected to have a low species diversity.
 b. would be expected to be in an early successional stage.
 c. would have a small species diversity but a large abundance of organisms.
 d. would have a high rate of colonization but a low rate of extinction.

20. The island recolonization experiment of Simberloff and Wilson showed that
 a. species diversity returns very slowly to an island after a disturbance.
 b. the species diversity was highest when disturbances were intermediate in frequency and severity.
 c. whereas the same numbers of species of arthropods returned to each island, the species composition was different, indicating the importance of chance events.
 d. islands closest to the mainland had the greatest numbers of arthropods recolonize, and their community composition and diversity were the same as prior to fumigation.

CHAPTER 49

ECOSYSTEMS

FRAMEWORK

This chapter describes energy flow and chemical cycling through ecosystems. Producers convert light energy into chemical energy, which is then passed, with a loss of energy at each level, from consumer to consumer and to detritivores. Energy makes a one-way trip through ecosystems. Chemical elements, such as carbon, oxygen, nitrogen, and phosphorus, are cycled in the ecosystem from reservoirs in the atmosphere, oceans, or soil through producers, consumers, and detritivores, and back to the reservoirs.

Burning fossil fuels, deforestation, dumping of toxic chemicals into the environment, and washing excess nutrients into lakes are human activities that have altered natural ecosystem dynamics and pose threats of future ecological disruption.

CHAPTER SUMMARY

An *ecosystem* is a community and its physical environment; it includes all the biotic and abiotic components of an area. The processes of energy flow and chemical cycling are described at the ecosystem level of organization. Most ecosystems are powered by energy from sunlight, which is transformed to chemical energy by autotrophs, passed to a series of heterotrophs in the organic compounds of food, and continually dissipated in the form of heat. Energy flows through the system; it is not recycled. Chemical elements, such as carbon and nitrogen, are cycled between the abiotic and biotic components of the ecosystem as autotrophs incorporate them into organic compounds and the processes of metabolism return them to the soil, air, and water.

Trophic Levels and Food Webs

The *trophic structure* of an ecosystem determines how energy and chemicals flow through different *trophic levels*. *Primary producers*, or autotrophs, usually use light energy to photosynthesize sugars for use as fuel in respiration and as building materials for other organic compounds. The *primary consumers* are herbivores, which eat plants and algae. *Secondary consumers* are carnivores that eat herbivores; *tertiary consumers* eat other carnivores. *Detritivores* consume organic wastes and dead organisms. *Omnivores* eat producers as well as consumers of various levels.

Plants are the main producers in terrestrial ecosystems; photosynthetic protists and cyanobacteria form the basis for most marine ecosystems. Chemosynthetic bacteria found around thermal vents deep in the seas depend on geothermal rather than solar energy.

Terrestrial herbivores include insects, snails, grazing mammals, and birds and mammals that eat fruit and seeds. In aquatic ecosystems, zooplankton, small invertebrates, and some fish are the primary consumers. Terrestrial secondary consumers include spiders, frogs, insect-eating birds, and carnivorous mammals. Secondary consumers in aquatic habitats include fish and larger invertebrates.

Fungi and bacteria are the most important decomposers in most ecosystems. Earthworms and such scavengers as crayfish, cockroaches, and bald eagles are also detritivores. Detritivores that feed on plant remains often form a link between producers and secondary consumers.

A *food chain* shows the transfer of food between trophic levels. Most ecosystems have complex, branching feeding relationships called *food webs*.

Energy Flow

The Global Energy Budget   The intensity of solar energy striking the earth varies by latitude and by region, depending on topography, cloud cover and dust in the air. Only a small portion of incoming solar radiation strikes algae and plants, and, of that, only 1% is converted to chemical energy by photosynthesis. Nevertheless, world-wide photosynthetic production is about 170 billion tons of organic material per year.

Primary Productivity   The rate of conversion of light to chemical energy is called the *primary productivity* of an ecosystem. *Net primary productivity* (NPP) is the *gross primary productivity* (GPP) minus the energy used by plants in their own cellular respiration (Rs). NPP = GPP – Rs. For most plants, 50% to 90% of gross primary productivity remains as net primary productivity, which is seen as plant growth and is the stored chemical energy available to consumers.

Gross primary productivity can be measured in aquatic habitats by comparing oxygen concentration in incubated dark and transparent bottles filled with water samples in which, respectively, just respiration or both respiration and photosynthesis have been occurring. In less productive waters, the level of radioactive carbon incorporated into plankton incubated with labeled bicarbonate is used to estimate net primary production.

Primary productivity can be expressed as energy per unit area per unit time ($J/m^2/yr$), or as *biomass* measured in terms of dry weight of organic material ($g/m^2/yr$). *Standing crop biomass* is the total biomass of plants in an ecosystem. Primary productivity is the rate at which new biomass is synthesized. Tropical rain forests are very productive ecosystems and, because of their large area, account for much of Earth's overall productivity. Estuaries and coral reefs also have high rates of productivity; deserts, tundra and oceans have low rates.

Productivity in terrestrial ecosystems is influenced by precipitation, heat, and light intensity as well as nutrients. A *limiting nutrient* is the one that restricts further productivity because it is not present in adequate amounts. Nitrogen or phosphorus is the limiting nutrient in many ecosystems.

Productivity in the seas is greatest in shallow waters and along coral reefs. The open ocean generally has low productivity because mineral nutrients are limited near the surface where light is available for photosynthesis. Phytoplankton communities are most productive where upwellings bring nitrogen, phosphorus, and other organic nutrients to the surface. In freshwater ecosystems, light intensity, temperature, and availability of minerals affect productivity.

Energy Transfers and Ecological Pyramids   *Secondary productivity* is the rate at which consumers in an ecosystem produce new biomass from organic material. Each trophic level sees a decline in productivity, partly because of a loss of energy to heat as each consumer converts organic fuel into its own molecules or into energy in cellular respiration. Only the chemical energy stored as growth or in offspring is available as food to higher trophic levels.

Herbivores cannot digest much of the cellulose they eat, and two-thirds of the energy they do absorb is used for cellular respiration. In a grass field, insects may convert only about 4% of the net primary productivity to secondary productivity. Carnivores can digest and absorb more of the organic compounds in their food, but often they use more of it for cellular respiration.

Ecological efficiency is a measure of the potential energy of one trophic level that makes it to the next level, often approximated at 10%. This progressive loss of energy can be represented as a *pyramid of productivity*. A *biomass pyramid* illustrates the standing crop biomass or total dry weight of organisms at each trophic level, an indication of the amount of chemical energy stored in organic compounds. This pyramid usually narrows rapidly from producers to the top trophic level, but some aquatic ecosystems have inverted biomass pyramids where zooplankton (primary consumers) outlive and outweigh the highly productive, but heavily consumed, phytoplankton. Although more biomass is present at any one time in the zooplankton, the energy pyramid is normal in shape.

Because of energy loss at each trophic level, most food webs are limited to three to five levels. Only about one-thousandth of the chemical energy produced in photosynthesis makes it to a tertiary consumer. The *pyramid of numbers* illustrates that higher trophic levels consist of small numbers of individuals.

Eating grain-fed beef as a main source of calories is an inefficient means for humans to obtain the energy trapped by photosynthesis.

Chemical Cycling

Chemical elements are passed between abiotic and biotic components of ecosystems through *biogeochemical cycles*. Plants and other autotrophs use inorganic nutrients to build new organic matter, which then is passed through the food chain. Atoms of organisms are returned to the atmosphere, water, or soil through respiration and the action of decomposers.

The route of a chemical cycle depends on the element involved. Gaseous carbon, oxygen, sulfur, and nitrogen have global cycles involving atmospheric reservoirs. Less mobile elements such as phosphorus, potassium, calcium, and the trace elements, have a more localized cycle in which soil is the main abiotic reservoir. Elements are found in four types of reservoirs or compartments: organic material in living organisms, available to other organisms; unavailable organic material in fossilized deposits; available inorganic elements, ions, and molecules in water, soil, or air; and unavailable elements in rocks.

The actual movement of elements through biogeochemical cycles is quite complex, with influx and loss of nutrients from ecosystems occurring in many ways. Ecologists have added radioactive tracers to chemical elements in several ecosystems to study biogeochemical pathways.

The Carbon Cycle   Carbon generally cycles at a fast rate, as plants take CO_2 from the atmosphere for photosynthesis and organisms release it in cellular respiration. One-seventh of atmospheric CO_2 is used each year by plants. Wood, coal, and petroleum can store carbon for long periods of time. The relatively low concentration of CO_2 in the atmosphere (0.03%) fluctuates slightly with seasonal changes in photosynthetic activity in the northern hemisphere. Fossil fuel combustion is increasing the amount of atmospheric CO_2.

In aquatic environments, CO_2 interacts with water and limestone to reversibly form bicarbonate, which acts as a carbon reservoir. The ocean reservoirs contain 50 times as much carbon as does the atmosphere and may continue to absorb some of the CO_2 from fossil fuel combustion.

The Nitrogen Cycle   Plants cannot assimilate atmospheric nitrogen. Only certain prokaryotes can fix or reduce N_2 into ammonia (NH_3), which can then be incorporated into amino acids and other nitrogenous organic compounds. Soil bacteria and symbiotic bacteria in root nodules fix nitrogen in terrestrial ecosystems; some cyanobacteria do so in aquatic ecosystems. Fertilizers also add a significant amount of nitrogen to agricultural soils and associated waters. In a process called *nitrification*, some soil bacteria oxidize ammonia to NO_2^- (nitrite) and NO_3^- (nitrate). Plants can absorb either ammonia or nitrate. Animals obtain nitrogen in organic form from plants or other animals. Some bacteria convert nitrate to N_2 to obtain their oxygen. This *denitrification* returns N_2 to the atmosphere.

Detritivores, in a process called *ammonification*, decompose nitrogenous compounds from organic wastes and dead organisms and return ammonia to the soil.

Most nitrogen cycling comes from nitrates recycled from organic compounds in soil and water by ammonification and nitrification. Nitrogen fixation and denitrification of atmospheric N_2 are minor processes in the nitrogen cycle.

The Phosphorus Cycle   Weathering of rock adds phosphorus to the soil in the form of PO_4^{3-}, which is absorbed by plants. Organic phosphate is transferred from plants to consumers and returned to the soil through the action of decomposers or by animal excretion. Humus and soil particles usually bind phosphorus, keeping it available locally for recycling. The phosphorus that leaches into the water table eventually travels to the sea, where sediments become incorporated into rocks. These rocks may eventually return to terrestrial ecosystems as a result of geological processes. The sedimentary cycle works within geological time, whereas the local cycle between soil, plants, and consumers operates in ecological time.

Variations in Nutrient Cycling Time   Nutrients cycle rapidly in a tropical rain forest; forest-floor litter contains only about 1% to 2% of the total organic material. In temperate forests, more nutrients are stored in the detritus. The rate of nutrient cycling in an ecosystem relates to decomposition rate, size of standing crop, soil chemistry, and frequency of fire.

Human Intrusions in Ecosystem Dynamics

Human activities and technology have affected the trophic structure, energy flow, and chemical cycling of most ecosystems.

Habitat Destruction and the Biodiversity Crisis   Natural ecosystems are cleared for agricultural, industrial, and residential development and clear-cutting of timber eliminates huge areas of forests. Nearly all large tracts of tropical forest will be gone in a decade or two if present practices continue. Wars and preparation for war also have severe ecological impacts. This destruction of natural habitat leads to the extinction of many species. The *biodiversity crisis* was addressed at the 1992 Earth Summit in Rio de Janeiro. A global treaty to preserve biodiversity was signed by almost all nations except the United States.

The extent of the biodiversity crisis is difficult to measure because only a fraction of species have been identified and described. Preserving endangered species is complicated by lack of knowledge about minimum viable population sizes and habitat size. Migratory species that range over several countries require cooperation in conservation measures. There is general concern that many potentially valuable species will become extinct before they are known to scientists.

Effects of Deforestation on Chemical Cycling: The Hubbard Brook Forest Study A team of scientists has looked at nutrient cycling in this forest ecosystem for 30 years. The mineral budget for each of six valleys was determined by measuring the influx of key nutrients in rainfall and their outflow through the creek that drained each watershed. About 60% of the precipitation exited through the stream; the rest was lost by transpiration and evaporation. Most minerals recycled within the forest ecosystem.

The effect of deforestation on nutrient cycling was measured for three years in a valley that was completely logged. Compared with a control, water runoff from the deforested valley increased 30%–40%; net loss of minerals, such as Ca^{2+}, K^+, and nitrate, was huge. The sixtyfold increase in nitrate in the creek created unsafe drinking water. Such *long-term ecological research (LTER)* is able to track interactions within natural ecosystems and the impact of human disruptions.

Agricultural Effects on Nutrient Cycling The harvesting of crops removes nutrients that would otherwise recycle into the soil. After depleting the organic and inorganic reserves of nutrients in a newly cleared area, crops require the addition of expensive synthetic fertilizers. Many nutrients are shunted into aquatic systems through sewage and runoff from fields, where they are lost to terrestrial systems and may stimulate excessive algal growth. Advances in sewage treatment may allow such nutrients to be safely returned to the land, reducing the need for fertilizer.

Accelerated Eutrophication of Lakes Lakes undergo a natural succession during which primary productivity increases as mineral nutrients are added gradually from runoff and recycled through the lake's food chain. Lakes eventually become *eutrophic*, or very productive and nutrient-rich. Sewage, factory wastes, and runoff of animal wastes and fertilizers from agricultural lands often accelerate this process. The rapid increase in nutrients can cause an explosive increase in producers. Oxygen shortages, due to plant and algal respiration at night, and the metabolism of decomposers that work on the accumulating organic material, kill off many fish and other lake organisms.

Poisons in Food Chains Many toxic chemicals dumped into ecosystems are nonbiodegradable; some may become more harmful as they react with other environmental factors. Organisms absorb these toxins from food and water and may retain them within their tissues. In a process known as *biological magnification*, the concentration of such compounds increases in each successive link of the food chain. DDT, a classic example of the problem of biological magnification, has been found in nearly every organism tested. High concentrations in birds of prey have interfered with successful egg shell production and reproduction.

The release of radioisotopes in nuclear accidents is a very serious environmental threat. These substances contaminate crops and natural ecosystems, increase birth defects and the incidence of cancer in animals, and are subject to biological magnification due to their long half-lives.

Intrusions in the Atmosphere The concentration of CO_2 in the atmosphere has been increasing since the Industrial Revolution as a result of the combustion of fossil fuel and wood. Atmospheric CO_2 increased 11% in the past 35 years.

If C_3 plants become able to outcompete C_4 plants with the increase in CO_2, species composition in natural and agricultural communities may be affected.

Through a phenomenon known as the *greenhouse effect*, CO_2 and water vapor in the atmosphere absorb infrared radiation reflected from Earth and re-reflect it back to Earth, causing an increase in temperature. The continued increase in atmospheric CO_2 concentration may have far-reaching consequences in climatic changes, in the amount and distribution of precipitation, and in rising sea levels due to the melting of the ice caps. Scientists use mathematical models to estimate the extent and consequences of increasing CO_2 levels. Paleoecologists study the effects of previous global warming trends to predict the effects of increasing temperatures. A treaty that set up a plan for stabilizing CO_2 emissions was signed at the Earth Summit in 1992, but changing the level of combustion of fossil fuel and wood in increasingly industrialized societies is a huge challenge.

A layer of ozone molecules (O_3) in the lower stratosphere absorbs damaging ultraviolet radiation, but has been gradually thinning since 1975, largely as a result of the accumulation of chlorofluorocarbons in the atmosphere. A seasonal "ozone hole" over Antarctica has been increasing in size in recent years. The effects of ozone depletion may include increased incidence of skin cancer and cataracts and unpredictable effects on phytoplankton, crops, and natural ecosystems.

Exotic species have been introduced into communities both accidentally by travelers and intentionally for agricultural or ornamental purposes. Examples of introduced species that have had negative impact include the rabbits in Australia, gypsy moths in America, and Africanized bees moving northward from Brazil. The release of genetically engineered organisms into natural environments has potential dangers of resistant crops becoming pests if they escape from agricultural fields or interbreeding with non-crop species and producing "superweeds."

Biotic Effects at the Biosphere Level

The Gaia hypothesis, as first proposed by Lovelock, maintained that the biosphere is a sort of superorganism that shapes the climate and atmosphere of Earth and is capable of self-regulating metabolism.

More recent proponents of the Gaia metaphor stress the feedback mechanisms inherent in natural ecosystems; the interconnectedness of trophic levels, energy cycling, and chemical cycles within and between ecosystems; and the unknown dangers of technological disruptions of these balances.

The Ecological Society of America has advocated a new research agenda, *The Sustainable Biosphere Initiative*, to encourage studies of global change, biological diversity, and maintenance of the productivity of natural and artificial ecosystems. The goal of this huge research initiative is to develop the ecological information needed to make intelligent decisions on the development, management, and conservation of natural resources.

STRUCTURE YOUR KNOWLEDGE

1. Two processes that emerge at the ecosystem level of organization are energy flow and chemical cycling. Develop a concept map that explains, compares, and contrasts these two processes.

2. From the nitrogen cycle pictured below, fill in the blanks with the names of organisms or compounds.

   a. _____

   b. _____

   c. _____

   d. _____

   e. _____

   f. _____

   g. _____

3. Describe three or four human intrusions in ecosystem dynamics that have detrimental effects.

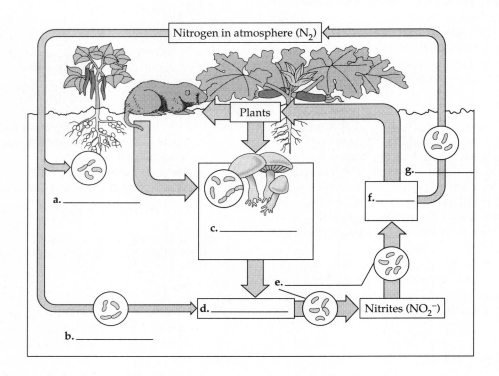

TEST YOUR KNOWLEDGE

MULTIPLE CHOICE: *Choose the one best answer.*

1. Which of the following organisms and trophic levels are mismatched?
   a. algae—producer
   b. phytoplankton—primary consumer
   c. earthworm—detritivore
   d. bobcat—secondary consumer

2. Chemosynthetic bacteria found around deep sea vents are examples of
   a. producers.
   b. detritivores.
   c. chemical cycling.
   d. secondary productivity.

3. In an ecosystem,
   a. energy is recycled through the trophic structure.
   b. energy is usually captured from sunlight by primary producers, passed to secondary producers in the form of organic compounds, and lost to detritivores in the form of heat.
   c. chemicals are recycled between the biotic and abiotic sectors, whereas energy makes a one-way trip through the food web.
   d. there is a continuous process by which energy is lost as heat, and chemical elements leave the ecosystem through runoff.

4. Primary productivity
   a. is equal to the standing crop of an ecosystem.
   b. is greatest in freshwater ecosystems.
   c. is the rate of conversion of light to chemical energy in an ecosystem.
   d. is all of the above.

5. Net primary productivity
   a. is less than gross primary productivity.
   b. takes into account the energy used by plants in respiration.
   c. can be measured in an aquatic habitat by the incorporation of radioactive carbon from labeled bicarbonate into phytoplankton.
   d. is all of the above.

6. Productivity in freshwater ecosystems is affected by
   a. temperature.
   b. light intensity.
   c. availability of nutrients.
   d. all of the above.

7. Secondary productivity
   a. is measured by the standing crop.
   b. is the rate of biomass production in consumers.
   c. is greater than primary productivity.
   d. is 10% less than primary productivity.

8. Which of the following is *not* true of an energy pyramid?
   a. Only about 10% of the energy in one trophic level is passed into the next level.
   b. Because of the loss of energy at each trophic level, most energy pyramids are limited to three to five steps.
   c. The energy pyramid of some aquatic ecosystems is inverted because of the large zooplankton primary-consumer level.
   d. Eating grain-fed beef is an inefficient means of obtaining the energy trapped by photosynthesis.

9. Biogeochemical cycles are global for
   a. elements that are found in the atmosphere.
   b. elements that are found mainly in the soil.
   c. carbon, nitrogen, and phosphorus.
   d. elements that have been marked with radioactive tracers.

10. Which of these processes is incorrectly paired with its description?
    a. nitrification—oxidation of ammonia in the soil to nitrite and nitrate
    b. nitrogen fixation—reduction of atmospheric nitrogen into ammonia
    c. denitrification—removal of nitrogen from organic compounds
    d. ammonification—decomposition of organic compounds into ammonia

11. Carbon cycles relatively rapidly except when it is
    a. dissolved in freshwater ecosystems.
    b. released by respiration.
    c. converted into sugars.
    d. stored in petroleum, coal, or wood.

12. The geological time scale of phosphorus cycling involves
    a. its incorporation into shells.
    b. the weathering of rock to add PO_4^{-2} to the soil.
    c. sedimentation to form rocks in the sea bed.
    d. the incorporation of phosphorus into fossils.

13. Which of the following was *not* shown by the Hubbard Brook Forest Study?
    a. Most minerals recycle within a forest ecosystem.
    b. Deforestation results in a large increase in water runoff.
    c. Mineral losses from a valley were great following deforestation.
    d. Following deforestation, mineral influx and outflow were nearly balanced, but water runoff increased dramatically.

14. The high concentrations of DDT found in birds of prey is an example of
    a. biological magnification.
    b. eutrophication.
    c. the Gaia hypothesis.
    d. chemical cycling.

15. The greenhouse effect
    a. could change global weather and lead to the flooding of coastal areas.
    b. could result in more C_4 plants in plant communities.
    c. causes an increase in temperature when CO_2 absorbs more sunlight entering the atmosphere.
    d. can do all of the above.

16. The Gaia hypothesis
    a. shows that atmospheric O_2 concentration is regulated by methane.
    b. is being researched by the Sustainable Biosphere Initiative.
    c. has shown that the biosphere is a superorganism, able to regulate the climate and atmospheric CO_2 concentration.
    d. may serve as a useful metaphor for the interactions between the biosphere and the physical parameters of Earth.

17. The open ocean and tropical rain forest are the two largest contributors to Earth's net primary productivity because
    a. both have high rates of net primary productivity.
    b. both cover large surface areas of the Earth.
    c. the ocean covers a large surface area and the tropical rain forest has a high rate of productivity.
    d. both a and b are correct.

18. To determine the gross primary productivity of a sample of phytoplankton using the light and dark bottle technique, one would
    a. subtract the decrease in O_2 in the dark bottle from the increase in O_2 in the light bottle.
    b. add the decrease in O_2 in the dark bottle to the increase in O_2 in the light bottle.
    c. average the O_2 concentrations in the two bottles.
    d. add the increase in O_2 in both bottles.

CHAPTER 50

BEHAVIOR

FRAMEWORK

This chapter introduces the complex and fascinating subject of animal behavior. The study of behavior integrates biochemistry, genetics, physiology, evolutionary theory, and ecology.

Behaviors can range from simple fixed action patterns in response to specific stimuli to insight learning in novel situations. Behaviors result from interactions among environmental stimuli, experience, and individual genetic makeup. The parameters of behavior are controlled by genetics and thus are acted upon by natural selection. Behavioral ecology focuses on the ultimate cause of reproductive fitness, which can be used to interpret foraging behavior, agonistic interactions, mating patterns, and altruistic behavior. Sociobiology extends evolutionary interpretations to human social interactions.

CHAPTER SUMMARY

Animal *behaviors* are observable movements or actions as well as unobservable physiological or brain changes in response to a stimulus. Behavior involves both innate and learned aspects, as illustrated by bird songs and human vocalizations.

The Evolutionary Logic of Behavioral Ecology

Behavioral ecology looks to evolution for explanations of animal behavior, with the underlying expectation that animals behave in ways that maximize their Darwinian fitness. This expectation of optimal behavior relies on the fact that behavior is influenced by genetics and therefore subject to natural selection. Behavior ecology allows researchers to frame questions about animal behavior in an evolutionary context, develop hypotheses, and make testable predictions.

Proximate Versus Ultimate Causation

The *ultimate causation* of a particular animal behavior relates to the evolutionary basis of the behavior—why it has been favored by natural selection. *Proximate causation* explains the immediate cause of a behavior in terms of the cues or stimuli that trigger it and the internal mechanisms that produce it. Many animals breed in the spring or early summer. The ultimate cause of this phenomenon is that breeding is most successful during this time. The proximate cause often is the stimulus of an increase in day length that results in reproductive behavior. Proximate causes produce behaviors that evolve because they result in increased fitness, which is the ultimate cause of behaviors.

Innate Components of Behavior

Ethology, the study of how animals behave in their natural environment, developed in the 1930s. The work of Lorenz, Tinbergen, and von Frisch provided evidence for the innate components of behavior and established the foundation of ethology.

Fixed Action Patterns The *fixed action pattern (FAP)*, or fixed motor pattern, was the fundamental concept to emerge from classical ethology. An FAP is a highly stereotyped, innate behavior that, once begun, is carried through to completion even if the activity becomes inappropriate. An FAP is triggered by a *sign stimulus* or *releaser*—some external sensory stimulus that, when perceived by the animal, initiates or "releases" a specific behavior.

The sign stimulus may be a feature of individuals of the same or another species. The releaser for attack behavior in stickleback fish is the red belly of the intruder. Any red colored object entering its territory may also act as a releaser and provoke the same response. FAPs often are associated with interactions between parents and young, such as the begging and feeding behavior of many bird species. Human behavior is usually based on more diverse, integrated information, which leads to intelligent action. A human infant's smile and grasp, however, are FAPs in response to visual or tactile stimuli.

Fixed action patterns are adaptive responses that have evolved due to selection pressures. Finer discrimination among sign stimuli has evolved in situations where there has been strong selective pressure, as in the ability in certain species of birds to recognize foreign, cuckoo eggs in their nests.

The mechanistic aspect of FAPs is illustrated by the repeated stereotyped behavior seen in some animals, such as a digger wasp performing its "drag and inspect" subroutine over and over when its prey, a paralyzed cricket it placed outside its nest, is experimentally moved.

The Nature of Sign Stimuli

The Nature of Sign Stimuli The experiments of ethologists have shown that sign stimuli usually involve just one or two simple characteristics of the object or organism. The pecking behavior of a herring chick, for example, can be released by a red spot swung horizontally at the end of a long object, as well as by the red spot on its parent's moving beak.

An animal's level of sensitivity to certain stimuli correlates closely with the type of sign stimuli to which it responds. Frogs are quite good at detecting movement, and it is the movement of objects that releases the frog's feeding behavior. Stronger responses often are elicited by exaggerated sign stimuli. The chick with the widest-gaping beak will be most likely fed first. Experiments with supernormal stimuli, such as giant models of eggs, have been used to illustrate these behavioral tendencies.

Ethologists have suggested that the use of simple cues in the animal's environment to release pre-programmed behavior (FAPs) ensures behavioral responses that require the least amount of processing and integrating of input.

Learning and Behavior

Learning is the modification of behavior as a result of experience and can affect even innate behaviors.

Nature Versus Nurture Early European ethologists were studying innate aspects of animal behavior, while American psychologists focused on how learning could influence behavior. A "nature versus nurture" debate developed concerning how much of an organism's behavior is innate or instinctive and how much is learned. Patterns of behavior are a consequence of both genetic and environmental inputs. Even FAPs improve with experience.

Learning Versus Maturation Improved performance of innate behaviors may result from neuromuscular maturation. Birds prevented from flying until older are able, upon release, to fly without practice or learning.

Habituation *Habituation*, a simple type of learning, is the loss of sensitivity to unimportant stimuli or to stimuli not associated with appropriate feedback. *Hydra* stop contracting in response to constant water currents; squirrels may stop responding to alarm calls not followed by an actual attack.

Imprinting With the phenomenon of *imprinting*, learning is closely associated with innate behavior. Imprinting is characterized by a critical period and the irreversibility of learning. Newly hatched ducks and geese instinctively follow the movement of an object, whether it be their mother, a ticking box, or Konrad Lorenz. A *critical period* is a limited time during which some type of imprinting may occur. Geese totally isolated during their first two days fail to imprint on anything. Adult birds appear to imprint on their young in the period following hatching. The length and timing of the critical period of sexual imprinting may vary. Male finches, after being placed with members of another closely related species during their critical period for sexual identity, readily mated with females of the second species, but reluctantly mated with their own species. The return of salmon to their home stream to spawn is an example of olfactory imprinting.

Song Development in Birds: A Study of Imprinting Learning and genetic programming may closely interact in vocal learning in young birds. In studies with experimentally raised sparrows, a bird learns to sing normally only if it can hear the song of its species during a critical period and then compare its own singing with the memory of that song. If a bird does not hear its own song, but only that of related species during its critical period, it does not learn that song. However, innate programming can be changed by experience. When sparrows older than 50 days are exposed to the singing of live adult birds of another species rather than to tape recordings, the length of their critical period for learning songs is longer and they are able to learn the song of the other species. People also have a critical period for learning vocalizations; learning foreign languages is easier up until the teen years.

Different species of birds show a great deal of diversity in their ability to learn songs. Some can develop nearly normal songs even when raised in isolation or deafened early in life. These diverse developmental patterns of song learning appear to involve a fairly flexible behavior system that is responsive to different selection pressures.

Classical Conditioning In *associative learning*, animals learn to associate one stimulus with another. *Classical conditioning*, described through the work of Pavlov in the 1900s with salivating dogs, illustrates the association of an irrelevant stimulus with a fixed physiological response. Classical conditioning in an animal's natural environment may involve the association of sharpened perceptions with release stimuli.

Operant Conditioning *Operant conditioning* refers to trial-and-error learning through which an animal associates a behavior with a reward or punishment. Skinner's work with rats is the best known laboratory study of operant conditioning. The association of good or bad tastes with food items is probably a common form of operant conditioning in nature. A food's nutritive value may lead to the programming of good or bad taste through natural selection.

Observational Learning Many vertebrates learn by watching the behavior of others. The spread of milk-bottle pecking through the English tit population illustrates this type of learning.

Play Despite its use of energy and the risk of injury, play behavior is common because it may confer adaptive advantages. Play that simulates stalking and attacking may allow animals to practice these behaviors. Play also maintains muscular and cardiovascular fitness, especially in young animals that do not yet engage in useful activities.

Insight *Insight learning*, sometimes called reasoning, involves the ability to perform correctly in a novel situation. A chimpanzee stacking boxes to reach a banana is a commonly cited example. Insight learning is most often observed in mammals, especially in primates.

The Ultimate and Proximate Bases of Learning Most animal behaviors can be explained as relatively fixed patterns in response to simple stimuli, which may be modified by simple kinds of learning. The capacity to learn confers a reproductive or survival advantage that can be acted upon by natural selection. The internal proximate mechanisms of learning are largely unknown, but some simple kinds of learning have been linked to biochemical or physiological changes.

Behavioral Rhythms

Feeding, sleeping, reproducing, and migrating are all regularly repeated behaviors that show a temporal rhythm. The safe and profitable execution of such behaviors is the explanation for the ultimate causation of these rhythms. The proximate mechanisms are the subject of much study.

Animals, as well as plants, show circadian rhythms of roughly 24 hours. To determine whether these rhythms are based on exogenous (external) cues or endogenous (internal) timers, researchers have placed animals into environments with no external cues and monitored their rhythmic behaviors. An endogenous, biological clock still requires exogenous cues to keep the behavior timed with the real world. The exogenous cue, sometimes called a *zeitgeber*, is usually light. Humans living in free-running conditions with no external time cues have a biological clock with a period of about 25 hours.

The role of endogenous factors in *circannual* behaviors, such as reproduction and hibernation, is not well understood. These behaviors are based in part on physiological and hormonal changes linked with changes in day length. One of the few studies of endogenous control of these long-term behaviors has shown that fat deposition in ground squirrels, associated with hibernation, occurs regularly, even in a constant environment.

Current hypotheses propose that the timing mechanism of the biological clock is based on regular molecular interactions.

Movement

The ultimate cause for oriented movements is to bring animals to environments for which they are adapted. Much research has focused on the proximate causes or mechanisms that animals use to detect and respond to environmental cues.

A *kinesis* is a change in activity rate in response to a stimulus. Although kinetic movements are randomly directed, they tend to maintain organisms in favorable environments because the animal's movements slow in response to favorable stimuli. A *taxis* is an oriented movement toward or away from a stimulus.

Migrations are long-distance, seasonal movements.

Three mechanisms are used in migration: Traveling from one familiar landmark to another is called piloting. Orientation occurs when an animal detects compass directions and moves in a straight line. Navigation involves the ability to detect compass direction and to determine present location relative to destination.

Some species of birds and other animals navigate by the heavens, using the sun or stars as directional cues. Many migrating animals have internal timing mechanisms and other means of adjusting to the changing positions of the sun and stars both throughout the day and along the migratory route.

Many bird species are able to continue migrating under clouds or through fog, probably because of their ability to detect and orient to Earth's magnetic field. Experimentally manipulating the magnetic field changes the migratory orientation of some species.

Foraging Behavior

Some animals are generalists, feeding on a large variety of items, whereas others are specialists with quite restricted diets. Specialists usually have morphological and behavioral adaptations specific for their food which make them extremely efficient at foraging.

Generalists often concentrate on an abundant food item, developing what we call a *search image* for it. When that food becomes less abundant, the animal appears to change its search image and concentrate on a new item. Search images allow for short-term specialization while maintaining the advantages of being a generalist.

Behavioral ecologists study foraging, with the assumption that natural selection favors animals that forage optimally, choosing foods that maximize energy intake over expenditure. Factors involved in an optimal foraging strategy include the distance to a food item; the time and energy required to pursue, catch, and handle the item; and the amount of usable energy and nutrients contained in it. Studies, such as those on prey selection by bluegill sunfish, have indicated that animals can modify their foraging behavior in ways that tend to maximize overall energy intake. This ability appears to be innate, although experience and physical maturation are thought to increase foraging efficiency.

Social Interactions

Social behavior involves interaction between two or more animals, usually of the same species. Cooperative prey capture and mating behavior are mutually beneficial, but animals are still operating in a manner that maximizes their own reproductive fitness. Many types of interactions are competitive, especially when they involve conflict over limited resources.

Agonistic Behavior *Agonistic behavior* involves a contest to determine which competitor gains access to a resource, such as food or a mate. The encounter may include a test of strength or, more commonly, threat displays and *ritual* behavior, which serve to avoid actual physical conflicts. A display of submission or appeasement by one of the competitors inhibits further aggressive activity and indicates surrender. Natural selection apparently favors the ending of a conflict without a violent combat that would reduce the reproductive fitness of the winner as well as the loser.

Dominance Hierarchies *Dominance hierarchies*, as illustrated by the "peck order" of hens, prevent continual combats by establishing which animals get first access to resources. In wolf packs, the alpha female controls mating of the others, allowing them to mate if there is an abundance of food and preventing it otherwise.

Territoriality A *territory* is an area established for feeding, mating, or rearing young from which conspecifics are excluded. A home range is the area in which an animal may roam, but a territory is the area that the animal defends. Territory size varies with species, function, and resource abundance. Agonistic behavior is used to establish and defend territories. Vocal displays, scent marks, and patrolling may be used to continually proclaim ownership.

Dominance systems and territoriality help to stabilize population density by ensuring that at least some individuals have a sufficient supply of a limited resource.

Mating Behavior

Courtship Most animals innately view conspecifics as threatening competitors to be driven off. This aversion must be overcome to accomplish mating. Complex courtship interactions, unique to each species, assure that individuals are not a threat and that they are of the proper species, sex, and physiological mating condition. These complex rituals often consist of a series of fixed action patterns, each released in sequence by the reciprocal behavior of the other individual involved.

Courtship behavior may also function as a basis for mate selection. Females, who usually have more parental investment in each offspring due to greater time and resource allocation, usually show more discrimination in choosing mates than do males. Sperm do not represent a large energy investment, and it may be reproductively advantageous for males to try to mate often and with a number of partners. Males often compete with each other for access to females. Intense courtship displays and secondary sexual characteristics may be used to attract females. A female's selection of a specific mate may be based on the competence he displays if the male provides parental care. Vigorous courtship displays and extreme secondary sexual characteristics also indicate good health. Especially if a male's only contribution to offspring is genetic, it is most beneficial to a female to choose a mate whose "performance" may indicate his genetic fitness.

Ritualized acts probably evolved from actions that had a more direct meaning. The behaviors of various species in the predaceous fly family Empididae show a progression from a male carrying a dead insect for the female to eat while he mates with her, to carrying a dead insect inside a silk balloon, to carrying simply an empty silk balloon.

Mating Systems   Many species have *promiscuous* mating, with no strong pair bonds forming. Longer-lasting relationships may be *monogamous* or *polygamous*. Polygamous relationships are most often *polygynous* (one male and many females), although a few are *polyandrous*.

The needs of young offspring are often an ultimate factor for the reproductive pattern of the parents. Newly hatched birds require more food than one parent may be able to supply. A male will leave more offspring, and thus increase his reproductive fitness, by helping to care for these young rather than by going off in search of more mates. With mammals, the female often provides all the food, and males are not monogamous.

Certainty of paternity also influences mating systems and parental care. With internal fertilization, the acts of mating and egg laying or birth are separated, and paternity is less certain than when eggs are fertilized externally. Male parental care is found much more frequently in species with external fertilization, where genetic contribution to the offspring is more certain.

Communication

Communication is the intentional transmission of information between individuals using visual, auditory, chemical, tactile, or electrical signals. A change in behavior of the "receiver" is an indication that communication has occurred. Behavioral ecologists argue that communications develop due to the benefits for the "sender," whereas ethologists assumed that communication evolved to maximize information transfer.

Pheromones are chemical signals commonly used by mammals and insects in reproductive behavior to attract mates and to release specific courtship behaviors. The trailing behavior of ants involves scents released by scouts.

Honeybees have highly complex communication systems, using ritualized dances to communicate the location of food sources. These dances were first described by von Frisch. Round dances are used when the food source is relatively close to the hive. Regurgitated nectar provides a scent to direct other bees to the food. The location of more distant food is communicated by waggle dances, in which the speed and vertical orientation of the dance on the comb, respectively, indicate the distance from the hive and the direction to the food relative to the horizontal angle to the sun.

Selfish Versus Altruistic Behavior

Altruistic behavior is difficult to explain in Darwinian terms of improving the reproductive success of the individual, but can be explained in terms of *inclusive fitness*, the ability of an individual to pass on its genes either by producing its own offspring or by helping close relatives produce their offspring. Natural selection can favor genes that produce altruistic behavior if the individual benefitting from an unselfish act also carries the altruistic genes. Cooperating individuals often are siblings or close relations, and thus are carrying many common genes. The coefficient of relatedness is a measure of the proportion of genes shared by two related individuals. *Kin selection* suggests that altruistic behavior is strongest between close relatives and less common as genetic relatedness decreases. This altruism among close relatives is a means of increasing inclusive fitness.

When altruistic behavior involves nonrelated animals, the explanation offered is *reciprocal altruism*; there is no immediate benefit for the altruistic individual, but some future benefit may occur when the helped animal may "return the favor." Reciprocal altruism often is used to explain altruism in humans and is primarily found only in species that form stable social groups.

Animal Cognition

Because it is impossible to know what goes on inside the minds of animals, most researchers have taken a mechanistic approach, called *behaviorism*, in the study of animal behavior. Recently, Griffin and others have argued that conscious thinking is an inherent part of animal behavior. According to *cognitive ethology*, cognitive ability forms an evolutionary continuum through many animal groups and has adaptive advantages that have been acted on by natural selection.

Human Sociobiology

In *Sociobiology*, published in 1975, Wilson elaborated the thesis that social behavior has an evolutionary basis and that behavioral characteristics are expressions of genes that have been acted on by natural selection. Wilson's work provided the basis for the modern approach of behavioral ecology. He also speculated on the evolutionary basis of certain social behaviors of humans.

The sociobiology debate continues the nature-versus-nurture controversy. In the example of human incest, the "nurture" side argues that the aversion to incest is not innate or else the taboos and laws against it would be unnecessary. The avoidance of incest must be a learned behavior, based on the experience that individuals who break the taboo are more likely to have defective children. The "nature" side argues that the occurrence of a behavior, such as the avoidance of incest, across many diverse cultures is evidence that it has an innate component.

Some sociobiologists, including Wilson, explain that cultural and genetic components of social behavior are linked together in a cycle of reinforcement. The development of cultural regulations of innate behaviors serves as an additional factor for the natural selection of that behavior. According to this view, human behavior represents an integration of genes and culture.

The parameters of human social behavior may be set by genetics, but the environment undoubtedly shapes behavioral traits just as it influences the expression of physical traits. Human behavior appears to be quite plastic. Our structured societies, with their regulations on behaviors, including behaviors that would enhance an individual's fitness, may be the one unique characteristic separating humans and other animals.

STRUCTURE YOUR KNOWLEDGE

1. How does the nature vs. nurture controversy apply to behavior?

2. Distinguish the following types of learning: imprinting, classical conditioning, operant conditioning, and insight learning.

3. Why are many interactions between members of the same species agonistic? What mechanisms reduce violent encounters?

4. What are the advantages and disadvantages of being a generalist or a specialist feeder?

5. How does the concept of Darwinian fitness apply to reproductive and altruistic behavior?

TEST YOUR KNOWLEDGE

MULTIPLE CHOICE: *Choose the one best answer.*

1. Behaviorism is the
   a. mechanistic study of the behavior of animals, focusing on stimulus and response.
   b. application of human emotions and thoughts to other animals.
   c. study of animal cognition.
   d. study of the origin and ultimate causes of animal behavior.

2. Proximate causes
   a. explain the evolutionary significance of a behavior.
   b. are immediate causes of behavior such as hunger or external stimuli.
   c. indicate that much of animal behavior is innate.
   d. are the focus of behavioral ecology.

3. A fixed action pattern
   a. shows that nature is the cause of animal behavior, not nurture.
   b. is an innate behavior, performed in its entirety in response to a sign stimulus.
   c. explains the development of bird songs.
   d. is a stereotyped behavior that has been used to explain imprinting in young animals.

4. Supernormal stimuli
   a. are innate releasing mechanisms between individuals of the same species.
   b. may elicit stronger responses or FAPs.
   c. are illustrated by the removal of the cricket from near a digger wasp's nest.
   d. are illustrated by the bobbing of a stick with a red spot past herring chicks.

5. Which of the following is an example of the evolution of a fixed action pattern with finer discrimination among sign stimuli?

a. a bluegill sunfish feeding on larger *Daphnia* when prey are abundant.
b. a chick pecking at the red spot on a parent's moving beak.
c. a bird recognizing and expelling an egg of a cuckoo from its nest.
d. a songbird learning its song after listening to a taped song of its species.

6. Learning
   a. cannot modify innate components of behavior.
   b. is a type of imprinting.
   c. must involve the use of insight.
   d. is a change in behavior as a result of experience or practice.

7. A critical period
   a. is the time right after birth when sexual identity is developed.
   b. usually follows the receiving of a sign stimulus.
   c. is a limited time during which imprinting can occur.
   d. is the period during which birds can learn to fly.

8. In classical conditioning,
   a. an animal associates a behavior with a reward or punishment.
   b. an animal learns as a result of trial and error.
   c. an irrelevant stimulus can elicit a response because of its association with a normal stimulus.
   d. a bird can learn the song of a related species if it hears only that song.

9. Insight learning
   a. only occurs in human beings.
   b. involves the ability to perform successfully in a novel situation.
   c. was studied by Skinner in his use of Skinner boxes.
   d. All of the above are correct.

10. Regularly repeated behaviors that show an approximate 24-hour rhythm are
    a. timed to the real world by the use of a biological clock.
    b. circadian rhythms.
    c. usually stopped in free-running conditions.
    d. a combination of learning and innate behavior.

11. Circannual behaviors
    a. are often linked to changes in day length.
    b. rely solely on endogenous cues.
    c. involve foraging, reproduction, and migration.
    d. do not occur in free-running conditions.

12. A kinesis
    a. is a randomly directed movement that is not caused by external stimuli.
    b. is a movement that is directed toward or away from a stimulus.
    c. is a change in activity rate in response to a stimulus.
    d. is illustrated by trout swimming upstream.

13. Which of the following is *not* true of long-distance migrations?
    a. Animals may use the stationary north star as a point of reference.
    b. Navigation using celestial bodies requires an internal clock to compensate for the daily movement of these objects.
    c. Birds that are able to continue migrating during fog and overcast weather may be sensing Earth's magnetic field.
    d. Migrations are the result of operant learning.

14. A dominance hierarchy
    a. may be established by agonistic behavior.
    b. determines which animals get first access to resources.
    c. helps to avoid potential injury of competitors.
    d. All of the above are correct.

15. An animal's territory may
    a. be larger than its home range.
    b. change in size with changing resource density.
    c. exclude both conspecifics and members of other species.
    d. be all of the above.

16. In a species in which females provide all the needed food and protection for the young,
    a. males are likely to be promiscuous.
    b. mating systems are likely to be monogamous.
    c. mating systems are likely to be polyandrous.
    d. males most likely will show sexual selection.

17. The ability of honeybees to fly directly to a food source, after having to wait several hours from the time of the waggle dance, indicates
    a. that the waggle dance provided directions relative only to the position of the hive.
    b. that bees have an internal clock that compensates for the movement of the sun during the elapsed time.
    c. that the bees must have been to that food source before.
    d. that bees are directed more by olfactory cues than by directional cues.

18. The concept of inclusive fitness explains
    a. optimal feeding behavior.
    b. sexual selection.
    c. kin selection and altruistic behavior.
    d. monogamous mating systems.

19. The coefficient of relatedness between brother and sister is
a. 0.5.
b. 0.25.
c. 0.125.
d. 1.0.

20. Sociobiology
a. explains the evolutionary basis of behavioral characteristics within animal societies.
b. applies evolutionary explanations to human social behaviors.
c. studies the roles of culture and genetics in social behavior.
d. does all of the above.

CHAPTER 46
ECOLOGY: DISTRIBUTION AND ADAPTATIONS OF ORGANISMS

Suggested Answers to Structure Your Knowledge

1. • Ecology is the study of the relationships of organisms to their abiotic and biotic environments, focusing on the abundance and distribution of species.

   • Areas of inquiry correspond to the levels of ecology and include the consideration of *organisms* and their physiological adaptations, the growth and regulation of *populations*, the interactions (such as predator-prey and competition) within *communities*, and the flow of energy and chemicals through *ecosystems*.

   • Experimental manipulation of variables is done both in the laboratory and in the field, along with observational measurements of various physical factors and kinds and numbers of organisms. Mathematical models and computer simulations are useful for predicting the effects of variables, especially in situations where experimentation is impractical or impossible.

   • Evolution, with the principles of natural selection and adaptation, is the key organizing theory applied to ecological study.

2. • Terrestrial biomes include the eight major world communities, which are identified on the basis of abiotic and biotic similarities.

   • Convergent evolution, the adaptations of different organisms to similar environments, accounts for similarities in life forms within each biome.

   • The areas of precipitation and temperature shown on the climograph are as follows:
   a. desert   d. temperate forest
   b. grassland   e. coniferous forest
   c. tropical forest   f. arctic and alpine tundra

Answers to Test Your Knowledge

Multiple Choice:

1. b	5. b	9. c	13. a
2. d	6. c	10. b	14. b
3. d	7. b	11. c	15. c
4. a	8. a	12. d	

Matching:

1. C	4. E	7. B
2. H	5. A	8. F
3. D	6. G	

CHAPTER 47
POPULATION ECOLOGY

Suggested Answers to Structure Your Knowledge

1.

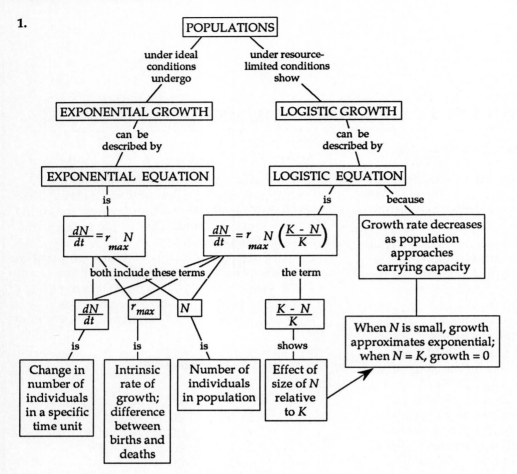

2. The age structure of a population affects population growth due to age specific fecundity, or the differential reproductive capabilities of various age groups. The larger the proportion of the population in age groups with high birth rates, the greater the rate at which the population will grow. Population growth rate is likely to continue to be high if the age structure is skewed toward younger age groups that have not yet reached the age of greatest fecundity. The shorter the generation time (age at which reproduction first begins), the greater the growth rate. The sex structure of a population will affect growth rate, depending on the mating patterns of the population. Survivorship influences growth rate in that the more individuals that live and reproduce, the more the population will grow. A population in which survival of offspring is usually low may show large fluctuations in population size depending on environmental conditions.

Answers to Test Your Knowledge

Multiple Choice:

1. a	5. a	9. d	13. a	17. b
2. b	6. a	10. d	14. b	18. c
3. c	7. d	11. c	15. a	19. a
4. b	8. c	12. a	16. b	20. d

CHAPTER 48
COMMUNITY ECOLOGY

Suggested Answers to Structure Your Knowledge

1.

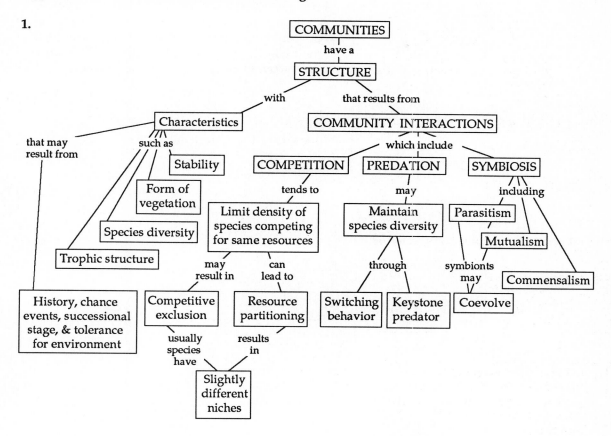

2. Succession is the gradual transition in species composition in a community, following a disturbance or creation of a new habitat. Each sere alters the environment somewhat and may facilitate the transition to the next stage. Early pioneers in the habitat may also inhibit the establishment of other species, until more competitive species are able to become established. Change in species composition also may be due to the gradual success of species most tolerant of the environment found in that habitat. Chance events, such as dispersion and allogenic disturbances, are important factors during succession. There is evidence that even communities that appear to be in final successional stages are in continual nonequilibrium, with species composition and diversity still changing.

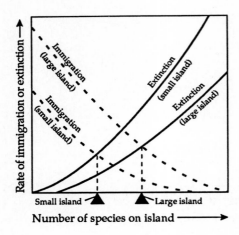

Rate of immigration or extinction →

Immigration (large island)

Immigration (small island)

Extinction (small island)

Extinction (large island)

Small island   Large island

Number of species on island →

3. The graph shows that the curves for immigration are inversely related to the number of species on

an island, and the curve for a large island is higher than that for a small island. The curve for extinction is directly related to an increase in species number, and the curve for a small island is higher that that of a large island. The intersection of these curves shows that the equilibrium point between immigration and extinction is lower for a small island than for a large island. The large island would have a greater species richness.

Answers to Test Your Knowledge

Multiple Choice:

1. c	5. b	9. a	13. c	17. d
2. d	6. a	10. c	14. d	18. b
3. b	7. a	11. a	15. a	19. a
4. d	8. b	12. b	16. b	20. c

CHAPTER 49
ECOSYSTEMS

Suggested Answers to Structure Your Knowledge

1.

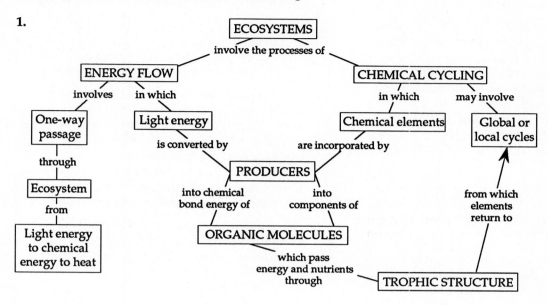

2. **a.** nitrogen-fixing bacteria in root nodules
   b. nitrogen-fixing soil bacteria
   c. decomposers
   d. ammonia (NH_3)
   e. nitrifying bacteria
   f. nitrates (NO_3^-)
   g. denitrifying bacteria

3. **a.** Deforestation can cause an increase in water runoff with an accompanying loss of soil and minerals. Clear-cutting of tropical forests results in loss of species diversity and a reduction in productivity of the area.
   b. Dumping of wastes and runoff from agricultural lands has resulted in eutrophication of many lakes, killing off fish and other organisms.
   c. Toxic chemicals introduced into the environment have been incorporated into the food chain. As a result of biological magnification, these substances pose threats particularly to top-level consumers.
   d. An increase in atmospheric CO_2 from fossil fuel and wood combustion adds to the greenhouse effect. The resulting increase in temperature may have far-reaching effects on climate and sea level. Changes in CO_2 levels may also affect species composition of C_3 and C_4 plants in natural communities and in agricultural crops.
   e. The release of chlorofluorocarbons into the atmosphere is breaking down the ozone layer and exposing organisms to harmful ultraviolet radiation.

Answers to Test Your Knowledge

Multiple Choice:

1. b	**5.** d	**9.** a	**13.** d	**17.** c
2. a	**6.** d	**10.** c	**14.** a	**18.** b
3. c	**7.** b	**11.** d	**15.** a	
4. c	**8.** c	**12.** c	**16.** d	

CHAPTER 50
BEHAVIOR

Suggested Answers to Structure Your Knowledge

1. There is controversy over how much of animal behavior is innate and genetically programmed and how much is a product of experience and learning. Fixed action patterns are clearly preprogrammed behavior, and many seemingly complex animal behaviors can be isolated into a series of FAPs. In other cases, genetics may set the parameters for an organism's behavior, but experience can modify behavior and learning is clearly evident.

2. *Imprinting* is a type of learning that occurs during a critical time in an organism's development. Innate behavior and learning are tightly coupled in imprinting.

   The concept of *classical conditioning* was developed by Pavlov. As classical conditioning is used in a laboratory study, an animal learns to associate an unrelated stimulus with a stimulus that elicits a behavioral or physiological response. Later, the unrelated stimulus alone can cause the behavior. In nature, classical conditioning is the type of learning in which an organism adds to its collection of the stimuli that it associates with particular cues or releasers.

   Operant conditioning, as studied by Skinner, is learning in which an animal comes to associate a positive or negative reward with a particular behavior. In a laboratory or circus, it can be used to teach animals all sorts of "tricks." In nature, it is a way in which an organism learns which foods to eat by the good or bad tastes of food.

   Insight learning involves the ability of an animal to reason and to figure out an appropriate behavior in a novel situation.

3. Members of the same species occupy the same niche and are thus competitors for resources such as food, territory, and mates. In agonistic encoun-

ters, one animal may establish its claim to these resources. Threat displays, appeasement gestures, dominance hierarchies, and territories all serve to reduce violent encounters between conspecifics.

4. A generalist is not as efficient in foraging for any one type of food, but is able to use more food items. The development of search images for specific prey may improve the efficiency of generalists. A specialist often has anatomical and behavioral adaptations for feeding efficiently on one type of food item. Should this item become scarce, however, the specialist is at a disadvantage.

5. According to the concept of Darwinian fitness, an animal's behavior should help to increase its reproductive success. The best strategy for reproductive behavior is based on the probability of producing the most offspring to carry one's genes into the next generation. Females with a high parental investment in each offspring maximize their fitness by discriminately choosing a mate. A male's reproductive fitness may be maximized by frequent mating or by helping to rear young if two parents are needed for offspring to survive. Altruistic behavior has been explained on the basis of kin selection; the inclusive fitness of an animal increases if it's altruistic behavior benefits related animals who may be carrying a high proportion of the same genes.

Answers to Test Your Knowledge

Multiple Choice:

1. a	6. d	11. a	16. a
2. b	7. c	12. c	17. b
3. b	8. c	13. d	18. c
4. b	9. b	14. d	19. a
5. c	10. b	15. b	20. d